Entomology in Focus

Volume 9

Insects are fundamentally important in the ecology of terrestrial habitats. What is more, they affect diverse human activities, notably agriculture, as well as human health and wellbeing. Meanwhile, much of modern biology has been developed using insects as subjects of study.

To reflect this, our aim with *Entomology in Focus* is to offer a range of titles that either capture different aspects of the diverse biology of insects or their management, or that offer updates and reviews of particular species or taxonomic groups that are important for agriculture, the environment or public health.

The series results from an agreement between Springer and the Entomological Society of Brazil (SEB) and as such may lean towards tropical entomology. The aim throughout is to provide reference texts that are simple in their conception and organization but that offer up-to-date syntheses of the respective areas, offer suggestions of future directions for research (and for management where relevant) and that don't shy away from offering considered opinions.

Adeney Freitas Bueno • Antônio Ricardo Panizzi
Editors

Stink Bugs (Hemiptera: Pentatomidae) Research and Management

Recent Advances and Case Studies from Brazil, Europe, and USA

Springer

SOCIEDADE ENTOMOLÓGICA DO BRASIL · 1972 ·

Editors
Adeney Freitas Bueno
Embrapa Soja
Londrina, PR, Brazil

Antônio Ricardo Panizzi
Embrapa Trigo
Passo Fundo, RS, Brazil

ISSN 2405-853X ISSN 2405-8548 (electronic)
Entomology in Focus
ISBN 978-3-031-69744-9 ISBN 978-3-031-69742-5 (eBook)
https://doi.org/10.1007/978-3-031-69742-5

This Springer imprint is published by the registered company Springer Nature Switzerland AG
The registered company address is: Gewerbestrasse 11, 6330 Cham, Switzerland

Foreword

The conception of this book originated from a Symposium on the taxonomy and economic importance of stink bugs (Heteroptera: Pentatomidae) in the Neotropics, organized by the distinguished Brazilian entomologists Antônio Ricardo Panizzi and Jocelia Grazia. The success of the Symposium at the XXVI Congress of Entomology, in Helsinki, Finland, 2022, led to the expansion of the original scope of the symposium to include developments in pentatomid research in the USA and Europe. Although the book has an international coverage, I use the progress of Brazilian entomology to illustrate the impact of the dramatic development of technology and the biological sciences in less than a century.

In 1940 the noted Brazilian entomologist Angelo Moreira da Costa Lima published the second of his 12-volume magnum opus *Insetos do Brasil*, a 350-page tome dedicated to the order Hemiptera (now the sub-order Heteroptera). While only 24 pages were on the Pentatomidae, the chapter contained an extensive bibliography that included very few works by Brazilian entomologists, thus highlighting the dearth of specialists in the country at the time.

However, the impact of Costa Lima, and a few of his contemporaries on the expansion of Brazilian entomology is remarkable, particularly in medical and agricultural entomology. After the 1950s' colleges of agronomy, veterinary, medicine, and natural history, all including entomology courses, proliferated in Brazil. Until introductory texts were published, young generations of entomologists had *"Insetos do Brasil"* as a key reference.

Eighty years or four generations after Costa Lima, Brazilian entomologists are in the forefront of the science. The "Sociedade Entomológica do Brasil" (SEB) (the Brazilian Entomological Society) is the leading professional organization in the Neotropics, and works by Brazilian entomologists appear in leading scientific publications worldwide.

Adeney Freitas Bueno and Antônio Ricardo Panizzi represent the fourth generation of Brazilian entomologists demonstrating the scientific excellence that Costa Lima pioneered. The present volume that they conceptualized reflects advances that have occurred in all areas of fundamental research and applications in control of pest pentatomids. The two Brazilian entomologists lead, as co-editors, an international cadre of some distinguished specialists from the USA and Europe in the production of this book on pest pentatomids, a fascinating and economically important family of Heteroptera.

These intervening 80 years between those four generations of entomologists produced some of the most dramatic advances in all realms of technology and science in the history of humankind. In technology a small example is a comparison of how Costa Lima's books were produced, and the present volume was probably produced.

Costa Lima's handwritten text was given to a secretary to typewrite the manuscript. The typescript was proofread by Costa Lima himself, as he insisted on doing with all his writings. Pages with corrections were re-typed and re-proofread. The corrected, typewritten text was passed on to the publisher to be typeset. Typesetting since the late 1880s' was done using linotype machines, a vast improvement over the previous letter-by-letter type setting. Yet the linotype operator was not infallible and again the printed text had to be proofread. Illustrations were India-ink drawings done by scientific illustrators (C. Lacerda in Costa Lima's books), or photos taken with film cameras that required a lengthy development process and usually reproduced poorly. Only then, when all corrections were made and all illustrations etched into metal plates were completed, the book could go into production.

By contrast, the present book was produced using advanced electronic technology. The co-editors contacted prospective chapter authors by email. Authors wrote their chapters using a word-processing application on a personal computer that automatically checked spelling and syntax errors. The electronically produced "manuscript" file was emailed to the editors who made comments on the file and emailed the comments back to the authors or directly to the publisher. All transfer of files was instantaneous. Graphs and illustrations were produced using graphics applications (apps). Photos were produced using digital cameras, easily transmitted through the internet, and printed with fidelity using laser printers. Finally, even the printing process was computerized allowing the publisher to produce a single copy or 1,000 copies at a time. A small example of the technological advances within only 80 years.

Equally impressive have been the advances in the biological sciences. Taxonomy, one of Costa Lima's strengths, passed through dramatic changes in the 1950s' from Sokal and Sneath's "Numerical Taxonomy" and the publication of Hans Hennig's "Grundzüge *einer Theorie der Phylogenetischen Systematik*" ushering in the era of phylogenetic systematics, or cladistics. Beyond that, advances in molecular biology allow the use of genetic markers to produce barcodes separating closely related species. These conceptual and technical advances are obviously only accessible to professionals in advanced research institutions. However, field entomologists and even the interested public can take advantage of advances in image recognition technology to closely identify insects from photos taken with iPhone cameras loaded with the appropriate application.

In agricultural entomology, Costa Lima set an example to future professionals with his work on the pink boll worm (*Pectinophora gossypiella* (Sounders, 1844)) and the coffee bean borer (*Hypothenemus hampei* (Ferrari, 1867)). However, his options for control recommendations were limited. Today, advances in all control tactics properly selected under principles of Integrated Pest Management (IPM) aim not only to reduce economic losses but also to protect the environment.

Well edited, multi-authored books, such as this one, represent mileposts in the progress of a specific branch of science. They give a snapshot of the state of science and highlight gaps in the knowledge that may inspire new lines of research. For future entomologists, 80 years from now, such books may be a valuable source of historical information. They may smile of how primitive we were still writing books because by the turn of the twenty-second century, using quantum computers and artificial intelligence, they may just feed some key words for the computer and AI will compile the literature, extract the key information, and format it into an elegant, clear text. This and much more, provided humankind is smart enough to contain and survive global warming or other man-made cataclysms.

Professor Emeritus Marcos Kogan
University of Illinois at Urbana-Champaign
Oregon State University
Corvallis, OR, USA

Preface

The XXVI International Congress of Entomology (ICE 2020) was planned to be held in Helsinki, Finland, in 2020. Frank Zalom (University of California, Davis), member of the Scientific Program Committee, invited Antônio Ricardo Panizzi (ARP) to organize a Symposium on advances in research of true bug (Heteroptera: Pentatomidae) pests in the Neotropics. After conversations with other Brazilian specialists, the invitation was accepted and Jocelia Grazia, (JG), a distinguished pentatomid taxonomist, joined ARP in the planning and organization of the symposium. Eight speakers were invited to address key topics in petatomid research, including taxonomy and geographical distribution of pest species; bioecology and host-plant interactions; feeding behavior and resulting damage; potential applications of gene silencing (RNAi) technology; managing stink bugs on cropping systems; biorational control using semiochemicals and vibrational communication; biological control; and insecticide resistance management to mitigate pentatomid pest species injury.

The Covid-19 pandemic of late 2019 forced the ICI 2020 to be postponed. This delay and restrictions on traveling and assembling conditions created problems for our planned symposium as some speakers and the co-organizer, JG, were not able to attend on the new date of the ICE. Despite these setbacks, the symposium was convened two years later, in 2022, with considerable success.

After the presentations, discussions, and conclusion, ARP was contacted by Lars Koerner, Senior Editor of Biomedicine and Life Sciences of Springer Nature with the suggestion to turn the symposium into a multi-authored book. One of the speakers, Adeney Freitas Bueno (AFB) participated in the talks and showed interest in collaborating on this endeavor. A considerably expanded book proposal was approved, with AFB and ARP as co-editors. The content of the book would now cover not only the topics presented in the Symposium on stink bugs of the neotropics but also research work on pest pentatomids carried out in the United States and Europe, as well. Therefore, the book now entitled *Stink Bugs (Hemiptera: Pentatomidae) Research and Management—Recent Advances and Case Studies from Brazil, Europe, and USA*, besides the expansion of the geographic scope, includes some additional research topics. From the original 8 Symposium topics, it increased to 17 chapters, as follows.

Chapter 1 covers recent advances and updates in the taxonomic, identification, and geographic distribution of pest stink bugs of soybean; Chap. 2 is dedicated to predatory stink bugs (Asopinae) as economically important biological control agents; Chap. 3 covers studies on life history and its implications for management of pest stink bugs; Chap. 4 is dedicated to a case study on the egg parasitoid *Telenomus podisi* Ashmead to control stink bugs in Brazil; Chaps. 5 and 6 deal with augmentative and conservation biological control, respectively; in Chap. 7, the potential of botanical oils as alternative insecticides to control stink bugs in agriculture is discussed; Chap. 8 covers the RNA interference (RNAi) as a promising alternative for pest control of heteropterans; Chap. 9 covers the potential challenges and major benefits of the *CRISPR* technology to manage pest stink bugs; Chap. 10 is dedicated to the Block technology as a host plant resistance tool to manage pest stink bugs on soybean; the present and the future of chemical control of stink bugs on soybean within integrated pest management (IPM) programs and spray technologies are presented in Chaps. 11 and 12, respectively; Chap. 13 discusses insecticide resistance management for stink bugs; in Chaps. 14, 15, and 16 case studies of the Neotropical brown stink bug *Euschistus heros* (F.) in Brazil, and the brown marmorated stink bug, *Halyomorpha halys* (Stål) in Europe and in the USA are covered, respectively; and, finally, as conclusion, Chap. 17 covers the perspectives on stink bugs research and management in agriculture.

With the new information on taxonomy and on bioecology of stink bugs (pentatomids) being generated, and the fast advances and innovations in technologies to manage pest species, the growing pressure for an updated text on this subject led us to produce this book. We hope that the content of the book will fulfill the expectations of the readers interested in the basic knowledge and advances in management of major stink bug pests of important commodities worldwide.

Londrina, PR, Brazil — Adeney Freitas Bueno
Passo Fundo, RS, Brazil — Antônio Ricardo Panizzi

Acknowledgments

During the long time this book was planned and come to an end with its publication, several people were involved and helped us to achieve this goal. We want to thank João Pildervasser, then Editor of Life Sciences and Neuroscience of Springer Nature (São Paulo, Brazil), who guided us during the steps taken to submit this project to the publisher. We are grateful to Arun Siva Shanmugam, Production Editor at Manapakkam, Chennai, India, who graciously supported us during the preparation and revision of proofs, indexes, and other details of the book production process. Adeney Freitas Bueno thanks the Conselho Nacional de Desenvolvimento Científico e Tecnológico (CNPq, Brasília, Brazil) for providing financial resources (grant 304052/2021-3) that made possible researching stink bugs and to attend several Congresses of Entomology which inspired organizing this work and the Brazilian Research Corporation, Embrapa Soja, for all the support to organize the book. Antônio Ricardo Panizzi thanks the Conselho Nacional de Desenvolvimento Científico e Tecnológico (CNPq, Brasília, Brazil) for providing financial resources to attend the International Congress of Entomology (ICE2022) in where this book originated after the presentation of our symposium on true bugs. We also thank the Sociedade Entomológica do Brasil (SEB) and Simon L. Elliot for the opportunity to include our book in the Entomology in Focus series sponsored by SEB. Finally, we want to thank the authors for their contributions and patience to answers all the queries during the preparation of the chapters. They are the reason why we could come up with such a valuable book.

Contents

1 **Soybean Stink Bugs: Updates on Classification, Taxonomy, and Distribution** 1
Jocelia Grazia and Cristiano Feldens Schwertner

2 **Predatory Stink Bugs of Economic Importance as Biological Control Agents** 29
Ricardo Brugnera and Jocelia Grazia

3 **Life History Studies and Their Implications for Management of Stink Bugs** 41
Antônio Ricardo Panizzi and Tiago Lucini

4 **The Use of *Telenomus podisi* to Manage Stink Bugs on Soybean: The Example of Brazil** 51
Adeney Freitas Bueno, Weidson Plauter Sutil, Yelitza Coromoto Colmenarez, and Leonardo Roswadoski

5 **Augmentative Biological Control of Stink Bugs Using Entomopathogens in Brazil** 65
Rodrigo Mendes Antunes Maciel, Raiza Abati, Nédia de Castilhos Ghisi, Everton Ricardi Lozano, and Michele Potrich

6 **Conservation Biological Control as an Alternative to Reduce Stink Bugs Outbreaks** 81
María Fernanda Cingolani

7 **Botanical Insecticides as an Alternative to Control Stink Bugs in Agriculture** 95
Everton Ricardi Lozano, Michele Potrich, Lucas Battisti, and Raiza Abati

8 **The Use of RNAi Against Stink Bugs** 117
Grazielle Celeste Maktura, Giovanna Vieira Guidelli,
Thiago Roncini Gomes da Costa, and Henrique Marques-Souza

9 **The CRISPR Technology to Manage Stink Bugs: Potential, Challenges and Major Benefits** 169
Renata Fuganti Pagliarini, Mayla Daiane Correa Molinari,
Alexandre Lima Nepomuceno, and Adeney Freitas Bueno

10 **Host Plant Resistance to Manage Pest Stink Bugs: The Block Technology on Soybean** 181
Tiago Lucini and Antônio Ricardo Panizzi

11 **The Present and Future of Chemical Control to Manage Stink Bugs in Brazil** 199
Renato Carvalho, Daniela Okuma, Oderlei Bernardi, and Ralf Nauen

12 **Application Technologies for Stink Bug Management in Different Cropping Systems** 213
Ulisses Rocha Antuniassi, Walter Boller, Clerison Regis Perini,
Glauber Renato Sturmer, Juliano de Bastos Pazini,
Vinadio Lucas Béga, Ivair Valmorbida, and Orcial Ceolin Bortolotto

13 **Insecticide Resistance Management of Stink Bugs** 241
Daniel Ricardo Sosa-Gómez and Oderlei Bernardi

14 **Case Study 1: *Euschistus heros* (F.) on Soybean in Brazil** 255
Adeney Freitas Bueno, Antônio Ricardo Panizzi, and Weidson
Plauter Sutil

15 **Case Study 2: *Halyomorpha halys* (Stål) in Europe** 271
Lara Maistrello

16 **Case Study 3: Current Pest Status and Management of the Invasive Brown Marmorated Stink Bug in the USA** 361
Laura J. Nixon, Anne L. Nielsen, and Tracy C. Leskey

17 **Perspectives on Pest Stink Bugs Research and Management in Agriculture** 383
Adeney Freitas Bueno and Antônio Ricardo Panizzi

Insect Index 395

Subject Index 399

Chapter 1
Soybean Stink Bugs: Updates on Classification, Taxonomy, and Distribution

Jocelia Grazia and Cristiano Feldens Schwertner

1 Introduction

The family Pentatomidae encompasses a monophyletic group of true bugs (Grazia et al. 2008), representing the third largest family within Heteroptera and one of the most diverse groups of hemimetabolous insects. The pentatomids are known as stink bugs or shield bugs and contain about 950 genera and more than 5,000 species in the world (Rider et al. 2018). Up to 10 subfamilies and more than 50 tribes are recognized, including several of them recognized as endemics (Table 1.1).

Stink bug species range from small to medium in size (1–20 mm), the body oval to elongate, antennae usually 5-segmented, scutellum always surpassing the middle of the abdomen, subtriangular to spatulate in shape (Grazia et al. 2008; Rider et al. 2018). There is an astonishing variation in color (Fig. 1.1), with several species brightly colored while others have more cryptic patterns, i.e. predominantly black, brown, green, or variegated, sometimes allowing them to camouflage with the substrate where the species live. Besides that, stink bugs represent the most common and abundant true bugs that release unpleasant odors, hence the name stink bugs.

Most of the stink bug species are phytophagous, poly- or oligophagous, feeding on the sap-fluid or reproductive parts (developing fruits and seeds) of the hosts (Rider et al. 2018). Only the species of the subfamily Asopinae are predacious, an evolutionary innovation within the family. Several species are related to

J. Grazia (✉)
Programa de Pós-Graduação em Biologia Animal, Universidade Federal do Rio Grande do Sul, Porto Alegre, RS, Brazil
e-mail: jocelia@ufrgs.br

C. F. Schwertner
Departamento de Ecologia e Biologia Evolucionária, Universidade de São Paulo, Diadema, SP, Brazil

A. F. Bueno, A. R. Panizzi (eds.), *Stink Bugs (Hemiptera: Pentatomidae) Research and Management*, Entomology in Focus 9,
https://doi.org/10.1007/978-3-031-69742-5_1

Table 1.1 Classification of the family Pentatomidae, with the number of genera and species and distribution of each recognized taxon. The classification follows Grazia et al. (2015), Rider et al. (2018), and Schuh and Weirauch (2020)

Subfamily	# Tribes	# Gen	# spp	Distribution
Aphylinae	–	2	3	Australian
Asopinae	–	66	299	World
Cyrtocorinae	–	4	11	Neotropical
Discocephalinae	2	81	325	Neotropical
Edessinae	–	15	338	Neotropical
Pentatominae	42	660	3,484	World
Phyllocephalinae	5	45	213	Afrotropical, Australian, Oriental, Palearctic
Podopinae	5	64	249	World
Stirotarsinae	–	1	1	Neotropical
Serbaninae	–	1	1	Oriental

agroecosystems, and the family includes some of the most important crop pests around the world (Panizzi et al. 2000; Grazia and Schwertner 2017).

2 Stink Bug Classification

True bugs (Heteroptera) represent a speciose group of hemimetabolous insects, with more than 45,000 species and 6,000 genera and high morphological and ecological diversification (Schuh and Weirauch 2020). Recognized as a suborder within Hemiptera, the unique characters of Heteroptera include the simply elongated sucking mouthparts (= rostrum) arising from the front of the head, the presence of a distinct gula (= post-buccal region), the forewing typically modified into hemelytron, scent glands openings on the abdominal terga in nymphs and on the metapleura in adults, and the enlarged and typical triangular scutellum (Schwertner et al. 2021). True bugs have a broad range in body length, including some of the largest living insects (Belostomatidae: *Lethocerus*, 130 mm), living in a remarkably wide range of habitats (Panizzi and Grazia 2015; Schuh and Weirauch 2020).

Seven infraorders are recognized within Heteroptera, each representing a monophyletic lineage (Schuh and Weirauch 2020). The family Pentatomidae is included in the infraorder Pentatomomorpha, which comprises more than 16,000 species; together with the infraorder Cimicomorpha comprises a clade also known as Terheteroptera (Weirauch et al. 2019) and includes almost 90% of the species diversity within Heteroptera. Among pentatomomorphans, the superfamily Pentatomoidea represents the lineage which includes the Pentatomidae and related families (Rider et al. 2018).

Classification of the family Pentatomidae has been considered in a state of flux since the current subfamilial and tribal arrangement being inconsistent with the evolutionary history of the group (see below). The current definition of the family was

Fig. 1.1 Stink bug diversity. (**a**) *Arocera apta* (Walker); (**b**) *Runibia decorata* (Dallas); (**c**) *Grazia tincta* (Distant); (**d**) *Mayrinia variegata* (Distant); (**e**) *Patanius vitattus* Rolston; (**f**) *Rhyncholepta grandicallosa* Bergroth; (**g**) *Euschistus picticornis* Stål; (**h**) *Lattinidea geographica* (F.); (**i**) *Eurystethus microlobatus* Ruckes; (**j**), *E. microlobatus*, nymphs and adults tending by ants, *Camponotus rufipes* F. and *C. melanoticus* Emery. (Photo credits: **a**, **c**-**h**, CF Schwertner; **b**, Aline Kazuno Hayachi)

stated by Leston (1953, 1954a, b), and further developed in the following years (McDonald 1966; Gross 1975, 1976; Rolston and McDonald 1979; Linnavuori 1982). However, these works differed widely in the proposed infrafamilial classification, strongly regional in scope, which hampered an effective unifying classification until today. The first phylogenetic scheme for the family was published by Gapud (1991), who recognized nine subfamilies within Pentatomidae. This classification was strongly supported by Grazia et al. (2008), nonetheless some authors still considered Serbaninae as part of the Pentatomidae (Rider et al. 2018). More recently, Cyrtocorinae was supported as a separate family (Roca-Cusachs et al. 2022). Moreover, the nominotypical subfamily Pentatominae as well as the Discocephalinae and Podopinae are supported as non-monophyletic groups in recent studies (Roell and Campos 2019; Genevcius et al. 2021; Roca-Cusachs et al. 2022; Sampaio and Campos 2023).

2.1 Incongruences of the Current Classification of Stink Bugs (Pentatomidae)

The classification of organisms represents our way of organizing information about the natural world, allowing storage, retrieval, and communication of information about organisms. Biological classifications are the baseline for comparative studies, and although it is desirable to be stable and universal, classifications are not static and are continuously subject to ongoing review. Taxa are described or reviewed in terms of validity and relationships, which allows us to improve classifications and consequently our knowledge about biodiversity and all the interactions and processes related to the species.

Evolutionary processes represent the main driver of the diversity of organisms at any level (biochemistry, ecology, genetics, morphology, etc.), making evolutionary relatedness the most informative basis for studying the diversity of organisms. Classification based on shared similarity (homology) represents the more informative, accurate, and predictive way to organize biological data (Ward et al. 2016; Sigwart 2022). In this context, taxa represent hypotheses that require testing and verification, a goal that has been achieved within a phylogenetic framework since the theoretical works of Willi Hennig (Rieppel 2016).

The most thorough phylogenetic analysis to date of the Pentatomidae supports the need for a complete re-evaluation of the current classification (Roca-Cusachs et al. 2022). Although not completely resolved (Fig. 1.2), the phylogenetic hypothesis clearly showed incongruences with the current subfamilial and tribal arrangements (Table 1.1). For instance, while the subfamilies Asopinae, Edessinae (which includes the stink pest *Edessa meditabunda* (F.) and Phyllocephalinae represent monophyletic groups, the subfamilies Discocephalinae, Pentatominae and Podopinae do not (Fig. 1.2).

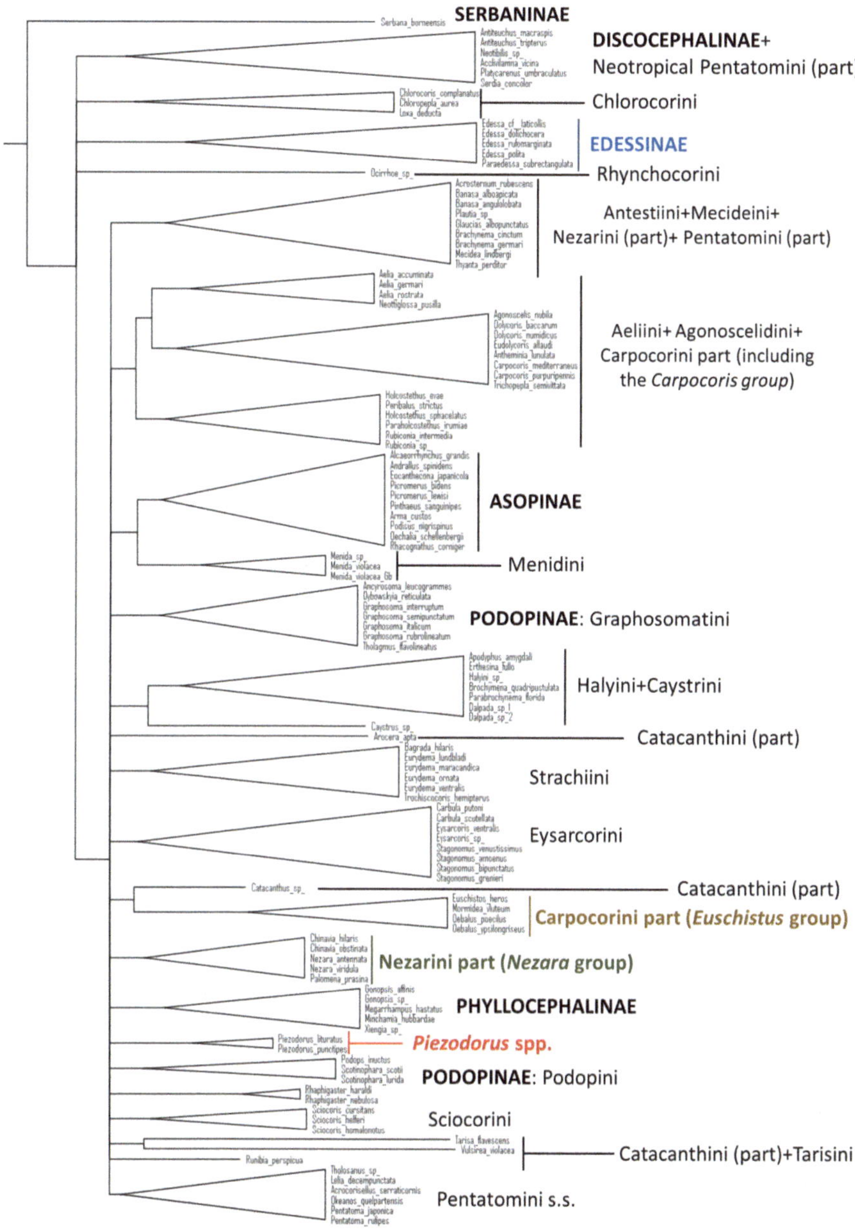

Fig. 1.2 Synoptical phylogeny depicting the phylogenetic classification of the stink bug family as discussed in this chapter, based on Roca-Cusachs et al. (2022). The phylogeny was derived from the published maximum likelihood analysis, all nodes weakly supported collapsed

Within the Pentatominae, which include most of the stink bugs of economic importance, several groups are recognized and supported as monophyletic (Fig. 1.2), however most of these groups do not correspond to the subfamily or tribe the species/genera are assigned (Genevcius et al. 2021; Roca-Cusachs et al. 2022). For instance, representatives of the tribe Carpocorini, which include the genus *Euschistus* Dallas and relatives, are separated into two different and unrelated clades: the type genus *Carpocoris* Kolenati and related genera (*Antheminia* Mulsant and Rey, *Dolycoris* Mulsantand Rey, etc.…) appeared closer to representatives of the Old World tribes Aeliini and Agonoscelidini than to the New World species currently included in the Carpocorini, the *Euschistus* group. The tribe Nezarini, which includes the genus *Nezara* Amyot and Serville and other economically important green stink bugs, clearly did not represent a monophyletic group, and a relationship with the representatives of the tribe Antestiini, Mecideini, and some Pentatomini is highly supported. The relationships among and within those monophyletic groups are not settled yet, and several studies have addressed this important gap over the years (i.e., Barcellos and Grazia 2003; Campos and Grazia 2006; Ferrari et al. 2010; Schwertner and Grazia 2012; Bianchi et al. 2017b; Roell and Campos 2019; Genevcius et al. 2021; Roca-Cusachs et al. 2022; Sampaio and Campos 2023).

At least three practical implications regarding these results must be highlighted. First, the current classification of the family Pentatomidae cannot be used to infer the relationship between suprageneric groups. Nonetheless, although still sketchy and not formally recognized in the current classification, relationships within the monophyletic groups supported may be based in recent published phylogenies which include species of economic importance (see ahead).

Second, some confusion may arise when we compare the classifications used by different authors in the last 20 years, which seem to be incongruent arrangements. For instance, Grazia et al. (2015) and Schuh and Weirauch (2020) excluded Serbaninae from Pentatomidae and recognized 9 subfamilies, while Rider et al. (2018) considered Serbaninae within the Pentatomidae, recognizing 10 subfamilies (see Table 1.1). More recently, Roca-Cusachs et al. (2022) also supported a classification with 9 subfamilies, but with Cyrtocorinae out, considered as a separated family, and Serbaninae in the Pentatomidae. The same level of incongruence may occur at the tribal level, depending on which group is lumped into one tribe or split into different tribes (i.e., Rider 2006; Ribes and Pagola-Carte 2013; Rider et al. 2018). Nonetheless, incongruences between classifications are usual in many groups of organisms (classifications are not static), both in the past and currently (i.e., Zhang et al. 2015; Santibáñez-López et al. 2020), and may be acknowledged as a period in which a better understanding of the taxa and its properties will emerge. An important recommendation to any comparative study involving organisms is always to make clear which classification one is following, phylogenetic or not since it will facilitate retrieving the information of the taxa in a broader context. Such recommendation is strongly reinforced here for stink bugs, because of the current knowledge of its classification.

Third, and related to the above, one can expect major changes in the classification of the stink bugs in the next couple of years, with a huge rearrangement of the groups currently assigned as subfamilies and tribes (Roca-Cusachs et al. 2022). For instance, there is high support for assigning the traditional subfamily Asopinae as a tribe within the subfamily Pentatominae, while the subfamily Podopinae needs to be split into three or four taxa, all recognized as a tribe as well (Roca-Cusachs et al. 2022). Other traditional taxa which include stink bugs of economic importance will be lumped (i.e., Antestiini and Nezarini) or split (i.e., Carpocorini). All these changes should be based in robust phylogenetic analysis, conducted under a hypothesis driven approach and including the best evidence available, reflecting a better understanding of the species and its properties.

3 Soybean Stink Bugs Around the World

More than three hundred species of phytophagous stink bugs are recorded on cultivated plants worldwide (Rider 2024; Schwertner and Cassis, unpublished), of which around 1/3 are found in commodity crops. Some of these species can damage extensive areas in different parts of the world, while others are considered secondary minor or local pests. At least one species, *Nezara viridula* (L.), is considered the main pest in major agricultural landscapes of the world, an example of the plasticity stink bugs can reach as pests (e.g., Esquivel et al. 2018). More recently, the invasive species *Bagrada hilaris* (Burmeister) and *Halyomorpha halys* (Stål) have put applied entomologists on alert whenever they were found (Bundy et al. 2018; Hamilton et al. 2018).

A seminal work reviewing information on stink bugs of economic importance was published by Panizzi et al. (2000). The authors compiled a list of references and summarized information from more than 85 species of phytophagous stink bugs of economic importance. Since then, information about the economically important species has been periodically published and/or revised (i.e., Koch et al. 2017; Grazia and Schwertner 2017; Sosa-Gómez et al. 2020).

For soybean, Panizzi et al. (2000) referred to thirteen main stink bug pest species and around fifteen less important ones. Seven species were recorded causing major damage to soybean, including the brown stink bugs, *Euschistus servus* (Say) and *Euschistus heros* (F.), the green stink bug, *Chinavia hilaris* (Say), the southern green stink bug, *Nezara viridula* (L.), the red banded stink bugs, *Piezodorus guildinii* (Westwood) and *Piezodorus hybneri* Gmelin, and the Neotropical brown-winged stink bug, *Edessa meditabunda* (F.). Minor damage to soybean was also cited for *Diceraeus furcatus* (F.) and *Euschistus tristigmus* (Say), which were considered of less importance at that time.

With the expansion of soybean cultivation over the last two decades, new challenges have arisen in managing pest insects (Bueno et al. 2021). Under a constantly changing landscape scenario, the relative importance of established and potential pest species is dynamic in soybean fields (i.e., Pilkay et al. 2015), and changes in species composition, distribution, and abundance have been recorded among different regions (i.e., Pezzini et al. 2019; Saldanha et al. 2024) and over the years (i.e., Horikoshi et al. 2021; Panizzi et al. 2022).

In an analysis of more than 300 papers published between 2000 and 2023, we found reference to at least 65 species of stink bugs associated with soybean worldwide (Table 1.2). Sixteen species were cited in five or more papers, representing ca. 85% of the records (Fig. 1.3). Most of the studies (64% of the papers) concerned species recorded in the Americas (Table 1.2), which reflects the importance of these countries to the rough world production of soybean (USDA 2024). The top three cited species were *E. heros* (cited in 19.4% of the papers), *N. viridula* (13.9%) and *P. guildinii* (13.9%). The two widely distributed stink bug pests, *H. halys,* and *N. viridula*, were cited in 20% of the studies. Only one species outside the Americas, *P. hybneri*, was among the top cited species; studies from Japan and South Korea are the main sources for this species, which is also recorded in the African and Indian continents (Ahmad et al. 2019).

Most of the species reviewed in Panizzi et al. (2000) were found in the literature published in the time span of our survey. The exceptions were *Chinavia acuta* (Dallas), *Chinavia marginata* (Palisot de Beauvois), and *Edessa ruformarginata* (De Geer). The species damaging soybean in Panizzi et al. (2000) are all included among the top cited species in our survey, and still are among the main species attacking this crop in the last 20 years. Nonetheless, our list also included at least one species which was not considered a soybean pest at all, *Halyomorpha halys*. This species was listed as a minor pest of fruits within its original distribution and currently is one of the threats to soybean in part of the US (Hamilton et al. 2018). Another example is the green belly stink bugs, *Diceraeus* spp., widely distributed in South America, and the growing importance of soybean in the last 20 years is well documented (i.e., Sosa-Gomez et al. 2020; Jacobi et al. 2022; Panizzi et al. 2022; Saldanha et al. 2024).

Other stink bugs species eventually collected in soybean, usually in low numbers, also appeared in our survey (Table 1.2), most of them recorded previously to the year 2000 (Panizzi and Slansky 1985; Link and Grazia 1987; Turnipseed and Kogan 1976; Zandigiacomo 1992). Unusual or unreported species cited in the soybean literature in the last 20 years included *Cholochrora permisilis* Horvath, *Cosmopepla lintneriana* (Kirkaldy), *Euschistus crassus* Dallas, *Mecidea major* Sailer, *Neotiglossa undata* (Say) and *Trichopepla semivittata* (Say) in North America (Suh et al. 2013; Koch and Pahs 2014; Pezzini et al. 2019), *Edessa loxdalii* Westwood and *Euschistus taurulus* Berg in South America (Lucini and Panizzi 2016; Hickmann et al. 2019).

Table 1.2 Diversity, current classification, and distribution of the soybean stink bugs recorded in the literature 2000–2023[a]. Classification follows Rider et al. (2018). Bold = species recorded in five or more references. P = cited in Panizzi et al. (2000). Distribution (Continents): AF = Africa, AS = Asia, CA = Central America, EU = Europe, OC = Oceania, NA = North America, SA = South America

Subfamily/Tribe	Species	P	Distribution
Edessinae	*Ascra bifida* Say, 1832		CA, NA
	Edessa loxdalii (Westwood, 1837)		CA, AS
	***Edessa meditabunda* (Fabricius, 1974)**	X	CA, AS
Pentatominae Aeliini	*Neotiglossa undata* (Say, 1831)		NA
Antestiini	*Plautia affinis* (Dallas, 1851)	X	OC
	Plautia stali Scott, 1874	X	AS, IN
Cappaeini	***Halyomorpha halys* (Stål, 1855)**	X	AF, AS, EU, NA, SA
Carpocorini	*Acledra bonariensis* (Stål, 1859)		SA
	Acledra fraternal (Stål, 1859)		SA
	Coenus delius (Say, 1831)		NA
	Cosmopepla lintneriana (Kirkaldy, 1909)		NA
	***Diceraeus furcatus* (Fabricius, 1775)**	X	SA
	***Diceraeus melacanthus* Dallas, 1851**	X	SA
	Dolycoris baccarum (Linnaeus, 1858)		EU
	Euschistus cornutus Dallas, 1851		SA
	Euschistus crassus (Dallas, 1851)		CA, NA
	Euschistus crenator (Fabricius, 1794)		CA, NA, SA
	***Euschistus heros* (Fabricius, 1798)**	X	SA
	Euschistus ictericus (Linnaeus, 1763)		NA
	Euschistus picticornis Stål, 1872		SA
	Euschistus quadrator Rolston, 1974		NA
	***Euschsitus servus* (Say, 1831)**	X	NA
	Euschistus taurulus Berg, 1878		SA
	***Eushistus tristigmus* (Say, 1831)**	X	NA
	***Euschistus variolarius* (Palisot de Beauvois, 1817)**		NA
	Holcostethus limbolarius (Stål, 1872)		NA
	Hymenarcis nervosa (Say, 1831)		NA
	Mormidea lugens (Fabricius, 1775)		NA
	Mormidea v-luteum (Lichtenstein, 1796)	X	SA
	Oebalus poecilus (Dallas, 1851)	X	SA
	Oebalus pugnax (Fabricius, 1775)		CA, NA, SA
	Oebalus ypsilogriseus (DeGeer, 1773)		AS
	Proxys albopunctulatus (Palisot de Beauvois, 1805)		AS
	Proxys punctulatus (Palisot de Beauvois, 1805)	X	CA, NA
	Tibraca limbativentris Stål, 1860	X	SA
	Trichopepla semivitatta (Say, 1831)		NA

(continued)

Table 1.2 (continued)

Subfamily/Tribe	Species	P	Distribution
Chlorocorini	*Loxa deducta* (Walker, 1867)	X	SA
	Mayrinia curvidens (Mayr, 1864)		SA
Eysarcorini	*Aspavia armigera* (Fabricius, 1781)	X	AF
Mecideini	*Mecidea major* Sailer, 1952		NA
Myrocheini	*Dictyotus coenosus* (Westwood, 1837)	X	OC
Nezarini	*Chinavia armigera* (Stål, 1859)	X	SA
	Chinavia aseada (Rolston, 1983)		SA
	Chinavia erythrocnemis (Berg, 1878)		SA
	***Chinavia hilaris* (Say, 1831)**	X	NA
	***Chinavia impicticornis* (Stål, 1872)**	X	SA
	Chinavia longicorialis (Breddin, 1901)		SA
	Chinavia nigridorsata (Breddin, 1901)	X	SA
	Chinavia obstinata (Stål, 1860)		SA
	Chinavia pengue (Rolston, 1983)		SA
	***Chinavia ubica* (Rolston,** 1983)		SA
	Chlorochroa persimilis Horvath, 1908		NA
	Nezara antennata Scott, 1874		AS
	***Nezara viridula* (Linnaeus, 1758)**	X	W
	Thyanta calceata (Say, 1831)		NA
	***Thyanta custator* (Fabricius, 1803)**	X	NA
	Thyanta pallidovirens (Stål, 1859)		NA
	***Thyanta perditor* (Fabricius, 1794)**	X	CA, NA, SA
Pentatomini	*Arvelius albopunctatus* (DeGeer, 1773)	X	CA, NA, SA
	Banasa dimidiata (Say, 1831)		NA
	Banasa euchlora Stål, 1872		NA
Piezodorini	***Piezodorus guildinii* (Westwood, 1837)**	X	CA, NA, SA
	***Piezodorus hybneri* (Gmelin, 1789)**	X	AF, AS, IN
	Piezodorus lituratus (Fabricius, 1794)		EU
	Piezodorus oceanicus (Montruzier, 1865)		AC
Strachiini	*Murgantia histrionica* (Hanh, 1834)		NA

[a]The search strategy for the survey included two online indexing sources, CABi Abstracts and Web of Science, using as initial queries the terms "Pentatomidae+ Soybean", "Pentatomidae+Glycine", "Stink bugs+Soybean" and "Stink bugs+ Glycine"; the search was oriented to retrieve information from the fields title, abstract and keywords, between 2000 and 2023. The abstract of each publication retrieved was checked, and only studies regarding stink bug-soybean interaction (field or/and laboratory experimental studies) were considered

4 The Most Common Species on Soybean

A detailed account of all species is out of the scope of this chapter. Nonetheless, in the next section, we provide updated information concerning the classification, identification, and current distribution of the most cited stink bugs associated with soybean in the last 20 years.

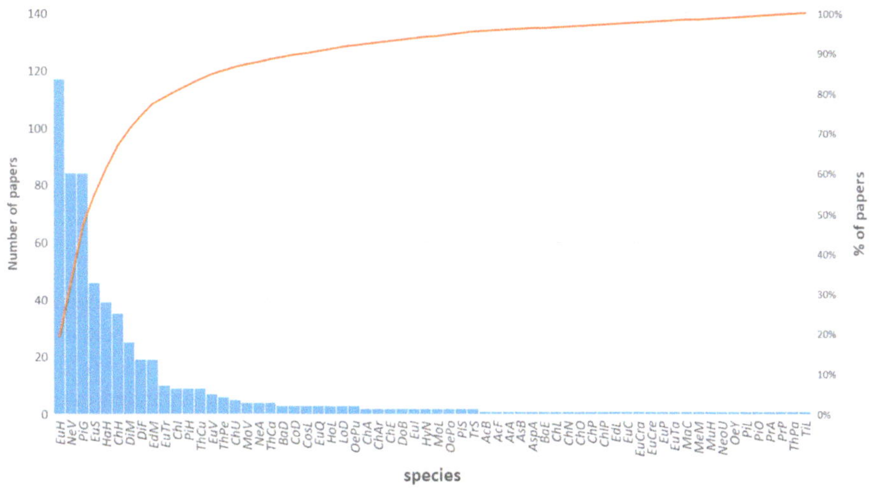

Fig. 1.3 Number and percentage of references citing each species of soybean stink bug in the literature, period 2000–2023. AcB, *Acledra bonariensis*; AcF, *Acledra fraterna*; ArA, *Arvelius albopunctatus*; AsB, *Ascra bifida*; AspA, *Aspavia armigera*; BaD, *Banasa dimidiata*; BaE, *Banasa euchlora*; ChAr, *Chinavia armigera*; ChA, *Chinavia asseada*; ChE, *Chinavia erythrocnemis*; ChH, *Chinavia hilaris*; ChI, *Chinavia impicticornis*; ChL, *Chinavia longicorialis*; ChN, *Chinavia nigridorsata*; ChO, *Chinavia obstinata*; ChP, *Chinavia pengue*, ChU, *Chinavia ubica*; ChlP, *Chlorochroa persimilis*; CoD, *Coenus delius*; CosL, *Cosmopepla lintneriana*; DiF, *Diceraeus furcatus*; DiM, *Diceraeus melacanthus*; DoB, *Dolycoris baccarum*; EdL, *Edessa loxdalii*, EdM, *Edessa meditabunda*; EuC, *Euschistus cornutus*; EuCra, *Euschistus crassus*; EuCre, *Euschistus crenator*; EuH, *Euschistus heros*; EuI, *Euschistus ictericus*; EuP, *Euschistus picticornis*; que, *Euschistus quadrator*; EuS, *Euschsitus servus*; EuTa, *Euschistus taurulus*, EuT, *Eushistus tristigmus*; EuV, *Euschistus variolarius*; HaH, *Halyomorpha halys*; HoL, *Holcostethus limbolarius*; HyN, *Hymenarcis nervosa*; LoD, *Loxa deducta*; MaC, *Mayrinia curvidens*; MeM, *Mecidea major*; MoL, *Mormidea lugens*; MoV, *Mormidea v-luteum*; MuH, *Murgantia histrionica*; NeoU, *Neotiglossa undata*, NeA, *Nezara antennata*, NeV, *Nezara viridula*; OeP, *Oebalus poecilus*; OePu, *Oebalus pugnax*; OeY, *Oebalus ypsilogriseus*; PiG, *Piezodorus guildinii*; PiH, *Piezodorus hybneri*; PiL, *Piezodorus lituratus*; PiO, *Piezodorus oceanicus*; PrA, *Proxys albopunctulatus*; PrP, *Proxys punctulatus*; PlS, *Plautia stali*; ThCa, *Thyanta* calceata, ThC, *Thyanta custator*; ThPa, *Thyanta pallidovirens*; ThP, *Thyanta perditor*; TiL, *Tibraca limbativentris*; TrS, *Trichopepla semivitatta*

4.1 Edessa meditabunda *(F.)*

The genus *Edessa* is the most speciose within Edessinae, which monophyly is strongly supported (Barcellos and Grazia 2003; Nunes et al. 2019; Roca-Cusachs et al. 2022). More recently some groups of species formerly included in *Edessa* have been recognized as distinct genera (Santos et al. 2015; Correia and Fernandes 2016). Several species groups within *Edessa* have been reviewed (e.g., Fernandes and Van Doesburg 2000; Silva et al. 2006; Silva and Fernandes 2012).

The brown-winged stink bug (Fig. 1.4a) is a common insect and widespread in South America (Fig. 1.5a). Identification of *E. meditabunda* can be difficult since the main characteristic used in its recognition (the dull brow hemelytra) is also found in other unrelated species of *Edessa*, i.e., *Edessa loxdalii* (Lucini and Panizzi 2016). There is no recent key published that allows the identification of this species among other edessines, and the only reliable character to recognize the species is the male genitalia.

4.2 Euschistus *Genus Group*

All genera are currently included in the *Euschistus* genus group and are currently treated within the tribe Carpocorini. The tribe Carpocorini is the most speciose tribe within Pentatomidae, with almost 130 genera and more than 500 species worldwide (Rider et al. 2018), and genera of economic importance can be separated into two broad unrelated monophyletic groups (Fig. 1.2): the *Carpocoris* group, which includes *Carpocoris* Kolenati and *Dolycoris* Mulsant and Rey, and the *Euschistus* group. The tribe did not represent a monophyletic group, and as addressed earlier in this chapter, a change in the classification of this group is expected in the following years.

The *Euschistus* group is restricted to the Central, North, and South American continents (Barão et al. 2020), and accounts for more than 1/3 of the species associated with soybean in our survey (Table 1.2).

The genus *Diceraeus* Dallas was recently recognized as a subgenus of *Dichelops* Spinola and includes five species distributed exclusively in South America (Barão et al. 2020), two of them considered of economic importance. Also known as the green belly stink bugs, both are among the most cited species in our survey (Table 1.2). *Diceraeus melacanthus* Dallas (Fig. 1.4c) has a widespread distribution (Fig. 1.5c), while *D. furcatus* (F.) (Fig. 1.4b) has expanded its distribution recently (Fig. 1.5b), see also Saldanha et al. 2024). Morphological identification of the species is addressed by Grazia (1978).

The genus *Euschistus* Dallas comprises more than 65 described species (Bianchi et al. 2017a; Paim et al. 2022), all restricted to the Western Hemisphere. Four species are among the most important economic stink bugs on soybean (Table 1.2). *Euschistus heros* (F.) (Figs. 1.4d and 1.5d) and *E. servus* (Say) (Figs. 1.4e and 1.6a) are key pests in this crop in South and North America, respectively. The species *Euschistus tristigmus* (Say) (Fig. 1.4f) and *E. variolarius* (Say) (Fig. 1.4g), both exclusively distributed in North America (Fig. 1.6b, c), were also included as the top cited species in our survey. Identification of the species level can be done following Rolston (1974, 1978, 1984), Esquivel et al. (2009), Bianchi et al. (2017a), Hickmann et al. (2019, 2021b) and Paim et al. (2022).

Traditionally, the North American species *E. servus* and *E. tristigmus* has been split into two subspecies (i.e., Froeschner 1988; McPherson and McPherson 2000), based on some minor morphological differences. Nonetheless, Rider and Swanson (2021) considered both as single species, with north and south forms. The synonymies of the subspecies were formally proposed by Paim et al. (2022).

Fig. 1.4 Soybean stink bugs. (**a**) *Edessa meditabunda* (F.); (**b**) *Diceraeus furcatus* (F.); (**c**) *Diceraeus melacanthus* Dallas; (**d**) *Euschistus heros* (F.); (**e**) *Euschistus servus* (Say); (**f**) *Euschistus tristigmus* (Say); (**g**) *Euschistus variolarius* (Say); (**h**) *Halyomorpha halys* (Stål). (Photo Credits. **a**, Ísis Meri Medri, iNaturalist website (photo 31209884); **b**, Gerónimo Martín Alonso, iNaturalist website (photo 7914161, CC-BY-NC-ND 4.0); **c**, Vinícius Rodrigues de Souza, iNaturalist website (photo 116543015); **d**, Ísis Meri Medri, iNaturalist website (photo 192705719); **e**, Mariko, iNaturalist website (photo 33184902 CC-BY-NC 4.0); **f**, Kevin, iNaturalist website (photo 321935880 CC-BY-NC 4.0); **g**, Kevin, iNaturalist website (photo 325638449 CC-BY-NC 4.0); **h**, Bill Keim, iNaturalist website (photo 342055502 CC-BY 4.0))

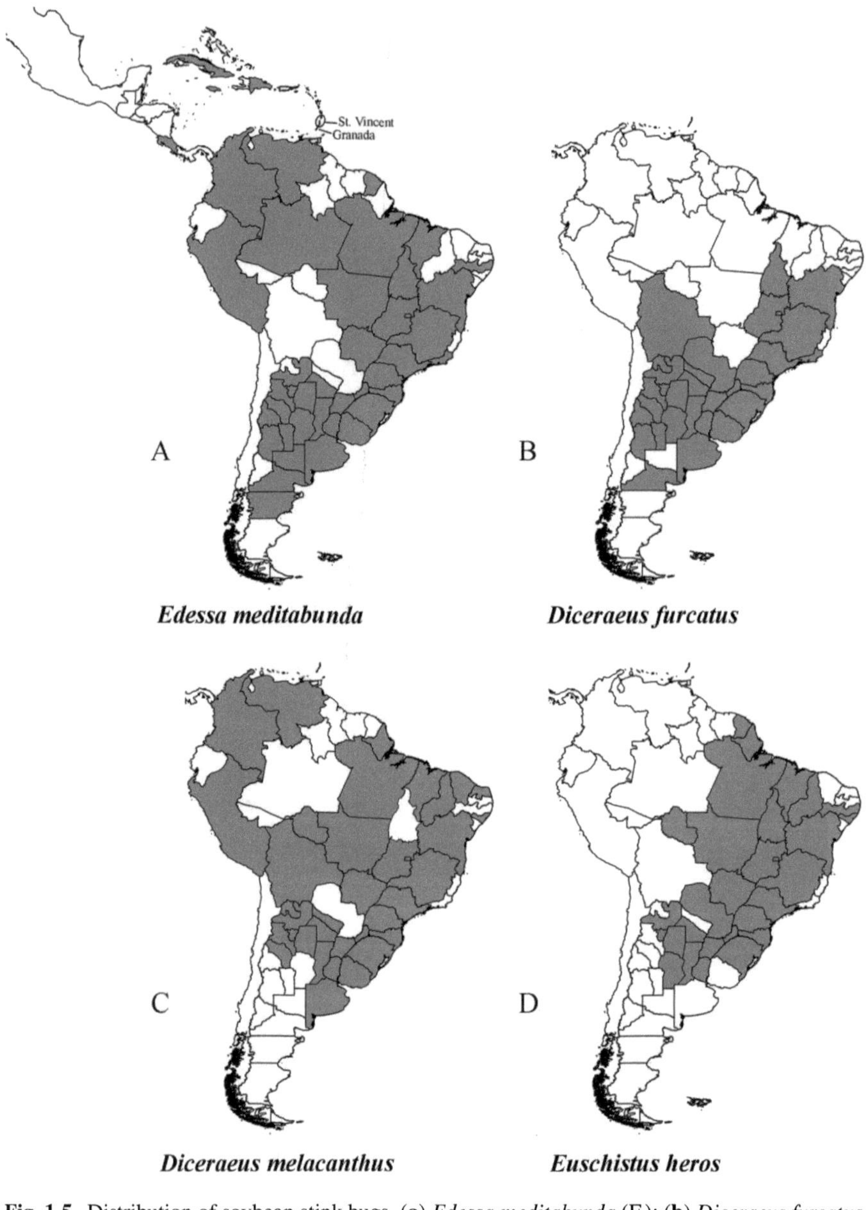

Fig. 1.5 Distribution of soybean stink bugs. (**a**) *Edessa meditabunda* (F.); (**b**) *Diceraeus furcatus* (F.); (**c**) *D. melacanthus* (Dallas); (**d**) *Euschistus heros* (F.)

Soares et al. (2018) recognized two populations strains of *E. heros* in Brazil with North-South allopatric distribution, in which individuals exhibit significant phenotypic, ecological, chromosomic, and genetic differences (Hickmann et al. 2021a, 2023). The southern strain is also recorded in Argentina (Balbi et al. 2023), where

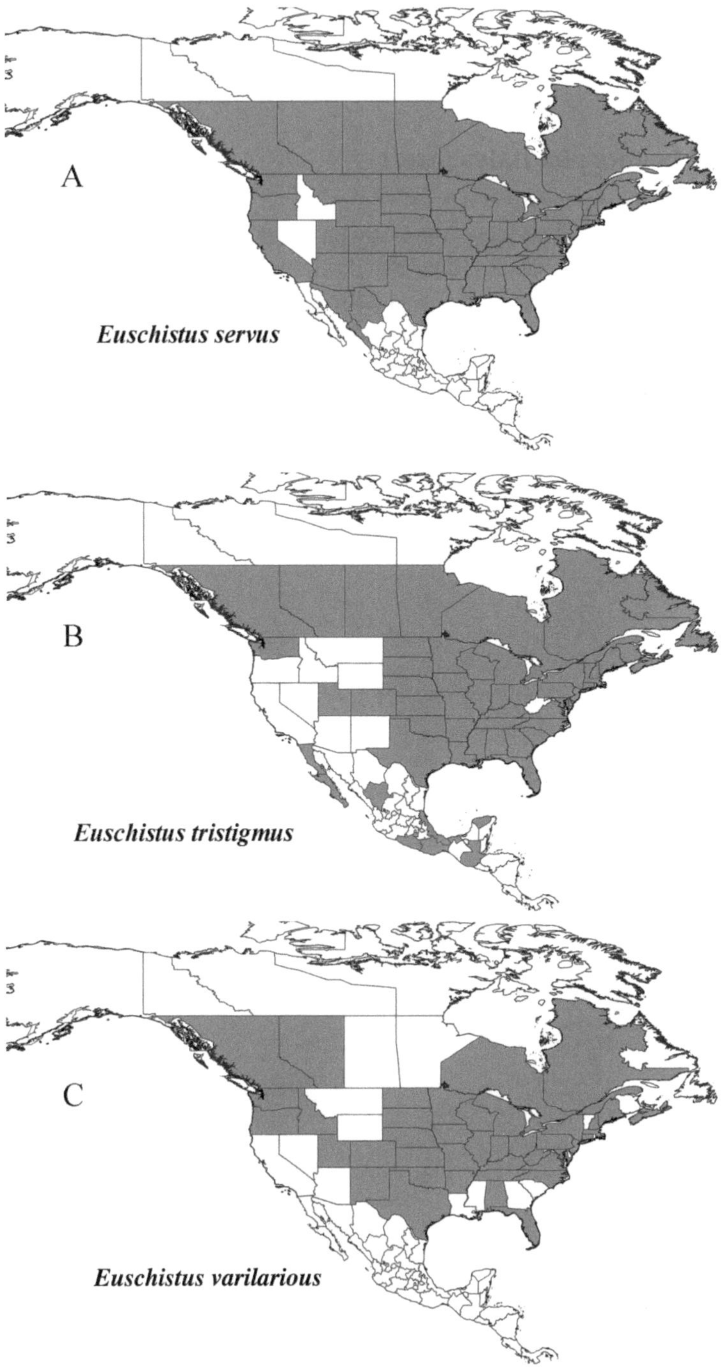

Fig. 1.6 Distribution of soybean stink bugs. (**a**) *Euschistus servus* (Say); (**b**) *Euschistus tristigmus* (Say); (**c**) *Euschistus variolarius* (Say)

E. heros has been recorded in soybean only recently (Saluso et al. 2011; Balbi et al. 2023).

4.3 Halyomorpha halys *(Stal)*

The genus *Halyomorpha* Mayr is currently part of the tribe Cappaeini Atkinson, and includes more than 35 species, distributed originally in the Afrotropical, Oriental, and Palearctic regions (Kment et al. 2021). The species *Halyomorpha halys* (Fig. 1.4h) is native to Southeast Asia and recently spread across most countries of the Northern Hemisphere (both in Europe and North America) and Chile in the Southern Hemisphere (Fig. 1.7a) (Hoebeke and Carter 2003; Lee et al. 2013; Lee 2015; Faundez and Rider 2017; Hamilton et al. 2018).

The taxonomy of *Halyomorpha* has been treated only on a regional basis, and the identification of the species in its native distribution can be difficult (Kment et al.

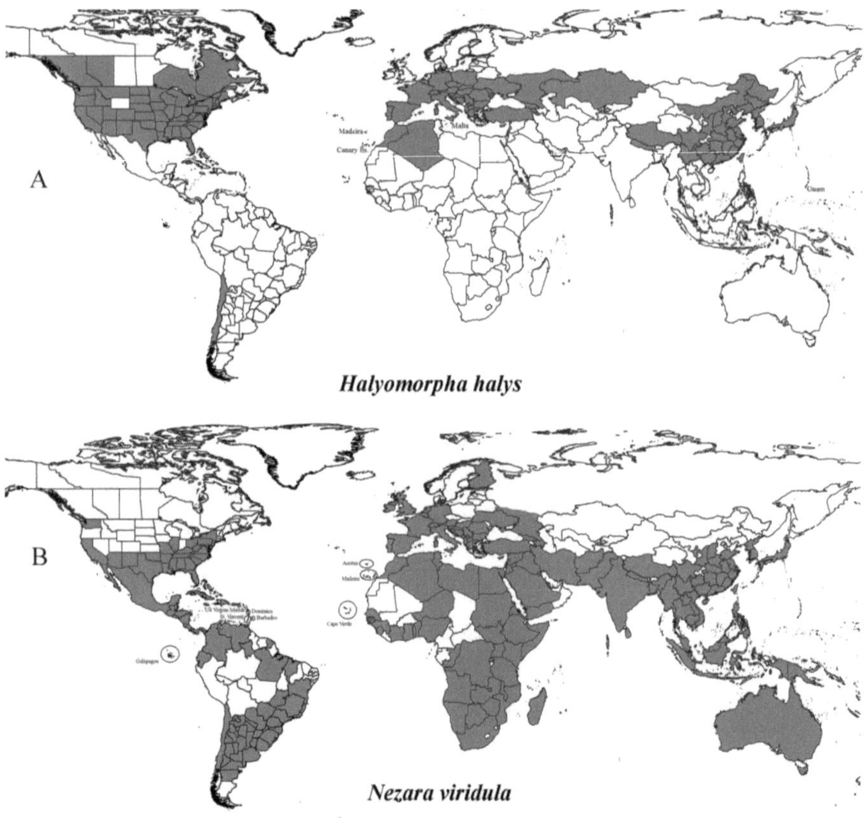

Fig. 1.7 Distribution of soybean stink bugs. (**a**) *Halyomorpha halys* (Stål); (**b**) *Nezara viridula* (L.)

2021; Salini et al. 2021). Species recorded in the Afrotropical were reviewed in Linnavuori (1982), while species of the Oriental region were reviewed by Ahmad and Zaidi (1989). A key to separate *H. halys* from similar species in Central Europe is presented by Wyniger and Kment (2010).

4.4 Nezara *Genus Group*

The *Nezara* group was first recognized by Linnavuori (1982) to include species of the genera *Aethemenes* Stål, *Acrosternum* Fieber, *Chinavia* Orian, and *Nezara* Amyot and Serville from the Afrotropical region. Schwertner (2005) tested the monophyly of the group and the relationship with other green stink bug genera included in the tribes Antestiini and Nezarini. The close relationship between the economically important genera *Chinavia*, *Chlorochroa* Stål, *Nezara*, and *Thyanta* Stål were supported more recently using combined morphological and molecular data (Genevcius et al. 2021; Roca-Cusachs et al. 2022; see also Fig. 1.2).

The tribe Nezarini comprises at least 26 genera and more than 250 species of green stink bugs that remain green after death (Rider et al. 2018). The monophyly of the tribe was not recovered by Roca-Cusachs et al. (2022), and as mentioned earlier in this chapter, the *Nezara* genus group is supported as part of a clade including also representatives of the tribes Antestiini and Mecideini.

The generic classification of these tribes is still confused given its wide distribution. The validity of the genus *Chinavia*, for example, was only recently settled (Schwertner 2005; Schwertner and Grazia 2007). Orian (1965) proposed the genus *Chinavia* to include nine species from Africa, Madagascar, and Mauritius, all removed from *Acrosternum*. Linnavuori (1972) did not agree with Orian and synonymized both genera, although Roche (1977) considered *Chinavia* a valid genus. Rolston (1983) discussed the classification of *Acrosternum* and considered *Chinavia* a subgenus, including all species from the Western Hemisphere included in *Acrosternum*. Schwertner (2005), based on morphological evidence, supported *Chinavia* as a distinct monophyletic group, more related to *Nezara* than to *Acrosternum*, thus considering *Chinavia* a valid genus. These results have been confirmed more recently based on molecular markers (Genevcius et al. 2021; Roca-Cusachs et al. 2022). Currently, the genus *Chinavia* comprises more than 80 species distributed in Africa (13 spp.) and in Central, North, and South America (72 spp.), including one of the most studied species regarding soybean in the US, the green stink bug *Chinavia hilaris* (Say) (Figs. 1.8a and 1.9a). Other species of *Chinavia* are cited on soybean (Table 1.2), usually collected in small numbers (i.e., Aquino et al. 2019; Saldanha et al. 2024) Two species, *Chinavia impicticornis* (Stål) (Figs. 1.8b and 1.9b) and *C. ubica* (Rolston) (Figs. 1.8c and 1.9c), were among the top cited species in our survey. An online electronic identification key to all species of *Chinavia* is available (Servino and Schwertner 2020), as well as a conspectus of the genus including images, diagnosis, host plants and distributional data for each species (Servino and Schwertner 2020).

Fig. 1.8 Soybean stink bugs. (**a**) *Chinavia hilaris* (Say); (**b**) *C. impicticornis* (Stål); (**c**) *Chinavia ubica* (Rolston); (**d**) *Nezara viridula* (L.); (**e**) *Thyanta custator* (F.); (**f**) *Thyanta perditor* (F.); (**g**) *Piezodorus guildinii* (Westwood); (**h**) *P. hybneri* (Gmelin). (Photo credits. **a**, Rob Van Epps, iNaturalist website (photo 162742783 CC-BY-NC 4.0); **b**, Ísis Meri Medri, iNaturalist website (photo 55554503); **c**, Vinícius Rodrigues de Souza, iNaturalist website (photo 149154374); **d**, Christopher Caine, iNaturalist website (photo 107190004 CC-BY-NC 4.0); **e**, Rob Van Epps, iNaturalist website (photo 301490662 CC-BY-NC 4.0); **f**, Carlos Alexandre Mattos Raposo, iNaturalist website (photo 334118572 CC-BY-NC 4.0); **g**, Tamara Moraes; **h**, Taewoo Kim, iNaturalist website (photo 11889442))

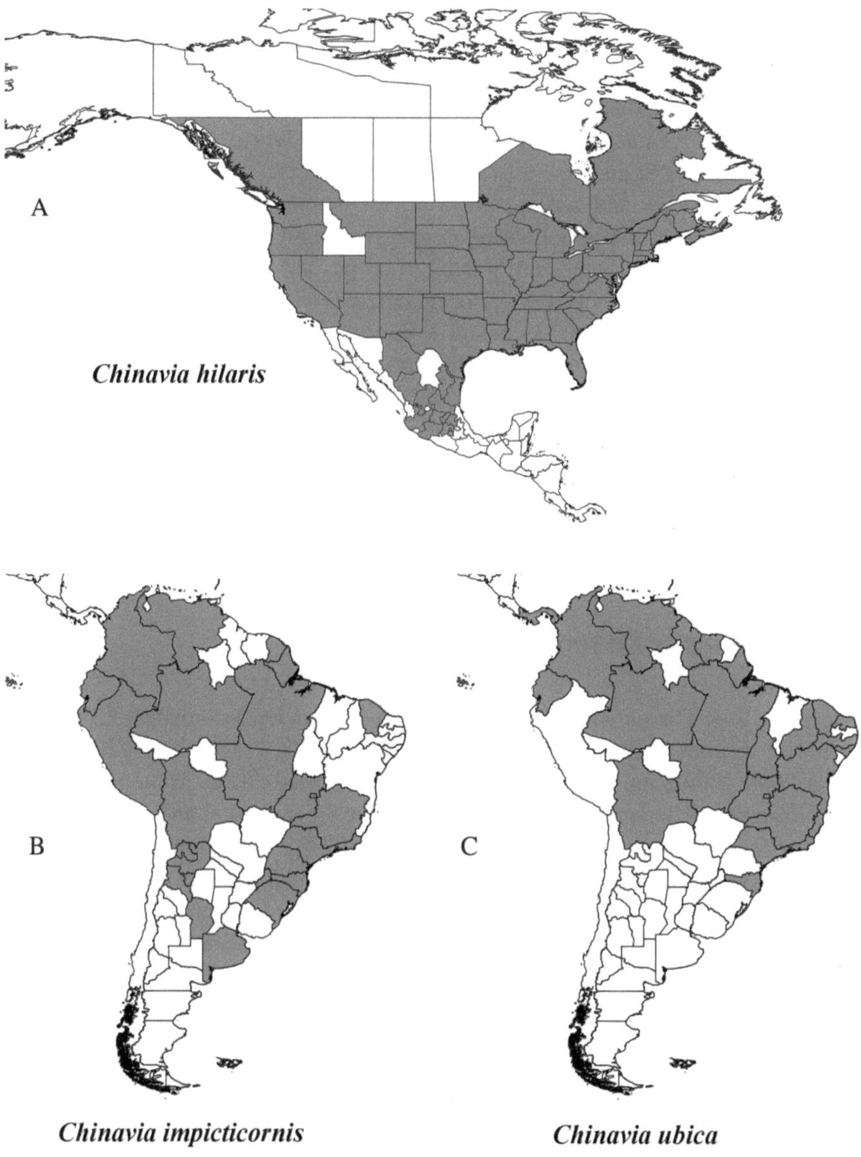

Fig. 1.9 Distribution of soybean stink bugs. (**a**) *Chinavia hilaris* (Say); (**b**) *C. impicticornis* (Stål); (**c**) *C. ubica* (Rolston)

The genus *Nezara* Amyot and Serville represent a monophyletic taxon (Ferrari et al. 2010; Roca-Cusachs et al. 2022), with *N. viridula* (Fig. 1.8d) widely distributed (Fig. 1.7b) and considered highly polyphagous (Esquivel et al. 2018). Also known as the southern green stink bug, this species is very similar to *N. antennata*

Scott, a species also found on soybean but restricted to Japan, and *N. yunnana* Zheg, restricted to the North of India. The three species represent a lineage separated to the African species within the genus (Ferrari et al. 2010). A key to identifying all *Nezara* species is available in Ferrari et al. (2010).

The genus *Thyanta* is also restricted to the New World, occurring widely in Central, North, and South America (Rider and Chapin 1992). An excellent revision, including all species and settling down all taxonomic problems concerning the genus and still the most important reference, was published by Rider and Chapin (1991, 1992). Several species are related to crops, and at least four species have been cited on soybean since 2000. *Thyanta custator* (F.) (Figs. 1.8e and 1.10a) was the more cited species in the US, while *T. perditor* (F.) (Figs. 1.8f and 1.10b) was the main species in South America. The similarities between the North American species, and the difficulties in identifying them, including the recognition of the subspecies within *T. custator* were addressed by Rider and Swanson (2021). Otherwise, although highly speciose in South America, the identification of *T. perditor* is more straightforward. Identification key to all species in Rider and Chapin (1991, 1992).

4.5 The Genus **Piezodorus** *Fieber*

The genus has a worldwide distribution, with 11 species recorded in all biogeographical regions (Linnavuori 1982; Froeschner 1988; Cassis and Gross 2002; Rider 2006; Salini and Viraktamath 2015; Moraes et al. 2023). Monophyly of the *Piezodorus* is well supported (Ahmad 1995), but the evolution and distribution of the species are still poorly studied (Moraes et al. 2023). At least three groups of species are recognized within the genus (Staddon and Ahmad 1995), with the important economic species included in the *hybneri* and *lituratus* groups (Table 1.2). The identification of these groups of species is straightforward (Ribes and Pagola-Carte 2013), but the separation of the species included in the *hybneri* group (*P. guildinii*, *P. hybneri* and *P. oceanicus* Montrouzier), are more subtle (Ahmad et al. 2019). A key to all species of *Piezodorus* is presented in Ahmad (1995). The most common species found on soybean include *P. guildinii* (Westwood) (Fig. 1.8g), recorded along all American continents (Fig. 1.11a), and *P. hybneri* (Gmelin) (Fig. 1.8h), distributed in the Old World tropics (Fig. 1.11b). More recently, *P. guildinii* was identified among a large sample of stink bugs from South Africa (Sonnekus et al. 2022; Moraes et al. 2023), challenging the current hypothesis of the origin of this species (Moraes et al. 2023).

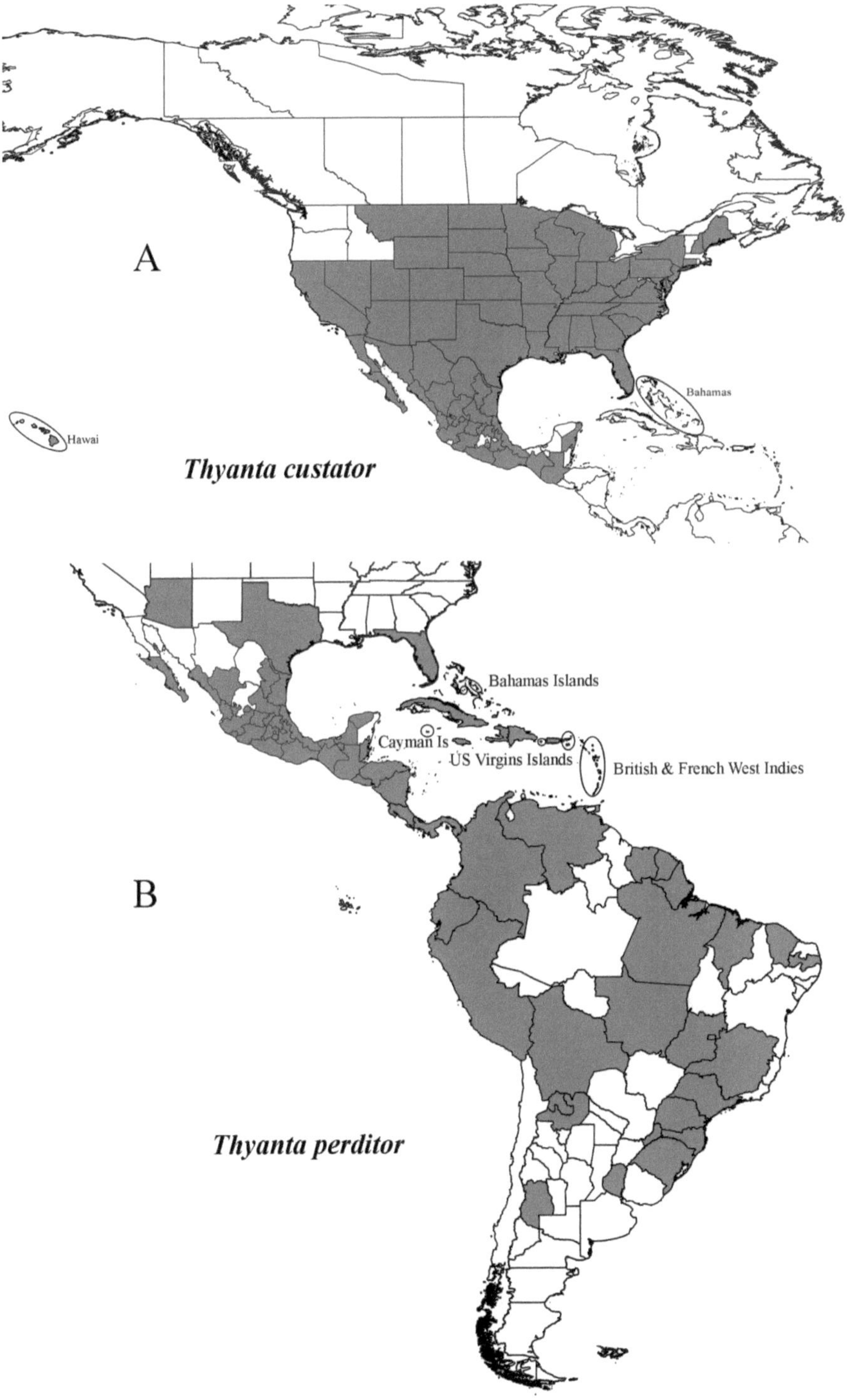

Fig. 1.10 Distribution of soybean stink bugs. (**a**) *Thyanta custator* (F.); (**b**) *Thyanta perditor* (F.)

Fig. 1.11 Distribution of soybean stink bugs. (**a**) *Piezodorus guildinii* (Westwood); (**b**) *P. hybneri* (Gmelin)

5 Concluding Remarks

Classification of the family Pentatomidae, which include some of the most important soybean pests around the world is in a state of flux. The current subfamilial and tribal arrangement is inconsistent with the evolutionary history of the group, hampering comparative studies. Recent analysis supports the recognition of several monophyletic groups, including the *Euschistus* and *Nezara* genus groups. Important steps have been taking to clear up and understand these taxa and its properties in a phylogenetic context. Major changes in the classification of the stink bugs in the next couple of years are expected.

At least 65 species of stink bugs associated with soybean worldwide were found in the literature of the last 20 years. Sixteen species were cited in five or more papers, representing ca. 85% of the records. The top three cited species were *E. heros*, *N. viridula,* and *P. guildinii* . The two widely distributed stink bug pests were *Halyomorpha halys,* and *Nezara viridula*. Other important species includes the green belly stink bugs (*Diceraeus furcatus* and *D. melacanthus*), the green stink bug (*Chinavia hilaris*), and the brown stink bugs (*Euschistus servus, E. tristigmus* and *E. variolarius*). Only one species outside the Americas, *Piezodorus hybneri*, was among the top cited species. Those 16 species are considered major or secondary pests in different countries. The remaining 49 stink bugs species associated with soybean are usually found in low numbers. Unusual or unreported species cited in the soybean literature included *Cholochrora permisilis* Horvath, *Cosmopepla lintneriana* (Kirkaldy), *Euschistus crassus* Dallas, *Mecidea major* Sailer, *Neotiglossa undata* (Say) and *Trichopepla semivittata* (Say) in North America, *Edessa loxdalii* Westwood and *Euschistus taurulus* Berg in South America.

Acknowledgments We acknowledge the National Research Council (CNPq) for the fellowship to JG (Proc. 310747/2020-1) and funding to CFS (Proc. n. 409988/2021-9). This work also benefited from a grant of the Fundação de Amparo a Pesquisa do Estado de São Paulo (FAPESP) to CFS (Proc. n. 2018/09469-5). We also acknowledge the authors of the photographs for kindly providing a license to illustrate this chapter.

References

Ahmad I (1995) A review of pentatomine legume bug genus *Piezodorus* Fieber (Hemiptera: Pentatomidae: Pentatominae) with its cladistic analysis. Proc Pak Congr Zool 15:329–358

Ahmad I, Zaidi RH (1989) A revision of the genus *Halyomorpha* Mayr (Hemiptera: Pentatomidae: Pentatominae: Carpocorini) from the Indo-Pakistan subcontinent with description of a new species from Potohar region of Pakistan and their cladistic analysis. Proc Pak Congr Zool 9:237–153

Ahmad I, McPherson JE, Navaid R, Bundy CS (2019) Systematic relationship between *Piezodorus guildinii* and *P. hybneri* (Hemiptera: Heteroptera: Pentatomidae) and diagnostic characters to separate species. Zootaxa 4613:443–462

Aquino MFS, Sujii ER, Borges M, Blassioli-Moraes MC, Laumann RA (2019) Diversity of stink bug adults and their parasitoids in soybean crops in Brazil: influence of a latitudinal gradient and insecticide application intensity. Environ Entomol 48:105–113

Balbi EI, Flores FM, Arneodo JD (2023) First genetic analysis of *Euschistus heros* F. (Hemiptera: Pentatomidae) from Argentina and its relative abundance in a newly invaded soybean region. Phytoparasitica 51:379–384

Barão KR, Ferrari A, Grazia J (2020) Phylogenetic analysis of the *Euschistus* group (Hemiptera: Pentatomidae) suggests polyphyly of *Dichelops* Spinola, 1837 with the erection of *Diceraeus* Dallas, 1851, stat. Rev. Austral Entomol 59: 770–783

Barcellos A, Grazia J (2003) Cladistics analysis and biogeography of *Brachystethus* Laporte (Heteroptera, Pentatomidae, Edessinae). Zootaxa 256:1–14

Bianchi FM, Barao KR, Grazia J (2017a) Review of the *sulcacitus* group of *Euschistus* (Pentatomidae: Pentatominae: Carpocorini) with description of the internal female genitalia and a new species. Zootaxa 4362:348–358

Bianchi FM, Depra M, Ferrari A, Grazia J, Valente VL, Campos LA (2017b) Total evidence phylogenetic analysis and reclassification of *Euschistus* Dallas within Carpocorini (Hemiptera: Pentatomidae: Pentatominae). Syst Entomol 42(2):399–409

Bueno ADF, Panizzi AR, Hunt TE, Dourado PM, Pitta RM, Gonçalves J (2021) Challenges for adoption of integrated pest management (IPM): the soybean example. Neotrop Entomol 50:5–20

Bundy S, Esquivel J, Panizzi AR, Eger J, Davis J, Walker A (2018) *Piezodorus guildinii* (Westwood). In: McPherson J (ed) Biology of invasive stink bugs and related species. CRC Press, Boca Raton, pp 425–451

Campos LA, Grazia J (2006) Análise cladística e biogeografia de Ochlerini (Hemiptera, Pentatomidae, Discocephalinae). Iheringia Ser Zool 96:147–163

Cassis G, Gross G (2002) Hemiptera: Heteroptera (Pentatomomorpha). In: Houston W, Wells A (eds) Zoological catalogue of Australia. Csiro Publishing, p xiv+737

Correia AO, Fernandes JAM (2016) *Grammedessa*, a new genus of Edessinae (Hemiptera: Heteroptera: Pentatomidae). Zootaxa 4107:541–565

da Silva VJ, Fernandes JAM (2012) A new species group in *Edessa* Fabricius, 1803 (Heteroptera: Pentatomidae: Edessinae). Zootaxa 3313:12–22

Esquivel JF, Anderson RM, Droleskey RE (2009) A visual guide for identification of *Euschistus* spp. (Hemiptera: Pentatomidae) in Central Texas. Southwest Entomol 4:485–488

Esquivel JF, Musolin DL, Jones WA, Rabitsch W, Greene JK, Toews MD, Schwertner CF, Grazia J, McPherson RM (2018) *Nezara viridula* (L.). In: McPherson J (ed) Biology of invasive stink bugs and related species. CRC Press, USA, Boca Raton, pp 351–424

Faúndez EI, Rider DA (2017) The brown marmorated stink bug *Halyomorpha halys* (Stål, 1855) (Heteroptera: Pentatomidae) in Chile. Arq Entomol 17:305–307

Fernandes JAM, van Doesburg PH (2000) The *E. cervus*-group of *Edessa* Fabricius, 1903 (Heteroptera: Pentatomidae: Edessinae). Zool Meded Leiden 74:151–165

Ferrari A, Schwertner CF, Grazia J (2010) Review, cladistics analysis and biogeography of *Nezara* Amyot and Serville (Hemiptera: Pentatomidae). Zootaxa 2424:1–41

Froeschner RC (1988) Family Pentatomidae Leach, 1815. The stink bugs. In: Henry TJ, Froeschner RC (eds) Catalog of the Heteroptera, or true bugs, of Canada and the continental United States. E.J. Brill, Leiden, New York, pp 544–607

Gapud VP (1991) A generic revision of the subfamily Asopinae, with consideration on its phylogenetic position in the family Pentatomidae and superfamily Pentatomoidea (Hemiptera-Heteroptera). Philipp Entomol 8:865–961

Genevcius BC, Greve C, Koehler S, Simmons RB, Rider DA, Grazia J (2021) Phylogeny of the stink bug tribe Chlorocorini (Heteroptera, Pentatomidae) based on DNA and morphological data: the evolution of key phenotypic traits. Syst Entomol 46:327–338

Grazia J (1978) Revisão do gênero *Dichelops* Spinola, 1837 (Heteroptera, Pentatomidae, Pentatomini). Iheringia Ser Zool 53:3–119

Grazia J, Schwertner CF (2017) Stink bug phylogeny classification, biology, and reproductive behavior. In: Cokl A, Borges M (eds) Stink bugs: biorational control based on communication processes. CRC Press, pp 226–246
Grazia J, Schuh RT, Wheeler WC (2008) Phylogenetic relationships of family groups in Pentatomoidea based on morphology and DNA sequences (Insecta: Heteroptera). Cladistics 24:932–976
Grazia J, Panizzi AR, Greve C, Schwertner CF, Campos LA, Garbelotto TA, Fernandes JAM (2015) Stink bugs (Pentatomidae). In: Panizzi AR, Grazia J (eds) True bugs (Heteroptera) of the Neotropics. Springer, Dordrecht, pp 681–756
Gross GF (1975) Handbook of the flora and fauna of South Australia. Plant-feeding and other bugs (Hemiptera) of South Australia. Heteroptera – part 1. Handbooks Committee, South Australian Government, Adelaide, Australia, p 250
Gross GF (1976) Handbook of the flora and fauna of South Australia. Plant-feeding and other bugs (Hemiptera) of South Australia. Heteroptera – part 2. Handbooks Committee, South Australian Government, Adelaide, Australia, p 250
Hamilton GC, Ahn JJ, Bu W-J, Leskey TC, Nielsen AL, Park Y-L, Rabitsch W, Hoelmer KA (2018) *Halyomorpha halys* (Stål). In: McPherson JE (ed) Invasive stink bugs and related species (Pentatomoidea): biology, higher systematics, semiochemistry, and management. CRC Press, pp 243–292
Hickmann F, Moraes T, Bianchi FM, Corrêa AS, Schwertner CF (2019) Integrating data to redescribe *Euschistus taurulus* Berg (Hemiptera: Pentatomidae). Zootaxa 4688:119–134
Hickmann F, Cordeiro EG, Soares PL, Aurélio MSL, Schwertner CF, Corrêa AS (2021a) Reproductive patterns drive the gene flow and spatial dispersal of *Euschistus heros* (Hemiptera: Pentatomidae). J Econ Entomol 114:2346–2354
Hickmann F, Savaris M, Correa AS, Schwertner CF (2021b) *Euschistus crenator* (Fabricius) (Hemiptera: Pentatomidae): a new invasive species on soybean fields in northern Brazil. Neotrop Entomol 50:497–503
Hickmann F, Cordeiro EMG, Aurélio MSL, Saldanha AV, Corrêa AS (2023) Host preference and survivorship of *Euschistus heros* strains on cotton and soybean. Entomol Exp Appl 171:564–571
Hoebeke ER, Carter ME (2003) *Halyomorpha halys* (Stål) (Heteroptera: Pentatomidae): a polyphagous plant pest from Asia newly detected in North America. Proc Entomol Soc Wash 105:225–237
Horikoshi RJ, Dourado PM, Berger GU, de S Fernandes D, Omoto C, Willse A, Martinelli S, Head GP, Corrêa AS (2021) Large-scale assessment of lepidopteran soybean pests and efficacy of Cry1Ac soybean in Brazil. Sci Rep 11:15956
Jacobi VG, Fernández PC, Zavala JA (2022) The stink bug *Dichelops furcatus*: a new pest of corn that emerges from soybean stubble. Pest Manag Sci 78:2113–2120
Kment P, Salini S, Rédei D, Rider D (2021) *Halyomorpha halys* fixed as the type species of the genus *Halyomorpha* (Hemiptera: Heteroptera: Pentatomidae). Acta Entomol Mus Natl Pragae 61:615–630
Koch RL, Pahs T (2014) Species composition, abundance, and seasonal dynamics of stink bugs (Hemiptera: Pentatomidae) in Minnesota soybean fields. Environ Entomol 43:883–888
Koch R, Pezzini DT, Michel AP, Hunt TE (2017) Identification, biology, impacts, and management of stink bugs (Hemiptera: Heteroptera: Pentatomidae) of soybean and corn in the Midwestern United States. J Integr Pest Manag 8:1–14
Lee DH (2015) Current status of research progress on the biology and management of *Halyomorpha halys* (Hemiptera: Pentatomidae) as an invasive species. Appl Entomol Zool 50:277–290
Lee DH, Shor BD, Joseph SV, Bergh JC, Leskey TC (2013) Review of the biology, ecology, and management of *Halyomorpha halys* (Hemiptera: Pentatomidae) in China, Japan, and the Republic of Korea. Environ Entomol 42:627–641
Leston D (1953) On the wing-venation, male genitalia and spermatheca of *Podops inuncta* (F.), with a note on the diagnosis of the subfamily Podopinae Dallas (Hem., Pentatomidae). J Soc Br Entomol 4:129–135

Leston D (1954a) Notes on the Ethiopian Pentatomoidea (Hem.) XII. On some specimens from southern Rhodesia, with an investigation of certain features in the morphology of *Afrius figuratus* (Germar) and remarks upon the genitalia in Amyotinae. Occ Papers Nat Mus South Rhodesia 19:678–686

Leston D (1954b) Notes on the Ethiopian Pentatomoidea (Hemiptera): XVII, Tessaratominae, Dinidorinae and Phylocephallinae of Angola. Public Cul Compan Diamantes Angola 24:11–22

Link D, Grazia J (1987) Pentatomídeos da região central do Rio Grande do Sul (Heteroptera). An Soc Entomol Brasil 16:115–129

Linnavuori RE (1972) Studies on African Pentatomoidea. Arq MusBocage 3:395–434

Linnavuori RE (1982) Pentatomidae and Acanthosomatidae (Heteroptera) of Nigeria and the Ivory Coast, with remarks on species of the adjacent countries in West and Central Africa. Acta Zool Fennica 163:1–176

Lucini T, Panizzi AR (2016) Three new host plants of *Edessa loxdalii* (Hemiptera: Heteroptera: Pentatomidae) and notes on its rearing in the laboratory. Fla Entomol 99:139–141

McDonald FJD (1966) The genitalia of North American Pentatomoidea (Hemiptera: Heteroptera). Quaest Entomol 2:7–150

McPherson JE, McPherson RM (2000) Stink bugs of economic importance in America North of Mexico. CRC Press, Boca Raton

Moraes T, Santos LMA, Schwertner CF, Corrêa AS (2023) Molecular insights on the historical dispersion of *Piezodorus guildinii* (Hemiptera: Pentatomidae) in Brazil. J Econ Entomol 116:2173–2183

Nunes B, Wallner A, Fernandes JA (2019) *Anisoedessa*, a new genus of Edessinae (Hemiptera: Heteroptera: Pentatomidae), and considerations on Edessinae relationships based on cladistic analysis. Arthropod Syst Phylogeny 77:215–237

Orian AJ (1965) A new genus of Pentatomidae from Africa, Madagascar and Mauritius (Hemiptera). Proc Royal Entomol Soc Lond Series B Taxonomy 34:25–29

Paim MR, Grazia J, Rider DA, Bianchi FM (2022) Revisiting Stål's thoughts: formalizing the *ictericus* group in *Euschistus* (*Euschistus*) (Hemiptera: Heteroptera: Pentatomidae). Zootaxa 5169:501–537

Panizzi AR, Grazia J (eds) (2015) True bugs (Heteroptera) of the Neotropics. Springer, Dordrecht

Panizzi AR, Slansky F Jr (1985) Review of phytophagous pentatomids (Hemiptera: Pentatomidae) associated with soybean in the Americas. Fla Entomol 68:184–214

Panizzi AR, McPherson JE, James DG, Javahery M, McPherson RM (2000) Stink bugs (Pentatomidae). In: Schaefer CW, Panizzi AR (eds) Heteroptera of economic importance. CRC Press, Boca Raton, pp 421–474

Panizzi AR, Lucini T, Aldrich J (2022) Dynamics in pest status of phytophagous stink bugs in the Neotropics. Neotrop Entomol 51:18–31

Pezzini DT, DiFonzo CD, Finke DL, Hunt TE, Knodel JJ, Krupke CH, McCornack B, Michel AP, Philips CR, Varenhorst AJ, Wright RJ, Koch RL (2019) Community composition, abundance, and phenology of stink bugs (Hemiptera: Pentatomidae) in soybean in the north central region of the United States. J Econ Entomol 112:1722–1731

Pilkay GL, Reay-Jones FP, Toews MD, Greene JK, Bridges WC (2015) Spatial and temporal dynamics of stink bugs in southeastern farmscapes. J Insect Sci 15:23

Ribes J, Pagola-Carte S (2013) Hémiptères Pentatomoidea Euro-Méditerranéens. Vol. 2: Pentatominae (suite). Faune de France 96. Fédér Franç Soc Scien Natur, Paris, p 423

Rider DA (2006) Family Pentatomidae. In: Aukema B, Rieger C (eds) Catalogue of the Heteroptera of the Palaearctic region, vol 5. Entomological Society, Amsterdam, pp 233–402

Rider DA (2024) Pentatomoidea home page: host index. North Dakota State University, Fargo. Available at: https://www.ndsu.edu/pubweb/~rider/Pentatomoidea/Hosts/host_index.htm

Rider DA, Chapin JB (1991) Revision of the genus *Thyanta* Stal, 1862 (Heteroptera: Pentatomidae) I. South America. J New York Entomol Soc 99:1–77

Rider DA, Chapin JB (1992) Revision of the genus *Thyanta* StDl, 1862 (Heteroptera: Pentatomidae) II. North America, Central America, and the West Indies. J New York Entomol Soc 100:42–98

Rider DA, Swanson DR (2021) A distributional synopsis of the Pentatomidae (Heteroptera) north of Mexico, including new state and provincial records. Zootaxa 5015:1–69
Rider DA, Schwertner CF, Vilímová J, Rédei D, Kment P, Thomas DB (2018) Higher systematics of Pentatomoidea. In: McPherson J (ed) Biology of invasive stink bugs and related species. CRC Press, Boca Raton, p 177
Rieppel O (2016) Phylogenetic Systematics. Haeckel to Hennig. CRC Press, Boca Raton, 402p
Roca-Cusachs M, Schwertner CF, Kim J, Eger J, Grazia J, Jung S (2022) Opening Pandora's box: molecular phylogeny of the stink bugs (Hemiptera: Heteroptera: Pentatomidae) reveals great incongruences in the current classification. Syst Entomol 47:36–51
Roche PJL (1977) Pentatomidae of the granitic islands of Seychelles (Heteroptera). Rev Zool Africaine 91:558–572
Roell T, Campos LA (2019) Phylogeny of Ochlerini (Hemiptera: Pentatomidae: Discocephalinae) and the evolution of the apical tarsomere in hind legs. Zool J Linnean Soc 185:1137–1149
Rolston LH (1974) Revision of the genus *Euschistus* in Middle America (Hemiptera, Pentatomidae, Pentatomini). Entomol Am 48:1–102
Rolston LH (1978) A new subgenus of *Euschistus* (Hemiptera: Pentatomidae). J New York Entomol Soc 86:102–120
Rolston LH (1983) A revision of the genus *Acrosternum* Fieber, subgenus *Chinavia* Orian, in the western hemisphere (Hemiptera: Pentatomidae). J New York Entomol Soc 91:97–176
Rolston LH (1984) Key to the males of the nominate subgenus of *Euschistus* in South America, with descriptions of three new species (Hemiptera: Pentatomidae). J New York Entomol Soc 92:352–364
Rolston LH, McDonald FJD (1979) Keys and diagnoses for the families of Western Hemisphere Pentatomoidea, subfamilies of Pentatomidae and tribes of Pentatominae (Hemiptera). J New York Entomol Soc 87:189–207
Saldanha AV, Horikoshi R, Dourado P, Lopez-Overejo RF, Berger GU, Martinelli S, Head GP, Moraes T, Corrêa AS, Schwertner CF (2024) The first extensive analysis of species composition and abundance of stink bugs (Hemiptera: Pentatomidae) on soybean crops in Brazil. Pest Manag Sci 80:3945–3956
Salini S, Viraktamath CA (2015) Genera of Pentatomidae (Hemiptera: Pentatomoidea) from South India–an illustrated key to genera and checklist of species. Zootaxa 3924:1–76
Salini S, David KJ, Pratheepa M (2021) Does India have the invasive brown marmorated stink bug, *Halyomorpha halys* (Stål). Curr Sci 120:268–269
Saluso A, Xavier L, Silva FAC, Panizzi AR (2011) An invasive pentatomid pest in Argentina: neotropical brown stink bug, *Euschistus heros* (F.) (Hemiptera: Pentatomidae). Neotrop Entomol 40:704–705
Sampaio VG, Campos LA (2023) *Paropsocolis*, a new pentatomid genus with two new species from Brazil (Heteroptera, Pentatomidae, Ochlerini). Zootaxa 5323:553–566
Santibáñez-López CE, Ojanguren-Affilastro AA, Sharma PP (2020) Another one bites the dust: taxonomic sampling of a key genus in phylogenomic datasets reveals more non-monophyletic groups in traditional scorpion classification. Invertebr Syst 34:133–143
Santos BT, Silva VJ, Fernandes JAM (2015) Revision of *Ascra* with proposition of the *bifida* species group and description of two new species (Hemiptera: Pentatomidae: Edessinae). Zootaxa 4034:445–470
Schuh RT, Weirauch C (2020) True bugs of the world (Hemiptera: Heteroptera): classification and natural history. Siri Scientific Press, Rochdale
Schwertner CF (2005) Phylogeny and classification of the green stink bugs of *Nezara group* of genera (Hemiptera, Pentatomidae, Pentatominae). Doctor Thesis, Universidade Federal do Rio Grande do Sul, RS, Brazil http://www.lume.ufrgs.br/handle/10183/1
Schwertner CF, Grazia J (2007) O gênero *Chinavia* Orian (Hemiptera, Pentatomidae, Pentatominae) no Brasil, com chave pictórica para os adultos. Rev Brasil Entomol 51:416–435
Schwertner CF, Grazia J (2012) Review of the Neotropical genus *Aleixus* Mcdonald (Hemiptera: Heteroptera: Pentatomidae: Procleticini), with description of a new species and cladistic analysis of the Tribe Procleticini. Entomol Am 118:252–262

Schwertner CF, Carrenho R, Moreira FF, Cassis G (2021) Hemiptera sampling methods. In: Santos JC, Fernandes GW (eds) Measuring arthropod biodiversity: a handbook of sampling methods. Springer, pp 289–313
Servino LM, Schwertner CF (2020) Electronic identification key and data base to the genus *Chinavia* (Heteroptera: Pentatomidae) with morphological and distributional data. Zootaxa 4809:195–196
Sigwart JD (2022) Taxonomic hierarchies as a tool for coping with the complexity of biodiversity. In: Wilkins JS, Zachos FE, Pavlinov IY (eds) Species problems and beyond. CRC Press, Boca Raton, pp 227–242
Silva EJE, Fernandes JAM, Grazia J (2006) Characterization of the group *Edessa rufomarginata* and description of seven new species (Heteroptera, Pentatomidae, Edessinae). Iheringia Ser Zool 96:345–362
Soares PL, Cordeiro EM, Santos FN, Omoto C, Correa AS (2018) The Reunion of two lineages of the Neotropical brown stink bug on soybean lands in the heart of Brazil. Sci Rep 8:2496
Sonnekus B, Slippers B, Hurley BP, Joubert E, Stiller M, Fourie G (2022) Diversity and molecular barcoding of stink bugs (Hemiptera: Pentatomidae) associated with macadamia in South Africa. Insects 13:601
Sosa-Gómez DR, Corrêa-Ferreira BS, Kraemer B, Pasini A, Husch PE, Delfino-Vieira CE, Reis-Martinez CBR, Negrao-Lopes IO (2020) Prevalence, damage, management and insecticide resistance of stink bug populations (Hemiptera: Pentatomidae) in commodity crops. Agric For Entomol 22:99–118
Staddon BW, Ahmad I (1995) Species problems and species groups in the genus *Piezodorus* Fieber (Hemiptera: Pentatomidae). J Nat Hist 29:787–802
Suh CP, Westbrook JK, Esquivel JF (2013) Species of stink bugs in cotton and other row crops in the Brazos River Bottom of Texas. Southwest Entomol 38:561–570
Turnipseed SG, Kogan M (1976) Soybean entomology. Annu Rev Entomol 21:247–282
USDA (U.S. Department of Agriculture) (2024) Soybean Explorer. USDA, Foreign Agriculture Service, https://ipad.fas.usda.gov/cropexplorer/cropview/Default.aspx [Access on 30 April 2024]
Ward PS, Brady SG, Fisher BL, Schultz TR (2016) Phylogenetic classifications are informative, stable, and pragmatic: the case for monophyletic taxa. Insect Soc 63:489–492
Weirauch C, Schuh RT, Cassis C, Wheeler WC (2019) Revisiting habitat and lifestyle transitions in Heteroptera (Insecta: Hemiptera): insights from a combined morphological and molecular phylogeny. Cladistics 35:67–105
Wyniger D, Kment P (2010) Key for the separation of *Halyomorpha halys* (Stål) from similar-appearing pentatomids (Insecta: Heteroptera: Pentatomidae) occurring in Central Europe, with new Swiss records. Mitt Schweiz Entomol Gesell 83:261–270
Zandigiacomo P (1992) Pest found in soybean crops in North-Eastern Italy. Inf Agrar 48:57–59
Zhang Q, Feild TS, Antonelli A (2015) Assessing the impact of phylogenetic incongruence on taxonomy, floral evolution, biogeographical history, and phylogenetic diversity. Am J Bot 102:566–580

Chapter 2
Predatory Stink Bugs of Economic Importance as Biological Control Agents

Ricardo Brugnera and Jocelia Grazia

1 Introduction

Known as predatory stink bugs, the Asopinae are currently a subfamily of Pentatomidae. The group is composed of 63 genera and more than 300 species (Rider et al. 2018), widely distributed around the world, with the Neotropical and Indo-Malayan regions being the most diverse. The classification of the group has undergone several modifications throughout its history, including the use of tribes, which are currently no longer used (Thomas 1992, 1994). Their monophyly has been supported in both molecular and morphological analyses, although their subfamily status has been questioned in recent studies (Roca-Cusachs et al. 2021). Among the main morphological characteristics that define the group are the robust labium used to capture and hold their prey, the presence of pseudoclaspers in the male genitalia (likely serving a function of coupling with the female genitalia during mating), and the phallus divided into the basal theca and thecal shield in the male's genitalia (Gapud 1991; Gapon and Konstantinov 2006; Roell et al. 2020).

Identifying Asopines can be a challenging task, as many genera have not undergone taxonomic revisions, and most of available identification keys present incongruences. Furthermore, the possible presence of cryptic species and chromatic polymorphisms are additional challenges to their identification (Schaefer 1996; Brugnera et al. 2019). Although slowly, the taxonomy of the group has improved around the world. The recently published catalog of types deposited in the Collection of the Natural History Museum of London by Roell et al. (2023) has provided valuable images of several genera and species, shedding light on the classification of Asopinae.

R. Brugnera (✉) · J. Grazia
Laboratório de Entomologia Sistemática, Instituto de Biociências, Departamento de Zoologia, Universidade Federal do Rio Grande do Sul, Porto Alegre, RS, Brazil

A. F. Bueno, A. R. Panizzi (eds.), *Stink Bugs (Hemiptera: Pentatomidae) Research and Management*, Entomology in Focus 9,
https://doi.org/10.1007/978-3-031-69742-5_2

Predatory stink bugs exhibit a mating behavior characterized by abdominal vibratory signals featuring species-specific temporal patterns and similar spectral traits, as noted in phytophagous stink bugs (Pentatominae) (Čokl 2008; Laumann et al. 2016). Females typically lay their eggs on the leaves of plant species that are more likely to harbor pests. The eggs are very diverse among genera, however, they exhibit similar morphotypes among related genera, facilitating some level of identification, as in *Podisus* Herrich-Schäffer, *Brontocoris* Thomas, and *Supputius* Stål in which the chorion is covered with spines and the aeromicropilary processes (structures that play a role in egg fertilization, and possibly in the gas exchange of the embryos) are very long, considered the longest found in Pentatomoidea (Javahery 1994; Brugnera et al. 2022). The nymphal stage consists of five instars, and its duration can vary depending on the species, food availability, and abiotic conditions. During the first instar, nymphs do not feed and only need water to survive. From the second instar onward, the nymphs begin to prey on other insects, often exhibiting gregarious behavior to attack their prey, thereby enhancing their chances of success (De Clercq 2008; Hyodo et al. 2014).

2 Predatory Stink Bugs as Biological Controllers

Predatory stink bugs play an important role in agroecosystems, exhibiting various morphological adaptations for detecting, capturing, and consuming pests that afflict several cultivated plants (Pires et al. 2015; De Clercq 2000). Species of Asopinae can function as either specific or generalist predators, with significant variation observed across genera (Grazia et al. 2015). Since many of these species also include plants in their diet, they are categorized as zoophytophagous (Torres et al. 2010). Their diverse feeding behaviors, intricate life cycles, and successful adaptation to various environments have, in recent decades, led to the recognition of their potential for biological pest control. These predators can efficiently regulate the populations of agricultural and forest pests without causing harm to the plants (Plata-Rueda et al. 2022).

Asopinae species inhabit a wide range of natural and agricultural environments, often found in shrublands and forests. In Brazil, many species are commonly associated with Myrtaceae plants. These stink bugs primarily target slow-moving, soft-bodied insects as their prey, including the larvae of Lepidoptera, Diptera, and Coleoptera. However, they can also feed on beetles and other true bugs with frequency (Schaefer 1996; De Clercq 2000). Predatory stink bugs use their rostrum to pierce and suck their prey. The labium encompasses mouthparts (stylets) that inject a saliva composed of various compounds, including non-proteinaceous venoms. This saliva induces paralysis, death, and initial (extra-oral) tissue digestion, while intermediary and final digestion, followed by nutrient absorption, takes place in the midgut (Martínez et al. 2016; Plata-Rueda et al. 2022).

Many asopinae species can be easily reared in the laboratory, with few differences in the methodology used, adapting to different sources of food, including artificial diets (Zou et al. 2015). To date, at least 28 species have been successfully reared under laboratory conditions (Brugnera et al. 2022).

Around the world, investigations point out the relevance of studying the biology and ontogeny of predatory stink bugs, thus better understanding their ecological role as predators and allowing a precise identification and management in Integrated Pest Management (IPM) programs (De Clercq 2000, 2008). As examples, in South America, *Podisus nigrispinus* (Dallas) was demonstrated to have more than 90% of predation effectiveness on the southern armyworm *Spodoptera eridania* (Cramer) (De Carvalho et al. 2020). In North America, *Podisus maculiventris* (Say) and *Perillus bioculatus* (F.) are known as important natural enemies of the Colorado potato beetle *Leptinotarsa decemlineata* (Say).

Utilized as an additional tool in pest management techniques, the predatory stink bugs holds promise for fostering more sustainable agriculture, thereby contributing to the health of both humans and the environments. Below, we outline the most common and promising species of predatory stink bugs for biological control in the US, Europe, and Brazil.

3 Main Species of Asopinae for Biological Control in USA, Europe and Brazil

3.1 USA

3.1.1 *Podisus maculiventris* (Say)

The spined soldier bug, known as *Podisus maculiventris* (Say) (Fig. 2.1a), stands out as the preeminent predatory stink bug inhabiting North America. It is naturally found in all regions of the United States, with its distribution extending southward to Mexico, the Bahamas, Haiti, and the Dominican Republic, and northward into Canada. With a length ranging from 10 to 14 millimeters, females of this species are typically larger than males. A distinguishing feature is the conspicuous dark spot located in the middle of their seventh abdominal sternite, while their humeri exhibit a distinctive thorn-like projection. (Thomas 1992). Noteworthy is the species' polyphagous nature, as it preys upon a wide array of more than 75 economically significant pests of eight different insect orders (McPherson 1980). This species thrives in a diverse range of environments, including natural ecosystems such as shrublands, woodlands, and streambanks, as well as agricultural settings like orchards and various field crops (De Clercq 2008). The study by Baek et al. (2013) showed that the optimal temperature for egg-to-adult development is 30.6 °C, which can be applied to mass rearing and to estimate its establishment potential into non-native locations. Instead its adaptability in different habitats, their introduction as a biocontrol agent against the Colorado potato beetle (*Leptinotarsa decemlineata* (Say)) in Europe (Greece, France, and Germany), and to control the fall webworm (*Hyphantria cunea* (Drury)) in Korea and Japan, was not successful (De Clercq 2008).

3.1.2 *Perillus bioculatus* (F.)

Perillus bioculatus (F.) (Fig. 2.1b), commonly known as the two-spotted stink bug, plays a pivotal role in controlling populations of its prey, which includes the Colorado potato beetle (*L. decemlineata*) and other plant-feeding insects (Hough-Goldstein and McPherson 1996). The two-spotted stink bug is recognizable by its striking coloration varying from red, orange and yellows with black, and marked by two distinctive semicircular black spots on the anterior half of pronotum. Adults measure between 10 and 15 millimeters in length and are voracious feeders on various life stages of their prey, from eggs to nymphs, larvae and adults. It is widely distributed in the United States, extending to Mexico and Canada. The species were successfully introduced in Europe and India to control the Colorado Potato Beatle (Jermy 1980; Prasad and Pal 2015); in Russia, Agasyeva et al. (2023) showed 98% of efficiency of predation with a rato of 1:10–1:15 (*P. bioculatus* - *L. decemlineata*). At least other six species of *Perillus* Stål occur in North America, some of them very similar to *P. bioculatus* (Thomas 1992).

Fig. 2.1 Principal species of Asopinae in use or with potential for biological control in United States, Brazil and Europe. (**a**) *Podisus maculiventris* (Say), ©Katja Schulz. (**b**) *Perillus bioculatus* (F.), ©Quinten Wiegersma. (**c**) *Stiretrus anchorago* (F.), ©Judy Gallagher. (**d**) *Arma custos* (F.), ©Gilles San Martin. (**e**) *Picromerus bidens* (L.), ©Denis Mariasis. (**f**) *Alcaeorrhynchus grandis* (Dallas), ©John P. Friel. (**g**) *Brontocoris tabidus* (Signoret), ©Ricardo Brugnera. (**h**) *Podisus nigrispinus* (Dallas), ©Vinícius Rodrigues de Souza. (**i**) *Supputius cincticeps* (Stål), ©Maurino André. (**j**) *Tynacantha marginata* (Dallas) feeding of *Edessa* sp., ©Fábio Luiz dos Santos

3.1.3 *Stiretrus anchorago* (F.)

The anchor stink bug, *Stiretrus anchorago* (F.) (Fig. 2.1c), is commonly found in all the east coast of United States, extending until Colombia (Castro-Huertas et al. 2022). Adults measure between 7 and 8 millimeters (Richman and Mead 2001). The species exhibits a preference to feed on eggs, larvae and adults of Coleoptera instead of Lepidoptera (Schaefer 1996), as *Epilachna varivestis* Mulsant, and *Microtheca ochroloma* Stål (Waddill and Shepard 1975; Balusu et al. 2017). As *Perillus bioculatus*, adults of the anchor stink bug exhibits a wide variety of morphotypes and colorations, ranging from shades of red, yellow, pale, to iridescent blue. The nymphs from the second instar have their abdomens entirely covered by dorsal plates, a unique pattern among known nymphs of predatory stink bugs (Brugnera et al. 2022). *Stiretrus* comprises seven other species distributed throughout the American continent, with particular importance of *Stiretrus decemguttatus* (Lepeletier & Serville), an important predator of cassidini beetles in Brazil (Paleari 2013; Gomes et al. 2019).

3.2 *Europe*

3.2.1 *Arma custos* (F.)

The species *Arma custos* (F.) (Fig. 2.1d), also known as the dock leaf bug, is widely distributed in the Paleartic Region, and one of the most recognized species of Asopinae regarding biological control (Thomas 1994; Bosio et al. 2014; Roca-Cusachs et al. 2020). It is easily reared in laboratory with natural and artificial preys, and plays an well adaptation to different environments, provident control of several pest insects of Lepidoptera, Coleoptera, Hymenoptera and Hemiptera orders (Zou et al. 2012, 2013; Liao et al. 2019). In some regions of China, for example, the mass rearing of *A. custos* can surpass the six million of individuals per year to control larvae of *Spodoptera* Guenée and *Helicoverpa* Hardwick species (Li et al. 2021). Advanced studies regarding its mitochondrial genome (Mu et al. 2022), morphology (Zhang et al. 2019), reproduction (Wu et al. 2021), and others have been developed. *Arma custos* is generally characterized by the body colorations dull brown without highlighted macula or spots, and obtuse to acute humeral angles; the male and female genitalia are detailed descripted by Zhao et al. (2018).

3.2.2 *Picromerus bidens* (L.)

The spiny shield bug, *Picromerus bidens* (L.) (Fig. 2.1e) is a common species of predatory stink bug, native and widely distributed in Europe and other parts of the Paleartic Region. The species also occurs in some regions of the United States and Canada, where is considered an alien species. In North America, it was first recorded in 1932, and it is suggested that the species was accidentally introduced through nursery stock or other horticultural plants (Javahery 1986; Swanson et al. 2013). It likes fresh, cool, and humid areas (Larivière and Larochelle 1989), mainly growing in moist bushes and forests more than 2 m above the ground (Cokl et al. 2011). *P. bidens* has an univoltine (one brood per year) life cycle with obligate embryonic diapause and overwinters primarily at the egg stage (Saulich and Musolin 2014). The species present as main characteristics the acute humeral spine and the yellow macula in the apex of scutellum (Thomas 1994).

3.3 Brazil

3.3.1 *Alcaeorrhynchus grandis* (Dallas)

The giant strong-nosed stink bug, *Alcaeorrhynchus grandis* (Dallas) (Fig. 2.1f), stands out as one of the largest representatives of Asopinae, with an impressive length of up to 25 mm. The dorsal coloration is variegated brown, and the humeral angles are distinctly bifid. Its range extends from the United States to Argentina, encompassing even the Galapagos Islands (Thomas 1992; Richman and Mead 2000). Females have the remarkable capacity to lay over 200 eggs in a single oviposition, and nymphs exhibit gregarious behavior throughout their developmental stages, acting together to preying on sizable targets such as defoliating caterpillars (Ribeiro et al. 2010; Brugnera et al. 2022). Records of this species near Trindade Island, Brazil, located more than 380 kilometers from the continent, suggest that *A. grandis* is capable of covering extensive distances through flight, being able for active migration (Alves et al. 2019).

3.3.2 *Brontocoris tabidus* (Signoret)

Although less common than other species, *Brontocoris tabidus* (Signoret) (Fig. 2.1g) has received attention in studies investigating its potential as a biological control agent. Over the past three decades, several researches suggest that the species shows promising performance for this method (Zanuncio et al. 2014; Pires et al. 2020, 2023). The genus *Brontocoris* Thomas includes only two species: *B. nigrolimbatus* (Spinola) found in Chile, Uruguay, and Argentina, and *B. tabidus* found in Brazil, Paraguay, and Argentina (Thomas 1992; Grazia et al. 2015). In Brazil, *B. tabidus* is a generalist predator commonly found controlling defoliating Lepidoptera caterpillars in *Eucalyptus* plantations (De Menezes et al. 2013). Among other prey types, *B. tabidus* is relevant in the predation of Hymenoptera larvae, such as leaf-feeding sawflies like *Haplostegus nigricrus* Conde and *Heteroperreyia hubrichi* Malaise, which are important defoliators of *Psidium guajava* L. (Guava) and *Schinus terebinthifolia* Raddi (Brazilian peppertree), respectively (Azevedo Pereira et al. 2008; McKay et al. 2019).

3.3.3 *Podisus nigrispinus* (Dallas)

The southern spined stink bug, *Podisus nigrispinus* (Dallas) (Fig. 2.1h) is one of the most common species of Asopinae in Brazilian regions and across most of South American countries. Its distribution spans from Colombia to Argentina, except Chile. A distinctive feature of this species is the presence of a light-colored margin along the anterolateral edges of the pronotum, and a protuberance posterior to the humeral angle, creating the impression of a double spine. The dorsal coloration of the body ranges from light brown to dark brown, with a paler ventral surface. Some individuals may exhibit red coloring on the ventral surface and the anterolateral margins of the pronotum (Thomas 1992). *Podisus nigrispinus* is found in diverse habitats, including woodlands, grasslands, and cultivated crops. As a generalist predator, it can feed of many pest insects, as caterpillars, beetles, stink bugs and aphids. This species readily adapts to laboratory rearing and is among the most studied species for biological pest control. Numerous in situ studies have demonstrated its potential for many pest insects, having over 90% predation effectiveness against the southern armyworm, *Spodoptera eridania* (Cramer) (De Carvalho et al. 2020).

3.3.4 *Supputius cincticeps* (Stål)

Supputius cincticeps (Stål) (Fig. 2.1i), is a common species found in the Atlantic Rainforest and Pampean regions of Brazil, Argentina, and Uruguay. Measuring among 8 to 10 cm, its body ranges from brown to light brown. It can be distinguished from other genera by the length of the second segment of the labium, which is longer than segments three and four combined (Thomas 1992). The nymphs exhibit polymorphism from the third instar, varying from black and yellow to entirely red (Brugnera et al. 2022). It is also a generalist feeder, preying on small caterpillars, beetles, membracids, aphids, and other insects. Several studies have been conducted on this species, focusing on its behavior (Silva et al. 2012), morphology (De Castro et al. 2013), life expectancy (Zanuncio et al. 2005), and predation capacity (Costa et al. 2019). It has been recognized as an important predator of the bronze bug (*Thaumastocoris peregrinus* Carpintero & Dellapé) in *Eucalyptus* plantations (Souza et al. 2012; Wilcken et al. 2014). For mass rearing, the use of *Musca domestica* (L.) or *Tenebrio molitor* (L.) in association with plant species, particularly from the Myrtaceae family is recommended (Zanuncio et al. 2014).

3.3.5 *Tynacantha marginata* (Dallas)

Commonly found in all regions of Brazil, the orange-belly predatory stink bug, *Tynacantha marginata* (Dallas) (Fig. 2.1j), is a promising candidate for application in biological control. Like other species, it can be easily reared under laboratory conditions, have generalist feeding habits, and exhibits a rapid life cycle compared to other species (Brugnera and Grazia 2018). Observations suggest that this species may have a preference for feeding on other true bugs, including species of economic importance such as *Nezara viridula* (L.) and *Edessa rufomarginata* (De Geer) (Sánchez et al. 2001), and *Piezodorus guildinii* (Westwood) (Panizzi and Smith 1976) although further study is needed to better understand its feeding behavior. The species is dorsally green, with the ventral surface and lateral margins varying from yellow to orange. The species becomes sexually dimorphic after a few weeks of adulthood when females change the color of their ventral surface to orange. The role of this feature is yet to be studied, however, considering that color change occurs when the period of oviposition is approaching (around the seventh week), the striking orange coloration may act as a defense against predators, prolonging the female's lifespan and enabling to lay more eggs. A single female of *Tynacantha marginata* can lay over 190 eggs in a period of 10 weeks (Moreira et al. 1995).

4 Conclusion

As a cosmopolitan group within Pentatomidae, asopine species are found across all zoogeographical regions. Throughout these regions, including those not mentioned in this chapter, there are species commonly occurring with potential for use in the biological control of both native and exotic/introduced pests, including other stink bug species. Furthermore, the dietary diversity and size range of these species enable their application for controlling various types of prey, either as the primary agent or in conjunction with other methods of biological control.

Apart from their predatory feeding habits, the key factor that makes asopines excellent candidates for effective biological control is their high adaptability to mass rearing. In general, the species (especially those mentioned here) can be easily reared in laboratory settings with minor protocol adjustments depending on the species, requiring low costs for necessary equipment, maintenance, and feeding. The use of common food sources such as *Tenebrio molitor* or *Musca domestica*, along with native plants, seems to work well for most species, ensuring good fertility, fecundity, and survival rates throughout rearing.

Despite advances in research over recent decades, further investment is needed to investigate the behavior, physiology, taxonomy, and phylogenetic relationships, which remain largely unknown for most species. For example, *Podisus fuscescens* (Dallas) and *Podisus sagitta* (F.) are relatively common species ocurring in large areas of the Neotropical Region, that possibly hold potential as biological control agents, but are still not well known. From this perspective, investing in research on local species from different regions of the world, in addition to revealing species relevant for biological control, may also provide an alternative to introducing exotic species for this purpose as seen with *Perillus bioculatus* in Europe and India. This approach minimizes the negative impacts associated with the introduction of exotic species, such as population control issues, competition with native species, and predation on non-target organisms, thereby preserving the balance of native populations.

References

Agasyeva I, Nefedova M, Ismailov V, Nastasy A (2023) Entomophages of the Colorado potato beetle, population dynamics of *Perillus bioculatus* Fabr. and its compatibility with biological and chemical insecticides. Agronomy 13:1496

Alves RJV, Costa LAA, Soares A, Silva NG, Pinto ÂP (2019) Open Ocean nocturnal insect migration in the Brazilian South Atlantic with comments on flight endurance. Peer J 7:e7583

Baek S, Son Y, Park Y (2013) Temperature-dependent development and survival of *Podisus maculiventris* (Hemiptera: Pentatomidae): implications for mass rearing and biological control. J Pest Sci 87:331–340

Balusu RR, Rhodes EM, Majumdar A, Cave RD, Liburd OE, Fadamiro HY (2017) Biology, ecology, and management of *Microtheca ochroloma* (Coleoptera: Chrysomelidae) in organic crucifer production. J Integr Pest Manag 8:14

Bosio G, Massombrio V, Chersi C, Scavarda G, Clarck S (2014) Spread of the ragweed leaf beetle, *Ophraella communa* LeSage, 1986 (Coleoptera Chrysomelidae), in Piedmont Region (northwestern Italy). Boll Soc Entomol Ital 146:17–30

Brugnera R, Grazia J (2018) External morphology of immature of *Tynacantha marginata* Dallas, 1851 (Hemiptera, Heteroptera, Pentatomidae). Zootaxa 4378:121–128

Brugnera R, Campos LA, Grazia J (2019) Morphology of immature stages, chromatic polymorphism of adults and natural history of *Oplomus catena* (Heteroptera: Pentatomidae: Asopinae). Neotrop Entomol 48:1046–1057

Brugnera R, Limberger GM, Campos LA, Grazia J (2022) The eggs and nymphs of predatory stink bugs (Hemiptera: Pentatomidae: Asopinae): what do we know? Zoology 151:125991

Castro-Huertas AV, Grazia J, Forero D, Fernandéz F, Schwertner CF (2022) Stink bugs (Hemiptera: Heteroptera: Pentatomidae) of Colombia: an annotated checklist of species. Zootaxa 5097:001–088

Čokl A (2008) Stink bug interaction with host plants during communication. J Insect Physiol 54:1113–1124

Čokl A, Žunič A, Virant-Doberle M (2011) Predatory bug *Picromerus bidens* communicates at different frequency levels. Cent Eur J Biol 6:431–439

Costa J, Torres L, Provance DW Jr, Brugnera R, Grazia J (2019) First report of predation by a stink bug (*Supputius cincticeps* Stål) on a walking-stick insect (*Cladomorphus phyllinus* Gray), with reflections on evolutionary mechanisms for camouflage. Acta Biol Parana 48:5–15

De Carvalho JR, Pratissoli D, de Araujo Junior LM, Damascena AP, Holtz AM, Dalvi LP, Vianna UR (2020) Predation behavior of *Podisus nigrispinus* on *Spodoptera eridania*. J Asia Pac Entomol 23:1279–1282

De Castro AA, Canevari GC, Pikart TG, Ribeiro RC, Serrão JE, Zanuncio TV, Zanuncio JC (2013) Salivary gland histology of the predator *Supputius cincticeps* (Heteroptera: Pentatomidae). Ann Entomol Soc Am 106:273–277

De Clercq P (2000) Predaceous stink bugs (Pentatomidae: Asopinae). In: Schaefer CW, Panizzi AR (eds) Heteroptera of economic importance. CRC Press, Boca Raton, pp 737–789

De Clercq P (2008) Predatory stink bugs (Pentatomidae: Asopinae). In: Capinera JL (ed) Encyclopedia of insects, Volume 2. Kluwer Academic Publishers, Dordrecht, pp 3042–3045

De Menezes CWG, Soares MA, De Assis J, Sebastião LM, Sady JMC, Dos Santos JB, Zanuncio JC (2013) *Brontocoris tabidus* (Heteroptera: Pentatomidae) preying on *Podalia walkeri* (Lepidoptera: Megalopygidae) on eucalypt plants in Brazil. Fla Entomol 96:261–263

Gapon DA, Konstantinov FV (2006) On the structure of the aedeagus of shield bugs (Heteroptera, Pentatomidae): III. Subfamily Asopinae. Entomol Rev 86:806–819

Gapud VP (1991) A generic revision of the subfamily Asopinae, with consideration of its phylogenetic position in the family Pentatomidae and superfamily Pentatomoidea (Hemiptera Heteroptera). Philipp Entomol 8:865–961

Gomes PAA, Hermes MG, Fernandes FR, Frieiro-Costa FA (2019) Tortoise beetles of an Atlantic Forest remnant in South Minas Gerais, Brazil: host plants and life history. J Nat Hist 55:15–60

Grazia J, Panizzi AR, Greve C, Schwertner CF, Campos LA, Garbelotto TA, Fernandes JAM (2015) Stink bugs (Pentatomidae). In: Panizzi AR, Grazia J (eds) True bugs (Heteroptera) of the Neotropics. Springer, Dordrecht, pp 681–756

Hough-Goldstein J, McPherson D (1996) Comparison of *Perillus bioculatus* and *Podisus maculiventris* (Hemiptera: Pentatomidae) as potential control agents of the Colorado potato beetle (Coleoptera: Chrysomelidae). J Econ Entomol 89:1116–1123

Hyodo D, Himuro C, Fujisaki K (2014) Prey size affects the costs and benefits of group predation in nymphs of the predatory stink bug *Andrallus spinidens* (Heteroptera: Pentatomidae). J Ethol 32:173–178

Javahery M (1986) Biology and ecology of *Picromerus bidens* (Hemiptera: Pentatomidae) in southeastern Canada. Entomol News 97:87–98

Javahery M (1994) Development of eggs in some true bugs (Hemiptera-Heteroptera). Part I. Pentatomoidea. Can Entomol 126:401–433

Jermy T (1980) The introduction of *Perillus bioculatus* into Europe to control the Colorado beetle. EPPO Bull 10:475–479

Larivière MC, Larochelle A (1989) *Picromerus bidens* (Heteroptera: Pentatomidae) in North America, with a world review of distribution and bionomics. Entomol News 100:133–146

Laumann RA, Čokl A, Blassioli-Moraes MA, Borges M (2016) Vibratory communication and its relevance to reproductive isolation in two sympatric stink bug species (Hemiptera: Pentatomidae: Pentatominae). J Insect Behav 29:643–665

Li W, Zhang C, Qin W, Xie H, Dong X, Li Z, Meng J, Guo X (2021) Safety assessment of fourteen insecticides to *Arma chinensis*. Hubei Agric Sci 60:89–92

Liao P, Miao SM, Xu RN, Liu CX, Chen GK, Wang MQ, Mao JJ, Zhang LS, Chen HY (2019) Evaluation of a new liquid artifcial diet of *Arma chinensis* Fallou (Hemiptera: Pentatomidae). Chin J Biol Control 35:9–14

Martínez LC, Fialho MCQ, Barbosa LCA, Oliveira LL, Zanuncio JC, Serrão JE (2016) Stink bug predator kills prey with salivary non-proteinaceous compounds. Insect Biochem Mol Biol 68:71–78

Mc Kay F, Dellapé G, Dyer K, Wheeler G (2019) New record of *Brontocoris tabidus* (Hemiptera: Pentatomidae) attacking larvae of *Heteroperreyia hubrichi* (Hymenoptera: Pergidae). Rev Soc Entomol Argent 78:22–25

McPherson JE (1980) A list of the prey species of *Podisus maculiventris* (Hemiptera: Pentatomidae). Gt Lakes Entomol 13:18–24

Moreira LA, Zanuncio JC, Picanço MC, Bruckner CH (1995) Tabelas de fertilidade e de esperança de vida de *Tynacantha marginata* Dallas (Heteroptera, Pentatomidae, Asopinae) alimentado com larvas de *Tenebrio molitor* (Coleoptera, Tenebrionidae) e folhas de *Eucalyptus urophylla* S. T. Blake. Rev Bras Zool 12:255–261

Mu Y, Zhang C, Zhang Y, Yang L, Chen X (2022) Characterizing the complete mitochondrial genome of *Arma custos* and *Picromerus lewisi* (Hemiptera: Pentatomidae: Asopinae) and conducting phylogenetic analysis. J Insect Sci 22:6

Paleari LM (2013) Developmental biology, polymorphism and ecological aspects of *Stiretrus decemguttatus* (Hemiptera, Pentatomidae), an important predator of cassidine beetles. Rev Bras Entomol 57:75–83

Panizzi AR, Smith JG (1976) Observações sobre inimigos naturais de *Piezodorus guildinii* (Westwood, 1837) (Hemiptera, Pentatomidae) em soja. An Soc Entomol Brasil 5:11–17

Pereira AIA, Curvêlo CRS, Pastori PL, Smith SR, Zanuncio JC (2008) Comportamento defensivo das larvas do *Symphyta* Neotropical *Haplostegus nigricrus* (Hymenoptera: Pergidae) expostas aos percevejos predadores *Podisus nigrispinus*, *Supputius cincticeps* e *Brontocoris tabidus* (Heteroptera: Pentatomidae). Rev Caatinga 21:167–171

Pires EM, Soares MA, Nogueira RM, Zanuncio JC, Moreira PSA, Oliveira MA (2015) Seven decades of studies with Asopinae predators in Brazil (1930–2014). Biosci J 31:1530–1549

Pires EM, Zanuncio JC, Nogueira RM, Soares MA, Oliveira MA (2020) Dispersal of the zoophytophagous predator *Brontocoris tabidus* and *Podisus nigrispinus* (Heteroptera: Pentatomidae) in an *Eucalyptus* plantation. Fla Entomol 103:168–171

Pires EM, Nogueira RM, Beber RC, Júnior WF, Soares MA, Serrão JE, Zanuncio JC (2023) Attack efcacy by *Brontocoris tabidus* and *Podisus nigrispinus* (Hemiptera: Pentatomidae) on *Thyrinteina arnobia* (Lepidoptera: Geometridae) with diferent pre-release food-free intervals. Phytoparasitica 51:967–973

Plata-Rueda A, Martínez LC, Zanuncio JC, Serrão JE (2022) Advances zoophytophagous stinkbugs (Pentatomidae) use in agroecosystems: biology, feeding behavior and biological control. J Pest Sci 95:1485–1500

Prasad CS, Pal R (2015) First record of two spotted stink bug, *Perillus bioculatus* (Fab.) from Meerut (U.P.) North India. Int J Environ Agric Res 1:9–11

Ribeiro RC, Lemos WP, Bernardino AS, Buecke J, Müller AA (2010) First occurrence of *Alcaeorrhynchus grandis* (Dallas) (Hemiptera: Pentatomidae) preying on defoliating caterpillars of oil palm in the state of Pará, Brazil. Neotrop Entomol 39:131–132

Richman DB, Mead FW (2000) Predatory stink bug, *Alcaeorrhynchus grandis* (Dallas). Pentatomidae). Institute of Food and Agricultural Sciences, University of Florida EENY-165, Hemiptera, pp 1–3

Richman DB, Mead FW (2001) Predatory stink bug, *Stiretrus anchorago* (Fabricius) (Insecta: Hemiptera: Pentatomidae). Institute of Food and Agricultural Sciences, University of Florida EENY-230, pp 1–3

Rider DA, Schwertner CF, Vilímová J, Rédei D, Kment P, Thomas DB (2018) Higher systematics of the Pentatomoidea. In: McPherson JE (ed) Invasive stink bugs and related species (Pentatomoidea): biology, higher systematics, semiochemistry, and management. CRC Press, Boca Raton, pp 25–201

Roca-Cusachs M, Kim J, Park J, Jung S (2020) Taxonomic review of the predatory stink bugs of the Korean Peninsula (Heteroptera: Pentatomidae: Asopinae), with a key to the Korean species and a discussion of their usefulness as biological control agents. J Asia Pac Entomol 23:113–123

Roca-Cusachs M, Schwertner CF, Kim J, Eger J, Grazia J, Jung S (2021) Opening Pandora's box: molecular phylogeny of the stink bugs (Hemiptera: Heteroptera: Pentatomidae) reveals great incongruences in the current classification. Syst Entomol 47:36–51

Roell T, Genevcius B, Campos LA (2020) Comparative morphology of clasping structures in predator stink bugs (Hemiptera: Pentatomidae: Asopinae): insights into their function and evolution. Arthropod Struct Dev 57:100949

Roell T, Lemaître VA, Webb MD, Campos LA (2023) An annotated and illustrated type catalogue of the predacious shield bugs (Hemiptera: Heteroptera: Pentatomidae: Asopinae) in the collection of the Natural History Museum, London. Zootaxa 5232:001–105

Sánchez MC, Borges B, Grazia J (2001) *Tynacantha marginata* (Heteroptera: Pentatomidae), depredador de la chinche verde hedionda (*Nezara viridula* L.) en Venezuela. Entomotropica 16:213–214

Saulich AK, Musolin DL (2014) Seasonal cycles in stink bugs (Heteroptera, Pentatomidae) from the temperate zone: diversity and control. Entomol Rev 94:785–814

Schaefer CW (1996) Bright bugs and bright beetles: Asopine Pentatomids (Hemiptera: Heteroptera) and their prey. In: Alomar O, Wiedemann RN (eds) Zoophytophagous Heteroptera: implications for the life history and integrated pest management. Proceeding of the Thomas Say Publications in Entomology, Entomol Soc Am Lanham, Maryland, pp 18–56

Silva RB, Corrêa AS, Lucia TMCD, Pereira AIA, Cruz I, Zanuncio JC (2012) Does the aggressiveness of the prey modify the attack behavior of the predator *Supputius cincticeps* (Stål) (Hemiptera, Pentatomidae)? Rev Bras Entomol 56:244–248

Souza GK, Pikart TG, Pikart FC, Serrão JE, Wilcken CF, Zanuncio JC (2012) First record of a native Heteropteran preying on the introduced *Eucalyptus* pest, *Thaumastocoris peregrinus* (Hemiptera: Thaumastocoridae), in Brazil. Fla Entomol 95:517–520

Swanson DR, Keller O, Rowley JD (2013) First record of the Palearctic predatory stink bug, *Picromerus bidens* (Heteroptera: Pentatomidae: Asopinae), in Michigan. Gt Lakes Entomol 46:231–234

Thomas DB (1992) Taxonomic synopsis of the asopinae Pentatomidae (Heteroptera) of the Western Hemisphere. Thomas Say Foundation Monograph, p 156

Thomas DB (1994) Taxonomic synopsis of the old world asopine genera (Heteroptera: Pentatomidae). Insecta Mundi 8:145–212

Torres JB, Barro EM, Coelho RR, Pimental RMM (2010) Zoophytophagous pentatomids feeding on plants and implications for biological control. Arthropod Plant Interact 4:219–227

Waddill V, Shepard M (1975) A comparison of predation by the pentatomids, *Podisus maculiventris* (Say) and *Stiretrus anchorago* (F.), on the Mexican bean beetle, *Epilachna varivestis* Mulsant. Ann Entomol Soc Am 68:1023–1027

Wilcken C, Barbosa L, Zache B, Firmino A, Zanuncio J, Junqueira L (2014) Biological control of the bronze bug, *Thaumastocoris peregrinus*, in *Eucalyptus* plantations in Brazil. Int For Rev 16:325

Wu S, Zeng W, Deng W, Li M, Hu W, Cai H, Li X, Xie P, Tan L, Zhou Z (2021) Egg cannibalism varies with sex, reproductive status, and egg and nymph ages in *Arma custos* (Hemiptera: Asopinae). Front Ecol Evol 9:705318

Zanuncio TV, Zanuncio JC, Serrão JE, Medeiros RS, Pinon TBM, Sediyama CAZ (2005) Fertility and life expectancy of the predator *Supputius cincticeps* (Heteroptera: Pentatomidae) exposed to sublethal doses of permethrin. Biol Res 38:31–39

Zanuncio JC, Tavares WS, Fernandes BV, Wilcken CF, Zanuncio TV (2014) Production and use of Heteroptera predators for the biological control of *Eucalyptus* in Brazil. Ekoloji 23:98–104

Zhang J, Zhang X, Liu C, Meng L, Zhou Y (2019) Fine structure and distribution of antennal sensilla of stink bug *Arma chinensis* (Heteroptera: Pentatomidae). Entomol Fenn 25:186–197

Zhao Q, Wei J, Bu W, Liu G, Zhang H (2018) Synonymize *Arma chinensis* as *Arma custos* based on morphological, molecular and geographical data. Zootaxa 4455:161–176

Zou D, Wang M, Zhang L, Zhang Y, Zhang X, Chen H (2012) Taxonomic and bionomic notes on *Arma chinensis* (Fallou) (Hemiptera: Pentatomidae: Asopinae). Zootaxa 3382:41–52

Zou DY, Wu HH, Coudron TA, Zhang LS, Wang MQ, Liu CX, Chen HY (2013) A meridic diet for continuous rearing of *Arma chinensis* (Hemiptera: Pentatomidae: Asopinae). Biol Control 67:491–497

Zou DY, Coudron TA, Wu HH, Gu XS, Xu WH, Zhang LS, Chen HY (2015) Performance and cost comparisons for continuous rearing of *Arma chinensis* (Hemiptera: Pentatomidae: Asopinae) on a zoophytogenous artificial diet and a secondary prey. J Econ Entomol 108:454–461

Chapter 3
Life History Studies and Their Implications for Management of Stink Bugs

Antônio Ricardo Panizzi and Tiago Lucini

1 Introduction

Stink bugs (Hemiptera: Heteroptera: Pentatomidae) are major pests of cultivated plants in many regions of the world. Their economic importance has been reviewed worldwide by Schaefer and Panizzi (2000), and in specific regions, such as in the America North of Mexico by McPherson and McPherson (2000). By being polyphagous, in a variable extent, they feed not only on cultivated but also on a vast array of uncultivated plants in order to reproduce and increase their success to build up populations.

Stink bugs are cosmopolitan, and perhaps in no other geographical region they achieve their peak of success as in the Neotropical Region. In this geographical region, because of mild winters, abundant sunshine, and favorable humidity, they can complete several generations and be active during the entire year, utilizing in sequence plants cultivated and non-cultivated (Panizzi 1997).

Integrated pest management (IPM) programs have been developed for several crops worldwide and their history of success has been well documented in the literature (e.g., Kogan 1998; Kogan and Jepson 2007; Peshin and Pimentel 2014; Grasswitz 2019; Green et al. 2020). In spite of that, we feel that many IPM programs could be improved and become more sustainable if we add to them additional details of the life history of the pest species been considered (see item 2 ahead for discussion of life history).

Regarding pest stink bugs, some examples of IPM considering major components of their life history can be found in the literature. For example, the phenology of the green stink bug, *Chinavia* (*Acrosternum*) *hilare* (Say), and of the southern green stink bug, *Nezara viridula* (L.) on soybean fields and on nearby wild plants

---

A. R. Panizzi (✉) · T. Lucini
Embrapa Trigo, Passo Fundo, RS, Brazil

A. F. Bueno, A. R. Panizzi (eds.), *Stink Bugs (Hemiptera: Pentatomidae) Research and Management*, Entomology in Focus 9,
https://doi.org/10.1007/978-3-031-69742-5_3

was recorded (Miner 1966). He found that bugs will complete one generation on the wild plants and then colonize the soybean fields. Other more recent studies will unveil the phenology of stink bugs including the influence of host plants (wild and cultivated), attractive plants used as traps, their overwintering behaviors either on less preferred plants or underneath crop residues or in other shelters, and the management of the landscape of farms areas to achieve efficient management strategies and tactics for stink bugs (e.g., Jones and Sullivan 1981, 1982; Hokkanen 1991; Panizzi 1997, 2000; Ehler 2000; Tillman et al. 2004; Mizell et al. 2008; Tillman 2014; Haseeb et al. 2016; Stahl et al. 2021).

In this chapter we present in a summarized way some examples of life history studies of major stink bug pests in the Neotropical Region as case studies. In this region, agriculture is practiced in a very intensive way, forming what is called the "green bridge". This consists of different or same crops cultivated in a continuum, providing abundant food and shelter for pest stink bugs during the entire year. This makes these pests to cause very significant reductions in grain yields of several commodities such as soybean, maize, wheat, cotton, and others. Therefore, there is a growing need to mitigate their impact on these crops. This may possibly be achieved by learning in greater details the activities of stink bugs throughout their life history to devise holistic integrated pest management programs.

2 Life History

Life history studies involve the comprehensive exploration of all the activities undertaken by a particular species over the course of an entire year. This includes learning which food sources are explored; on which food resources the insects will reproduce and have their offspring to develop and reach adulthood, completing generations in sequence; which food resources are used in a limited way (less preferred feeding plants) providing water and nutrients for sustainability only; which shelters are used during unfavorable biotic and abiotic conditions; on which time of the year insects are more suitable to the action of parasites and predators; and so on. In short, the idea is to have a complete view of what is happening with the insect biology at any time of the year. This comprehensive knowledge contributes significantly not only to promote enhanced sustainability in the management of insect populations but also to develop more effective Integrated Pest Management (IPM) programs.

In Table 3.1, we present and give a short description of the major components of the life history of pest stink bugs in the Neotropics. These components include generations completed per year, cultivated and non-cultivated (wild) plants explored as hosts or only as associated, diapause (oligopause) occurrence underneath crop residues and/or other shelters, and the incidence of natural enemies, mostly parasites which are more specific than predators and entomopathogenic agents.

Regarding pest stink bugs in the Neotropical Region, which will be here analyzed, not many studies on their life history have been published, considering the number of generations per year, plants (wild and cultivated) used, and diapause

Table 3.1 Main components of the life history of pest stink bugs in the Neotropics with explanation of their characteristics

Components	Description
Generations completed on	*Crops* (preferred plants in where the majority of generations are completed per year). *Wild plants* (in general, less preferred plants in where occasional generations are completed; most serve as sources of nutrients/water for maintenance or as shelters).
Plants utilized	*Host plants* (plants that provide nutrients that allow nymph development to reach adulthood and on which several generations are completed). *Associated plants* (plants used temporarily as sources of nutrients/water and as shelters that do not allow nymph development).
Diapause (oligopause)	*Overwinter strategy* (arrested development during unfavorable conditions shown by some stink bug species).
Natural enemies (parasites)	*Egg* (multiple egg parasitoids of the families Scelionidae attack eggs). *Nymph* (late instar nymphs are attacked by dipteran tachinid flies). *Adult* (commonly attacked by several species of dipteran tachinid flies [Diptera] and hymenopterous [Encyrtidae]).

(olipopause) occurrence. Panizzi (1997) described the major components of the life history of the pest species *Nezara viridula* (L.), *Piezodorus guildinii* (Westwood), and *Euschistus heros* (F.) in northern Paraná state, Brazil. The first two species complete 5–6 generations per year. Three generations are completed on the major host plant, soybean. *Nezara viridula* which is greatly polyphagous complete another additional 3 generations on wild plants. *Piezodorus guildinii*, which is less polyphagous, complete 2 more generations on wild plants. In contrast to these two former species, *E. heros*, that also complete 3 generations on soybean, instead of colonizing wild plants after soybean harvest, it looks for shelter under crops residues and stay there in partial hibernation (oligopause) until the next spring, when beans and soybean become available again. This condition for these three pest stink bugs, reported over 25 years ago, nowadays changed substantially.

As the cultivation systems in the Neotropics drastically changed, with the generalized used of non-tillage cultivation systems and addition of multiple crops cultivated in sequence year-round, the dynamics of stink populations also changed. The former number one pest, *N. viridula*, entered in decline, and its abundance was drastically reduced, becoming a secondary pest. The reasons to explain that was presented and discussed in a review article published by Panizzi and Lucini (2016). Two other species gain momentum. The so-called green belly stink bug, *Diceraeus melacanthus* (Dallas), a former minor pest of maize (Ávila and Panizzi 1995), and the Neotropical brown stink bug, *E. heros*, which was also of minor importance, became the two most important stink bug pests of major commodities in the Neotropics (Panizzi et al. 2022).

Studies on life history of stink bugs offers numerous beneficial outcomes. Among these advantages is the capacity to gain insights into how the role of natural enemies can be enhanced by promoting the nowadays very important considered, the conservation biological control. This research gives details on the role of alternative hosts in sustaining populations of natural enemies; how these natural enemies move from crop plants and alternative hosts and vice-versa; and how effective they are to mitigate the populations of pest stink bugs (Panizzi and Lucini 2023).

3 Pest Stink Bugs Abundance

The abundance of pest stink bugs is affected by several variables depending on the food availability, favorability of the environment considering biotic and abiotic factors, and the characteristic of the landscape. We will give a general view and discuss the behavior of pest stink bugs populations in the Neotropics. In Fig. 3.1, we depict the curve of the populations considering the major factors that influence its increase or decrease along the year.

Starting on the spring in October, stink bug populations that came from hosts (cultivated or non-cultivated) or associated plants and from shelters start to increase in abundance. The peak of the population is reached during summer, primarily on reproductive plants, mostly of soybean, but also on other cultivated crops. As the crops are harvested, these stink bugs move to adjacent plants and colonize plants of crops cultivated in sequence (commonly referred to as 2nd crops). As these crops are harvested in mid-autumn, populations move to other host and associated plants or to shelters and wait for the next spring in September to start their annual cycle

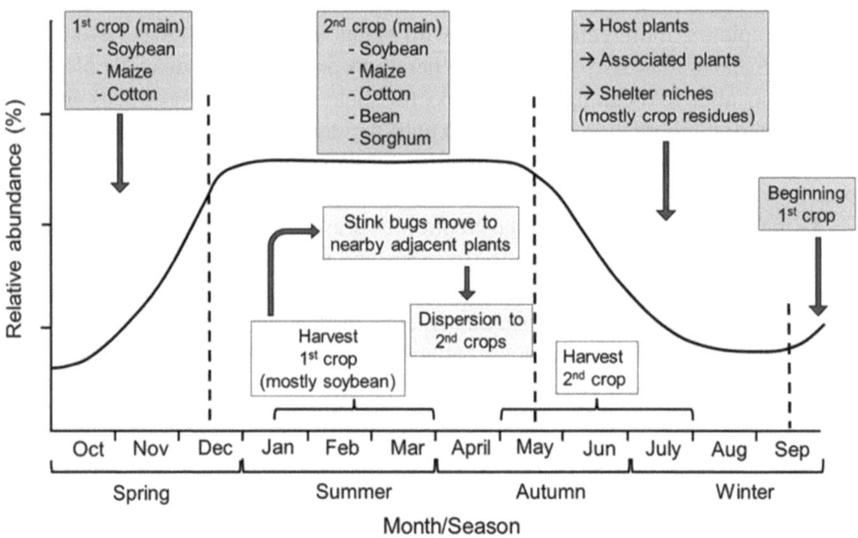

Fig. 3.1 General dynamics of pest stink bug populations along the year in the Neotropics exploring major commodity crops and less preferred associated plants. Populations peak on crops, which are harvest at maturation forcing stink bugs to move to adjacent plants. These plants may allow further reproduction acting as hosts, or just offer some nutrients/water for maintenance or serving as shelter. (From Panizzi and Lucini 2023 with permission)

again (Fig. 3.1). This pattern of movement and colonization highlights the seasonal dynamics of stink bug populations as they adapt to changing plant availability and environmental conditions throughout the year. This adaptability makes them to be considered the most important pest complex of the major commodities in the Neotropics.

3.1 Cultivated Plants

The most important component of the life history of stink bugs in the Neotropics, which explains the tremendous abundance of these pest insects is the availability of food, provided mainly by crop plants. In Fig. 3.2, the increase of area dedicated to cultivated plants (grains and fiber) in Brazil in the last 20 years is shown as an example. The expansion of this area increased from around 46 million hectares in the year 2006 to 78 million hectares in 2023 (IBGE 2023). This means a 70% increase in the cultivated area in this time frame, which is estimated to keep growing. As mentioned before, crops cultivated in sequence form a "green bridge" of food plants, which are explored by stink bugs favoring their abundance.

The majority of cultivated plants play a pivotal role as hosts for stink bugs, i.e., plants that support the full lifecycle of stink bugs, from nymph development to adult reproduction. Some, however, only serve as associated plants, i.e., serving to provide nutrients for maintenance and shelter during specific periods. For example, the green belly stink bug, *Diceraeus melacanthus* (Dallas) feeds on maize and wheat seedlings, causing severe damage. Despite this, they do not lay their eggs on these

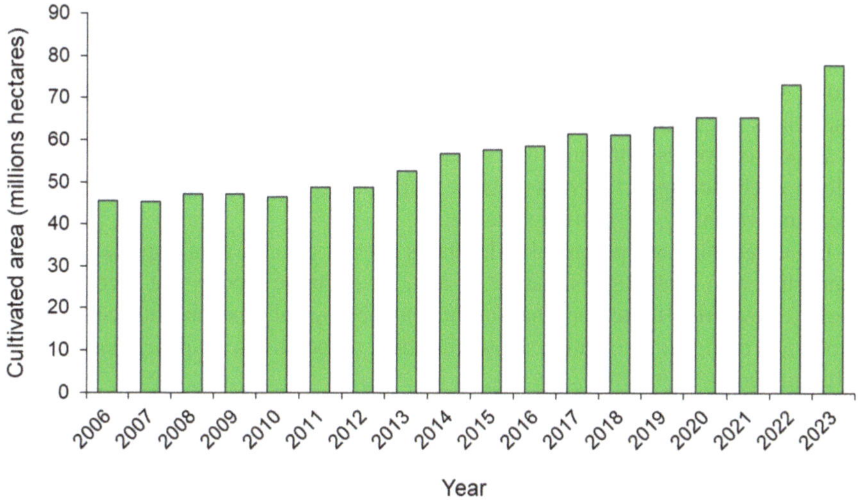

Fig. 3.2 Gradually increase in the use of land allocated to cultivated crops (grains and fibers) in the major area of the Neotropics (Brazil) in the last two decades. (Source: IBGE 2023)

seedlings, and when nymphs are present, they are unable to complete their development while feeding on these plants.

Later in the season they might feed on plants in the reproductive stage, such as on wheat ears, but no reproduction occurs. On soybean, however, they feed on pods and adults lay eggs and nymphs complete their development by feeding on them (Chocorosqui and Panizzi 2008). This distinction in behavior highlights the varying interactions between stink bugs and different plant species, where certain plants serve as suitable hosts for their complete life cycle, while others merely provide sustenance and shelter at specific stages of their development.

3.2 *Non-cultivated Plants*

In addition to the cultivated plants, temporary wild plants, mostly weeds, and perennial wild plants (trees) have an important role in supporting pest stink bug populations.

In the first case, weed plants play the most important role. With the adoption of non-cultivated (non-tillage) systems, these plants became very abundant in crop fields and also nearby them. Despite of that, their contribution to pest stink bugs biology has been negligible, as recently pointed out (Panizzi and Lucini 2022).

The number of weed plants serving as hosts, or which are simply associated with some main pest stink bugs in the Neotropics is shown in Fig. 3.3. In general, the total number of weeds explored by each species varies depending on their polyphagous habits in a range of over one to two dozen (14–20 weed species in the case analyzed).

Interesting to note that for the species *D. melacanthus* and for *E. heros*, 23 and 16% of the weeds, respectively, serve as hosts, and the majority of them, 77 and 84%, serve as associated plants. In contrast, for *N. viridula*, which is highly polyphagous, 40% of the weeds serve as hosts and 60% as associated. For *P. guildinii* the situation is intermediated with 50% of the weeds serving as host and 50% serving as associated weeds. Although this variability occurs among stink bug species, the role of weeds to support their abundance is germane and cannot be overseen, as has been, unfortunately, the current situation.

In the second case, perennial wild plants (trees), also have an important role in supporting pest stink bug abundance. In fact, the endemic species that become pests usually move from these trees that for the long ecological time have hosted them, in a long and complex association. However, these associations are poorly known and little studied.

Perennial wild trees are, in a variable extent, an important component of the agricultural landscape. As the use of land for crops became more intense, the pressure over the land covered with trees reached dangerous levels, in some cases with the complete elimination of them to give place to crops. This was particular true in the large areas of the Brazilian savannahs in the Central-West. In other areas, perennial trees adjacent to crop fields is more common, and they may host stink bugs (see

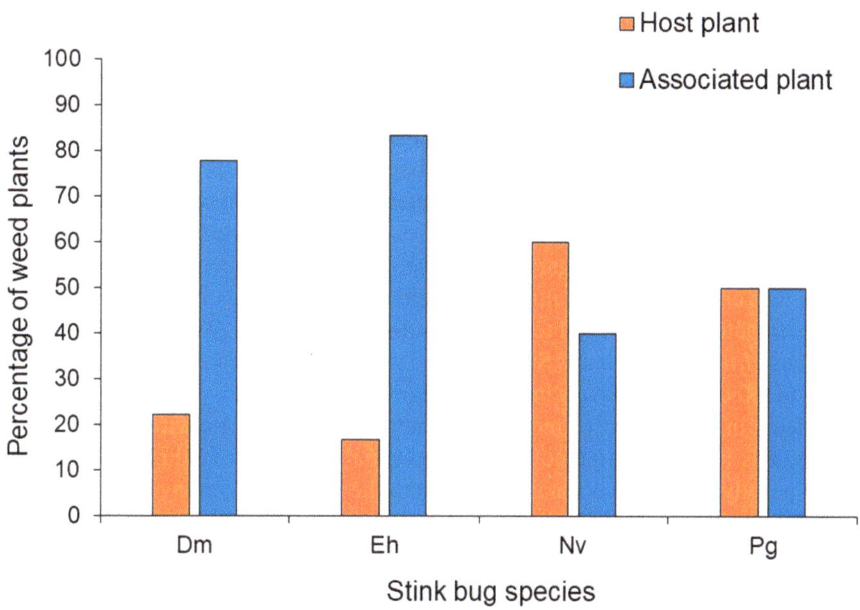

Fig. 3.3 Number (%) of weed plants found in row crop fields or in pasture lands in the Neotropics ranked as hosts (allow nymph survivorship and/or adult reproduction) or as associated plants (provide nutrients/water/shelter only) of selected stink bug pest species. Dm = *Diceraeus melacanthus* (number of weed plants registered = 18); Eh = *Euschistus heros* (n = 18); Nv = *Nezara viridula* (n = 20); and Pg = *Piezodorus guildinii* (n = 14). (Modified from Panizzi and Lucini 2022)

examples of wild trees hosting pest stink bugs in Costa et al. 1995; Panizzi and Grazia 2001; Zerbino et al. 2015; Marsaro et al. 2020).

4 Agricultural Systems Landscape and IPM for Stink Bugs

Stink bugs exhibit significant mobility, with adults frequently shifting between various host plant species over lifecycles. Consequently, agricultural fields might be surrounded by a rich and diverse array of alternative host and associated plants that could potentially heighten the presence of pests, leading to increased crop damage. The role of uncultivated land bearing temporary and perennial plants affecting the distribution and abundance of pests insects, as well as their natural enemies, on nearby crops has long been recognized (van Emden 1965a, b). Since that time, a myriad of integrated pest management (IPM) programs has been conceived and implemented (examples can be found on the IPM literature previously cited).

The majority of these IPM programs have been used by farmers worldwide with great success. However, few have taken in consideration seriously the effect of the mosaic of the vegetation (of cultivated or non-cultivated plants), that surrounds crops been managed, on the pest's species targeted. Therefore, a pressing question emerges: How can we further enhance the efficacy and sustainability of IPM programs? We believe that the implementation of IPM programs could be improved if greater attention be dedicated in considering the landscape, identifying the factors linked to elevated stink bug damage in agricultural products.

By delving into the life history of stink bugs and their interactions with the landscape, we can identify key factors that influence their population dynamics, movement patterns, and the timing of their life stages. This knowledge empowers us to make informed decisions regarding the planning and implementation of IPM programs. For example, Rice et al. (2017) reported that presence of larger forest edges nearby tomato fields have led to the most substantial damage of brown marmorated stink bug *Halyomorpha halys* Stål. It can be inferred that landscape configuration is likely to affect damage levels in adjacent agricultural fields.

Indeed, in the context of modern agriculture and evolving environmental challenges, it is imperative to recognize the potential of the surrounding landscape to shape pest dynamics. This entails deeper into the intricate interplay between crops and their ecological surroundings. The recognition that pest populations are not isolated entities but are intimately connected to the mosaic of plant species and ecosystems surrounding them opens doors to innovative and holistic approaches to pest management. Such an approach not only mitigates the risks associated with pest outbreaks but also promotes the well-being of natural enemies and biodiversity, contributing to more sustainable and resilient agricultural practices. In light of these insights, fine-tuning IPM programs to account for the broader landscape not only enables the precise targeting of specific pests but also nurtures the cultivation of a healthier and more balanced ecosystem.

5 Concluding Remarks

In this chapter we present a brief view of the implications of life history studies for the design and implementation of effective integrated pest management (IPM) programs, completer and more sustainable for management of pest stink bugs.

As agriculture moves toward more intense systems of cultivation, and pressure grows on the use of land for cultivation, replacing natural vegetation by crops, the development and use of the IPM programs here in referred becomes mandatory. This applies to any pest species, and in particular for the case of stink bugs in the Neotropics. The reason for that is because, not only they are major pests, but because they put pressure on several crops during most, or if not, total time of the year. This has several impacts, such as the use of agrichemicals (insecticides, fungicides, nematicides, herbicides, just to mention the main ones) that are dumped on the fields, spoiling "mother nature".

To implement these IPM programs refereed, a balance on the mosaic composition of the landscape at the farm levels must be seek. In a recent paper, Panizzi and Lucini (2023) presented a theoretical scheme of farm landscape considering areas under crop cultivation, areas with weeds and other wild temporary vegetation, and areas with perennial vegetation. They also present and discuss how stink bugs move from one area to other, and how the intensity of movement from one area to the other may vary. This exercise of learning how pest stink bugs behave in a certain agroecosystem set the ground to develop effective and holistic IPM programs. Needless to mention that as diversity and complexity is added to the agroecosystem, the greater stability is achieved and less likely pest outbreaks will occur (e.g., Altieri 1994; Bianchi et al. 2006; Letourneau et al. 2011).

In conclusion, the intricate relationship between agricultural systems landscapes and IPM is pivotal for modern stink bug control. Successful pest management necessitates a comprehensive understanding of how stink bugs interact with their environment and how the landscape shapes their behavior. This knowledge forms the bedrock for the development of effective, environmentally sustainable, and resilient pest management strategies that are indispensable for contemporary agriculture.

References

Altieri MA (1994) Biodiversity and pest management in agroecosystems. Haworth Press, New York

Ávila CJ, Panizzi AR (1995) Occurrence and damage by *Dichelops* (*Neodichelops*) *melacanthus* (Dallas) (Heteroptera: Pentatomidae) on maize. An Soc Entomol Brasil 24:193–194

Bianchi FJJA, Booij CJH, Tscharntke T (2006) Sustainable pest regulation in agricultural landscapes: a review on landscape composition, biodiversity and natural pest control. Proc R Soc B 273:1715–1727

Chocorosqui VR, Panizzi AR (2008) Nymph and adult biology of *Dichelops melacanthus* (Dallas) (Heteroptera: Pentatomidae) feeding on cultivated and non-cultivated host plants. Neotrop Entomol 37:353–360

Costa EC, Bogorni PC, Bellomo VH (1995) Percevejos coletados em copas de diferentes espécies florestais. Pentatomidae-1. Rev Cienc Flor 5:123–128

Ehler LE (2000) Farmscape ecology of stink bugs in northern California. Thomas Say Publication in Entomology: Memoirs, Entomological Society of America, Lanham

Grasswitz TR (2019) Integrated pest management (IPM) for small-scale farms in developed economies: challenges and opportunities. Insects 10(179):1–20

Green KK, Stenberg JA, Lankinen Å (2020) Making sense of integrated pest management (IPM) in the light of evolution. Evol Appl 13:1791–1805

Haseeb M, Gordon T, Legaspi J, Kanga L (2016) Integrated pest management of the southern green stink bug, *Nezara viridula*, on tomato using trap and refuge crops in North Florida. Proceedings of the 52nd Caribbean Food Crops Society meeting, Guadeloupe, July 10–16

Hokkanen HMT (1991) Trap cropping in pest management. Annu Rev Ent 36:119–138

IGBE – Instituto Brasileiro de Geografia e Estatística (2023) Levantamento sistemático da produção agrícola. Available at https://www.ibge.gov.br/estatisticas/economicas/agricultura-e-pecuaria/9201-levantamento-sistematico-da-producao-agricola.html. Accessed 23 Oct 2023

Jones WA Jr, Sullivan MJ (1981) Overwintering habitats, spring emergence patterns, and winter mortality of some South Carolina Hemiptera. Environ Entomol 10:409–414

Jones WA Jr, Sullivan MJ (1982) Role of host plants in population dynamics of stink bug pests of soybean in South Carolina Hemiptera. Environ Entomol 11:867–875

Kogan M (1998) Integrated pest management: historical perspectives and contemporary developments. Annu Rev Ent 43:243–270

Kogan M, Jepson P (2007) Perspectives in ecological theory and integrated pest management. Cambridge University Press, Cambridge

Letourneau DK, Armbrecht I, Rivera BS, Lerma JM, Carmona EJ, Daza MC, Escobar S, Galindo V, Gutierrez C, Lopez SD, Mejia JL, Rangel AMA, Rangel JH, Rivera L, Saavedra CA, Torres AM, Trujillo AR (2011) Does plant diversity benefit agroecosystems? A synthetic review. Ecol Appl 21:9–21

Marsaro AL Jr, Panizzi AR, Sargiorato AC, Lucini T (2020) The invasive plant *Pittosporum undulatum* Ventenat (Pittosporaceae) hosting pest-stink bugs in southern Brazil. Entomol Communic 2:ec02025

McPherson JE, McPherson RM (2000) Stink bugs of economic importance in America North of Mexico. CRC Press, Boca Raton

Miner FD (1966) Biology and control of stink bugs on soybeans. Agric Exp Sta Univ Arkansas Bull 708:1–40

Mizell RF, Riddle TC, Blount AS (2008) Trap cropping system to suppress stink bugs in the southern coastal plain. Proc Florida State Horticultural Soc 121:377–382

Panizzi AR (1997) Wild hosts of pentatomids: ecological significance and role in their pest status on crops. Annu Rev Ent 42:99–122

Panizzi AR (2000) Suboptimal nutrition and feeding behavior of hemipterans on less preferred plant food sources. An Soc Entomol Brasil 29:1–12

Panizzi AR, Grazia J (2001) Stink bugs (Heteroptera, Pentatomidae) and a unique host plant in the Brazilian subtropics. Iheringia Ser Zool 90:21–35

Panizzi AR, Lucini T (2016) What happened to *Nezara viridula* (L.) in the Americas? Possible reasons to explain populations decline. Neotr Ent 45:619–628

Panizzi AR, Lucini T (2022) The overlooked role of weed plants affecting pest stink bug (Hemiptera: Heteroptera: Pentatomidae) bioecology in the Neotropics. Arthropod Plant-Interact 16:1–14

Panizzi AR, Lucini T, Aldrich JR (2022) Dynamics in pest status of phytophagous stink bugs in the Neotropics. Neotr Ent 51:18–31

Panizzi AR, Lucini T (2023) Life history studies of stink bugs: much-needed research to support their conservation biological control. BioControl. https://doi.org/10.1007/s10526-023-10214-5

Peshin R, Pimentel D (2014) Integrated pest management. Experiences with implementation, global overview, vol 4. Springer, Dordrecht

Rice KB, Troyer RR, Watrous KM, Tooker JF, Fleischer SJ (2017) Landscape factors influencing stink bug injury in mid-Atlantic tomato fields. J Econ Entomol 110:94–100

Schaefer CW, Panizzi AR (2000) Heteroptera of economic importance. CRC Press, Boca Raton

Stahl JM, Wilson H, Straser RK, Maccaro JJ, Daane KM (2021) Irrigated trap crops impact key hemipteran pests in organic pistachio orchard. Arthropod Plant Interact 15:949–959

Tillman PG (2014) Physical barriers for suppression of movement of adult stink bugs into cotton. J Pest Sci 87:419–427

Tillman PG, Schomberg H, Phatak S, Mullinix B, Lachnicht S, Timper P, Olson D (2004) Influence of cover crops on insect pests and predators in conservation tillage cotton. J Econ Entomol 97:1217–1232

van Emden H (1965a) The effect of uncultivated land on the distribution of cabbage aphid (*Brevicoryne brassicae*) on an adjacent crop. J Appl Ecol 2:171–196

van Emden H (1965b) The role of uncultivated land in the biology of crop pests and beneficial insects. Sci Hort 17:121–136

Zerbino MS, Altier N, Panizzi AR (2015) Seasonal occurrence of *Piezodorus guildinii* on different plants including morphological and physiological changes. J Pest Sci 88:495–505

Chapter 4
The Use of *Telenomus podisi* to Manage Stink Bugs on Soybean: The Example of Brazil

Adeney Freitas Bueno, Weidson Plauter Sutil, Yelitza Coromoto Colmenarez, and Leonardo Roswadoski

1 Introduction

Telenomus podisi Ashmead is an egg parasitoid wasp from the family Scelionidae, which parasitizes eggs of both phytophagous and predaceous stink bugs (pentatomids), including a number of important agricultural pests. Adults are ~1.0 mm long (Fig. 4.1) and develop (from egg to adult) inside the host's eggs, lasting ~17, 105, and 120 h for the egg, larva, and pupa stages (Fig. 4.2), respectively, under 25 °C (Bueno et al. 2012).

The parasitoid is well distributed in Brazil. Its occurrence has been reported from the Midwest (Medeiros et al. 1997) to the South of the country (Moreira and Becker 1986), illustrating the parasitoid adaption to different climatic conditions (Torres et al. 1997). Despite evidences that *T. podisi* may be composed of more than one species (Bowers 2015), we will be providing in this chapter the most important information published in the literature grouped as *T. podisi*.

A. F. Bueno (✉)
Embrapa Soja, Londrina, PR, Brazil
e-mail: adeney.bueno@embrapa.br

W. P. Sutil · L. Roswadoski
Universidade Federal do Paraná, Curitiba, PR, Brazil

Y. C. Colmenarez
CABI-UNESP-FEPAF, Botucatu, SP, Brazil

A. F. Bueno, A. R. Panizzi (eds.), *Stink Bugs (Hemiptera: Pentatomidae) Research and Management*, Entomology in Focus 9,
https://doi.org/10.1007/978-3-031-69742-5_4

Fig. 4.1 *Telenomus podisi* parasitizing *Euschistus heros* eggs. (Photo: Adair V. Carneiro)

2 The Importance of *T. podisi*

Telenomus podisi has been responsible for more than 80% of natural parasitism recorded for eggs of the Neotropical brown stink bug, *Euschistus heros* (F.) (Pacheco and Corrêa-Ferreira 2000). This stink bug is the most important soybean pest in Brazil, responsible from 50% to 60% of the insecticide use in the country (Bueno et al. 2023). One single female of *T. podisi* can parasitize from 104 to 211 eggs of *E. heros* during its lifetime (Pacheco and Corrêa-Ferreira 2000; Silva et al. 2018), and more than 70% parasitism of the eggs of *E. heros* has been observed in the field (Bueno et al. 2020). Moreover, *T. podisi* can parasitize the eggs of several other pentatomid species (Yeargan 1982) as, for example, *Diceraeus melacanthus* Dallas, which parasitism reaches ~100 eggs (Taguti et al. 2019). Thus, this parasitoid has been considered one of the most important natural enemies not only of *E. heros* (Silva et al. 2018) but also of different other stink bug species (Koppel et al. 2009; Laumann et al. 2010; Bueno et al. 2012, 2020). As one of the most-studied egg parasitoids, this species is considered as an effective biocontrol agent for the key stink bug pests that attack soybean crops (Hoback et al. 2024). Consequently, the parasitoid was officially registered to be used in Brazil against stink bugs, being currently available in the market as a commercial bioinsecticide to control *E. heros* (Bueno et al. 2022). Despite parasitoids' availability in the market and its confirmed potential and importance to stink bug control, some important challenges still need to be faced, which still limit the release of *T. podisi* (Pinto and Parra 2002), notably in extensive crops such as soybean.

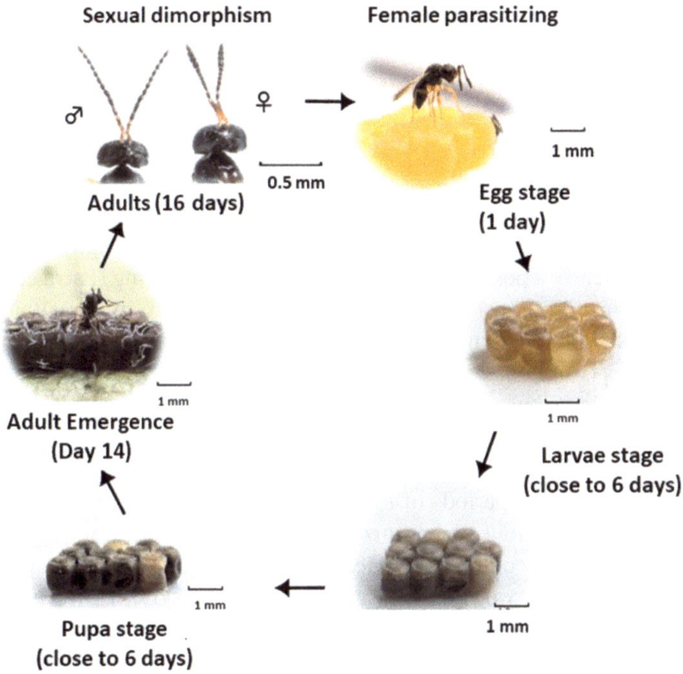

Fig. 4.2 Biological cycle of *Telenomus podisi* parasitizing eggs of *Euschistus heros*. (Adapted from Bueno et al. 2022)

3 *Telenomus podisi* Mass Rearing

To apply *T. podisi* to manage stink bugs we need to have the parasitoid available in large scale. Therefore, an efficient mass rearing protocol is of crucial importance to the success of applied release of *T. podisi* in the field. Consequently, mass rearing of the parasitoid is a critical step to successfully release it as part of an augmentative biological control (ABC) program of *E. heros* (Parra 2010; Parra and Coelho 2022). An efficient and economic mass rearing protocol that offers *T. podisi* in large numbers and at low costs is crucial. It will allow the ABC program to be competitive with other pest control tools (Macedo et al. 2006; Bernardo et al. 2008; Carvalho et al. 2008; Colinet and Boivin 2011; Queiroz et al. 2017) and, therefore, massively adopted by soybean farmers (Corrêa-Ferreira and Moscardi 1993; van Lenteren and Tommasini 2002).

Despite *T. podisi* mass rearing being studied for more than 20 years (Peres and Corrêa-Ferreira 2004), with significant advances recently reported (Parra et al. 2023), this critical step of ABC program still poses challenges to *T. podisi* adoption as an applied management strategy. Theoretically, there are three different strategies in which *T. podisi* can be reared (Parra 2010): (1) on the natural host, (2) on factitious hosts (both in vivo), and (3) on an artificial diet (in vitro). So far, to the best of our knowledge, *T. podisi* has only been reared in vivo, with both pros and cons of this strategy.

An advantage of using natural host eggs for *T. podisi* mass-rearing is the possibility of storing such host eggs in liquid nitrogen at −196 °C, for up to 6 months, as reported for both *E. heros* (Favetti et al. 2014) and *Piezodorus guildinii* (Westwood) (Cingolani et al. 2018) eggs, with no harm to their quality. *Telenomus podisi* requires host eggs with good quality and sufficient nutritional value (Ramos 2020). Therefore, hosts of good quality will provide enough nutrients, assuring a high rate of parasitism and consequently a positive impact on the parasitoid rearing (Parra et al. 2023).

Differently from the host that can be stored, the storage of parasitoids at low temperatures is still a challenge that would benefit biological control adoption for different reasons (van Lenteren and Tommasini 2003). Firstly, it would help to both decrease production costs (Colinet and Boivin 2011) and keep production flow to provide a steady and sufficient supply of insects for releasing in ABC programs. Secondly, storage of the parasitoids would build up reserve supplies of biocontrol agents to compensate for periods of low production or unexpectedly high demand. Lastly, storage would allow synchronized field releases of natural enemies during critical stages of pest outbreaks (McDonald and Kok 1990; Leopold 1998; Venkatesan et al. 2000; van Lenteren and Tommasini 2003). However, adults of *T. podisi* can stand storage only for a few days. Adult wasps of the parasitoid can be stored for only 6 days at 5 °C with no detectable injury to its biology or parasitism capacity (Silva et al. 2019) before being released in the field. Therefore, due to the lack of a more efficient storage methodology for parasitoid adults (Colinet and Boivin 2011), *T. podisi* must be produced shortly before their intended use (Macedo et al. 2006). Recently, Roswadoski (2024) reported that adult wasps of *T. podisi* can be stored inside capsules with food supply for up to 15 days at 25 °C before releasing without no injury to parasitism capacity, adding some extra days to the shelf life of the parasitoid. Nevertheless, more research and development in this field is still required.

On the other hand, one negative aspect of using natural host eggs for mass rearing *T. podisi* might be the labor-intensive work demanded in rearing the stink bugs in the laboratory to have the required number of host eggs to mass rear the parasitoid (Hayashida et al. 2018; Parra and Coelho 2022). When reared on natural diet composed of fresh green bean pods (*Phaseolus vulgaris* L.); dry soybean seeds (*Glycine max* L.); raw shelled peanuts (*Arachis hypogaea* L.); sunflower seeds (*Helianthus annuus* L.); and water (Silva et al. 2008), parasitoid colony maintenance and diet replacement must be performed from twice to three times a week what can be too labor intensive and, therefore, expensive (Hayashida et al. 2018; Parra and Coelho 2022).

A strategy to mitigate this limitation and reduce the costs of stink bug egg production is the development of artificial diets to support mass rearing and production of those insects (Parra 2009). However, the lack of suitable and cost-effective artificial diets to completely replace the natural beans still hampers pentatomid species rearing and colony maintenance (Mendoza et al. 2016).

The development of artificial diets for the rearing of *E. heros* has been reported by several authors (Panizzi et al. 2000; Fortes et al. 2006; Siqueira 2007; Mendoza et al. 2016; Hayashida et al. 2018). However, despite the progress made in these studies, adjustments are still needed to improve the efficiency of pentatomid egg production. Mendoza et al. (2016) provided an artificial diet with lyophilized material that could meet the nutritional requirements of *E. heros* to produce insects of quality comparable to field insects. However, the diet requires bean pod lyophilization, which can be expensive and time-consuming. Furthermore, acquiring and maintaining an industrial freeze dryer is an economic disadvantage for small biofactories. Considering all this, Hayashida et al. (2018) proposed a cheaper diet than the initially one proposed by Mendoza et al. (2016) without lyophilized components, thus reducing costs of *E. heros* rearing and consequently of *T. podisi* rearing. Nevertheless, this diet proposed by Hayashida et al. (2018) still needs to be improved with preservatives to avoid contamination and the necessity of diet replacement. Industry in Brazil (for instance, Vivus, Piracicaba, SP, Brazil) just started rearing *E. heros* on artificial diet, producing enough eggs to release the parasitoid in 40,000 hectares of soybean per day (considering 6,500 parasitoids per hectare). However, the innovations incorporated in the artifical diet were not published and still kept as industrial secrets.

Telenomus podisi mass rearing can be performed at temperatures between 21 and 30 °C, although the highest daily and accumulated parasitism rates and viability occur at 24 °C. Therefore, 24 °C is the temperature adopted at *T. podisi* mass-rear facilities. Fresh and cryopreserved *E. heros* eggs can perform successfully as parasitoid hosts (Parra et al. 2023). However, it is relevant to point out that the quality control protocols for both *T. podisi* rearing and release are crucial issues that need further development since *T. podisi* use in ABC is relatively new in Brazil. The parasitoid official registration and commercialization in Brazil was only done after 2019 (MAPA 2019).

One challenging aspect of *T. podisi* rearing is the risk of parasitoid fitness losses in long-term production systems. It has been recorded for other egg parasitoid species that over the generations at constant ideal conditions in rearing facilities, foraging and flying abilities of adult wasps were reduced after several generations (Naranjo-Guevara et al. 2020). The establishment of varying rearing conditions (different regimes of light, temperature, and humidity, for example) and the introduction of wild individuals (periodically) into laboratory rearing have been suggested as possible solutions to this problem (Colmenarez et al. 2022).

As newly released parasitoids usually have to face field conditions with daily thermal fluctuations and periods of extremely high or low temperatures (Bannerman and Roitberg 2014, Colinet et al. 2015), Castellanos et al. (2019) proposed rearing *T. podisi* under fluctuating temperature conditions of 30 ± 2 °C during the day (12 h) and 20 ± 2 °C at night (12 h). This varing temperature is considered more suitable for parasitoid mass-rearing since such conditions resulted in fitness benefits (shorter developmental time, longer female longevity, higher fecundity/fertility) and reduced the costs of parasitoid production by approximately 23.5%.

The insect colony refreshment with individuals collected from natural populations is also relevant (Colmenarez et al. 2022). However, this procedure should be done with caution since taxonomic knowledge is required to avoid colony contamination with other parasitoid species (Bowers 2015) or fungal and bacterial pathogens from the field, among other possible adverse effects to the parasitoid colony maintenance (Colmenarez et al. 2022).

4 Recommendations for the Applied Used of *Telenomus podisi* to Manage Stink Bugs on Soybean

The egg parasitoid *T. podisi* is an efficient biocontrol to be released on soybean to manage stink bugs (Bueno et al. 2020). Despite its efficiency and the several commercial products available in the market being sold and used by farmers in Brazil, essential challenges still need to be addressed to successfully release this parasitoid in the field inside ABC programs (Bueno et al. 2022).

Among the most crucial considerations to enable *T. podisi* success (Bueno et al. 2022), the most important ones are:

4.1 *The Number (Density) of* T. podisi *to Be Released*

Releasing the ideal density of parasitoids is very important for the success of *T. podisi*. The official recommendations from *T. podisi* registration in Brazil are from two to three releases of 6,500 parasitoids/ha. The weekly releases should sum up to 13,000 to 19,500 wasps/ha (Bueno et al. 2022). Following these recommendations, Bueno et al. (2020) recorded more than 70% parasitism of *E. heros* eggs after three releases of 6,500 *T. podisi* at weekly intervals, while the natural parasitism rate was only close to 10%.

Differently from Bueno et al. (2022), Weber et al. (2022), using a computation model, based on biological parameters of the parasitoid, recommended three to four releases of 5,000 female parasitoids per hectare. If considering a sex ratio of 0.54 (Silva et al. 2018), it is similar to 30,000–40,000 wasp/ha, a number almost 100% higher than the current official recommendation of the registered biopesticides with *T. podisi*. Since some variations are still been seem in the published results, it is still important to test different numbers of parasitoids and better define the field recommendation to be followed by farmers.

4.2 *Release Technology of* T. podisi *in the Field Considering Dispersal, Release Frequency, Predation, and the Best Time*

Telenomus podisi can be released as adult wasps or pupae close to adult emergence (Bueno et al. 2022). Releasing parasitoid pupae has been preferred by industries that sell this biocontrol agent, mainly due to the mechanization possibilities, which facilitate the operation in the field. Automation of the process reduces the need for labor and the operational costs involved. Usually, these pupae are bulk distributed strategically in the field at equidistant points (minimum of 50 points per hectare) (Bueno et al. 2022). However, to release pupae is a practice based upon previous experiences with another egg parasitoid, *Trichogramma pretiosum* Riley, which release is easily programmed to occur only hours before parasitoid emergence.

However, for *T. podisi*, males emerge from 17 to 24 h before females (Yeargan 1980). Thus, released *T. podisi* female pupae will stay longer in the field before the adult emergence and can be susceptible to different biotic and abiotic causes of mortality, reducing parasitoid emergence. Brazil has a diverse fauna, including many ant species and other predators. These predators can prey on exposed and susceptible natural enemies, reaching 100% predation within a few hours after release (Parra 2014).

In addition to predation, among the abiotic causes of mortality, the temperature is highlighted as one of the most critical threats to parasitoids (Frazer and Mcgregor 1992) because it can affect not only survival but also several biological parameters (Denis et al. 2011). When *T. podisi* pupae were released in soybean crops during the vegetative period, before complete plant development and soil shading between rows, those pupae exposed directly to the sunlight between rows had significantly reduced adult emergence compared to pupae shaded under plants. Parasitoid emergence reduced from 76% to close to only 20% emergency (Braz et al. 2021).

Those causes of mortality are even more relevant for egg parasitoids when released as immobile pupae, unable to search for shelter to protect themselves from predators or adverse temperatures (Braz et al. 2021). Therefore, releases in the late afternoon should favor adult emergence (Grande et al. 2021). Also, Roswadoski (2024) suggested that *T. podisi* should be released as fed adults inside capsules (Fig. 4.3). Besides increasing the shelf life of the bioinsecticide based on the parasitoid, releasing fed adults can reduce pupa mortality, minimizing the adverse effects caused by extreme temperatures and predation since released adults can immediately look for shelter to protect themselves from those causes of mortality.

In addition, it is also relevant to release the parasitoid when the vulnerable stage of the host egg is available in the field. Therefore, one of the biggest challenges for successfully managing stink bugs using *T. podisi* is releasing the parasitoid at the right time. *Telenomus podisi* adults reaches 80% of their lifetime parasitism around 14–16 days (Silva et al. 2018). Furthermore, *T. podisi* showed similar parasitism capacity on *E. heros* eggs at 24, 48, 72, and 96 h of embryonic development (Queiroz

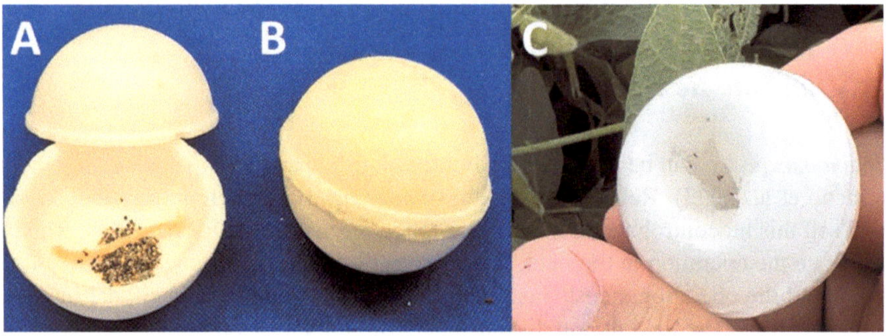

Fig. 4.3 Release of adults of *Telenomus podisi* inside capsules with a solid diet. (**a**) *Telelenomus podisi* pupae with the solid diet at the releasing capsule preparation. (**b**) Releasing capsule read to be used in the field from 2 to 15 days after the emergence of first parasitoid adults (Photos: Leonardo Roswadoski) and (**c**) *Telenomus podisi* being released in the field after opening the capsules. (Photo: Adeney Freitas Bueno)

et al. 2019). Therefore, the success of *T. podisi* depends on a release strategy that allows adult parasitoids 14–16 days old to match host eggs aged up to 4 days old in the field. The official recommendation establishes that the releases should be made weekly (Bueno et al. 2022). However, considering that *T. podisi* only reaches 80% of its lifetime parasitism around 14 days after parasitism (Silva et al. 2018), field release intervals of 2 weeks could also be tested in order to improve the parasitoid efficacy in the field. Moreover, rigorous monitoring of stink bugs in the field is crucial, in addition to a fast response of the bioindustry to release the parasitoid in the area right after the parasitoid is bought by the farmer.

The minimum number of points per area at which the parasitoid needs to be released is another decisive requirement needing observation when releasing parasitoids in the field. The official recommendation is 50 points per hectare (Bueno et al. 2022). Differently, Weber et al. (2022) determined that the release of *T. podisi* should be at points or strips spaced 25 m.

Regardless of the release as pupae or adult, it is necessary to consider that *T. podisi* are small insects, highly influenced by the wind direction (Fig. 4.4). Since the wind influences the dispersion of parasitoids in the field (Corbett and Rosenheim 1996), the release method should preferably be for a perimeter distribution of the released insects, considering the prevailing wind direction in the area, whenever possible (Pomari-Fernandes et al. 2018).

4.3 *Stink Bug Density and Climatic Conditions at the Time of Release*

Unlike insecticides that kill stink bugs, *T. podisi* will prevent stink bug outbreaks by its egg parasitism. Therefore, parasitoid release should start at the first stink bug adults and their first eggs in the field (Bueno et al. 2022). That is a challenge,

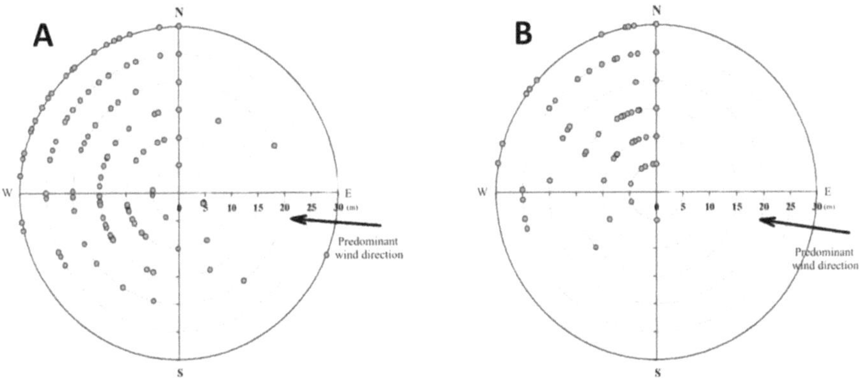

Fig. 4.4 The effect of the predominant wind direction on the dispersal of the egg parasitoid *Telenomus remus* after releases performed at soybean stages of development V3 (**a**) and R3 (**b**). Each point represents a parasitized egg recovered at different distances (from 5 to 30 m) from the released point, which was the cross of the axes. (Adapted from Pomari-Fernandes et al. 2018)

considering that soybean pest sampling with the beating cloth is the alternative most recommended by research (Bueno et al. 2023). The use of traps to capture adults may be an option to help determining the ideal time to release the parasitoids. However, the correlation between adults captured in traps and the egg presence in the field still needs to be established. Today, *T. podisi* is released when farmers visually observe the first adults and eggs in the field, which is subjective and dependable on visual perception to succeed.

The climatic conditions, especially rain and extreme temperatures, can negatively impact the success of biological control using egg parasitoids (Braz et al. 2021; Grande et al. 2021). Therefore, the recommendation is to release egg parasitoids at milder times of the day, with more favorable climatic conditions for the emergence and survival of parasitoids. In this scenario, the release of *T. podisi* in the early morning or late afternoon favors parasitism since it is less affected by the absence of light (Grande et al. 2021). With the release of fed adults inside capsules, as suggested by Roswadoski (2024), farmers can wait up to 15 days for better weather conditions after the emergence of the first parasitoids inside the capsules, without any reduction of parasitoid parasitism in the field. This makes the parasitoid release more flexible and, it is very convinient for delaying the release for a couple of days, when the weather conditions are not favorable.

4.4 *Interactions of* T. podisi *with Other Pesticides Used in Crop Management*

Despite the high potential of biological control using *T. podisi* releases against stink bugs, additional control strategies (chemical or biological) might still be necessary. Thus, understanding the threats that pesticides pose for *T. podisi* and the knowledge of selective pesticides are essential to allow the use of both chemical and biological controls in combination in IPM programs (Torres and Bueno 2018).

Active ingredients of the insect growth regulators group (IGRs), such as chlorfluazuron, teflubenzuron, novaluron, and lufenuron, are relatively more selective to *T. podisi* (Stecca et al. 2018). In contrast, pyrethroids such as bifenthrin, beta-cyfluthrin, zeta-cypermethrin and organophosphates such as chlorpyrifos and acephate were among the most harmful pesticides to this parasitoid, especially to adults, which is generally the most susceptible parasitoid stage (Hassan et al. 1985; Carmo et al. 2020; Silva et al. 2018; Stecca et al. 2018). Therefore, soybean farmers must avoid broad-spectrum insecticide applications around 10 days before and 15 days after *T. podisi* releases. Fungicides and herbicides are generally less harmful to beneficial organisms when compared to insecticides (Bueno et al. 2017; Zantedeschi et al. 2018).

4.5 *Size of the* T. podisi *Release Field*

The size of the field influences *T. podisi* and all other macro biologicals. Macro organisms move from plant to plant or place to place, and their migration to crops near the released field is always possible. Adults of *T. podisi* can move to areas where they were not released. Also, adults of *E. heros* can migrate to areas where the parasitoid was released. Therefore, the success of *T. podisi* release with high parasitism does not eliminate the need for stink bug control using other strategies since the parasitoids will kill the eggs but adults from nearby areas can migrate to the area where the parasitoid was released, reaching or surpassing stink bug economic thresholds of 2 bugs/meter (or beath cloth) (Bueno et al. 2015), highlighting the importance of other compatible control alternatives in IPM.

The smaller the size of the field, the higher the chances of either *T. podisi* or *E. heros* migrating and frustrating biological control with the released parasitoids. The minimum size for *T. podisi* releases still needs to be determined in future research.

5 Concluding Remarks

Further studies are still needed to precisely determine optimal release rates, release times and frequencies, the number of release points, the best stage and equipment for releases, and other aspects, such as how large the fields should be to achieve efficient pest control releasing *T. podisi*. Despite these needs, *T. podisi* release against stink bugs is currently considered an efficient management strategy besides being commercially available in Brazil. The adoption by farmers of parasitoid releases with other IPM recommendations is allowing the reduction of synthetic insecticide use and prioritizing the use of the most selective pesticides, which will trigger a more favourable environment for the success of biological control. Among the most selective insecticides to natural enemies, biological insecticides (entomopathogens) stand out, followed by the insecticide growth regulators (IGRs) group (Torres and Bueno 2018). Unfortunately, no IGRs are available against stink bugs. However, more recently, entomopathogens (*Metarhizium anisopliae* and *Beauveria bassiana*) have been officially registered and are commercially available in Brazil for stink bug control. *Telenomus podisi*, in combination with those entomopathogens, promises to be a viable and sustainable solution to stink bug management for soybean fields. However, they still need to be tested in conjunction in future trials.

References

Bannerman JA, Roitberg BD (2014) Impact of extreme and fluctuating temperatures on aphid-parasitoid dynamics. Oikos 123:89–98

Bernardo U, Iodice L, Sasso R, Pedata PA (2008) Effects of cold storage on *Thripobius javae* (=*T. semiluteus*) (Hymenoptera: Eulophidae). Biol Sci Technol 18:921–933

Bowers KR (2015) *Telenomus podisi*: one species, or more? Master's Thesis, Graduate School of The Ohio State University, Ohio, United States of America, 67p

Braz EC, Bueno AF, Colombo AF, Queiroz AP (2021) Temperature impact on *Telenomus podisi* emergence in field releases of unprotected and encapsulated parasitoid pupae. Neotrop Entomol 50:462–469

Bueno AF, Sosa-Gómez DR, Côrrea-Ferreira BS, Moscardi F, Bueno RCOF (2012) Inimigos naturais das pragas da soja. In: Hoffmann-Campo CB, Côrrea-Ferreira BS, Moscardi F (eds) Soja: Manejo Integrado de Insetos e Outros Artrópodes-praga. Embrapa, Brasília, pp 493–630

Bueno AF, Bortolotto OC, Pomari-Fernandes A, França-Neto JB (2015) Assessment of a more conservative stink bug economic threshold for managing stink bugs in Brazilian soybean. Crop Prot 71:132–137

Bueno AF, Carvalho GA, Santos AC, Sosa-Gomez DR, Silva DM (2017) Pesticide selectivity to natural enemies: challenges and constraints for research and field recommendation. Cienc Rural 47:e20160829

Bueno AF, Braz ÉC, Favetti BM, França-Neto JB, Silva GV (2020) Release of the egg parasitoid *Telenomus podisi* to manage the Neotropical brown stink bug, *Euschistus heros*, in soybean production. Crop Prot 137:105310

Bueno AF, Parra JRP, Colombo FC, Colmenarez YC, Narde BVF, Pereira FF (2022) Manejo de pragas com parasitoides. In: Meyer MC, Bueno AF, Mazaro SMM, Silva JC (eds) Bioinsumos na Cultura da Soja. Embrapa, Brasília, DF, Brasil pp 417–434

Bueno AF, Colmenarez YC, Carnevalli RA, Sutil WP (2023) Benefits and perspectives of adopting soybean-IPM: the success of a Brazilian programme. Plant Health Cases, pp 1–16

Carmo ED, Bueno AF, Bueno RC (2020) Pesticide selectivity for the insect egg parasitoid *Telenomus remus*. BioControl 55:455–464

Carvalho JS, Vacari AM, De Bortoli SA, Viel SR (2008) Efeito do armazenamento de pupas de *Cotesia flavipes* (Cameron, 1891) (Hymenoptera: Braconidae) em baixa temperatura. Bol Sanid Veg 34:21–26

Castellanos NL, Bueno AF, Haddi K, Silveira EC, Rodrigues HS, Hirose E, Oliveira EE (2019) The fitness and economic benefits of rearing the parasitoid *Telenomus podisi* under fluctuating temperature regime. Neotrop Entomol 48:934–948

Cingolani MF, Greco NM, Liljesthröm GG (2018) Cold-storage of *Piezodorus guildinii* (Hemiptera: Pentatomidae) eggs for rearing *Telenomus podisi* (Hymenoptera: Platygastridae). J Agric Sci Technol 17:1507–1516

Colinet H, Boivin G (2011) Insect parasitoids cold storage: a comprehensive review of factors of variability and consequences. Biol Control 58:83–166

Colinet H, Sinclair BJ, Vernon P, Renault D (2015) Insects in fluctuating thermal environments. Annu Rev Entomol 60:123–140

Colmenarez YC, Babendreier D, Wurst FRF, Vásquez freytez CL, Bueno AF (2022) The use of *Telenomus remus* (Nixon, 1937) (Hymenoptera: Scelionidae) in the management of *Spodoptera* spp.: potential, challenges and major benefits. CABI Agric Biosci 3:5

Corbett A, Rosenheim JA (1996) Quantifying movement of a minute parasitoid *Anagrus epos* (Hymenoptera: Mymaridae), using fluorescent dust marking and recapture. Biol Control 6:35–44

Corrêa-Ferreira BS, Moscardi F (1993) Técnicas de armazenamento de ovos do percevejo-verde visando à multiplicação do parasitoide *Trissolcus basalis* (Wollaston). Pesq Agropec Bras 28:1247–1253

Denis D, Pierre JS, van Baaren J, van Alphen JJ (2011) How temperature and habitat quality affect parasitoid lifetime reproductive success a simulation study. Ecol Model 222:1604–1613

Favetti BM, Butnariu AR, Doetzer AK (2014) Storage of *Euschistus heros* (Fabricius) eggs (Hemiptera: Pentatomidae) in liquid nitrogen for parasitization by *Telenomus podisi* Ashmead (Hymenoptera: Platygastridae). Neotrop Entomol 43:291–293

Fortes P, Magro SR, Panizzi AR, Parra JR (2006) Development of a dry artificial diet for *Nezara viridula* (L.) and *Euschistus heros* (Fabricius) (Heteroptera: Pentatomidae). Neotrop Entomol 35:567–572

Frazer BD, McGregor RR (1992) Temperature-dependent survival and hatching rate of eggs of seven species of Coccinellidae. Can Entomol 124:305–312

Grande MLM, Queiroz AP, Gonçalves J, Hayashida R, Ventura MU, Bueno AF (2021) Impact of environmental variables on parasitism and emergence of *Trichogramma pretiosum*, *Telenomus remus* and *Telenomus podisi*. Neotrop Entomol 50:605–614

Hassan SA, Bigler F, Blaisinger P, Bogenschütz H, Brun J, Chiverton P, Van Zon AQ (1985) Standard methods to test the side-effects of pesticides on natural enemies of insects and mites developed by the IOBC/WPRS Working Group 'Pesticides and Beneficial Organisms'. Eppo Bull 15:214–255

Hayashida R, Bueno AF, Hermel AO, Hirakuri MH, Silva FAC, Roggia S (2018) *Euschistus heros* (Hemiptera: Pentatomidae) fitness on artificial diets: an approach to optimize mass rearing of *Telenomus podisi* (Hymenoptera: Platygastridae) for augmentative biological control. J Econ Entomol 111:1605–1613

Hoback WW, Ramos G, Hayashida R, Santos DM, Alvarez DL, Oliveira RC (2024) Optimizing the release pattern of *Telenomus podisi* for effective biological control of Euschistus heros in soybean. Insects 15:192

Koppel AL, Herbert DA Jr, Kuhar TP, Kamminga K (2009) Survey of stink bug (Hemiptera: Pentatomidae) egg parasitoids in wheat, soybean, and vegetable crops in Southeast Virginia. Environ Entomol 38:375–379

Laumann RA, Moraes MC, Silva JP, Vieira AM, Silveira SD, Borges M (2010) Egg parasitoid wasps as natural enemies of the neotropical stink bug *Dichelops melacanthus*. Pesq Agropec Bras 45:442–449

Leopold RA (1998) Cold storage of insects for integrated pest management. In: Hallman GJ, Denlinger DL (eds) Temperature sensitivity in insects and application in integrated pest management. Westview Press, Boulder, pp 235–267

Macedo LPM, Maracajá PB, Bezerra CES (2006) Técnicas de armazenamento de ovos de insetos para programas de controle biológico. Rev Verde 1:1–19

Mapa (2019) Instrução Normativa da Especificação de Referência. https://www.gov.br/agricultura/pt-br/assuntos/insumos-agropecuarios/insumos-agricolas/agrotoxicos/produtosfitossanitarios/arquivos-especificacao-de-referencia/INSDAn252019.especifreferDOU.pdf. Accessed 16 Dec 2022

McDonald RC, Kok LT (1990) Post refrigeration viability of *Pteromalus puparum* (Hymenoptera: Pteromalidae) prepupae within host chrysalids. J Entomol Sci 25:409–413

Medeiros MA, Schimidt FVG, Loiácono MS, Carvalho VF, Borges M (1997) Parasitismo e predação em ovos de *Euschistus heros* (Fab.) (Heteroptera: Pentatomidae) no Distrito Federal. Brasil An Soc Ent Bras 26:397–401

Mendoza AC, Rocha ACP, Parra JRP (2016) Lyophilized artificial diet for rearing the Neotropical *Euschistus heros* (Hemiptera: Pentatomidae). J Insect Sci 16:41

Moreira GRP, Becker M (1986) Mortalidade de *Nezara viridula* (Linnaeus, 1758) (Heteroptera: Pentatomidae) no estágio de ovo na cultura da soja: II Parasitoides. An Soc Ent Bras 15:291–308

Naranjo-Guevara N, Santos LAO, Barbosa NCCP, Castro ACMC, Fernandes AO (2020) Long-term mass rearing impacts performance of the egg parasitoid *Telenomus remus* (Hymenoptera: Platygastridae). J Entomol Sci 55:69–86

Pacheco DJP, Corrêa-Ferreira BS (2000) Parasitismo de *Telenomus podisi* Ashmead (Hymenoptera: Scelionidae) em populações de percevejos pragas da soja. An Soc Ent Bras 29:295–302

Panizzi AR, Parra JRP, Santos CH, Carvalho DR (2000) Rearing the Southern green stink bug using artificial diet and artificial plant. Pesqui Agropec Bras 35:1709–1715

Parra JRP (2009) A evolução das dietas artificiais e suas interações em ciência e tecnologia. In: Panizzi AR, Parra JRP (eds) Bioeologia e Nutrição de Insetos: Base para o Manejo Integrado de Pragas. Embrapa, Brasília, Brasil, pp 91–174

Parra JRP (2010) Mass rearing of egg parasitoids for biological control programs. In: Consoli F, Parra JRPP, Zucchi R (eds) Egg parasitoids in agroecosystems with emphasis on *Trichogramma*. Progress in biological control, vol 9. Springer, Dordrecht, pp 325–342

Parra JRP (2014) Biological control in Brazil: an overview. Sci Agr 71:345–355

Parra JRP, Coelho A (2022) Insect rearing techniques for biological control programs, a component of sustainable agriculture in Brazil. Insects 13:105

Parra LM, Carvalho JR, Hoback WW, Oliveira RC (2023) Optimizing mass rearing of the egg parasitoid, *Telenomus podisi*, for control of the brown stink bug, *Euschistus heros*. Insects 14:435

Peres WAA, Corrêa-Ferreira BS (2004) Methodology of mass multiplication of *Telenomus podisi* Ash. and *Trissolcus basalis* (Woll.) (Hymenoptera: Scelionidae) on eggs of *Euschistus heros* (Fab.) (Hemiptera: Pentatomidae). Neotrop Entomol 33:457–462

Pinto AS, Parra JRP (2002) Liberação de inimigos naturais, p. 325–342. In: Parra JRP, Botelho PSM, Corrêa-Ferreira BS, Bento JMS (eds) Controle Biológico no Brasil: Parasitoides e Predadores. Manole, São Paulo, SP, Brasil, p 635

Pomari-Fernandes A, Bueno AF, Bortoli SA, Favetti BM (2018) Dispersal capacity of the egg parasitoid *Telenomus remus* Nixon (Hymenoptera: Platygastridae) in maize and soybean crops. Biol Control 126:158–168

Queiroz AP, Bueno AF, Pomari-Fernandes A, Grande MLM, Bortolotto OC, Silva DM (2017) Low temperature storage of *Telenomus remus* (Nixon) (Hymenoptera: Platygastridae) and its factitious host *Corcyra cephalonica* (Stainton) (Lepidoptera: Pyralidae). Neotrop Entomol 46:182–192

Queiroz AP, Favetti BM, Hayashida R, Grande ML, Neiva MM, Panizzi AR, Bueno AF (2019) Effect of the ages of parasitoid and host eggs on *Telenomus podisi* (Hymenoptera: Platygastridae) parasitism. Neotrop Entomol 48:974–982

Ramos GS (2020) Capacidade de dispersão e parâmetros biológicos inerentes à criação massal de *Telenomus podisi* Ashmead (Hymenoptera: Platygastridae) para o controle de pentatomídeos-praga em soja. Master's Thesis, São Paulo State University, Botucatu, 122p

Roswadoski L (2024) Dieta para adultos de *Telenomus podisi* Ashmead, 1893 (Hymenoptera: Scelionidae) visando a liberação em campo de parasitoides alimentados e copulados. Master's Thesis, Federal University of Parana, Curitiba, PR, Brasil, 122p

Silva CC, Laumann RA, Blassioli MC (2008) *Euschistus heros* mass rearing technique for the multiplication of *Telenomus podisi*. Pesqui Agropec Bras 43:575–580

Silva GV, Bueno AF, Neves PMOJ, Favetti BM (2018) Biological characteristics and parasitism capacity of *Telenomus podisi* (Hymenoptera: Platygastridae) on eggs of *Euschistus heros* (Hemiptera: Pentatomidae). J Agric Sci 10:210–220

Silva GV, Bueno AF, Favetti BM, Neves PMOJ (2019) Use of low temperature storage to preserve host and parasitoid to improve the rearing of *Telenomus podisi* (Hymenoptera: Platygastridae) on *Euschistus heros* (Hemiptera: Pentatomidae) eggs. Neotrop Entomol 48:126–135

Siqueira F (2007) Adição de frutos de *Ligustrum lucidum* ait. (Oleaceae) em dieta artificial seca para a criação do percevejo marrom *Euschistus heros* (F.) (Heteroptera: Pentatomidae). Master's Thesis, Universidade Estadual do Paraná, Paraná, Curitiba, 43p

Stecca CS, Bueno AF, Pasini A, Silva DM, Andrade K, Zirondi Filho DM (2018) Impact of insecticides used in soybean crops to the egg parasitoid *Telenomus podisi* (Hymenoptera: Platygastridae). Neotrop Entomol 47:281–291

Taguti EA, Gonçalvez J, Bueno AF, Marchioro ST (2019) *Telenomus podisi parasitism on Dichelops melacanthus and Podisus nigrispinus eggs at different temperatures. Fla Entomol 102:607–613*

Torres JB, Bueno AF (2018) Conservation biological control using selective insecticides: a valuable tool for IPM. BioControl 126:53–64

Torres JB, Pratissoli D, Zanuncio JC (1997) Exigências térmicas e potencial de desenvolvimento dos parasitóides *Telenomus podisi* Ashmead e *Trissolcus brochymenae* (Ashmead) em ovos do percevejo predador *Podisus nigrispinus* (Dallas). An Soc Entomol Brasil 26:445–453

van Lenteren JC, Tommasini MG (2002) Mass production, storage, shipment and quality control of natural enemies. In: Albajes R, Gullino ML, van Lenteren JC, Elad Y (eds) Mass production, storage, shipment and quality control of natural enemies, integrated pest and disease management in greenhouse crops. Springer, Dordrecht, pp 276–294

van Lenteren JC, Tommasini MG (2003) Mass production, storage, shipment and release of natural enemies. In: van Lenteren JC (ed) Quality control and production of biological control agents: theory and testing procedures. CABI, Wallingford, pp 181–189

Venkatesan T, Singh SP, Jalali SK (2000) Effect of cold storage on cocoons of *Goniozus nephantidis* Muesebeck (Hymenoptera: Bethylidae) stored for varying periods at different temperature regimes. J Entomol Res 24:43–47

Weber ID, Garcia AG, Bueno AF, Oliveira RC, Godoy WAC (2022) Release strategies of *Telenomus podisi* for control of *Euschistus heros*: a computational modeling approach. Pest Manag Sci 78:4544–4556

Yeargan KV (1980) Effects of temperature on developmental rate of *Telenomus podisi* (Hymenoptera: Scelionidae). Ann Entomol Soc Am 73:339–342

Yeargan KV (1982) Reproductive capability and longevity of the parasitic wasps *Telenomus podisi* and *Trissolcus euschisti*. Ann Entomol Soc Am 75:181–183

Zantedeschi R, Grützmacher AD, Pazini JDB, Bueno AF, Machado LL (2018) Selectivity of pesticides registered for soybean crop on *Telenomus podisi* and *Trissolcus basalis*. Pesq Agropec Trop 48:52–58

Chapter 5
Augmentative Biological Control of Stink Bugs Using Entomopathogens in Brazil

Rodrigo Mendes Antunes Maciel, Raiza Abati, Nédia de Castilhos Ghisi, Everton Ricardi Lozano, and Michele Potrich

1 Introduction

Among the ecofriendly strategies studied for stink bug control, the use of biological control, especially using entomopathogens, should be highlighted, mainly because it is a natural control tool that can be used on large scales (van Lenteren et al. 2018). The use of augmentative biological control (ABC) in soybean has been used to control both caterpillars (Moscardi 1999) and stink bugs (Corrêa-Ferreira and Moscardi 1996). For the control of *Euschistus heros* (F.), the use of biological control agents is on the rise. Among biological control agents, the possible commercial choices for control are especially entomopathogenic fungi and one egg parasitoid (*Telenomus podisi* Ashmead discussed in more details in Chap. 4).

Currently, three species of entomopathogenic fungi are commercialized: *Beauveria bassiana* (Bals.-Criv.) Vuill. (Ascomycetes: Clavicipitaceae), *Metarhizium anisopliae* (Metsch.) (Ascomycetes: Clavicipitaceae) and *Cordyceps fumosorosea* (Wize) (Hypocreales: Cordycipitaceae). However, few isolates originate commercial products (Potrich et al. 2022).

The growth in the adoption of the use of entomopathogenic fungi occurs year after year and with prospects for continuity (Mascarin et al. 2019). Therefore, the prospecting of new isolates (Dalla-Nora et al. 2020; Silva-Santana et al. 2021) also shows growth and it is necessary for offering more efficient and diversified products on the market. Furthermore, for more efficient and sustainable management, the association of entomopathogens with other control strategies inside integrated pest

R. M. A. Maciel · R. Abati
Programa de Pós-Graduação em Ciências Biológicas (Entomologia), Universidade Federal do Paraná, Curitiba, PR, Brazil

N. de Castilhos Ghisi · E. R. Lozano · M. Potrich (✉)
Universidade Tecnológica Federal do Paraná, Dois Vizinhos, PR, Brazil

A. F. Bueno, A. R. Panizzi (eds.), *Stink Bugs (Hemiptera: Pentatomidae) Research and Management*, Entomology in Focus 9,
https://doi.org/10.1007/978-3-031-69742-5_5

management (IPM) is essential, such as the use of entomopathogenic fungi in conjunction with other control strategies, such as parasitoids and even compatible chemical insecticides.

2 Growth and Perspectives of Biological Control of Insects on Soybean

Augmentative biological control (ABC) is the most used biological control (BC) method in agriculture and, currently, the global biocontrol market corresponds to more than 5.2 billion dollars, with prospects of reaching more than 10 billion dollars by 2027 (Borsari and Vieira 2022). The commercialization of ABC agents begins with the rearing (macro biologicals) or mass production (micro biologicals) of the natural enemies, however, after these processes, adjustments are still needed for application/release, that is, defining the most appropriate methodology to be used. Methodologies for releasing natural enemies can be: inundative, inoculative and seasonal inoculative (Parra 2014).

Inundative is used for both entomopathogens and entomophages. This method is very similar to the way synthetic chemical insecticides are used, and when the economic threshold (ET) is achieved or surpassed, after field monitoring, sufficient amounts of the biocontrol agent are released/sprayed, capable of suppressing the pest population in a short period of time.

Inoculative release is applied to long-term control, as reduced quantities of the control agent are released, aiming for it to proliferate and establish itself in the environment, thus being able to carry out long-lasting and effective control, returning the environment as close to a natural balance as possible (Panizzi and Corrêa-Ferreira 1997; Weber et al. 2021). Seasonal inoculative release has the same principle as the previous methodology, however, in order to control the insect pest in a shorter, previously established period, the releases occur periodically. This method is used mainly in annual crops, which have insect pests at a certain stage of the plant, the release of control agents occurs before or concomitantly with the increase in the population incidence of the pest insect (Corrêa-Ferreira and Moscardi 1996).

In Brazilian agriculture, the main ABC programs are based on inundative releases/applications, mainly because they are used in short-cycle annual monocultures. Among the most successful programs of ABC programs, it is important to highlight the use of the baculovirus *Anticarsia* to control soybean caterpillars, in which the program at its peak covered more than two million hectares (Moscardi 1999), and with entomopathogenic fungi, mainly *M. anisopliae* to control leafhoppers and borers, both pests of sugarcane (Li et al. 2010). Also in sugarcane, the use of parasitoids *Cotesia flavipes* Cameron (Hymenoptera: Braconidae) and *Trichogramma galloi* Zucchi (Hymenoptera: Trichogrammatidae) to control the sugarcane borer *Diatraea saccharalis* F. (Lepidoptera: Crambidae) has also been successful (Bezerra et al. 2021).

Currently, BC agents are used as tools that make up IPM, except for the sugarcane leafhopper and borer control program, where only on-farm production of *M. anisopliae* is used on large-scale fields (Faria et al. 2023). Over time, BC rose due to the need for new tools to control insect pests in agriculture, generating demand for them, thus culminating in a significant increase in the number of producing companies, number of registered products and, consequently, availability of bioinsecticides to farmers. One of the main crops with use of such products is soybean, mainly for stink bug control. Despite the great demand for new tools to control *E. heros*, the only effective, and, therefore, commercialized control agents, are entomopathogenic fungi and egg parasitoid *T. podisi*. In recent years there has been an increase in the number of registered products based on entomopathogenic fungi and *T. podisi* and companies that produce them, thus increasing the stink bug ABC market.

2.1 Augmentative Biological Control for Stink Bugs

The use of augmentative biological control agents to control stink bugs, especially *E. heros*, must be analyzed from the perspective of the insect's morphology and behavior. *Euschistus heros* (F.) is a phytophagous stink bug that feeds mainly on the endosperm of soybean (Lucini et al. 2021; Tessmer et al. 2022). Thus, products that need to be ingested are not eligible for choice as a control tool since ABC control agents do not present translocation characteristics. Thus, the remaining ABC agents are: nematodes, predators, fungi and parasitoids, of which the first two are not still commercialized on large scales, thus making it impossible to use them on soybean fields, restricting the options to entomopathogenic fungi and *T. podisi* (Mascarin et al. 2019; Bueno et al. 2022). Both options are effective tools, however there is a greater number of entomopathogenic in terms of products available on the market and more established in terms of production, storage and application technology. In this scenario, it is one of the main ABC tools for controlling stink bugs.

The entomopathogenic fungi used in ABC belong to the order Hypocreales and are concentrated in two families, Cordicypitaceae and Clavicipitaceae, and allocated to these families, the genera *Beauveria*, *Cordyceps* and *Metarhizium* stand out. These fungi are naturally present in the environment and can occasionally cause widespread infection of insect pests, leading to their death, in addition to being used as bio-inputs in insect management (Potrich et al. 2022).

The first entomopathogenic fungus used in agriculture was *M. anisopliae*, in 1880 in Russia, to control beetles that affected several crops. *Beauveria bassiana* was also studied in that period, however, as a problem that caused the disease of white muscardine in *Bombyx mori* L. (Lepidoptera: Bombycidae), thus devastating silk production in Europe.

Cordyceps fumosorosea is a fungus used more recently, with focus and success in the management of whiteflies and leafhoppers (Inglis et al. 2001; Lacey et al. 2015).The mode of action of this fungi is similar for the three main species previously mentioned, starting with the contact of the conidium with the insect cuticle and, subsequently, release of the appressorium, a structure responsible for the mechanical force of rupture of the insect cuticle, in addition to enzymes, such as proteases, lipases and chitinases. After the rupture of the insect cuticle, in contact with the hemocele, the proliferation of blastospores begins. The growth through these structures confers an important infective characteristic, since the insect's immune system does not recognize it, without any implications on the part of the hematocytes, growth occurs at an accelerated rate (Sosa-Gómez and Alessandro 2019; Islam et al. 2021).

During the colonization of the insect's body, mycotoxins are released, which cause behavioral and physiological changes, ultimately causing the death of the insect. After the death of the insect, it becomes a reservoir of the entomopathogenic fungus and, when under favorable environmental conditions, external sporulation occurs through the insect's orifices and dissemination into the environment, potentially infecting other organisms (Islam et al. 2021).

The general mode of action of fungi is shared, however, it has specificities. The *B. bassiana* fungus is characterized by its white color (Fig. 5.1a) and, during the host colonization process, this fungus is capable of producing and releasing toxins, from secondary metabolites, causing toxicity to the infected insect. The toxins released are beauvericin, bassianin, beauverolide, oxalic acid, among others. Furthermore, these metabolites can promote the growth of the fungus and inhibit the development of other microorganisms present in the insect's body (Zimmermann 2007; Islam et al. 2021).

The entomopathogenic fungus *M. anisopliae* has dark green color as its main visual characteristic (Fig. 5.1b), and is capable of producing destruxins during the colonization process, molecules that are known to be neurotoxic (Alves 1998; Qasim et al. 2020). The toxins are produced after the fungus penetrates and infects the host's body and, due to the depolarization of the membrane due to the opening of calcium channels, leads to irregular contraction of muscle cells and, consequently, paralysis of the insect (Islam et al. 2021). Furthermore, destruxins can cause cytotoxic and immunosuppressive effects in insects (Sosa-Gómez and Alessandro 2019; Potrich et al. 2022).

Cordyceps fumosorosea, which is the most recent fungus to be used in ABC, shows rapid growth and produces the first white colonies, which over time change to shades of pink or purple (Fig. 5.1c) (Zimmermann 2007; Ozdemir et al. 2021). This fungus produces toxins that aid in the host's infection process, such as the IF8 protein and beauvericin (Wang et al. 2021).

The particularities of each species of entomopathogenic fungus associated with the contact mode of action, increase the diversity of target insects that the control agent can manage, in addition to giving them the ability to control insect pests of different orders. *Beauveria bassiana* is reported to be capable of infecting more than 700 hosts (Imoulan et al. 2017) and *M. anisopliae* is reported to be infective to at least 200 insect species (Parra 2014; van Lenteren et al. 2018). *Cordyceps fumosorosea*, as it is the most recent fungus being massively produced and market in

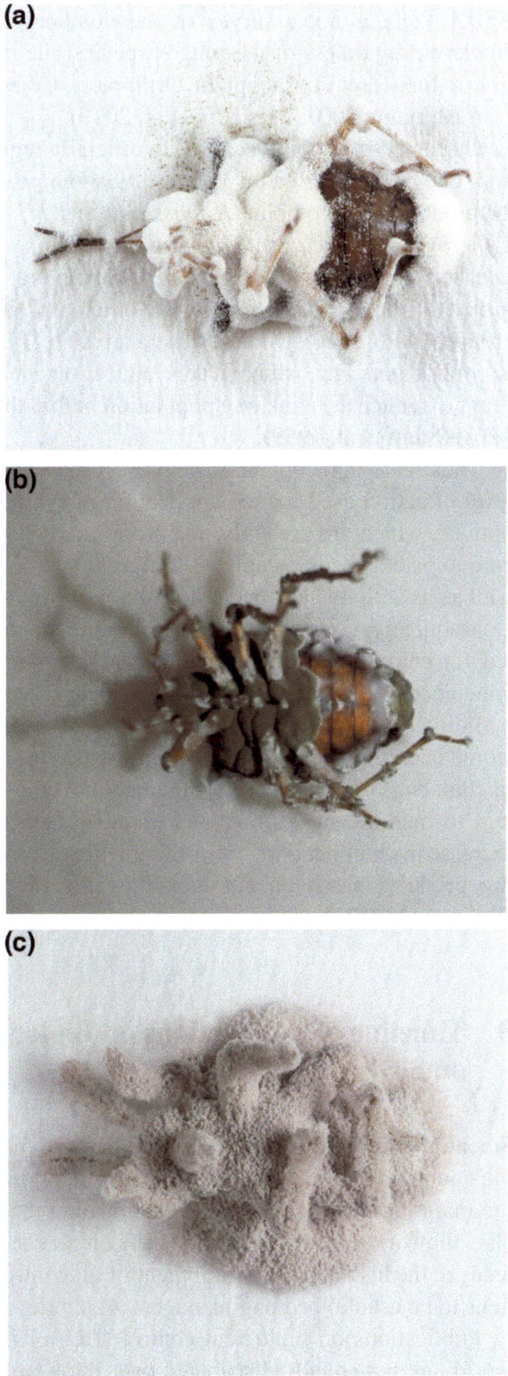

Fig. 5.1 *Euschistus heros* adults infected with entomopathogenic fungi (**a**) Infected by *Beauveria bassiana* (Photo: Adair V. Carneiro); (**b**) Infected by *Metarhizium anisopliae* (Photo: Rodrigo A. Maciel) (**c**) Infected by *Cordyceps fumosorosea*. (Photo: Adair V. Carneiro)

Brazil, there is not a survey of the number of hosts it is capable of infecting. However, regardless of the fungus species, the main orders of insect pests susceptible to fungi are: Thysanoptera, Orthoptera, Coleoptera, Lepidoptera and Hemiptera (Zimmermann 2007; Mascarin et al. 2019).

Currently, in Brazil there are six officially registered products based on *B. bassiana*, one product based on *M. anisopliae*, no product based on *C. fumosorosea* and 20 products that combine *B. bassiana* and *M. anisopliae* for the control of both *E. heros* and *D. melacanthus*. Also, there is one product formulated with *B. bassiana*, for *Scaptocoris castanea* Perty (Hemiptera: Cydnidae) and three products formulated from *M. anisopliae*, for the other stink bugs that are prominent on soybean crops (MAPA 2024). However, there are no ETs defined to use those bioinsecticides to control stink bug, which makes unclear the ideal time to apply entomopathogenic fungi to reduce the stink bug population before they reach the economic injury level (EIL) (Stern et al. 1959).

Thus, entomopathogenic fungi are used subjectively or considering the same level of action used for the application of synthetic chemical insecticides, both of options, which are certainly inappropriate. It is known that the mode of action of entomopathogens is completely different from traditional chemical insecticides, as well as its activity and speed of action in the management of stink bugs in the field. Consequently, it is of fundamental importance that appropriated ETs are also studied for entomopathogenic fungi, in order to improve the efficiency and increase adherence of this control tool in the management of stink bugs in IPM programs.

Furthermore, there are numerous isolates of the three fungal species mentioned stored in entomopathogen collections, which have not yet been tested for the control of stink bugs on soybean or formulated into commercial products. Thus, it is important to measure the pathogenicity and virulence of different isolates in order to increase the number of agents available for stink bug control, in addition to improving products based on entomopathogenic fungi through more virulent isolates (Bueno et al. 2022).

3 Timeline of Publications on Biological Control of Insects on Soybean

Scientometric analyzes allow us to understand the history of research over the years, the countries that required the most effort and investment for research at that time, the main journals responsible for publishing the results, among other factors, they also allow us to identify research gaps on the theme. In this sense, in order to better analyze the history, the development of biocontrol products based on entomopathogens to be used in bed bug management, a brief detailed analysis was carried out.

Publications on biological control (BC) of insects on soybean, all around the world, are not equally distributed over the years in the period from 1973 to 2024, totaling 745 publications. These publications were collected after a search conducted in the *Web of Science Core Collection* (WoS) using the keywords ("*Glycine*

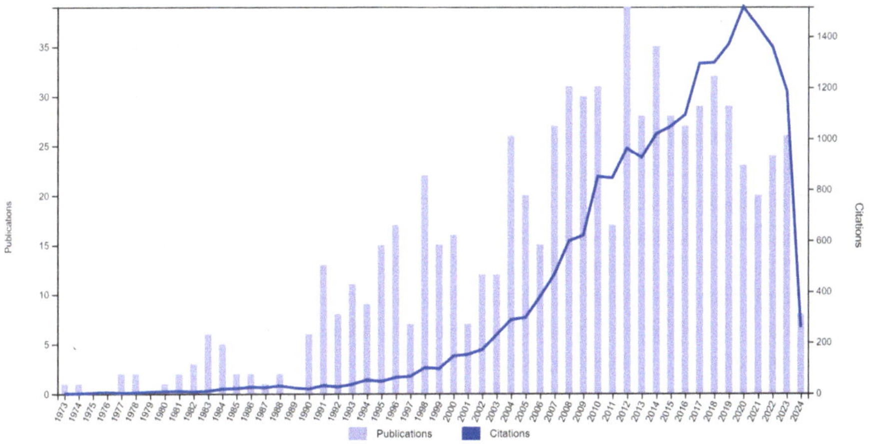

Fig. 5.2 Relationship between publications, citations and average citations/year of publication related to biological control of insects on soybean

max" OR "soybean" OR "soy") AND ("biological control" OR "entomopathogenic fungi" OR "entomopathogenic bacteria" OR "entomopathogenic virus" OR "entomopathogenic nematode" OR "predator*" OR "parasitoid*") and followed by manual refinement according to the selection criteria. The beginning of this period has few publications, showing an increase from 1989 which had accelerated until today (April 2024) (Fig. 5.2). Most publications referring to BC of insects on soybean are scientific articles (93.8%). Only small percentage of the publications are distributed over reviews (3.7%), protocols (2.0%), book chapters (0.5%) and miscellaneous (less than 2%). The H factor of all publications was 69.

Studies with BC on soybean, starting in 1973, are linked to the prohibition of several products based on chlorinated hydrocarbons, that have caused environmental problems around the world. However, the most accurate reporting of the impacts of these products was in the US, described by Rachel Carson in Silent Spring (Carson 1962), and the 1972 IPM standardization (Parra 2014). But, despite the need for new control strategies after the banning of the products and the popular call for the reduction of products that are harmful to the environment, the growth curve of studies on BC in the soybean crop is discreet between 1973 and the beginning of the decade of 90 (Parra 2014).

The main reason is that, after the ban on hydrocarbons, there is an increase in pyrethroids, carbamates and organophosphates, considered safer than the old products used, thus being more accepted in the general society. However, pesticides tend to select resistant populations of pest insects, generating the need to seek control strategies that manage resistance and are efficient in controlling insects (Sosa-Gómez et al. 2020; Stacke et al. 2020). In the analysis of articles listed by scientometrics, until the early 1990s, studies on BC were based on control or on the evaluation of the potential of BC agents, such as bacteria and viruses, restricting studies to laboratory tests, in order to determine BC efficiency.

Although studies related to insect BC on soybean began in 1973, the first publications on stink bug control with entomopathogenic fungi began in the 90s and represent only a small portion of publications, approximately 1.1% (Table 5.1) of the total publications evaluated here. Following this analysis, it appears that from 1990 to 2024 there is a growing increase in studies of augmentative BC with entomopathogens and entomophages. The increase in these studies results from the growth of soybean productive areas, consequently, increase of insect pests and resistant populations of pesticides, resulting the need to provide efficient control agents for field use (van Lenteren and Bueno 2003; Bale et al. 2008; Parra 2014), safer than that already used. Besides that, during 2000's there are an increase in the number of biological control agents' products registered and available in the market, reaffirming the increase of studies in this area.

Differently from the works published until 1990, an increase in the use of biotechnologies related to BC is observed, such as the molecular identification of control agents, or even the use of these agents in plant transgenic processes. In this sense, despite verifying in the last decade that BC studies have focused on ABC and on biotechnology tools that improve control agents, it is still possible to verify studies with natural, conservative, or classic BC. This fact occurs because these studies provide knowledge for the development of control agents used in ABC.

3.1 The Impact of Research on Biological Control of Insects on Soybean

The importance assigned to an article's study is measured mainly by the number of citations it has, in the sense that the number of citations reflects that the work was consulted, and the information contained useful for the scientific community. Among the tools which reflect the importance of the journal, the main one is the impact factor by the journal, which is the sum of citations that the journal received in the two preceding years, divided by the number of journal publications, in the same period. The impact factor is used to qualify the best journals in an area, giving even greater weight of importance to the articles published in these journals. Both publication and journal impact factor deal with journals and articles individually.

The journals with the most citations in the field of BC of insects on soybean are demonsrated in Fig. 5.3. To determine this ranking, the relevance of each journal is measured by frequency (number of publications represented in the figure by node size - entire circle), centrality (strength of the connection between two or more groups of nodes, influence represented in the figure by the purple halo), half-life (number of years a publication has received half of its citations since publication), Burst (citation explosion represented in the figure by the red center) and impact factor (Chen 2014).

Journal of Economic Entomology is the scientific journal with the highest number of citations of published works (498). In addition, this journal is given an

Table 5.1 Publications relating biological control with entomopathogenic fungi and soybean stink bugs

Title	Citation	Journal	Fungi	Stink bug	Country
Lethal concentration and sporulation by contact and direct spray of the entomopathogenic fungus *Beauveria bassiana* on different stages of *Nezara viridula* (Heteroptera: Pentatomidae)	Portilla et al. (2022)	Journal of Fungi	*B. bassiana*	*N. viridula*	USA
Selection and characterisation of *Beauveria bassiana* fungus and their potential to control the brown stink bug	Silva-Santana et al. (2022)	Biocontrol Science and Tecnology	*B. bassiana*	*E. heros*	Brazil
Virulence of *Metarhizium brunneum* (Ascomycota: Hypocreales) strains against stink bugs *Euschistus heros* and *Dichelops furcatus* (Hemiptera: Pentatomidae)	Resquín-Romero et al. (2020)	Journal of Economic Entomology	*M. brunneum*	*E. heros* and *D. furcatus*	Paraguay
Effectiveness of *Beauveria bassiana* against *Piezodorus guildinii* (Hemiptera: Pentatomidae), a key pest of soybeans in the neotropics	Parys and Portilla (2020)	Biocontrol Science and Technology	*B. bassiana*	*P.guildini*	USA
Biological parameters of *Euschistus heros* (F.) (Heteroptera: Pentatomidae) and its susceptibility to entomopathogenic fungi when fed on different diets	Oliveira et al. (2016)	Brazilian Archives of Biology and Technology	*B. bassiana* and *M. anisopliae*	*E. heros*	Brazil
The fungistatic and fungicidal effects of volatiles from metathoracic glands of soybean-attacking stink bugs (Heteroptera: Pentatomidae) on the entomopathogen *Beauveria bassiana*	Lopes et al. (2015)	Journal of Invertebrate Pathology	*B. bassiana*	*E. heros, C. ubica, D. melacanthus* and N. *viridula* gland extracts	Brazil
Laboratory and field studies on the infection of stink bugs, *Nezara viridula, Piezodorus guildinii, and Euschistus heros* (Hemiptera: Pentatomidae) with *Metarhizium anisopliae* and *Beauveria bassiana* in Brazil	Sosa-Gómez and Moscardi (1998)	Journal of Invertebrate Pathology	*B. bassiana* and *M. anisopliae*	*N. viridula, P. guildinii,* and *E. heros*	Brazil

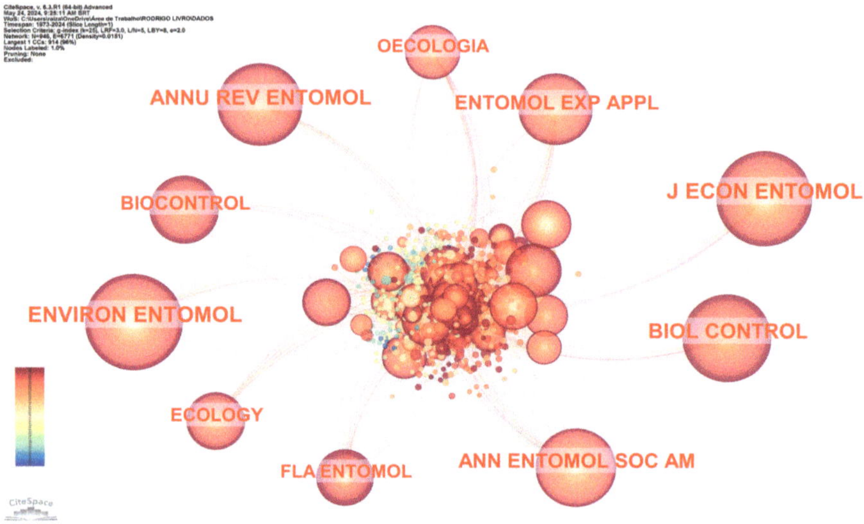

Fig. 5.3 Scientific journal with the most cited publications about biological control on soybean

explosion of citations between the years 1973 and 2003. The journal has 88 published articles about CB on soybean, one of which provides information on entomopathogenic fungi and stink bugs. The article entitled "Virulence of *Metarhizium brunneum* (Ascomycota: Hypocreales) strains against stink bugs, *Euschistus heros* and *Dichelops furcatus* (Hemiptera: Pentatomidae)" published by Resquín-Romero et al. (2020) in this journal, demonstrates the potential of three strains of *M. brunneaum* in controlling eggs, nymphs and adults. This journal has an economic focus and, by publishing and establishing a work by classical BC as the most cited work, it implies how much the BC profitably favors the soybean crop, a view that is often made difficult because it is not measurable by unit currency. However, the environmental value added by this type of control is high and at the same time invaluable.

Frequency is an important factor for illustrating the range of information that a scientific journal passes on to its reader, through the number of publications. However, another factor to be observed is centrality, which reflects the journal's influence on a given subject, thus demonstrating the engagement with other scientific journals in the area. In this sense, the scientific journal with the highest centrality value is *Anais da Sociedade Entomologica do Brasil* (0.12), actually called *Neotropical Entomology*, attributing a relevant influence on the BC theme. However, the purple halo is barely visible because its thickness varies according to the strength of the centrality of intermediations. Furthermore, these results demonstrate the importance of soybean in Brazil and the concern with more sustainable management strategies since this is an originally Brazilian scientific journal.

In addition to frequency and centrality, which are assigned based on scientometric analyzes with the CiteSpace program, the impact factor is another criterion used to rank journals. The impact factor is a tool to facilitate the choice of scientific journals and help the rise of small scientific journals. The scale is determined from the number of citations of articles in a newspaper over a two-year period. In this ranking index, the scientific journal *Annual Review of Entomology*, whose Journal Citation Report (JCR) Impact Factor is 15 (2024) stands out as the highest ranked among the 15 most important in BC.

3.2 Relationship Between Research and the Country's Socioeconomic Status

Publications on biological control are distributed in 52 different countries. However, approximately 83% of the publications are concentrated in five countries, with the United States of America being the country that most published articles on BC (391), followed by Brazil (181), China (40), Canada (39) and Argentina (28) (Fig. 5.4).

Analyzing the publications related to BC with entomopathogenic fungi on stink bugs on soybean, it is noted that Brazil stands out (Fig. 5.5), demonstrating once again the search for more sustainable management strategies in the country that is today the largest soybean producer in the world.

The United States of America and Brazil, in addition to being the countries that most publish about BC on soybean, are the countries with the highest centrality values, 0.69 and 0.28 respectively, visualized through the purple halos (Fig. 5.6). The value of centrality reflects how much the country works on the subject in partnership with other countries and highlights the influence on the academic community within the scope of BC on soybean.

The first two placed in terms of number of publications and centrality are also the only ones to present explosions of citations (represented by the red rings in the

Fig. 5.4 Geographic distribution of general publications on biological control of insects

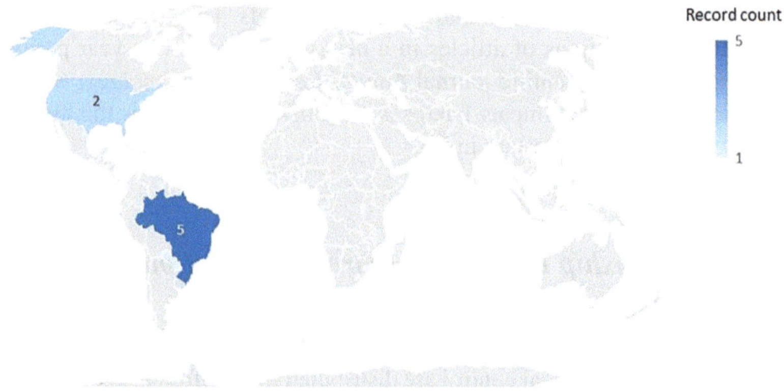

Fig. 5.5 Geographic distribution of soybean publications on biological control with entomopathogenic fungi against stink bugs

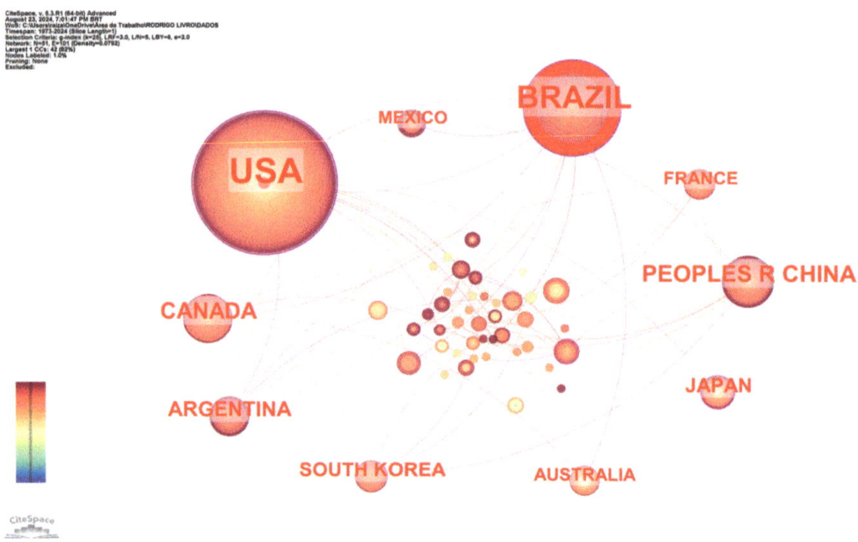

Fig. 5.6 Frequency, centrality and burst of bibliographic citations by countries of publications relating biological control on soybean

respective country nodes) (Fig. 5.6). The red rings demonstrate that there was a significant increase in publications in each time interval. The explosion of citations referring to the USA (13.3) occurred between 1980 and 1994. In Brazil (9.1), the explosion of citations has occurred since 2016 (Fig. 5.6) and is still persistent today, for this reason the red rings are in the outermost part of the node, it is likely that the burst value of citations will increase, consequently increasing centrality. Possibly, the increase in citations in Brazil was determined by *Helicoverpa armigera* Hübner (Lepidoptera: Noctuidae) outbreak in 2013. Research on soybean were focus on to verify products and agents to manage the insect pest as soon as possibly aiming to reach the best control tool, since transgenic technologies and synthetic chemical insecticides are not efficient to control this insect pest.

USA leads frequency and centrality metrics on published works related to BC of insects on soybean. This country's leadership can be attributed to the first BC studies that were carried out in the state of Florida (van Lenteren et al. 2018), in addition, until 2020, USA was positioned as the country with the largest production of soybean in recent decades (USDA 2024). So, there were more efforts and incentive in study BC of insects in the last years.

Other important factors that may be linked to the importance of these countries in the metrics evaluated in this study may be related to North America leading the sales of biopesticides worldwide, moreover, the use of BC agents is on the rise in Latin America (van Lenteren et al. 2018). As in Brazil, the use of BC agents is negatively correlated with the use of synthetic chemical products. Since 2000s until the present time, the use of synthetic chemical insecticides has an average dose of 200 g/ha, a dose approximately 10 times lower compared to the 60s. In contrast, from 2000 to the present, there has been an increase in the number of registered biological products from 2 to 484.

The consolidation of these factors is the result of investment in research and the quality of studies carried out by these countries. In the USA, another point that collaborates with the increase in the use of BC agents on soybean is the increase in the production of organic soybean, from 2002 to 2012 the planted area increased from 0.5 million hectares to 1.3 million hectares. This increase directly reflects the use of BC, as this cultivation system does not allow the use of synthetic insecticides (Hartman et al. 2011).

From product selection to marketing, as well as the efficiency, selectivity, and biosafety testing phases, are conducted by educational or research institutions. Thus, understanding the most relevant institutions and relating them to the BC's scenario in the native country of these institutions indicates the segment that the market should cover in the coming years, in addition to providing confidence to the products present in the market. As an example, in Brazil, several commercialized biological products have the seal of the *Empresa Brasileira de Pesquisa Agropecuária* (EMBRAPA) as a way of guaranteeing product innovation and quality.

4 Final Considerations

With the verification of the status of research on BC on soybean and knowing the economic importance of soybean culture for the economy of countries and the need for more sustainable management strategies for its cultivation, the trend is that research and publications continue to increase in the next years. Besides there are still gaps to be filled, mainly regarding the action of entomopathogenic fungi on stink bugs, to research on the compatibility of biological control agents with each other and with other products, as well as their efficiency with other products, whether biological, alternative, or synthetic. In addition, greater investment in research on strategies to increase the shelf life of BC agents is necessary, without changes in the efficiency of the products, which can thus support studies for new strategies for the integrated management of pests on soybean.

References

Alves SB (1998) Controle Microbiano de Insetos. FEALQ, Piracicaba

Bale JS, van Lenteren JC, Bigler F (2008) Biological control and sustainable food production. Philoshophical Trans Royak Soc 363:761–776

Bezerra JIM, Molter A, Rafikov M, Frighetto DF (2021) Biological control of the chaotic sugarcane borer-parasitoid agroecosystem. Ecol Modell 450:109564

Borsari ACP, Vieira LC (2022) Mercado e perspectivas dos bioinsumos no Brasil. In: Meyer MC, Bueno AF, Mazaro SM, Silva JC d (eds) Bioinsumos na Cultura da Soja, 1ª ed. Embrapa, Brasília, pp 39–52

Bueno AF, JRP P, Colombo FC, Colmenarez YC, BVF N, Pereira FF (2022) Manejo de pragas com parasitoides. In: Meyer MC, Bueno AF, Mazaro SM, Silva JC d (eds) Bioinsumos na Cultura da Soja, 1ª ed. Embrapa, Brasília, pp 417–434

Carson RL (1962) Silent Spring. Boston: Cambridge, Mass., Houghton Mifflin, 1962

Chen C (2014) The CiteSpace manual. Coll Comput Informatics 103:1003–1022

Corrêa-Ferreira BS, Moscardi F (1996) Biological control of soybean stink bugs by inoculative releases of *Trissolcus basalis*. Entomol Exp Appl 79:1–7

Dalla-Nora D, Piovesan BC, Bellé C, Stackeb RS, Balardin RR, Guedes JVC, Michaud JP, Jacques RJS (2020) Isolation and evaluation of entomopathogenic fungi against the neotropical brown stink bug *Euschistus heros* (F.) (Hemiptera: Pentatomidae) under laboratory conditions. Environ Sci Biol 31:22–34

Faria M, Mascarin GM, Butt T, Lopes RB (2023) On-farm production of microbial entomopathogens for use in agriculture: Brazil as a case study. Neotrop Entomol 52:122–133

Hartman GL, West ED, Herman T (2011) Crops that feed the World 2. Soybean-worldwide production, use, and constraints caused by pathogens and pests. Food Secur 3:5–17

Imoulan A, Hussain M, Kirk PM, Meziane AE, Yao Y-J (2017) Entomopathogenic fungus Beauveria: Host specificity, ecology and significance of morpho-molecular characterization in accurate taxonomic classification. J Asia Pac Entomol 20:1204–1212

Inglis GD, Goettel MS, Butt TM, Strasser H (2001) Use of hyphomycetous fungi for managing insect pests. In: Butt TM, Jackson C, Magan N (eds) Fungi as biocontrol agents: progress, problems and potential. CABI, Wallingford, pp 23–69

Islam W, Adnan M, Shabbir A et al (2021) Insect-fungal-interactions: a detailed review on entomopathogenic fungi pathogenicity to combat insect pests. Microb Pathog 159:105122

Lacey LA, Grzywacz D, Shapiro-Ilan DI, Frutos R, Brownbridge M, Goettel MS (2015) Insect pathogens as biological control agents: back to the future. J Invertebr Pathol 132:1–41

Li Z, Alves SB, Roberts DW et al (2010) Biological control of insects in Brazil and China: history, current programs and reasons for their successes using entomopathogenic fungi. Biocontrol Sci Technol 20:117–136

Lopes RB, Laumann RA, Blassioli-Moraes MC, Borges M, Faria M (2015) The fungistatic and fungicidal effects of volatiles from metathoracic glands of soybean-attacking stink bugs (Heteroptera: Pentatomidae) on the entomopathogen *Beauveria bassiana*. J Invertebr Pathol 132:77–85

Lucini T, Panizzi R, Bueno ADF (2021) Evaluating resistance of the soybean block technology cultivars to the Neotropical brown stink bug, *Euschistus heros* (F.). J Insect Physiol 131:104228

Mascarin GM, Lopes RB, Delalibera Í Jr, Fernandes EKK, Luz C, Faria M (2019) Current status and perspectives of fungal entomopathogens used for microbial control of arthropod pests in Brazil. J Invertebr Pathol 165:46–53
Moscardi F (1999) Assessment of the application of baculoviruses for control of Lepidoptera. Annu Rev Entomol 44:257–289
Oliveira DGP, Dudczak AC, Alves LFA, Sosa-Gomez DR (2016) Biological parameters of *Euschistus heros* (F.) (Heteroptera: Pentatomidae) and its susceptibility to entomopathogenic fungi when fed on different diets. Brazilian Arch Biol Technol 59:e16150141
Ozdemir IO, Tuncer C, Goskel O (2021) Molecular characterisation and efficacy of entomopathogenic fungi against the green shield bug *Palomena prasina* (L.) (Hemiptera: Pentatomidae) under laboratory conditions. Biocontrol Sci Technol ISSN 31:1298–1313
Panizzi AR, Corrêa-Ferreira BS (1997) Dynamics in the insect fauna adaptation to soybean in the tropics. Trends Parasitol 1:71–88
Parra JRP (2014) Biological control in Brazil: An overview. Sci Agric 71:420–429
Parys KA, Portilla M (2020) Effectiveness of *Beauveria bassiana* against *Piezodorus guildinii* (Hemiptera: Pentatomidae), a key pest of soybeans in the neotropics. Biocontrol Sci Technol 30:451–461
Portilla M, Zhang M, Glover JP, Reddy GVP, Johnson C (2022) Lethal concentration and sporulation by contact and direct spray of the entomopathogenic fungus B*eauveria bassiana* on different stages of *Nezara viridula* (Heteroptera: Pentatomidae). J Fungi 8:1164
Potrich M, Neves PMOJ, Lozano ER, Maciel RMA, Colombo FC, Abati R (2022) Manejo de pragas com fungos entomopatogênicos. In: Meyer MC, Bueno AF, Mazaro SM, Silva JC (eds) Bioinsumos na Cultura da Soja, 1ª ed. Embrapa, Brasília, pp 401–416
Qasim M, Ul S, Islam W et al (2020) Toxicon Characterization of mycotoxins from entomopathogenic fungi (*Cordyceps fumosorosea*) and their toxic effects to the development of asian citrus psyllid reared on healthy and diseased citrus plants. Toxicon 188:39–47
Resquín-Romero G, Cabral-Antúnez C, Sarubbi-Orue H, Garrido-Jurado I, Valverde-Garcia P, Schade M, Butt TM (2020) Virulence of *Metarhizium brunneum* (Ascomycota: Hypocreales) strains against stink bugs *Euschistus heros* and *Dichelops furcatus* (Hemiptera: Pentatomidae). J Econ Entomol 113:2540–2545
Silva-Santana MF, Alves LFA, Ferreira TT, Bonini AK (2021) Selection and characterisation of *Beauveria bassian*a fungus and their potential to control the brown stink bug. Biocontrol Sci Technol 32:90–102
Silva-santana MF, Francisco L, Alves A et al (2022) Selection and characterisation of *Beauveria bassiana* fungus and their potential to control the brown stink bug. Biocontrol Sci Technol ISSN 32:90–102
Sosa-Gómez DR, Alessandro CPD (2019) Hongos entomopatógenos del phylum ascomycota. In: Lastra CCL, Lecuona RE (eds) Micopatología de artrópodos, 1ª ed. INTA Ediciones, Buenos Aires, p 233
Sosa-Gómez DR, Moscardi F (1998) Laboratory and field studies on the infection of stink bugs, *Nezara viridula, Piezodorus guildinii,* and *Euschistus heros* (Hemiptera: Pentatomidae) with *Metarhizium anisopliae* and *Beauveria bassiana* in Brazil. J Invertebr Pathol 71:115–120
Sosa-Gómez DR, Corrêa-ferreira BS, Kraemer B, Pasini A, Husch PE, Vieira CED, Martinez CBR, Lopez ION (2020) Prevalence, damage, management and insecticide resistance of stink bug populations (Hemiptera: Pentatomidae) in commodity crops. Agric Forest Entomol 22:99–118
Stacke RF, Godoy DN, Halberstadt SA, Bronzatto ES, Giacomelli T, Hettwer BL, Muraro DS, Guedes JVC, Bernardi O (2020) Inheritance of lambda-cyhalothrin resistance, fitness costs and cross-resistance to other pyrethroids in soybean looper, *Chrysodeixis includens* (Lepidoptera: Noctuidae). Crop Prot 131:105096
Stern VM, Smith RF, Bosch R v d, Hagen KS (1959) The integrated control concept. J Agric Sci 29:81–101
Tessmer MA, Kuhn TM d A, Appezzato-da-Glória B, Lops JRS, Erler G, Bonani JP (2022) Histology of damage caused by *Euschistus heros* (F.) nymphs in soybean pods and seeds. Neotrop Entomol 51:112–121

USDA (2024) Foreign Agricultural Service – U.S. Department of Agriculture. World agricultural production, Circular Series. https://fas.usda.gov/data/world-agricultural-production-03082024

van Lenteren J, Bueno V (2003) Augmentative biological control of arthropods in Latin America Augmentative biological control of arthropods in Latin. BioControl 48:123–139

van Lenteren JC, Bolckmans K, Köhl J, Ravensberg WJ, Urbaneja A (2018) Biological control using invertebrates and microorganisms: plenty of new opportunities. BioControl 63:39–59

Wang Z-L, Pan H-B, Li M-Y, Wu W, Yu XP (2021) Comprehensive insights into host-pathogen interaction between brown planthopper anda fungal entomopathogen by dual RNA sequencing. Pest Manag Sci 77:4903–4914

Weber ID, Lopes WR, Paula LCB, Albernaz-Godinho KC, Czepak C (2021) Tachinids associated with lepidopteran pests in agricultural crops in south-central region of Goiás, Brazil. BioControl 66:625–637

Zimmermann G (2007) Review on safety of the entomopathogenic fungi *Beauveria bassiana* and *Beauveria brongniartii*. Biocontrol Sci Technol 17:553–596

Chapter 6
Conservation Biological Control as an Alternative to Reduce Stink Bugs Outbreaks

María Fernanda Cingolani

1 Introduction

Natural environments and healthy ecosystems provide several benefits to humans, known as ecosystem services (Daily 1997). In this sense, the biological interactions among species that spontaneously occurs in nature contribute to one of these ecosystem services, the regulation of pests. However, current intensive farming practices that include large monocultures and an excessive use of insecticides generate and maintain high levels of disturbance in the agroecosystem, in the detriment of those natural biological interactions. That situation limit resources for natural enemies, tending to lower their populations and favoring rapid recolonization and population growth by pests (Chaplin-Kramer et al. 2011).

Conservation biological control of pests (CBC) is a strategy that relies on naturally occurring enemies that are presumable well adapted to the target pest. Its implementation requires the modification of the environment or existing practices, made to protect and enhance those beneficial organisms, in order to reduce the effect of pests (Eilenberg et al. 2001). This may be done either by eliminating adverse factors or providing improved conditions for colonization and survival (DeBach 1964). Unlike the other biological control strategies (i.e., classical, inoculation and inundation), in the case of CBC natural enemies are not released but are protected and provided with resources to enhance their fitness and ultimate impact on pests. In comparison, conservation is a better option than classical biological control, which implies the importation of non-indigenous biological control agents, when importation of natural enemies is precluded or when the candidate natural

M. F. Cingolani (✉)
Centro de Estudios Parasitológicos y de Vectores/CEPAVE/CONICET – FCNyM, UNLP, La Plata, Argentina
e-mail: fernandacingolani@cepave.edu.ar

A. F. Bueno, A. R. Panizzi (eds.), *Stink Bugs (Hemiptera: Pentatomidae) Research and Management*, Entomology in Focus 9,
https://doi.org/10.1007/978-3-031-69742-5_6

enemy lacks the degree of host specificity required for environmental safety. Conservation biological control can be also better than inoculation and inundation biological control, when the release of large numbers of insectary produced agents is not economical.

Entomophagous natural enemies of pests (i.e., parasitoids and predators) require hosts or preys, sources of nectar and/or free water, and refugia for completing their life cycles. Most of these components increase, on average, in habitats with greater vegetational diversity and temporal stability (Lichtenberg et al. 2017). In this sense, the approach of CBC encompasses a variety of management practices that limit the use of pesticide and actively provide food and shelter to specified natural enemies that help control specified pests (Barbosa 1999). In general, CBC strategies do not exercise pest control directly, as is the purpose of cultural control and good farming practices (Eilenberg et al. 2001); instead, the aim of CBC tactics is to promote the abundance and diversity of natural enemies which are already present in the agroecosystem (Vargas et al. 2022). In this sense, CBC becomes a preventive measure within an integrated pest management approach (Begg et al. 2017). Moreover, the addition of resources back to simplified agriculture systems contribute to numerous other ecosystem services such as erosion control, reduced runoff, aesthetic benefits, increased revenue, nutrient management, pollination services and soil health (Philips et al. 2014).

Particularly with respect to stink bug (Hemiptera: Pentatomidae) pests, Panizzi and Lucini (2023) pointed out that life history studies are fundamental in order to enhance CBC. Traits of life history include number of generations per year, nymph and adult bioecology on host and associated plants, and identification of natural enemies (parasitoids, predators, and entomopathogens) impacting the stink bugs pests year-round (see also Chap. 3 for further discussion). Among the natural enemies of these pests, egg parasitoids (Hymenoptera: Scelionidae) are the main ones, but also the parasitoids of adult hosts and predators are of importance. Even though there is a high diversity of these native parasitoids (Cingolani 2012), the difficulty in rearing these insects and their susceptibility to insecticides (Corrêa-Ferreira et al. 2010) make conservation biological control using natural populations of parasitoids the better option (Blassioli-Moraes et al. 2022). As an applied example, Bueno et al. (2011) compared the pest infestation rates of stink bugs in soybean fields considering three different pest management strategies: prophylactic use of pyrethroids, use of insecticides more selective for natural enemies (insect-growth regulators, acephate), and biological control. The authors found that the prophylactic use of insecticides did not result in higher productivity than in fields under biological control.

Providing limited resources to benefit natural enemy communities, such as alternative food and hosts, microclimates, and nesting and overwintering habitat, natural enemies' populations may be enhanced, although this is not always consistent with pest suppression or with the increase of crop yields (Vargas et al. 2022). Shields et al. (2019) give a good historical summary of the development of CBC and illustrates its methods, current situation and challenges. For instance, in the mid-1900s it was discovered that some parasitoids need obligatory alternative hosts to persist through the year, while other parasitoids require many species of alternative hosts when the main pest is scarce. Also, the importance of nectar and the right microclimate conditions for parasitoids was demonstrated. In particular, the availability of additional resources such as extra-floral substances were pointed out as necessary for parasitoids to successfully reduce pest populations. However, reduced use of pesticides was one of the first CBC tools that became mainstream. Only in subsequent decades, research involving the influence of non-crop habitat within-field on natural enemy communities became more common.

2 Conservation Biological Control Strategies

Conservation biological control involves manipulation of the environment to enhance the survival, fecundity, longevity, and behavior of natural enemies to increase their effectiveness. This may be achieved by mitigating harmful conditions or enhancing favorable ones. According to Landis et al. (2000) conservation practices can be further categorized as those that focus on reducing mortality, providing supplementary resources, controlling secondary enemies, or manipulating host plant attributes to the benefit of natural enemies. Moreover, this conservation practices may occur at different spatial scales (i.e., within-crop, within-farm, or landscape levels), and also at different temporal scales (i.e., short versus long term). Understanding the key elements of diversity needed by natural enemies as resources is crucial to selectively enhance their efficiency. The spatial arrangement and temporal availability of such resources may boost early season activity of natural enemies within the crop.

The general objectives pursued by CBC strategies focus on providing: alternative food sources that improve the survival and fecundity of natural enemies; optimal overwintering habitats and oviposition sites that maximizes survival and reproduction; refugia from pesticides applications outside the treated area; and alternative preys or hosts that encourage generalist natural enemies' establishment within the crop before the arrival and seasonal increase of pests. In the following sections, some examples of CBC strategies applied for stink bugs control are detailed.

2.1 *Strategies Aimed at Selective Use of Pesticides*

Multiple field experiments have demonstrated that pesticide applications may not significantly reduce pest density. Moreover, they do affect natural enemies in most cases (Janssen and van Rijn 2021). The development of protocols for a reduction on the frequency of applications with broad-spectrum pesticides, and the detection of selective products is of paramount importance.

Cropped plants can tolerate certain injury level caused by phytophagous insects, with no economically significant yield reduction. This economic threshold is the insect's population level or extent of crop damage at which the value of the crop destroyed exceeds the cost of controlling the pest. Above this pest's density a control treatment will provide an economic return. Cautious application of selective insecticide, based on economic thresholds, will contribute to maximize its effects on pests, with the minimum impact on beneficial organisms. In particular, it was well demonstrated that for the control of stink bugs, natural enemies represent an important mortality factor that can keep the damage caused by these pests below the economic threshold levels without additional control actions (Bueno et al. 2023). In this way, for controlling these phytophagous insects, it is recommended to use pesticides only when pest numbers are equal to or surpass two stink bugs per meter (Bueno et al. 2015). Although this economic threshold was developed for the rational use of chemical insecticides without considering the use of biocontrol agents, the simple adoption of the existing economic threshold contributes to a significant reduction in insecticide use, providing ecological and financial benefits. For example, the adoption of integrated pest management and its economic threshold in soybeans in Paraná State (Southern Brazil) during six consecutive growing seasons resulted in a reduction of around 40% of insecticides applications to control stink bugs and a 75 kg higher average yield than fields under conventional pest management strategies (Bueno et al. 2023).

An important concept to consider is the selectivity of pesticides, defined as the ability of a product to control the targeted pest while also causing the least possible impact on beneficial organisms such as predators, parasitoids, bees, and entomopathogen (Bueno et al. 2023). Determining insecticides' selectivity is crucial for achieving CBC, for the case when biological control is not sufficient and insecticides applications are required. This selectivity is a combination of differential insecticide's toxicity to pest than to their natural enemies (physiological selectivity), and differential probability of pesticide contact with the pest and their natural enemies through differences in behavior between them (ecological selectivity) (Ripper et al. 1951). Physiological selectivity of a given pesticide towards a specific natural enemy is determined by the insect sensitivity to the pesticide, and capacity for detoxification, sequestration and excretion. Ecological selectivity is related to the delivery approach through careful application to differentially access the target pest. For example, systemic pesticides may be available only to leaf-feeding herbivores and have no effect on many natural enemies that frequent the leaf surface. Therefore, the use of nonselective insecticides can be improved by functionally achieved ecological selectivity (Torres and Bueno 2018), that involve spatial and temporal separation of pesticide and natural enemies.

When evaluating pesticides selectivity, besides mortality caused by direct contact with toxic substances, secondary effects affecting life history characteristics, population demography, and behavior of natural enemies must also be considered. Another aspect that must be taken into account is the impact of different exposure methods and routes of toxicant entry (i.e., direct topical exposure, direct exposure to residues, indirect exposure by ingestion of intoxicated hosts, preys, nectar, sap or pollen) (see Ruberson et al. 1998 for detailed discussion on this regard). Moreover, toxicant also can move across trophic levels via treated preys or hosts, and can accumulate in predators or parasitoids.

Broadly speaking, pesticides can exert two different types of effects on natural enemies: lethal effects expressed as acute or chronic mortality arising from contact with a pesticide; and sublethal effects expressed as chronic change in the insect's life history attributes, such as its fecundity, developmental time, egg viability, consumption rates, behavior, etc. The outcomes of these two types of effects on natural enemies' populations may be quite similar or vastly different. Efficacy for biological control of the individual natural enemies that survive sublethal doses of pesticides may be damaged due to the effect of the pesticide on a variety of life history attributes others than survival. An interesting detailed list of these attributes is given in Ruberson et al. (1998) (Table 6.1).

Another important aspect to consider when evaluating pesticides is that the susceptibility among natural enemies' life stages and sexes to these compounds often varies considerably. Data on stage-specific susceptibility of natural enemies that have stages protected from the external environment, such as endoparasitoids, provide unavoidable information for timing pesticide applications in order to minimize the impact on natural enemies. In the same way, gender-specific data are important for understanding changes in population structure and population dynamics following pesticide applications (Ruberson et al. 1998).

As a corollary, consciously considering when and where to apply selective pesticides can reduce their impact on natural enemies of pests. In particular, regarding the impact of pesticides on two ubiquitous natural enemies of stink bugs, the egg parasitoids *Telenomus podisi* (Ashmead) and *Trissolcus basalis* (Wollaston) (Hymenoptera: Scelionidae), it was demonstrated that neurotoxic products do not show physiological selectivity for these species (Stecca et al. 2018). Also, some fungicides showed to be not selective. However, growth-regulating insecticides were classified as selective. Abbate et al. (2022) provide an extensive list of active ingredients that when applied at doses recommended for stink bug control do not affect *T. podisi* larvae. In another interesting work, Werdin-González et al. (2013) evaluated toxicological and behavioral effects of four essential oils that were previously detected as a potential phytoinsecticide against *Nezara viridula* L. (Hemiptera: Pentatomidae), on its main natural enemy, the parasitoid *T. basalis*. The authors found that essential oils of two Lamiaceae species, *Origanum vulgare* L. and *Thymus vulgaris* L., were highly selective when used as fumigant and did not change parasitoid behavior. Moreover, residues were harmless after one week, with no sublethal effects against the parasitoid. Turchen et al. (2016) assessed the toxicity of essential oil from *Piper aduncum* L. (Piperales: Piperaceae) against *Euschistus heros* (F.) (Hemiptera: Pentatomidae) and on the performance of their egg parasitoids. Although the evaluated compounds were toxic to the stink bug, they exhibited lower toxicity for the parasitoids, as there was no effect on parasitism and emergence of wasps. Similar results were observed by Abudulai and Shepard (2003) and Mitchell et al. (2004), when testing azadirachtin. However, contrasting results were observed in the bioassay with azadirachtin and *Annona* extract, as there was a lower parasitism rate, when the egg host was treated, indicating deterrence of oviposition

Table 6.1 Crucial aspects to consider when designing conservation biological control strategies

Target	Approaches	Variables/proxies to be evaluated
Selection of species of natural enemies to preserve		
	Relative importance within system	Abundance within system
		Hosts/preys' range
		Parasitism/predation rate
		Synchronization with the pest in the crop
		Interactions with other parasitoids/predators
		Ability to tolerate pesticides
		Ability to use alternative food items
Reduction of pesticide-induced harm to natural enemies		
Assessing susceptibility of natural enemies to pesticides		
	Evaluation of pesticides considering particular life stages/sexes of the natural enemies	Active, exposed/concealed, protected stages
		Gender-specific variability in susceptibility
	Evaluation of pesticides considering particular life history parameters of the natural enemies	Survival
		Longevity
		Developmental time
		Fecundity/fertility
		Searching behavior
		Dispersal ability
		Parasitization/consumption rates
		Population growth/reduction
	Evaluation of pesticides considering particular routes of pesticide entry	Direct topical contact
		Contact with residues on substrate
		Ingestion of toxified host/prey tissues
		Ingestion of toxified plant products (e.g., nectar, pollen, sap)
	Pesticide formulations and rates	Range of recommended rates
		Possible dilution in environment
Habitat manipulations that favor natural enemies at the expense of pests		
Assessing influence of plant patch structure and diversity		

(continued)

Table 6.1 (continued)

Target	Approaches	Variables/proxies to be evaluated
	Selected plant species	Potential weed status of selected plants
		Potential alternate host for crop pathogens
		Potential toxicity to livestock
		Perennially/self-sowing annual
		Seed commercial availability and cost
	Size, number and shape of plant patches; distance to cropped plants	Response of natural enemies to plant chemical signals
		Response of natural enemies to physical attributes of plants
		Plant toxicity to natural enemies
		Plants as sources of food for natural enemies
		Plants as sites for natural enemies' mating/oviposition
		Plants as sources of alternative hosts/preys
		Plants as refuge/hibernation sites for natural enemies
		Effect of plant structure on microclimatic conditions
		Plants as habitats for other phytophagous species
		Promotion of intraguild predation
		Promotion of pathogens for the natural enemies

Modified from Gurr et al. (1998) and Ruberson et al. (1998)

(Smaniotto et al. 2013; Turchen et al. 2016) and negative sublethal effects on parasitism and population.

In it turns, Danieli (2010) evaluated the effects of some insecticides commonly used in soybean culture, on several insects that predate on stink bugs eggs and nymphs, such as species of the genus *Geocoris*, *Orius*, *Podisus*, *Nabis* and some reduvids (Hemiptera), coccinellids (Coleoptera) and hymenopterans. They all were negatively affected and their population's recovery was slowed down. Chlorantraniliprole was also verified as harmful to several predators of stink bugs (coleopterans, ants, dipterans and spiders) and acephate was verified as harmful for carabideans, ants and dipterans (Zucoloto 2017). Moreover, in field evaluation, both pesticides were associated to a decrease in parasitism of *E. heros* eggs (Zucoloto 2017).

Other research contributing to the evaluation of insecticides on natural enemies of stink bugs found alterations in behavioral parameters of *T. basalis* exposed to deltamethrin. These sublethal effects influenced foraging ability in insecticide-treated parasitoids due to a change in their response to host cues, decreasing *T. basalis* efficacy in controlling *N. viridula* (Salerno et al. 2002).

2.2 Strategies Aimed at Habitat Management

While natural habitat has been shown to increase pest control in many systems, Tscharntke et al. (2016) identified several hypotheses for when and why natural habitat fail to support biological pest control. For example, pest populations may have no effective natural enemies in the region, or natural habitat may be a greater source of pests than natural enemies, or crops may provide more resources for natural enemies than does natural habitat. Chaplin-Kramer et al. (2011) conducted a meta-analysis of 46 landscape-level studies, and found that natural enemies have a strong positive response to landscape complexity, with specialist enemies responding more strongly to landscape complexity at smaller scales than generalist ones. However, the positive response of natural enemies does not necessarily translate into pest control, since pest abundances showed no significant response to landscape complexity.

The effects of organic farming and plant diversification on abundance, local diversity (communities within fields), and regional diversity (communities across fields) of pollinators, predators, herbivores, and detritivores was analyzed by Lichtenberg et al. (2017) using a global metadataset. The analyzes pointed out that in-field plant diversification exerted the strongest effects on pollinators and predators, suggesting these management schemes can provide temporal and spatial stability of ecosystem service provisioning without augmenting herbivore populations.

Adding diversity to a simplified system provides some limiting resource to beneficial arthropods. This involves, for example, planting different crop and noncrop plants, as most beneficial insects feed on nectar and pollen at some point in their life. In the frame of CBC, after the acquisition of resources, natural enemies should disperse into adjacent cropping systems (Philips et al. 2014). Besides, suitable flower species must enhance the survival and fecundity of natural enemies but in addition they also need to be highly attractive and thus frequently visited (Foti et al. 2017). For example, the effects of four flowering plants species on the fecundity and olfactory attractiveness of the egg parasitoid of stink bugs, *T. basalis* was tested. It was found that access to buckwheat and basil flowers increased the parasitoid offspring, but only the former was attractive to this parasitoid (Foti et al. 2017).

Several studies have demonstrated also that soybean fields located near forest fragments showed a higher richness of predatory mites and arachnids than fields located at greater distances. This higher richness of predators resulted also in lower herbivory and greater number of full soybean pods (higher yield) (Bortolotto et al. 2022). Similar results were found by González et al. (2017), who sampled arthropods on soybean plants on nine landscapes with different amounts of forest cover. They reported that landscapes with high forest cover presented higher richness and abundance of natural enemies and higher levels of biological control on *Diceraeus furcatus* (Fabricius) (Hemiptera: Pentatomidae) eggs. In general, natural enemies showed more species and individuals at short distances from the forest than at larger distances and, conversely, stink bug abundance was lower at the closest distance to the forest. Based on these results, the authors suggest that maintaining remnants of forest in agricultural landscapes can be effective for conservation of arthropod biodiversity while contributing to biological control of stink bugs in soybean fields.

Considering natural enemies' behavioral mechanisms that may be influenced by habitat manipulation, it was demonstrated that some semiochemicals have the potential to attract stink bug parasitoids in the laboratory and field, e.g., plant volatiles (Blassioli-Moraes et al. 2005, 2008; Michereff et al. 2015, 2016), host sex pheromones (Borges et al. 1998) and host defensive compounds (Laumann et al. 2009). However, the application of semiochemicals under field conditions increased the recruitment of egg parasitoids but it did not increase parasitism rate (Vieira et al. 2013, 2014).

The use of flowering plants in combination with semiochemicals, for the design of attract-and-reward strategies (a biological control approach that combines semiochemicals to improve immigration of beneficial organisms into crops, with nectar plants to maintain their populations) may solve this problem. Indeed, the effect of several flowering plants on the increase of offspring of egg parasitoids of stink bugs is well documented (Foti et al. 2017; Rahat et al. 2005; Wäckers 2004).

Ayelo et al. (2021), in a review article provide extensive examples of field application of semiochemicals in CBC to increase the number of natural enemies attracted to a crop. The authors remark that although it is well demonstrated that semiochemicals are useful to increase the number of attracted natural enemies, the effect of this increase of natural enemies is barely reflected in the literature into a reduction of pest populations and crop damage below economic threshold levels and an increase in crop yield. In line with this assessment, Lee et al. (2022) denote that application of methyl salicylate, an herbivore-induced plant volatile commercially sold for attracting natural enemies, has increased the abundance of natural enemies in numerous food crops but its impact on the reduction of pests is not as frequently demonstrated. Although several predatory species (thrips, green lacewing adults, coccinellids) were attracted to experimental plots where methyl salicylate was applied, predation on sentinel *Halyomorpha halys* Stål (Hemiptera: Pentatomidae) eggs was not increased. In another work, Morrison et al. (2018) found that the presence of the *H. halys* aggregation pheromone did not influence the frequency of predation, parasitism, or number of remaining undeveloped eggs from sentinel egg masses. Moreover, supplemental semiochemicals (methyl salicylate) did not affect the predation and parasitism of egg masses.

When manipulating habitats, shelters provide natural enemies with suitable microclimate, alternative prey or host, non-prey food, and protection from intraguild predators and pesticides (Gontijo 2019). This contributes to increase the abundance and diversity of these beneficial organisms. However, when designing shelters for natural enemies, whether made of synthetic material or natural structures, it should be considered that these shelters might also affect the biocontrol agents by hindering their movement and diverting their attack.

Nezara viridula parasitism by naturally occurring egg parasitoids was measured in tomato plots with and without a margin of flowering buckwheat, a plant species that attracts the parasitoid *T. basalis* and whose nectar increases this wasp's fecundity (Foti et al. 2017). The results of this research showed that another parasitoid of *N. viridula* eggs, *Ooencyrtus telenomicida* (Vassiliev) (Hymenoptera: Encyrtidae), was repelled by the odor of buckwheat flowers, unlike *T. basalis*. The latter parasitoid species parasitized more egg masses when flower margins were present, and egg parasitism correlated positively with proximity to the buckwheat margin. However, egg parasitism by *O. telenomicida* occurred almost exclusively in the control plots. Based on these results, the authors conclude that buckwheat strips can influence intraguild competition among these two natural enemies of *N. viridula*.

3 Final considerations

Increasing global demand for food requires an ecologically sustainable intensification of food production (Shields et al. 2019). Therefore, ecosystem services are crucial and must be protected and restored, and technologies based on ecological principles must be boosted.

Among them, CBC as a way to prevent crop losses by pests results of paramount importance. As it was mentioned, the excessive use of pesticides with high toxicity and broad action spectrum, mostly neurotoxic ones, is recognized as the main cause of ecological imbalances in agroecosystems (Carvalho et al. 2019). Alternatively, the use of selective pesticides and minimizing their applications taking in consideration economic thresholds is recognized as a practicable and efficient strategy. Many of these ecologically safer pesticides are products derived from the secondary metabolism of plants. In recent years, the number of botanical insecticides published in Latin America has increased considerably; however, few are commercialized (Alves and Ascari 2019) due to difficulties of standardization and competition with multinationals. Knowledge about the safety of botanical insecticides for beneficial species is growing, demonstrating that botanical pesticides can be compatible with biological control.

Another mitigation strategy against negative consequences of human modification of ecosystems is the promotion of plant diversity, as it is demonstrated that increased plant diversity favors higher trophic levels, mainly natural enemies of herbivores (Barnes et al. 2020; Moreira et al. 2016 and references therein). However, it is not just the increased plant diversity per se that favors natural enemies, and that is why bioecological aspects of the tritrophic crop-pest-natural enemy interactions must be considered in deep when designing CBC tools based on habitat management. A good understanding of the ecology of parasitoids and predators, both inside and outside the cultivated habitat is needed, identifying those resources that are necessary for their survivorship and reproduction. The implementation of strategies to enhance the amount and quality of critical habitat for natural enemies will lead to the maintenance of their populations that may help to regulate pests in the agroecosystem.

To achieve biological control of pests through conservation tactics, it is necessary to integrate extensive knowledge (Table 6.1), and several key issues must be addressed. In this regard, Landis et al. (2000) proposed the following crucial aspects: the selection of the most appropriate plant species; the predator/parasitoid behavioral mechanisms which are influenced by the manipulation; the spatial scale over which the habitat enhancement operates, with implications for the area, shape and spacing of resources and refugia for predators and parasitoids; the negative aspects associated with adding new plants to an agroecosystem, such as the use of the plant resources by the pest being targeted; the degree of uptake by the agricultural/horticultural community of the proposed habitat changes.

A good amount of research supports the positive response of natural enemies to CBC strategies (Begg et al. 2017 and references therein), although unfortunately it is also denoted that such strategies do not always benefit natural enemy populations and that they often fail to result in effective pest regulation or improved crop yield, impacting negatively on degree of uptake by growers of the proposed strategies. Begg et al. (2017) provide an extensive and detailed analysis of the multiplicity of aspects that are being studied, and contribute to CBC as a complex strategy for pest suppression. The authors point out that the design and implementation of CBC tactics demand the integration of knowledge on nutritional and chemical ecology, molecular genetic approaches, natural enemy dispersal and habitat connectivity, trophic interactions, landscape effects, impact assessment, and societal and economic aspect. Given such complexity, it is not surprising that the design and adoption of reliable management prescriptions is not an easy task. Anyway, the high potential of CBC to contribute to the production of healthier foods and support more environmentally friendly agriculture is undeniable.

References

Abbate S, Silva H, Ribeiro AF, Bentancur O, Castiglioni E (2022) Effectiveness of some insecticides against soybean stink bugs andside-effects on *Telenomus podisi* (Ashmead) and generalist predators. Int J Trop Insect Sci 42:1813–1824

Abudulai M, Shepard BM (2003) Effects of neem *Azadirachta indica* A. Juss on *Trissolcus basalis* Wollaston (Hymenoptera: Scelionidae), a parasitoid of *Nezara viridula* L. Hemiptera: Pentatomidae. J Entomol Sci 38:386–397

Alves DS, Ascari J (2019) Use of natural chemical products for pest control. In: Souza B, Vázquez LL, Marucci RC (eds) Natural enemies of insect pests in Neotropical agroecosystems. Biological control and functional biodiversity. Springer Nature, Cham, pp 479–488

Ayelo PM, Pirk CWW, Yusuf AA, Chailleux A, Mohamed SA, Deletre E (2021) Exploring the kairomone-based foraging behaviour of natural enemies to enhance biological control: a review. Front Ecol Evol 9. https://doi.org/10.3389/fevo.2021.641974

Barbosa P (1999) Conservation biological control. Academic, San Diego

Barnes AD, Scherber C, Brose U, Borer ET, Ebeling A, Gauzens B, Giling DP, Hines J, Isbell F, Ristok C, Tilman D, Weisser WW, Eisenhauer N (2020) Biodiversity enhances the multitrophic control of arthropod herbivory. Sci Adv 6:eabb6603

Begg GS, Cook SM, Dye R, Ferrante M, Franck P, Lavigne C, Lövei GL, Mansion-Vaquie A, Pell JK, Petit S (2017) A functional overview of conservation biological control. Crop Prot 97:145–158

Blassioli-Moraes MCB, Laumann RA, Pires CSS, Borges M (2005) Induced volatiles in soybean and pigeon pea plants artificially infested with the Neotropical brown stink bug, *Euschistus heros*, and their effect on the egg parasitoid, *Telenomus podisi*. Entomol Exp Appl 115:227–237

Blassioli-Moraes MCB, Pareja M, Laumann RA, Hoffmann-Campos CB, Borges M (2008) Response of the parasitoid *Telenomus podisi* to induced volatiles from soybean damaged by stink bug herbivory and oviposition. J Plant Int 5:1–13

Blassioli-Moraes MCB, Venzon M, Silveira LCP, Gontijo LM, Togni PHB, Sujii ER, Haro MM, Borges M, Furtado Michereff MF, Santos de Aquino MF, Laumann RA, Caulfield J, Birkett M (2022) Companion and smart plants: scientific background to promote conservation biological control. Neotrop Entomol 51:171–187

Borges M, Medeiros MA, Mori K, Zarbin PHG, Ferreira JTB (1998) Field responses of stink bugs to the natural and synthetic pheromone of the Neotropical brown stink bug, *Euschistus heros* (Heteroptera: Pentatomidae). Phys Entomol 23:202–207

Bortolotto OC, Hoshino AT, Pazini J d B, Barroso G, Hata FT (2022) Manejo de pragas com artrópodes. In: Meyer MC, Bueno AF, Mazaro SM, da Silva JC (eds) Bioinsumos na cultura da soja. Parte V – Manejo de pragas. Embrapa, Brasília, pp 435–450

Bueno AF, Batistela MJ, Bueno RCOF, França-Neto JB, Naime Nishikawa MA, Filho AL (2011) Effects of integrated pest management, biological control and prophylactic use of insecticides on the management and sustainability of soybean. Crop Prot 30:937–945

Bueno AF, Bortolotto OC, Pomari-Fernandes A, França-Neto JB (2015) Assessment of a more conservative stink bug economic threshold for managing stink bugs in Brazilian soybean. Crop Prot 71:132–137

Bueno AF, Sutil WP, Jahnke SM, Carvalho GA, Cingolani MF, Colmenarez YC, Corniani N (2023) Biological control as part of the soybean integrated pest management (IPM): potential and challenges. Agronomy 13:2532

Carvalho GA, Grützmacher AD, Passos LC, de Oliveira RL (2019) Physiological and ecological selectivity of pesticides for natural enemies of insects. In: Souza B, Vázquez LL, Marucci RC (eds) Natural enemies of insect pests in neotropical agroecosystems. Biological control and functional biodiversity. Springer Nature, Cham, pp 469–478

Chaplin-Kramer R, O'Rourke EO, Blitzer EJ, Kremen C (2011) A meta-analysis of crop pest and natural enemy response to landscape complexity. Ecol Lett 14:922–923

Cingolani MF (2012) Parasitismo de huevos de *Piezodorus guildinii* (Hemiptera: Pentatomidae) por *Trissolcus basalis* y *Telenomus podisi* (Hymenoptera: Scelionidae) en el noreste de la provincia de Buenos Aires. PhD Thesis, Universidad Nacional de La Plata, Buenos Aires

Corrêa-Ferreira BS, Alexandre TM, Pellizzaro EC, Moscard F, Bueno A de F (2010) Práticas de manejo de pragas utilizadas na soja e seu impacto sobre a cultura. Embrapa Soja, Londrina, Circular Técnica 78, 16p

Daily G (1997) Introduction: what are ecosystem services. Island Press, Washington, DC

Danieli T (2010) Eficácia de inseticidas utilizados no controle de lagartas desfolhadoras e impacto na artropodofauna da soja RR. Moura Lacerda University Center, Agric Eng, Ribeirão Preto

DeBach P (1964) Biological control of insect pests and weeds. Cambridge University Press, Cambridge

Eilenberg J, Hajek A, Lomer C (2001) Suggestions for unifying the terminology in biological control. BioControl 46:387–400
Foti MC, Rostás M, Peri E, Chung Park K, Slimani T, Wratten SD, Colazza S (2017) Chemical ecology meets conservation biological control: identifying plant volatiles as predictors of floral resource suitability for an egg parasitoid of stink bugs. J Pest Sci 90:299–310
Gontijo LM (2019) Engineering natural enemy shelters to enhance conservation biological control in field crops. Biol Control 130:155–163
González E, Salvo A, Valladares G (2017) Arthropod communities and biological control in soybean fields: forest cover at landscape scale is more influential than forest proximity. Agr Ecos & Environ 239:359–367
Gurr GM, van Emden HF, Wratten SD (1998) Pesticides and conservation of natural enemies in pest management. In: Barbosa P (ed) Conservation biological control. Academic, pp 155–183
Janssen A, van Rijn P (2021) Pesticides do not significantly reduce arthropod pest densities in the presence of natural enemies. Ecol Lett 24:2010–2024
Landis DA, Wratten SD, Gurr GM (2000) Habitat management to conserve natural enemies of arthropod pests in agriculture. Annu Rev Entomol 45:175–201
Laumann RA, Aquino MFS, Blassioli Moraes MC, Pareja M, Borges M (2009) Response of the egg parasitoids *Trissolcus basalis* and *Telenomus podisi* to compounds from defensive secretions of stink bugs. J Chem Ecol 35:8–19
Lee JC, Flores SM, Velasco GK, Skillman VP (2022) Methyl salicylate can benefit ornamental pest control, and does not alter per capita predator consumption at close-range. Front Ecol Evol 9. https://doi.org/10.3389/fevo.2021.788187
Lichtenberg EM, Kennedy CM, Kremen C et al (2017) A global synthesis of the effects of diversified farming systems on arthropod diversity within fields and across agricultural landscapes. Glob Change Biol 23:4946–4957
Michereff MFF, Michereff Filho M, Blassioli-Moraes MC, Laumann RA, Diniz IR, Borges M (2015) Effect of resistant and susceptible soybean cultivars on the attraction of egg parasitoids under field conditions. J Appl Entomol 193:207–216
Michereff MFF, Borges M, Aquino MFS, Laumann RA, Mendes Gomes ACM, Blassioli-Moraes MC (2016) The influence of volatile semiochemicals from stink bug eggs and oviposition damaged plants on the foraging behaviour of the egg parasitoid *Telenomus podisi*. Bull Entomol Res 106:663–671
Mitchell PL, Gupta R, Singh AK, Kumar P (2004) Behavioral and developmental effects of neem extracts on *Clavigralla scutellaris* (Hemiptera: Heteroptera: Coreidae) and its egg parasitoid, *Gryon fulviventre* (Hymenoptera: Scelionidae). J Econ Entomol 97:916–923
Moreira X, Abdala-Roberts L, Rasmann S, Castagneyrol B, Mooney KA (2016) Plant diversity effects on insect herbivores and their natural enemies: current thinking, recent findings, and future directions. Curr Opin Insect Sci 14:1–7
Morrison WR, Blaauw BR, Nielsen AL, Talamas E, Leskey TC (2018) Predation and parasitism by native and exotic natural enemies of *Halyomorpha halys* (Stål) (Hemiptera: Pentatomidae) eggs augmented with semiochemicals and differing host stimuli. Biol Control 121:140–150
Panizzi AR, Lucini T (2023) Life history studies of stink bugs: much-needed research to support their conservation biological control. BioControl. https://doi.org/10.1007/s10526-023-10214-5
Philips CR, Rogers MA, Kuhar TP (2014) Understanding farmscapes and their potential for improving IPM programs. J Integ Pest Manag 5:C1–C9
Rahat S, Gurr GM, Wratten SD, Mo J, Neeson R (2005) Effect of plant nectars on adult longevity of the stinkbug parasitoid, *Trissolcus basalis*. Int J Pest Manag 51:321–324
Ripper WE, Greenslade RM, Hartley GS (1951) Selective inseticides and biological control. J Econ Entomol 44:448–458
Ruberson J, Nemoto H, Hirose Y (1998) Pesticides and conservation of natural enemies in pest management. In: Barbosa P (ed) Conservation biological control. Academic, pp 207–220

Salerno G, Colazza S, Conti E (2002) Sub-lethal effects of deltamethrin on walking behaviour and response to host kairomone of the egg parasitoid *Trissolcus basalis*. Pest Manag Sci 58:663–668

Smaniotto LF, Gouvea A de, Potrich M, Silva ERL da, Silva J da, Pegorini CS (2013) Seletividade de produtos alternativos a Telenomus podisi Ashmead (Hymenoptera: Scelionidae). Semina: Ciências Agrárias 34(6Supl1):3295–3306

Shields MW, Johnson AC, Pandey S, Cullen R, González- Chang M, Wratten SD, Gurr GM (2019) History, current situation and challenges for conservation biological control. Biol Control 131:25–35

Stecca CS, Bueno AF, Pasini A, Silva DM, Andrade K, Zirondi Filho DM (2018) Impact of insecticides used in soybean crops to the egg parasitoid *Telenomus podisi* (Hymenoptera: Platygastridae). Neotrop Entomol 47:281–291

Torres JB, Bueno AF (2018) Conservation biological control using selective insecticides – a valuable tool for IPM. Biol Control 126:53–64

Tscharntke T, Karp DS, Chaplin-Kramer R, Batáry P, DeClerck F, Gratton C, Hunt L, Ives A, Jonsson M, Larsen A, Martin EA, Martínez-Salinas A, Meehan TD, O'Rourke M, Poveda K, Rosenheim JA, Rusch A, Schellhorn N, Wanger TC, Wratten S, Zhang W (2016) When natural habitat fails to enhance biological pest control – five hypotheses. Biol Cons 204(Part B):449–458

Turchen LM, Piton LP, Dall'Oglio EL, Butnariu AR, MJB P (2016) Toxicity of *Piper aduncum* (Piperaceae) essential oil against *Euschistus heros* (F.) (Hemiptera: Pentatomidae) and non-effect on egg parasitoids. Neotrop Entomol 45:604–611

Vargas G, Rivera-Pedroza LF, García LF, Jahnke SM (2022) Conservation biological control as an important tool in the Neotropical Region. Neotrop Entomol 52:134–151

Vieira CR, Blassioli-Moraes MC, Borges M, Pires CSS, Sujii ER, Laumann RA (2013) Cis-Jasmone indirect action on egg parasitoids (Hymenoptera: Scelionidae) and its application in biological control of soybean stink bugs (Hemiptera: Pentatomidae). Biol Control 64:75–82

Vieira CR, Blassioli-Moraes MC, Borges M, Pires CSS, Sujii ER, Laumann RA (2014) Field evaluation of (E)-2-hexenal efficacy for behavioral manipulation of egg parasitoids in soybean. Biocontrol 59:525–537

Wäckers FL (2004) Assessing the suitability of flowering herbs as parasitoids food sources: flower attractiveness and nectar accessibility. Biol Control 29:301–314

Werdin-González JO, Laumann RA, da Silveira S, Blassioli Moraes MC, Borges M, Ferrero AA (2013) Lethal and sublethal effects of four essential oils on the egg parasitoids *Trissolcus basalis*. Chemosphere 92:608–615

Zucoloto GC (2017) Impacto de acefato e clorantraniliprole sobre organismos de superfície de solo e *Telenomus podisi* em soja. Monograph, Moura Lacerda University Center, Ribeirão Preto

Chapter 7
Botanical Insecticides as an Alternative to Control Stink Bugs in Agriculture

Everton Ricardi Lozano, Michele Potrich, Lucas Battisti, and Raiza Abati

1 Introduction

Due to the global demand for reducing the use of synthetic chemical insecticides in agriculture, research into effective and environmentally safe techniques for the integrated management of pests, especially stink bugs, is prominent, among which the use of applied biological control stands out, with macroorganisms and microorganisms and natural products derived from plants (Parra and Coelho 2019; Isman 2020). Botanical insecticides are potential "green alternatives" for improving efficiency in agricultural production, as they have different modes of action, most of which are more economically viable, biodegradable, more environmentally friendly, and more selective to non-target organisms compared to synthetics products, suggesting greater safety for human health and the environment (Souto et al. 2021).

The properties of botanical derivatives result from substances originating from the secondary metabolism of plants, developed throughout the evolutionary process as a strategy for adaptation, defence against pathogens and herbivores, and attraction of pollinators, among others (Boncan et al. 2020; Matsuura and Fett-Neto 2015; Shen et al. 2022; Treutter 2006; Zaynab et al. 2018). These metabolites are

E. R. Lozano (✉) · M. Potrich
Laboratório de Controle Biológico (LABCON), Universidade Tecnológica Federal do Paraná, Dois Vizinhos, PR, Brazil

L. Battisti
Programa de Pós-graduação em Agroecossistemas (PPGSIS), Universidade Tecnológica Federal do Paraná, Dois Vizinhos, PR, Brazil

R. Abati
Programa de Pós-Graduação em Ciências Biológicas (Entomologia), Universidade Federal do Paraná, Curitiba, PR, Brazil

A. F. Bueno, A. R. Panizzi (eds.), *Stink Bugs (Hemiptera: Pentatomidae) Research and Management*, Entomology in Focus 9,
https://doi.org/10.1007/978-3-031-69742-5_7

exclusive to some plant species, which are produced throughout the plant body and therefore can be found in different parts, such as leaves, flowers, fruits, seeds, roots and stems (Shen et al. 2022; Zaynab et al. 2018).

Among the main known secondary metabolites, phenolic compounds, terpenes, flavonoids and alkaloids stand out. Phenolic compounds are responsible for protecting the plant against attacks by insects and microorganisms and ultraviolet radiation and water stress, among other functions (Chowdhary et al. 2021). Terpenes also have insecticidal properties and protect plants against abiotic stress (Boncan et al. 2020). Flavonoids have the function of protecting the plant from pollution and chemical compounds, thermal and water stress, UV radiation, and pathogens (Shen et al. 2022; Treutter 2006). Alkaloids mainly act against pathogens and insects (Matsuura and Fett-Neto 2015).

These metabolites are formed by different chemical components, which may vary in the same plant species, considering the part of the plant used, the collection time, seasonality, the stage of development of the plant, temperature, ultraviolet radiation, water availability during development, availability of nutrients in the soil, cultivation altitude, atmospheric pollution, mechanical stimuli or pathogen attack, among others (Berini et al. 2018; Boncan et al. 2020; Gobbo-Neto and Lopes 2007).

The use of plant-based products, especially extracts and essential oils, to control insect pests and diseases has re-emerged after the era of synthetic chemicals, as a viable technological innovation strategy in the context of the development of more sustainable and regenerative agriculture. This is evidenced by the increasing registration of botanical insecticides in developed countries (Isman 2020) and, still, in an incipient form, in Brazil, according to data from the Ministério da Agricultura Pecuária e Abastecimento (MAPA 2023). Without doubt, technology will play an increasingly important role both in increasing efficiency in the use of these inputs and in developing alternatives that efficiently replace inputs based on non-renewable resources. It is also important to highlight that, in the current scenario, the prospects for the use of botanical insecticides are promising in the context of integrated pest management (IPM), going beyond the boundaries of use by small producers, since botanical insecticides associated with other control strategies can contribute to reduce the doses and applications of synthetic chemical insecticides (Isman 2015). This is because they have a broad spectrum of action, which can cause insect repellence, food and oviposition repellence, development changes, growth inhibition, and acute toxicity (Isman 2017), and in some cases, they have an effect equivalent to synthetic insecticides (Koul and Walia 2009; Siegwart et al. 2015). Although the worldwide publication of works relating to botanical insecticides is evident, commercialisation and field application, proportionally, are still incipient. The trend is for the global bioinsecticide market (microorganisms and botanical insecticides) to reach 20% of the insecticide market by 2025, with botanicals reaching a range of 7% (Isman 2015).

2 Toxicity of Botanical Insecticides to Phytophagous Stink Bugs

Although botanical diversity is enormous, there are still few plant species explored to control insect pests. In a recent study, Turchen et al. (2020) analysed 2,500 publications between 1945 and 2019 referring to the effects of botanical insecticides and found that 13 botanical families prevailed in these studies, with emphasis on the Meliaceae family. Regarding the target insect species, Coleoptera, Diptera, Lepidoptera and Hemiptera stood out. Although work on Hemiptera is prominent for pentatomids, information is still limited.

In this context, to seek information about botanical insecticides and pentatomides, a careful search for articles was carried out in the indexers Web of Science (http://www.webofscience.com), Scopus (https://www.scopus.com/), PubMed (https://www.ncbi.nlm.nih.gov/pubmed/) and Academic Google (https://scholar.google.com/). The Boolean script used was TOPIC ("botanical insecticide*" "plant extract*" "essential oil*" "biopesticide*" "bioinsecticide*" "alternative control") AND "bug*" "pentatomid*" "Euschistus" "Piezodorus" "Nezara" "Dichelops" "Diceraeus" "Edessa" "Scaptocoris" to select articles that evaluated the toxicity of insecticides for phytophagous stink bugs published until September 2023. After the search, the articles were analysed if they presented data related to the control of pentatomids with botanical insecticides. A total of 18 articles were considered (Fig. 7.1, Table 7.1).

The exclusion criteria were as follows: (1) articles without data on lethal and/or sublethal effects; (2) articles that evaluated toxicity with insects belonging to other families; (3) articles that evaluated toxicity with predatory pentatomids; and (4) non-peer-reviewed studies.

Based on the data collection, the articles found came from six countries, namely Brazil (7), the USA (5), Indonesia (3), India (2), Argentina and Italy (1 article each), with the works published in the last 18 years, from 2003 to 2023 (Fig. 7.1a).

2.1 Main Species of Phytophagous Stink Bugs

Among the phytophagous stink bugs, pentatomids stand out, being the largest family of hemimetabolous insects, with around 800 genera and 4,700 species described (Panizzi and Grazia 2015; Triplehorn and Jonnson 2016). Understanding these insects is crucial for developing sustainable pest control approaches and preserving the balance between agricultural and natural ecosystems. After surveying the articles, seven species of phytophagous stink bugs belonging to two families and two subfamilies were identified as being used as a model for toxicological studies with botanical insecticides (Table 7.1). The aforementioned stink bugs are species that stand out due to their impact on natural and agricultural ecosystems. *Halyomorpha halys* Stål (Pentatomidae: Pentatominae), also known as the brown marmorated

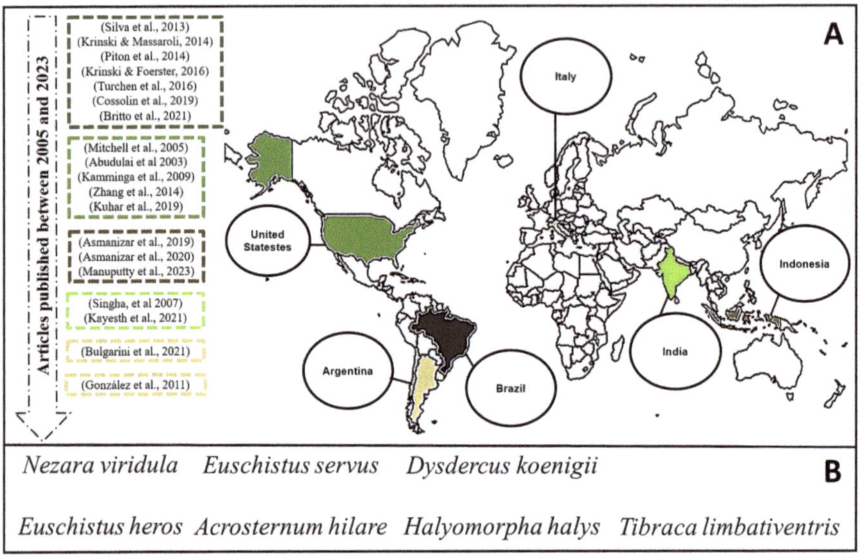

Fig. 7.1 Articles that evaluated the toxicity of botanical insecticides for phytophagous stink bugs published until September 2023. (**a**) Articles found in the data search and their respective countries; (**b**) Species of phytophagous stink bugs used as a toxicological study model. The colour gradient represents the number of articles published (dark green more articles—Brazil; light green fewer articles—Argentina and Italy)

stink bug, has expanded globally, causing significant damage to crops by feeding on a wide range of cultivated plants (Zhu et al. 2012). Chemical control is the main method of controlling this insect (Kuhar and Kamminga 2017).

Another notable example is *Tibraca limbativentris* Stål (Pentatomidae: Pentatominae), known as the rice stem bug, which can cause economic damage to rice, wheat, soybean and tomato, among others (Garbelotto and Campos 2014; Grazia et al. 2015). Furthermore, *Dysdercus koenigii* F. (Pyrrhocoridae: Pyrrhocorinae), known as the cotton bug or red cotton bug, stands out for its preference for feeding on cotton fruit and seeds, representing a direct threat to cotton production. This insect is highly mobile, which makes chemical control difficult (Shah et al. 2022); therefore, research into alternative forms of control is relevant.

Another economically important species is *Acrosternum hilare* Say (Pentatomidae: Pentatominae), also called the green stink bug. It is recognised for its ability to adapt to different habitats and feed on a variety of host plants, but its food preference is for soybean crops, for which it is a main insect pest (Panizzi and Slansky 1985; Koppel et al. 2011).

The neotropical brown stink bug, *Euschistus heros* F. (Pentatomidae: Pentatominae), is a predominant insect pest species in soybean crops and is the most abundant in soybean-producing regions in Brazil (Silva et al. 2012; Garbelotto and Campos 2014; Grazia et al. 2015). In soybean, adults and nymphs of *E. heros* feed

Table 7.1 Species, family, subfamily, main economically damaged crops and main places of occurrence of stink bug species used as a model for toxicological studies with botanical insecticides in articles published between 2005 and 2023

Species	Family	Subfamily	DC[a]	GD[b]
Dysdercus koenigii	Pyrrhocoridae	Pyrrhocorinae	Cotton Soybean	South and Central America
Tibraca limbativentris	Pentatomidae	Pentatominae	Rice Corn Pastures	South America (Brazil, Argentina and Paraguay)
Acrosternum hilare			Soybean Cotton Bean	South and Central America
Halyomorpha halys			Apple Peach Tomato	Asia (China, Japan, Korea and Taiwan), North America (USA and Canada) and Europe
Euschistus servus			Soybean Bean Corn	North America (USA and Canada)
Euschistus heros			Soybean Bean Corn	South America and Central America
Nezara viridula			Soybean Bean Apple Peach	North America (United States and Canada), Central and South America, Europe and Asia

[a]DC – main economically damaged crops
[b]GD – geographic distribution (main regions)

on pods and grains, causing significant losses in the production of this crop (Sosa-Gómez et al. 2014).

Nezara viridula L. (Pentatomidae: Pentatominae), known as the southern green stink bug or green stink bug, is a species of stink bug recognised by its bright green colour (Garbelotto and Campos 2014). This insect feeds on several plant species, especially soybean, being an agricultural pest of significant importance, where adults and nymphs cause damage similar to that caused by other stink bugs (Panizzi 1997; Sosa-Gómez et al. 2014).

2.2 *Lethal and Sublethal Effects of Botanical Insecticides on Phytophagous Stink Bugs*

Some plant species can have different effects on stink bugs, which can reduce the damage these insects have to agricultural productivity. Among these effects are repellence, immediate mortality of nymphs or adult insects, and sublethal effects, such as reduction of progeny or malformation. Table 7.2 presents a compilation of information regarding the species of stink bugs studied, their life stage, the effects

Table 7.2 Species of phytophagous stink bugs, botanical insecticides, formulation, concentration, life stage of the insects, effects of the identified botanical insecticides on the insect, country of origin of the study and reference

Species	Insecticide plant	For	Con	Application	Life stage	Effects	Country	Reference
Nezara viridula	*Azadirachta indica*	AS	0.1, 0.5, 1.0, 5.0, 20.0 (%)	Nymph immersion and feeding preference for treated beans	4th instar nymphs	It interrupted nymph molting, caused defects in the morphology of adults, increased the development time until adulthood, and decreased the longevity of females. It caused feeding repellence of the treated pods.	USA	Abudulai et al. (2003)
Nezara viridula	*Jatropha curcas*	GE	0.015 0.0312 0.0625 0.125 0.2505 (%)	Contact feeding (residual toxicity)	Eggs Nymphs Adults	The crude extract of *J. curcas* seed showed residual and ingestion toxicity, causing high mortality (100% of insects) at crude extract concentrations of 0.5% and 0.25%, without production of eggs or progeny.	Indonesia	Asmanizar et al. (2020)

Nezara viridula	*Annona muricata* (seed), *Alpinia galanga* (rhizome), *Jatropha curcas* (seed), *Citrus aurantifolia* (peel), *Carica papaya* (leaf), *Isotoma longifora* (leaf), *Lantana camara* (leaf), *Tagetes erecta* (leaf), *Ageratum conyzoides (plant)*, *Cymbopogon nardus* (leaf), *Cymbopogon nardus* (clump), *Ocimum citriodorum* (leaf), *Piper betle* (leaf) and *Murraya koenigii* (leaf)	GE	0.25 0.5 (%)	Contact feeding (residual toxicity)	Adults	Crude extracts of *A. muricata* (seed), *A. galanga* (rhizome) and *J. curcas* (seed) at concentrations of 0.5% and 0.25% showed insect mortality of 80–100% and exhibited residual toxicity against *N. viridula* adults.	Indonesia	Asmanizar et al. (2019)

(continued)

Table 7.2 (continued)

Species	Insecticide plant	For	Con	Application	Life stage	Effects	Country	Reference
Nezara viridula	*Origanum vulgare* and *Thymus vulgaris*	EO	0.625 1.25 6.25 12.5 (μg egg^{-1})	Topic Fumigation Contact	Eggs Nymphs (second instar) Adults	Oregano and thyme EOs showed an ovicidal effect dependent on concentration, with the percentage of hatching inhibition (at 12.5 μg egg^{-1}) being 99.2% and 99.7%, respectively. Both oils (oregano and thyme) showed insecticidal effects on *N. viridula* nymphs and adults using the fumigation method. The LC_{50} values for oregano were 26.8 and 285.6 μg mL^{-1} for nymphs and adults, respectively. As for thyme, it was 8.9 μg mL^{-1} for nymphs and 219.2 μg mL^{-1} for adults. Significant mortality of nymphs and adults was observed. For contact in a glass bottle, the LC_{50} for oregano was 1.7 and 169.2 μg cm^{-2} for nymphs and adults, respectively; for thyme, they were 3.5 and 48.8 μg cm^{-2}, respectively. Both EOs induced a repellent effect on the second instar at 40, 20 and 10 μg cm^{-2}. In adults, at 160, 80 and 40 μg cm^{-2}, the oils showed repellence activity.	Argentina	González et al. (2011)

Nezara viridula	*Allium sativum*, *Cymbopogon winterianus*, *Azadirachta indica* (leaf), *Azadirachta indica* (EO)	GE	15 15 2.5 0.15 (%)		Nymphs (third instar)	Significant mortality was observed for the GEs and EOs evaluated.	Indonesia	Manuputty et al. (2023a)
Nezara viridula	*Azadirachta indica*	AS HE	5, 2,5, 1.25, 0.62, 0.31 0.15 (%)	Contact (topical application)	4th instar nymphs	Deformity of the oral apparatus and death.	India	Singha et al. (2007)
Euschistus heros	*Ficus carica* (branches and leaf)	GE	1.5 10.0 15.0 20.0 30.0 ($mg\ mL^{-1}$ and $\mu g\ cm^{-2}$)	Contact	Nymphs (third instar)	Insecticidal activity was observed for third instar nymphs for extracts of *F. carica* twigs and leaves. Psoralen and Bergapten were identified by phytochemical analysis as compounds in the extract. Computational predictions revealed that these molecules had different abilities to physically interact with *E. heros* acetylcholinesterases, playing relevant roles in toxicity.	Brazil	Britto et al. (2021)

(continued)

Table 7.2 (continued)

Species	Insecticide plant	For	Con	Application	Life stage	Effects	Country	Reference
Euschistus heros	*Piper aduncum*	EO	20 40 60 80 100 (mg mL^{-1})	Contact		*Piper aduncum* essential oil was toxic to *E. heros* adults with an LD_{50} of 36.23 mg per insect and an LD_{90} of 50.52 per insect. Cytological changes, such as tissue disruption, increased population of mitochondria, and depletion of glycogen and lipids, occurred in the cells of the fat body, causing intracellular stress in the fat body and compromising important processes for the insect's survival.	Brazil	Cossolin et al. (2019)
Euschistus heros	*Piper aduncum* (leaf)	GE	0.5 1 2 4 8 (%)	Immersion Contact	Eggs Nymphs Adults	In immersed eggs, all concentrations above 1% rendered more than 70% of the eggs unviable. However, in pulverised eggs, only 19% were invalidated. In the nymph bioassay, concentrations of 8 and 4% caused deaths of 72 and 52%, respectively. In adults, all concentrations tested significantly reduced insect survival and reproduction.	Brazil	Piton et al. (2014)
Euschistus heros	*Annona* spp. (*A. crassiflora* and *A. coriácea)* Azadirachtin	GE CP	2 (%)	Field application		After the seventh day of application, both GE and CP were efficient in controlling the *E. heros* population.	Brazil	Silva et al. (2013)

Euschistus heros	*Piper aduncum* (leaf)	EO		Contact Immersion Topic	Eggs Nymphs Adults	In the egg bioassay, both exposure methods were efficient for non-viable eggs, with an emphasis on the immersion method. For nymphs, greater EO toxicity was observed, with lower concentrations necessary to cause insect death. For adults, there was a reduction in insect survival and, consequently, a reduction in the number of individuals in the next generation.	Brazil	Turchen et al. (2016)

(continued)

Table 7.2 (continued)

Species	Insecticide plant	For	Con	Application	Life stage	Effects	Country	Reference
Euschistus servus *Acrosternum hilare*	Azadirachtin Pyrethrins	CP	318 L ha^{-1} (152 mg, 97 mg [Active Ingredient] L^{-1})	Sprayed on bean pods and food preference (tomatoes with treatments)	Nymphs (fifth instar)	Exposure of *E. servus* nymphs and adults to insecticide treatments resulted in significant mortality after 72 h. Exposure of *A. hilare* to insecticide treatments resulted in significant mortality in all bioassays, with nymphs being more susceptible than adults. A preference for foods not treated with both principles was observed after 24 h of exposure. Pyrethrins significantly repelled *E. servus* and *A. hilare* adults compared to the control. Azadirachtin did not have a significant repellence effect on either species.	USA	Kamminga et al. (2009)
Dysdercus koenigii	*Catharanthus roseus* (leaf)	HE	0.00625 a 10 (%)	Exposure to volatiles (1 mL concentrations)	Nymphs (fifth instar)	The survival of the nymphs was not affected by the volatiles present in the extract. However, such compounds influenced the growth and development of fifth instar nymphs. Thus, no normal adult could develop in treatments of 5% or more.	India	Kayesth et al. (2021)

Tibraca limbativentris	*Piper* spp. (leaf) (*P. aduncum*, *P. gaudichaudianum*, *P. malacophyllum*, *P. marginatum* and *P. tuberculatum*)	EO	0.25 0.5 1.0 2.0 4.0 (%)	Insertion (10 s in treatment solutions)	Eggs	All essential oils tested showed an ovicidal effect. The LC_{50} values indicated that both younger and older eggs were susceptible to different oils.	Brazil	Krinski and Foerster (2016)
Tibraca limbativentris	*Annona mucosa* (leaf and seed) and *Annona crassiflora* (seed)	CME	0.5 1.0 2.0 4.0 8.0 (%)	Topical (2 μL on the back)	Nymphs (2nd instar)	High toxicity in the second instar of *T. limbativentris* nymphs, with mortality exceeding 75% at an extract concentration of 1.0%, within the first 24 h after application. *Annona crassiflora* seed extracts presented intermediate toxicity among all the extracts tested, and nymph mortality exceeded 80% at the highest concentration after 120 h of application of the extract. Extracts from *A. mucosa* leaves had low toxicity, with mortality not exceeding 40%, even 120 h after application.	Brazil	Krinski and Massaroli (2014)
Halyomorpha halys	*Rosmarinus officinalis* and *Mentha piperita*	EO	10 2 (%)	Oral	Nymphs (2nd and 4th instar) Adults	Repellence. There were fewer nymphs and adults in fruits and leaves treated with EO.	USA	Kuhar et al. (2019)

(continued)

Table 7.2 (continued)

Species	Insecticide plant	For	Con	Application	Life stage	Effects	Country	Reference
Halyomorpha halys	*Cymbopogon citratus*, *Allium sativum*, *Rosmarinus officinalis*, *Pelargonium hortorum*, *Gaultheria procumbens*, *Mentha spicata* and *Mentha pulegium*	EO	–	Exposure to volatiles	Adults	A repellent effect was observed for all EOs evaluated.	USA	Zhang et al. (2014)
Halyomorpha halys	*Zingiber officinale*, *Syzygium aromaticum*, *Chrysopogon zizanioides* and *Curcuma longa*	EO	1 3 6 12 25 50 (%)	Oral	Adults	All EOs tested were repellents at concentrations greater than 3%, regardless of gender or exposure time. Turmeric and clove were the most repellent for EOs, while ginger and vetiver showed, on average, medium to low repellence.	Italy	Bulgarini et al. (2021)

Legend: *For* formation, *Con* Concentration, *GE* gross extract, *HE* hexane extract, *EO* essential oil, *CME* chloroform-methanolic extracts, *ES* Ethanolic solution, *AS* Aqueous suspension, *CP* Commercial product observed, the botanical insecticide used, its formulation and concentration, and the country where the study was carried out. Among the plants studied for stink bug control are species from the Piperaceae family, which have shown positive results. In a study of adults of *E. heros*, *Piper aduncum* essential oil presented an LD_{50} of 36.23 mg/insect in contact exposure and caused cytological changes in the fat body, which can compromise vital processes in these insects (Cossolin et al. 2019). The essential oil of *P. aduncum* can cause the unviability of *E. heros* eggs, the mortality of nymphs and adults, and a reduction in progeny (Turchen et al. 2016). In addition to the essential oil, the acetone extract of *P. aduncum* has the potential to make *E. heros* eggs unviable, cause mortality in nymphs and adults, and reduce reproduction (Piton et al. 2014). In addition to the species *P. aduncum*, other species of the same genus, such as *P. gaudichaudianum*, *P. malacophyllum*, *P. marginatum* and *P. tuberculatum*, also present ovicidal potential for stink bug species *T. limbativentris* (Krinski and Foerster 2016).

The efficiency of the essential oil of *Piper* species is possibly due to the presence of components classified as phenylpropanoids, especially dilapiol. Although most of the articles found did not evaluate biochemical and systemic effects, the mode of action of essential oils in general is known to occur in the central nervous system of insects and may be related to the inhibition of acetylcholinesterase, which affects the synapses of octopamine and GABA receptors (Jankowska et al. 2018). Furthermore, diapiol, as it is formed with dioxy methylene rings, can cause inhibition of the insect's cytochrome P450, which is responsible for phase I of detoxification (Durofil et al. 2021). Hypothetically, inhibition of P450 can lead to increased toxicity to other substances (chemical or natural), considering its role in cellular detoxification, which would make the insect more susceptible. This opens up the possibility of new research evaluating the combined use of *Piper* spp. oils with other products.

Some species of the Meliaceae family also have insecticidal potential for pentatomids, especially *Azadirachta indica*, which has already been shown to be lethal for *E. heros* (Silva et al. 2013), *E. servus*, *A. hilare* (Kamminga et al. 2009) and *N. viridula* (Abudulai et al. 2003; Manuputty et al. 2023b; Singha et al. 2007). In addition to mortality, *A. indica* can interrupt nymphs' ecdysis, cause defects in the morphology of the wings and scutellum of adults, increase the development time until adulthood, and decrease the longevity of *N. viridula* females (Abudulai et al. 2003). Furthermore, it can cause deformities in the mouth parts of these insects (Singha et al. 2007). *Azadirachta indica* can also cause food repellence in *N. viridula* (Abudulai et al. 2003), *E. servus* and *A. hilare* (Kamminga et al. 2009).

Azadirachtin is a triterpene belonging to the limonoid class, which is present in the seeds of *A. indica* and is considered mainly responsible for causing morphological and reproductive defects, deregulating growth, and inhibiting insect feeding (Mordue 2004; Kilani-Morakchi et al. 2021). This occurs because azadirachtin prevents cell division, blocks the transport and release of neurosecretory peptides, and prevents sperm formation. Furthermore, it inhibits protein synthesis in cells, directly interfering with midgut cells, reproduction and the fat body, which are responsible for detoxification processes (Mordue 2004; Yu et al. 2022). However, little caution is needed regarding the use of natural products based on *A. indica*, as there is evidence that commercial products based on azadirachtin can have lethal and sublethal effects in non-target organisms, as suggested by Battisti et al. (2020) and Smaniotto et al. (2013), who evaluated the selectivity of different neem-based products on the egg parasitoid *Telenomus podisi* Ashmead (Hymenoptera: Scelionidae).

Another species belonging to the Meliaceae family that presents insecticidal potential for *N. viridula* is *Melia volkensii*. After exposing nymphs to the ethanolic solution of the fruit, morphological deformations of the wing, scutellum, pronotum, legs and antennae of the insects that reached the stage were observed in adults. Furthermore, when nymphs were exposed to 50 g/L of the solution, immediate mortality was observed (Mitchell et al. 2005). However, unlike *A. indica*, the majority component responsible for the effect on insects is salanin, a type of limonoid. Although the mode of action has not yet been defined, salanin, as well as other triterpenes, deregulates insect growth and causes antifeedant effects (Jaoko et al. 2020). This may explain the aforementioned morphological deformities since during the nymph phase, the demand for nutrients is high, considering the stink bug's growth and ecdysis.

Origanum vulgare and *Thymus vulgaris* essential oils belonging to the Lamiaceae family can cause repellence of nymphs and adults of *N. viridula* (González et al. 2011). Furthermore, the essential oils of these two plants can have an ovicidal effect and be lethal to *N. viridula* nymphs and adults exposed both by the fumigation method and by contact (González et al. 2011). *Origanum vulgare* and *T. vulgaris* essential oils have a repellent action, especially due to the major oxygenated components: carvacrol and thymol present in the secondary metabolite of these plants (De-Oliveira et al. 2021). In addition to repellence, carvacrol and thymol have a high toxicological potential for stink bugs because they are saturated compounds, which have simple carbon–carbon bonds outside the benzene ring, and because of the presence of functional compounds of hydroxyl groups on the benzene ring, which facilitate penetration into the cuticle of these insects (Gaire et al. 2019). However, there is still no concrete definition of their mode of action in insects. One possibility is that they act on the central nervous system of insects, potentiating the chloride channels controlled by ligands in the nervous system. Furthermore, thymol can block octopamine receptors, causing insect death (Youssefi et al. 2019). Essential oils from other Lamiaceae, such as *Rosmarinus officinalis*, *Mentha piperita* and *Mentha pulegium*, can also cause repellence in *H. halys* (Zhang et al. 2014; Kuhar et al. 2019).

Annona species (Annoneaceae) have shown high lethality potential to stink bugs in different types of extracts from seeds. Among them, the lethality to *E. heros* of the crude extract of *A. crassiflora* and *A. coriacea* stood out in a field trial (Silva et al. 2013). The crude extract of *A. muricata* seeds also causes high mortality in *N. viridula*, with an LC_{50} estimated at 0.038% (Asmanizar et al. 2019). Finally, chloroform-methanolic extracts of *A. mucosa* and *A. crassiflora* seeds are lethal to *T. limbativentris* second instar nymphs (Krinski and Massaroli 2014). The use of seeds from the genus *Annona* is due to the high concentrations of acetogenins, one of the major components of the secondary metabolites of this plant, which cause lethality to insects. The mode of action of acetogenins has not been well studied; however, it is estimated that they act as in mammals, inhibiting mitochondrial complex I and inducing programmed cell death by suspending ATP production (Bermejo et al. 2005, Da Silva Costa et al. 2016). From this perspective, the most affected systems may be the muscular and nervous systems, considering the high energy demands of the cells. This may explain the high lethality potential observed.

In addition to these plant species, other species have been highlighted for possible stink bug control. Among these, the crude extract of *Jatropha curcas* Euphorbiaceae seeds causes lethality to adults of *N. viridula* (Asmanizar et al. 2019, 2020) and the inhibition of progeny (Asmanizar et al. 2020). Botanical extracts of *Allium sativum* Amaryllidaceae and *Cymbopogon winterianus* Poaceae were also lethal to *N. viridula* third instar nymphs (Manuputty et al. 2023b).

In a recent study, 3rd instar nymphs of *E. heros* were exposed to different ethanolic extracts of *Ficus carica* (Moraceae), where they presented an LC_{50} of 5.9 m/mL for extracts extracted from the twigs and an LC_{50} of 14.1 mg/mL for extracts extracted from the leaves (Britto et al. 2021). By analysing these compounds, the presence of the main compounds psoralen and bergapten, classified as furanocoumarins, was verified. Although information about this extract is scarce, according to the same authors, computational predictions indicate that these components may act on the acetylcholinesterase enzyme. *Dysdercus koenigii* had altered development after exposure of 5th instar nymphs to hexane extract of *Catharanthus roseus* (Apocynaceae) at a concentration above 5%, although it did not interfere with the survival of the newly nymphs exposed (Kayesth et al. 2021), thus demonstrating the ability of this extract to cause changes in insect development.

Finally, *Cymbopogon citratus* Poaceae, *Pelargonium hortorum* Plantae, *Gaultheria procumbens* Ericaceae, *Allium sativum* Amaryllidaceae (Zhang et al. 2014), *Zingiber officinale*, *Curcuma longa* Zingiberaceae, *Syzygium aromaticum* Myrtaceae and *Chrysopogon zizanioides* Poaceae (Bulgarini et al. 2021) showed repellence potential for *H. halys*. However, studies evaluating these plants detrimental effects on stink bugs are still scarce.

Although botanical products are often associated with more sustainable agricultural practices, it is important to highlight that the way they are used can also pose risks to ecosystems (natural and agricultural) and living beings, even more so considering their large-scale use. It must be considered that, in the same way that they trigger toxic effects in insect pests, they can cause lethal and sublethal effects in non-target organisms, such as parasitoids, predators and pollinators, even in humans. It is known that many metabolic processes in insects are similar to those in humans. From this perspective, research evaluating the selectivity of plant-based products over non-target organisms is extremely important. For this reason, an increase in the number of articles published in recent decades has been noted, but even with this growth, the need for more reliable information on the topic is evident.

In general, there is a lack of data confirming the potential of plant-based products for stink bug control. Knowing that the components of secondary metabolites vary based on biotic and abiotic factors, the importance of more in-depth studies on the majority components of plant extracts and essential oils, as well as the quantity present, is highlighted so that they act on insects and can be used in bioinsecticide formulations.

3 Final Considerations

In this chapter we present a general perspective on studies evaluating the toxicity of botanical insecticides to phytophagous pentatomid bugs. The negative impacts resulting from the excessive use of synthetic insecticides to control insect pests have led to a re-evaluation and, consequently, an increase in studies with products of botanical origin, especially since the 1990s. Despite the importance of this topic, few countries have invested resources for this purpose. Brazil has the highest number of published articles (n = 7), followed by the USA (n = 4), Indonesia (n = 3), Argentina (n = 1), India (n = 1) and Italy (n = 1) (Fig. 7.1a). In addition to the small number of studies on the topic, there is also a lack of standards or protocols in the studies. Most studies did not use a positive control group (synthetic chemical insecticide) in their methodology; thus, the results often cannot be compared. In such a context, it is important to highlight the need to standardise methodological protocols for evaluating the toxicity of botanical insecticides to phytophagous pentatomids.

Most of the results found were restricted to mortality/survival of nymphs and adults, ovicidal effects, repellence effects and morphological changes. No study analysed here evaluated the mechanisms of action of botanical insecticides on phytophagous pentatomids, a fundamental factor in defining control strategies in IPM. Another important point to be highlighted is the lack of semi-field or field studies, limiting the use of these products by the producer.

Furthermore, considering the use of botanical insecticides to control insect pests in general, there is a huge gap when it comes to dissemination/education strategies and incentives for their use by farmers. It is known that in addition to technical effectiveness, it is important to understand the barriers that hinder the use of these products, such as costs, availability, residual effects and even knowledge, which vary in different regions and cultures. Table 7.3 lists some points about the main research gaps on the effects of botanical insecticides on phytophagous pentatomids, in addition to describing suggestions for future directions for improving the population control of these insects.

In summary, botanical insecticides are an excellent alternative to synthetic chemicals, as several compounds have lethal and sublethal effects on phytophagous pentatomids, in addition to generally being more selective to non-target organisms and less toxic in environmental terms compared to synthetic ones. However, even though they are more selective, it is extremely important to evaluate the impact of these products on non-target organisms. Another point to consider is that botanical insecticides can be used together with biological control agents, which can increase the control effect, but investment in research is necessary to understand this possible effect.

Table 7.3 Overview of research gaps in the literature on the effects of botanical insecticides on phytophagous pentatomids and suggestions for future directions

Research gaps	Future directions
Lack of uniform methodologies and specific protocols to evaluate toxic effects (lethal and sublethal);	Develop specific and standardised protocols to evaluate toxic effects (lethal and sublethal) with the inclusion of a positive control in the laboratory, semi-field and field methodology;
Lack of studies evaluating the mechanisms of action of botanical insecticides;	Increase knowledge about the mechanisms of action of botanical insecticides on phytophagous pentatomids;
Lack of inclusion of a positive control (chemical insecticide) in most studies;	Intensify the dissemination of botanical insecticides to farmers to increase acceptance/adoption of this control method;
Few studies carried out in field or semi-field conditions;	Expand research into plant species and their effects on phytophagous pentatomids.
Lack of studies evaluating the effectiveness of botanical insecticides on phytophagous pentatomids over several generations, considering that some species can develop resistance over time;	
Lack of research on dissemination and education strategies to encourage the use of botanical insecticides by farmers.	

References

Abudulai M, Shepard BM, Mitchell PL (2003) Antifeedant and toxic effects of a neem (*Azadirachta indica* A. Juss)-based formulation Neemix® against *Nezara viridula* (L.) (Hemiptera: Pentatomidae). J Entomol Sci 38:398–408

Asmanizar A, Aldywaridha A, Sumantri E, Lubis RM, Siregar D, Maimunah F (2019) Evaluation of potential plant crude extracts against green stink bug *Nezara viridula* Linn. (Hemiptera: Pentatomidae). UKM J Artic Repos 24:15–24

Asmanizar A, Aldywaridha A, Sumantri E, Lubis RM (2020) Potency of *Jatropha curcas* seed crude extract againts *Nezara viridula* (Hemiptera: Pentatomidae). Social Behav Sci. https://doi.org/10.15405/epsbs.2020.03.03.11

Battisti L, Warmling JV, Vieira CF, Oliveira DHR, Lima YRA, Potrich M, Bueno AF, Lozano ER (2020) Side effects of organic products on *Telenomus podisi* (Hymenoptera: Platygastridae). J Econ Entomol 113:1694–1701

Berini JL, Brockman SA, Hegeman AD, Reich PB, Muthukrishnan R, Montgomery RA, Forester JD (2018) Combinations of abiotic factors differentially alter production of plant secondary metabolites in five woody plant species in the boreal-temperate transition zone. Front Plant Sci 9. https://doi.org/10.3389/fpls.2018.01257

Bermejo A, Figadère B, Zafra-Polo MC, Barrachina I, Estornell E, Cortes D (2005) Acetogenins from Annonaceae: recent progress in isolation, synthesis and mechanisms of action. Nat Prod Rep 22:269–303

Boncan DAT, Tsang SSK, Li C, Lee IHT, Lam HM, Chan TF, Hui JHL (2020) Terpenes and terpenoids in plants: Interactions with environment and insects. Int J Mol Sci 21:7382

Britto IO, Araújo SHC, Toledo PFS, Lima GDA, Salustiano IV, Alves JR, Mantilla-Afanador JG, Kohlhoff M, Oliveira EE, Leite JPV (2021) Potential of *Ficus carica* extracts against

Euschistus heros: Toxicity of major active compounds and selectivity against beneficial insects. Pest Manag Sci 77:4638–4647
Bulgarini G, Bortolini S, Maistrello L (2021) Repellent activity of essential oils on adults of *Halyomorpha halys* (Heteroptera: Pentatomidae) in different physiological-behavioural phases. J Appl Entomol 145:575–586
Chowdhary V, Alooparampil S, Pandya R, Tank JG (2021) Physiological function of phenolic compounds in plant defense system. In: Phenolic compounds—chemistry, synthesis, diversity, non-conventional industrial, pharmaceutical and therapeutic applications. IntechOpen, London, pp 185–205
Cossolin JFS, Pereira MJB, Martínez LC, Turchen LM, Fiaz M, Bozdoğan H, Serrão JE (2019) Cytotoxicity of *Piper aduncum* (Piperaceae) essential oil in brown stink bug *Euschistus heros* (Heteroptera: Pentatomidae). Ecotoxicology 28:763–770
Da-Silva-Costa M, De-Paula SO, Martins GF, Zanuncio JC, Santana AEG, Serrão JE (2016) Multiple modes of action of the squamocin in the midgut cells of *Aedes aegypti* larvae. PLoS One 11:1–13
De-Oliveira AA, França LP, Ramos AS, Ferreira JLP, Maria ACB, Oliveira KMT, Araújo-Jr ES, Silva JN, Branches ADS, Barros GA, Silva NG, Tadei WP, Amaral ACF, Andrade-Silva JR (2021) Larvicidal, adulticidal and repellent activities against *Aedes aegypti* L. of two commonly used spices, *Origanum vulgare* L. and *Thymus vulgaris* L. S Afr J Bot 140:17–24
Durofil A, Radice M, Blanco-Salas J, Ruiz-Téllez T (2021) *Piper aduncum* essential oil: A promising insecticide, acaricide and antiparasitic. A review. Parasite 28:1–19
Gaire S, Scharf ME, Gondhalekar AD (2019) Toxicity and neurophysiological impacts of plant essential oil components on bed bugs (Cimicidae: Hemiptera). Sci Rep 9:1–12
Garbelotto TA, Campos LA (2014) Série Zoológica: Pentatominae do Sul de Santa Catarina. Sociedade Brasileira de Zoologia, Curitiba, 80 pp
Gobbo-Neto L, Lopes NP (2007) Plantas medicinais: Fatores de influência no conteúdo de metabólitos secundários. Quim Nova 30:374–381
González JOW, Gutiérrez MM, Murray AP, Ferrero AA (2011) Composition and biological activity of essential oils from Labiatae against *Nezara viridula* (Hemiptera: Pentatomidae) soybean pest. Pest Manag Sci 67:948–955
Grazia J, Panizzi AR, Greve C, Schwertner CF, Campos LA, Garbelotto TA, Fernandes JAM (2015) Stink bugs (Pentatomidae). In: Panizzi AR, Grazia J (eds) True bugs (Heteroptera) of the neotropics. Springer, pp 681–757
Isman MB (2015) A renaissance for botanical insecticides? Pest Manag Sci 71:1587–1590
Isman MB (2017) Bridging the gap: Moving botanical insecticides from the laboratory to the farm. Ind Crop Prod 110:10–14
Isman MB (2020) Botanical insecticides in the twenty-first century-fulfilling their promise? Annu Rev Entomol 65:233–249
Jankowska M, Rogalska J, Wyszkowska J, Stankiewicz M (2018) Molecular targets for components of essential oils in the insect nervous system—a review. Molecules 23:1–20
Jaoko V, Nji C, Taning T, Backx S, Mulatya J, Abeele JVD, Magomere T, Olubayo F, Mangelinckx S, Werbrouck SPO, Smagghe G (2020) The phytochemical composition of *Melia volkensii* and its potential for insect pest management. Plan Theory 9:1–11
Kamminga KL, Herbert DA, Kuhar TP, Malone S, Doughty H (2009) Toxicity, feeding preference, and repellency associated with selected organic insecticides against *Acrosternum hilare*, and *Euschistus servus* (Hemiptera: Pentatomidae). J Econ Entomol 102:1915–1921
Kayesth S, Kumar S, Shazad M, Gupta KK (2021) No diminished fitness of red cotton bug *Dysdercus koenigii* Fabricius induced by the volatiles in the hexane extract of *Catharanthus roseus*. Int J Trop Insect Sci 41:389–399
Kilani-Morakchi S, Morakchi-Goudjil H, Sifi K (2021) Azadirachtin-based insecticide: Overview, risk assessments, and future directions. Front Agron 3. https://doi.org/10.3389/fagro.2021.676208

Koppel AL, Herbert DA, Kuhar TP, Malone S, Arrington M (2011) Efficacy of selected insecticides against eggs of *Euschistus servus* and *Acrosternum hilare* (Hemiptera: Pentatomidae) and the egg parasitoid *Telenomus podisi* (Hymenoptera: Scelionidae). J Econ Entomol 104:137–142
Koul O, Walia S (2009) Comparing impacts of plant extracts and pure allelochemicals and implications for pest control. CAB Rev Perspect Agric Vet Sci Nutr Nat Resour 4:1–30
Krinski D, Foerster LA (2016) Toxicity of essential oils from leaves of Piperaceae species in rice stalk stink bug eggs, *Tibraca limbativentris* (Hemiptera: Pentatomidae). Ciênc Agrotecnol 40:676–687
Krinski D, Massaroli A (2014) Nymphicidal effect of vegetal extracts of *Annona mucosa* and *Anonna crassiflora* (Magnoliales, Annonaceae) against rice stalk stink bug, *Tibraca limbativentris* (Hemiptera, Pentatomidae). Rev Bras Frutic 36:217–224
Kuhar TP, Kamminga K (2017) Review of the chemical control research on *Halyomorpha halys* in the USA. J Pest Sci 90:1021–1031
Kuhar TP, Morehead JA, Formella AJ (2019) Applications of kaolin protect fruiting vegetables from brown marmorated stink bug (Hemiptera: Pentatomidae). J Entomol Sci 54:401–408
Manuputty C, Patty JA, Noya S (2023a) Efektivitas beberapa ekstrak tanaman terhadap nimfa instar III *Nezara viridula* L. (Hemiptera: Pentatomidae) pada polong kacang panjang effectiveness some plant extracts of third instar nymphs *Nezara viridula* L. (Hemiptera). J Agrosilvopast Tech 2:7–44
Manuputty C, Patty JA, Noya SH (2023b) Effectiveness some plant extracts of third instar nymphs *Nezara viridula* L. (Hemiptera: Pentatomidae) attack on string bean. Pod J Agrosilvopast Tech 2:37–44
MAPA (2023) Ministério da Agricultura e Pecuária [WWW Document]. URL https://www.gov.br/agricultura/pt-br
Matsuura HN, Fett-Neto AG (2015) Plant alkaloids: main features, toxicity, and mechanisms of action. Plant Toxins:1–15. https://doi.org/10.1007/978-94-007-6728-7_2-1
Mitchell PL, Thielen JB, Stell FM, Fescemyer HW (2005) Activity of *Melia volkensii* (Meliaceae) extract against southern green stink bug (Hemiptera: Heteroptera: Pentatomidae). J Agric Urban Entomol 21:131–141
Mordue AJ (2004) Present concepts of the mode of action of azadirachtin from neem. In: Koul O, Wahab S (eds) Neem: today and in the new millennium. Kluwer Academic Publishers, pp 229–242
Panizzi AR (1997) Wild hosts of pentatomids: ecological significance and role in their pest status on crops. Annu Rev Entomol 42:99–122
Panizzi AR, Grazia J (2015) Introduction to true bugs (Heteroptera) of the neotropics. In: Panizzi AR, Grazia J (eds) True bugs (Heteroptera) of the neotropics. Springer, Dordrecht, pp 3–22
Panizzi AR, Slansky F (1985) Review of phytophagous pentatomids (Hemiptera: Pentatomidae) associated with soybean in the Americas. Florida Entomol 68:184–214
Parra JRP, Coelho A (2019) Applied biological control in Brazil: From laboratory assays to field application. J Insect Sci 19:1–6
Piton LP, Turchen LM, Butnariu AR, Pereira MJB (2014) Natural insecticide based-leaves extract of *Piper aduncum* (Piperaceae) in the control of brown stink bug on soybean. Ciênc Rur 44:1915–1920
Shah SIA, Saeed R, Jaleel W, Sarwar ZM, Azad R, Akbar R, Jan MT, Ghram HA, BM AL-S, Ahmad Z, Khan KA, Hussain S (2022) Ovipositional response of *Antilochus coquebertii* and *Dysdercus koenigii* in different media under controled conditions. J King Saud Univ Sci 34:102252
Shen N, Wang T, Gan Q, Liu S, Wang L, Jin B (2022) Plant flavonoids: Classification, distribution, biosynthesis, and antioxidant activity. Food Chem 383:132531
Siegwart M, Graillot B, Lopez CB, Besse S, Bardin M, Nicot PC, Lopez-Ferber M (2015) Resistance to bio-insecticides or how to enhance their sustainability: A review. Front Plant Sci 6:1–19

Silva FAC, Silva JJ, Depieri RA, Panizzi AR (2012) Feeding activity, salivary amylase activity, and superficial damage to soybean seed by adult *Edessa meditabunda* (F.) and *Euschistus heros* (F.) (Hemiptera: Pentatomidae). Neotrop Entomol 41:386–390

Silva VP, Pereira MJB, Turchen LM (2013) Efeito de extratos vegetais no controle de *Euschistus heros* (F.) (Hemiptera: Pentatomidae) em lavoura de soja na região sudoeste do Estado de Mato Grosso. Braz J Agric 88:185

Singha A, Thareja V, Singh AK (2007) Application of neem seed kernel extracts result in mouth-part deformities and subsequent mortality in *Nezara viridula* (L.) (Hem: Pentatomidae). A J Appl Entomol 131:197–201

Smaniotto LF, De-Gouvea A, Potrich M, Da-Silva ERL, Da-Silva J, Pegorini CS (2013) Seletividade de produtos alternativos a *Telenomus podisi* Ashmead (Hymenoptera: Scelionidae). Semin Agrar 34:3295–3306

Sosa-Gómez DR, Corrêa-Ferreira BS, Hoffmann-Campo CB, Corso IC, Oliveira LJ, Moscardi F, Panizzi AR, Bueno AF, Hirose E, Roggia S (2014) Manual de identificação de insetos e outros invertebrados da cultura da soja. Embrapa Soja, Londrina, 100 pp

Souto AL, Sylvestre M, Tölke ED, Tavares JF, Barbosa-Filho JM, Cebrián-Torrejón G (2021) Plant-derived pesticides as an alternative to pest management and sustainable agricultural production: prospects, applications and challenges. Molecules 26. https://doi.org/10.3390/molecules26164835

Treutter D (2006) Significance of flavonoids in plant resistance: a review. Environ Chem Lett 4:147–157

Triplehorn CA, Jonnson NF (2016) Estudo dos Insetos. Tradução Noveritis do Brasil, 2nd edn. Cengage Learning, São Paulo

Turchen LM, Piton LP, Dall'Oglio EL, Butnariu AR, MJB P (2016) Toxicity of *Piper aduncum* (Piperaceae) essential oil against *Euschistus heros* (F.) (Hemiptera: Pentatomidae) and non-effect on egg parasitoids. Neotrop Entomol 45:604–611

Turchen LM, Cosme-Júnior L, Guedes RNC (2020) Plant-derived insecticides under meta-analyses: status, biases, and knowledge gaps. Insects 11:1–13

Youssefi MR, Tabari MA, Esfandiari A, Kazemi S, Moghadamnia AA, Sut S, Dall'Acqua S, Benelli G, Maggi F (2019) Efficacy of two monoterpenoids, carvacrol and thymol, and their combinations against eggs and larvae of the West Nile vector *Culex pipiens*. Molecules 24:1–11

Yu H, Yang X, Dai J, Li Y, Veeran S, Lin J, Shu B (2022) Effects of azadirachtin on detoxification-related gene expression in the fat bodies of the fall armyworm, *Spodoptera frugiperda*. Environ Sci Pollut Res 30:42587–42595

Zaynab M, Fatima M, Abbas S, Sharif Y, Umair M, Zafar MH, Bahadar K (2018) Role of secondary metabolites in plant defense against pathogens. Microb Pathog 124:198–202

Zhang QH, Schneidmiller RG, Hoover DR, Zhou G, Margaryan A, Bryant P (2014) Essential oils as spatial repellents for the brown marmorated stink bug, *Halyomorpha halys* (Stål) (Hemiptera: Pentatomidae). J Appl Entomol 138:490–499

Zhu G, Bu W, Gao Y, Liu G (2012) Potential geographic distribution of brown marmorated stink bug invasion (*Halyomorpha halys*). PLoS One 7. https://doi.org/10.1371/journal.pone.0031246

Chapter 8
The Use of RNAi Against Stink Bugs

Grazielle Celeste Maktura, Giovanna Vieira Guidelli, Thiago Roncini Gomes da Costa, and Henrique Marques-Souza

1 Introduction

The use of RNA molecules to control agricultural pests and diseases represents a sustainable and highly specific alternative for crop protection (Mamta and Rajam 2017; Christiaens et al. 2020). The control of Pentatomidae pests has been mainly based on the use of agrochemicals, such as neonicotinoids, phenylpyrazole, organophosphates, and pyrethroids, requiring from four to eight applications to obtain effective control (Takeuchi and Endo 2012; Tibola et al. 2021). Although chemical molecules represent the most successful method to prevent losses from pests and diseases, frequent development of insecticide resistance (Zerbino and Panizzi 2019) and its prejudicial effects on the environment and non-target species (Mamta and Rajam 2017) have urged for the development of more specific and eco-friendly control tools to integrated pest management (IPM) and achieve a safe, sustainable, and high level of crop protection (Panizzi 2013; Bueno et al. 2020).

The RNAi technology works through the mechanism of gene silencing mediated by double stranded RNAs (dsRNAs) molecules that are processed into single-strand small interfering RNA (siRNA) by the RNA Induced Silencing Complex (RISC), which binds to the target complementary mRNA, leading to its degradation (Mello and Conte 2004) or blocking protein translation (Samir et al. 2016). The use of

G. C. Maktura
HMS Agroscience, Campinas, SP, Brazil

G. V. Guidelli
Agrivalle Brasil, Indaiatuba, SP, Brazil

T. R. G. da Costa · H. Marques-Souza (✉)
Department of Biochemistry and Tissue Biology, University of Campinas, Campinas, SP, Brazil
e-mail: hmsouza@unicamp.br

A. F. Bueno, A. R. Panizzi (eds.), *Stink Bugs (Hemiptera: Pentatomidae) Research and Management*, Entomology in Focus 9,
https://doi.org/10.1007/978-3-031-69742-5_8

RNAi technology in agriculture started with the discovery that plant-expressing dsRNA could result in insect pest mortality through specific gene silencing (Price and Gatehouse 2008). Since then, studies emerged aiming to propose solutions for controlling the entire range of agricultural pest species (Mamta and Rajam 2017), such as Coleoptera (Willow and Veromann 2021), Diptera (Maktura et al. 2021), Lepidoptera (Terenius et al. 2011) and Hemiptera (Jain et al. 2021).

Despite being conserved in eukaryotes, the efficiency of gene silencing through RNAi varies significantly among different insect orders and even in species within the same order (Joga et al. 2016; Singh et al. 2017; Zhang et al. 2022) (Fig. 8.1). Hemipterans and lepidopterans appear to be most recalcitrant to RNAi, compared to the highly sensitive coleopteran group (Vogel et al. 2018; Guan et al. 2018). This efficiency variation depends on many factors, such as the presence of specific genes of the core RNAi machinery, target tissue and gene, delivery method, presence of dsRNases, refractoriness, and dsRNA length and stability (Dias et al. 2020; Joga et al. 2016; Cooper et al. 2019). Understanding these variations is important to guide research and development programs to target primarily the most sensitive pests but also to guide the development of enabling biotechnological solutions for the most commercially important insect pests.

Although stink bugs are among the most recalcitrant species to RNAi, the technology has been successfully used to knockdown target genes of different species

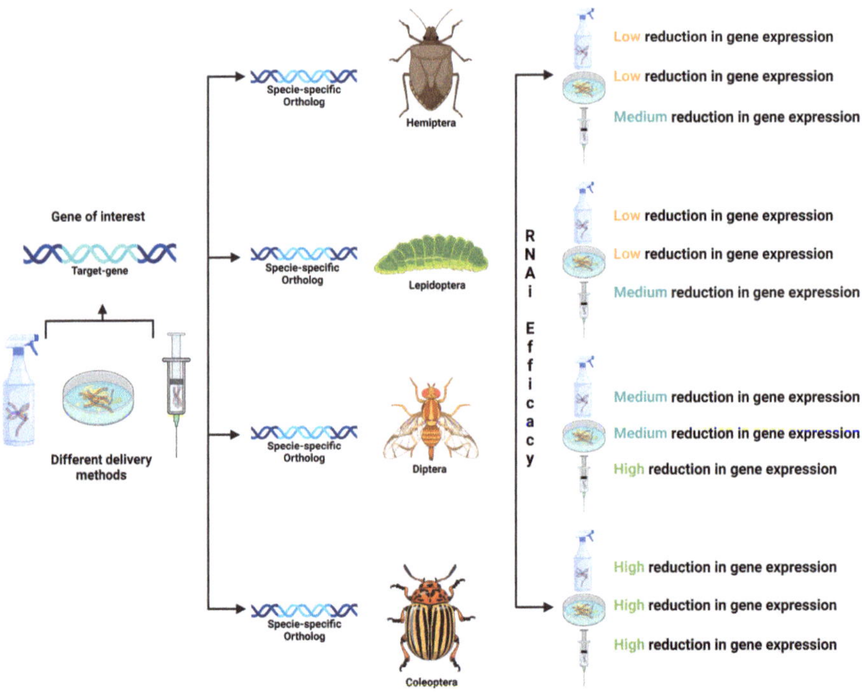

Fig. 8.1 Variation of RNAi efficiency on gene silencing in different insect orders (Created with BioRender.com)

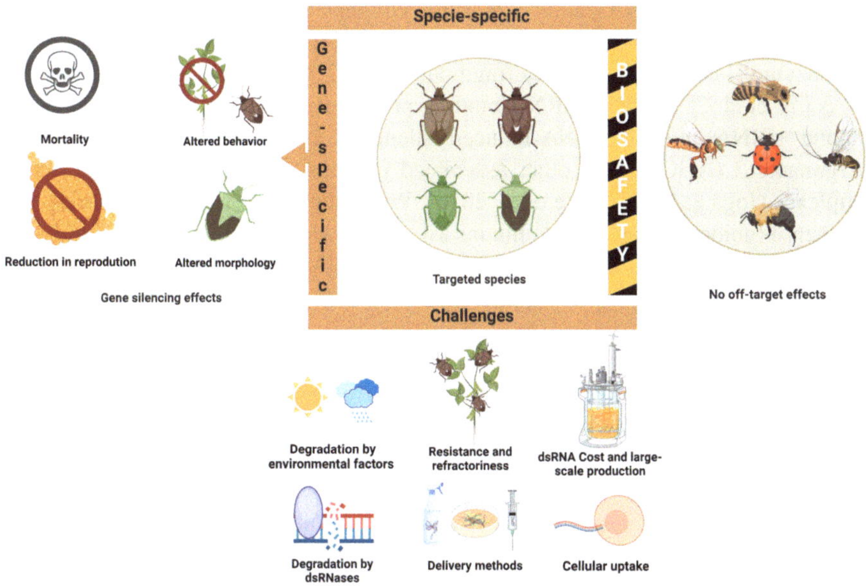

Fig. 8.2 Overview of the RNAi technology regarding delivery methods, specific phenotypic effects, safety for non-target organisms and challenges related to RNAi for pest control (Created with BioRender.com)

such as the Neotropical brown stink bug *Euschistus heros* (F.) (Fishilevich et al. 2016) and the brown marmorated stink bug *Halyomorpha halys* Stål (Lu et al. 2017). In this chapter, we review the entire body of information on RNAi in stink bugs and discuss the specificities of the core RNAi machinery genes in stink bugs that might present advantages or limitations to specific species, the success rate of target genes and biological processes to date and present the most important aspects of the technology that should be considered for the development of successful products for field control of stink bug agricultural losses (Fig. 8.2).

2 RNAi Machinery

As a primitive eukaryote mechanism against viruses and transposable elements, RNAi is capable of regulating gene expression in a sequence-specific manner using noncoding RNAs (Kim and Rossi 2008; Zhou and Rana 2013). This basic biological process was later transformed into a powerful tool to silence specific genes in a wide range of eukaryote species (Zhao and Guo 2022). In general, gene silencing via RNAi, as a cellular process, works through three different pathways: microRNA (miRNAs), PIWI-Interacting RNA (piRNA) and small interfering RNA (siRNA), the last being the pathway employed by the RNAi technology (Obbard et al. 2009; Joga et al. 2016).

RNAi through siRNA pathway is triggered when a gene-specific dsRNA molecule enters the cell of the target organism, which is recognized by the ribonuclease III enzyme Dicer-2 (DCR-2) and fragmented into 21–24 bp molecules, called siRNAs (Ye et al. 2019). The dsRNA-binding protein R2D2 incorporates the DCR-2-bound siRNA into the RNA-induced silencing complex (RISC), formed by Argonaute 2 (AGO-2) and other associated proteins, which unwinds the siRNA duplexes (Joga et al. 2016; Ye et al. 2019). While the sense strand is degraded, the antisense (guide siRNA) strand remains associated with the RISC complex and specifically binds to its complementary messenger RNA (mRNA) molecules, causing their cleavage and, consequently, post-transcriptional gene silencing (Li et al. 2011; Joga et al. 2016).

The successful use of RNAi in a variety of Pentatomidae species indicates that these insects have an active RNAi machinery, as reported in other hemipteran families (Jain et al. 2020). RNAi-related genes have been found in transcriptomic analysis of *Piezodorus guildinii* (Westwood) (Schvartzman et al. 2022), *E. heros* (Cagliari et al. 2020a) and *H. halys* (Sparks et al. 2014), indicating that stink bugs have a complete RNAi machinery, sufficient to allow gene silence, as described below.

2.1 Core RNAi Machinery

The efficiency of RNAi varies across different species and depends on several factors, including the presence of essential components within the RNAi machinery (Cagliari et al. 2020a). For this reason, the characterization of the core RNAi machinery in the target pest is essential to evaluate the potential of this technology in target organisms (Fig. 8.3).

As already pointed out, RNAi occurs through three different pathways: miRNA, siRNA and piRNA. The miRNAs are small single-stranded RNAs that act as endogenous regulators of gene expression, whereas siRNAs are involved in the genome protection against exogenous dsRNA (Joga et al. 2016; Samir et al. 2016). In both cases miRNA and siRNAs are produced by an endonuclease, the RNase III-enzyme Dicer (Kutter and Svoboda 2008). In turn, PIWI-associated RNAs are produced in a dicer-independent manner and act in embryonic development, maintenance, and protection of germline integrity against transposon (Kutter and Svoboda 2008; Huang et al. 2014). For all three pathways, RNA silencing occurs via the association of the non-coding RNA molecules with an RNA-induced silencing complex (RISC) (Kutter and Svoboda 2008; Clark and Rager 2020) (Fig. 8.3).

Proteins related to the miRNA, siRNA and piRNA pathways were identified in both *E. heros* and *P. guildinii* transcriptomes, with high similarity to *H. halys* homologues or, in a lower occurrence level, to *Nezara viridula* L. (Table 8.1).

- *miRNA pathway*

In the miRNA pathway, DCR-1, Drosha, Exportin5, Pasha, Loquacious, AGO-1 were found in *E. heros* and *P. guildinii*. For *E. heros,* DCR-1 presented the

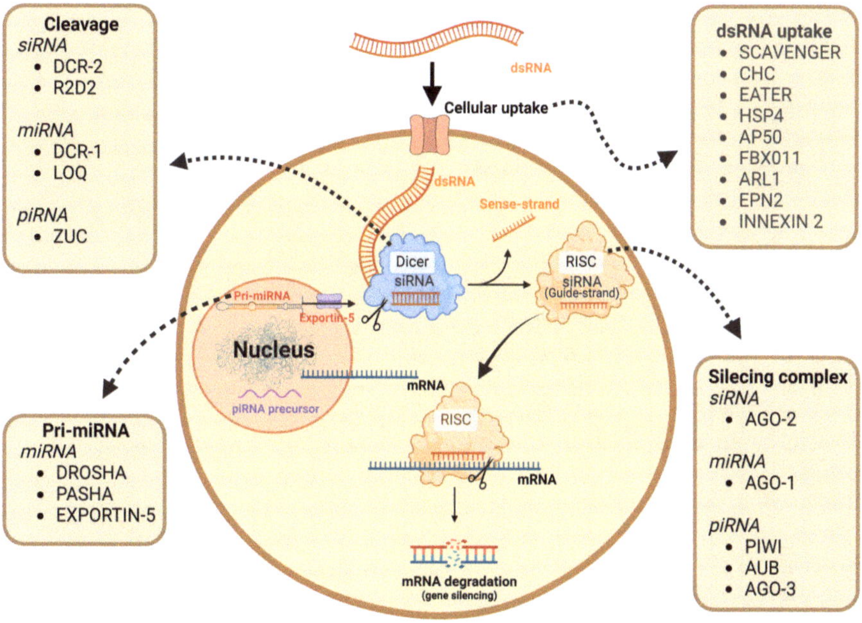

Fig. 8.3 Major phenomenon and corresponding genes related to miRNA, piRNA and siRNA pathways and dsRNA cellular uptake (Created with BioRender.com)

conserved domains PAZ, two RNaseIII and Double-stranded RNA-binding domain (dsRBD), with no presence of helicase domain (Cagliari et al. 2020a), whereas in *P. guildinii* was observed only a partial sequence, containing two RNaseIII and a dsRBD domain (Schvartzman et al. 2022). The Drosha showed two Ribonuclease III domain in the two stink bug groups, but *E. heros* protein also presented a Ribonuclease III C terminal domain (Cagliari et al. 2020a), while *P. guildinii* showed a dsRBD domain (Schvartzman et al. 2022). Regarding Loquacious, two dsRBD conserved domains were found for both species. Finally, for AGO-1 proteins were identified PAZ and PIWI conserved domains. However, only one isoform was described in *P. guildinii* transcriptome, whereas three additional isoforms were described for *E. heros* (Cagliari et al. 2020a; Schvartzman et al. 2022).

- *siRNA pathway*

In the siRNA pathway, DCR-2, R2D2 and AGO-2 were found to be expressed in *E. heros* and *P. guildinii* and in samples of salivary glands of *Chlorochroa ligata* (Say), *Euschistus conspersus* (Uhler), *Thyanta pallidovirens* (Stål), *H. halys* and *N. viridula* (Marshall et al. 2023). While DCR-2 was identified as two isoforms in *E. heros*, only one copy was identified in *P. guildinii*. The DCR-2 isoform 1, common to both species, showed conservation of the N-terminal helicase domains, PAZ domain, two RNaseIII domains and one dsRBD (Cagliari et al. 2020a, Schvartzman et al. 2022). In turn, *E. heros* DCR-2 isoform 2 exhibited two RNaseIII domains and

Table 8.1 Comparison of genes related to core RNAi machinery identified in *Euschistus heros* and *Piezodorus guildinii*

		Euschistus heros		*Piezodorus guildinii*	
Gene ID	Homologue ID	Comparison	% Identity	Comparison	% Identity
miRNA					
DCR-1	AVK59457.1 – *N. viridula*	E = 0.0 – bits = 2109	91	E = 0.0 – bits = 1321	94
AGO-1 isoform 1	AVK59466.1 – *N. viridula*	E = 0.0 – bits = 1924	99.89	E = 0.0 – bits = 1257	99.84
AGO-1 isoform 3	AVK59466.1 – *N. viridula*	E = 0.0 – bits = 1923	99.89	n.a	n.a
AGO-1 isoform 4	AVK59466.1 – *N. viridula*	E = 0.0 – bits = 1924	99.89	n.a	n.a
AGO-1 isoform 5	XP_014287705.1 – *H. halys*	E = 0.0 – bits = 1877	99	n.a	n.a
Loquacious	XP_014274312.1 – *H. halys*	E = 0.0 – bits = 521	96	E = 0.0 – bits = 739	95
Drosha	XP_014278529.1 – *H. halys*	E = 0.0 – bits = 2366	91	E = 0.0 – bits = 1442	93
Pasha	XP_014282581.1 – *H. halys*	E = 0.0 – bits = 1078	89	E = 0.0 – bits = 1210	87
Exportin-5	XP_014280932.1 – *H. halys*	E = 0.0 – bits = 2420	98	E = 0.0 – bits = 2405	86
siRNA					
DCR-2 isoform 1	XP_014275310.1 – *H. halys*	E = 0.0 – bits = 2795	83	E = 0.0 – bits = 2808	84
DCR-2 isoform 2	XP_014275311.1 – *H. halys*	E = 0.0 – bits = 852	88	n.a	n.a
AGO-2 isoform 1	AVK59468.1 – *N. viridula*	E = 0.0 – bits = 565	80	E = 0.0 – bits = 2808	81
AGO-2 isoform 2	AVK59468.1 – *N. viridula*	E = 0.0 – bits = 1516	75	n.a	n.a
R2D2	XP_014288218.1 – *H. halys*	E = 0.0 – bits = 559	82	E = 2.73e-170 – bits = 582	77
piRNA					
AGO-3	XP_014276831.1 – *H. halys*	E = 0.0 – bits = 1595	85	E = 0.0 – bits = 1028	86
Aubergine (*AUB*)	XP_014275927.1– *H. halys*	E = 0.0 – bits = 1676	96	E = 0.0 – bits = 1714	97
PIWI	XP_014270559.1 – *H. halys*	E = 0.0 – bits = 1172	63	E = 0.0 – bits = 1195	65
Zucchini (*ZUC*)	XP_014288409.1 – *H. halys*	E = 1e-152 – bits = 432	86	E = 3.52e-167 – bits = 468	92

a Ribonuclease III C terminal domain (Cagliari et al. 2020a). For both species, the R2D2 protein was found to possess dsRBD conserved domains. Two isoforms of AGO-2 were found in *E. heros* transcriptome, while only one was identified in *P. guildinii* data. In AGO-2 isoform 1, common to both species, PAZ and PIWI conserved domains were identified, while in *E. heros* AGO-2 isoform 2, only PIWI domain was found (Cagliari et al. 2020a, Schvartzman et al. 2022).

- *piRNA pathway*

In the piRNA pathway, Zucchini (ZUC), AGO-3, AUB and PIWI proteins were found, with the last three containing all PAZ and PIWI conserved domains in both species. The presence of a nuclease-conserved domain in ZUC was described in *E. heros* transcriptome, but no additional information about its domains was described in *P. guildinii* data (Cagliari et al. 2020a).

The presence of DCR, Loquacious, AGO and AUB proteins was also described to be found in *Murgantia histrionica* (Hahn) transcriptome, although no more details about its isoforms, domains or homologues were described (Sparks et al. 2017). This analysis shows that stink bugs possess the major genes involved in RNA silencing pathways, although slight differences in isoform numbers and/or domains have been described in the transcriptome of different species.

2.2 Auxiliary RISC Factors

In addition to having the core RNAi machinery genes, the presence of auxiliary RISC protein also influences the susceptibility of the target insect to RNAi (Xu et al. 2019; Christiaens et al. 2020) (Fig. 8.3). RISC is the multiprotein complex responsible for incorporating the siRNA or miRNA guide-strand and performing the recognition of complementary mRNA, allowing its binding to siRNA/miRNA and causing post-transcriptional gene silencing (Tang 2005). Highlighting the siRNA pathway, in this process the mature RISC complex, also known as holo-RISC, is formed when AGO-2 converts the siRNA duplexes into single-stranded siRNA (Kim et al. 2007). Its composition also includes DCR-2, R2D2, in addition to auxiliary factors in its formation and maturation, such as CLIP-associating protein (CLIP-1), Tudor-SN (TSN), Vasa intronic gene (VIG), Fragile X related protein (FXMR1), component 3 promoter of RISC (C3P0), Translin and Translin-associated factor-X (TRAX) proteins (Liang et al. 2015; Schvartzman et al. 2022).

At least 20 RISC complex-associated genes have been found to be expressed in *E. heros* and *P. guildinii*, including all the described above (Cagliari et al. 2020a; Schvartzman et al. 2022) (Table 8.2). The nucleases Armitage (ARMI), Homeless (SPINDLE-E), Maelstrom and HEN-1 were also identified in both species, in addition to methyltransferase PRMT, all involved in the piRNA pathway. However, only a partial sequence of HEN-1 was identified in *P. guildinii*, whereas no conserved domain was found for this protein in *E. heros* (Cagliari et al. 2020a, Schvartzman et al. 2022). Helicases p68, Gemin 3 homolog, Belle and PRP16 were also expressed

Table 8.2 Comparison of identified genes related to RISC in *Euschistus heros* and *Piezodorus guildinii*

		Euschistus heros		*Piezodorus guildinii*	
Gene ID	Homologue ID	Comparison	% Identity	Comparison	% Identity
Tudor-SN (TSN)	XP_014284230.1 – *H. halys*	E = 0.0 – bits = 2031	76	E = 0.0 – bits = 1660	75
Translin	XP_014290495.1 – *H. halys*	E = 1e-154 – bits = 434	85	E = 8.5e-162 – bits = 513	86
Similar to translin associated factor-X (TRAX)	XP_014289754.1 – *H. halys*	E= 3e-162 – bits = 456	85	E = 1.50e-160 – bits = 613	88
Armitage	XP_014289817.1 – *H. halys*	E = 0.0 – bits = 1110	96	E = 0.0 – bits = 2098	95
Homeless (SPINDLE-E)	XP_014286769.1 – *H. halys*	E = 0.0 – bits = 2707	89	E = 0.0 – bits = 2285	88
Maelstrom	XP_014290039.1 – *H. halys*	E = 0.0 – bits = 694	79	E = 0.0 – bits = 449	83
HEN1	XP_014284423.1 – *H. halys*	E = 0.0 bits = 656	70	E = 0.0 – bits = 1264	67
PRP16, mut6 homolog	XP_014279344.1 – *H. halys*	E = 0.0 – bits = 2423	96	E = 0.0 – bits = 2428	99
Clp1 homolog (kinase)	XP_014275582.1 – *H. halys*	E = 0.0 – bits = 2731	94	E = 0.0 – bits = 2821	96
Elongator complex protein 1 (ELP-1)	XP_014290480.1 – *H. halys*	E = 0.0 – bits = 2045	82	E = 0.0 – bits = 2104	85
GLD-1 homolog	XP_014290348.1 – *H. halys*	E = 0.0 – bits = 1073	87	E = 0.0 – bits = 1058	86
ACO-1 homolog	XP_014275296.1 – *H. halys*	E= 0.0 – bits = 1660	92	E = 0.0 – bits = 1709	94
Vasa intronic gene (VIG)	XP_014292052.1 – *H. halys*	E = 0.0 – bits = 644	96	E = 0.0 – bits = 782	97
Staufen (STAU)	XP_014282526.1 – *H. halys*	E = 0.0 – bits = 956	96	E = 0.0 – bits = 1323	96
RNA helicase Belle	XP_014279436.1 – *H. halys*	E = 0.0 – bits = 1377	97	E = 0.0 – bits = 340	98
Protein arginine methyltransferase 7 (PRMT)	XP_014292128.1 – *H. halys*	E = 0.0 – bits = 726	84	E = 0.0 – bits = 749	87
Gawky	XP_014288686.1 – *H. halys*	E = 0.0 – bits = 2803	97	E = 0.0 – bits = 2830	99
Similar to fragile X mental retardation syndrome related protein 1 (FXMR1)	XP_969396 E – *T. castaneum*	n.a	n.a	E = 0.0 – bits = 613	71
Gemin 3 homolog	EFA00789 – *T. castaneum*	n.a	n.a	E = 0.0 – bits = 427	50
p68 Helicase	NP_001164095 – *T. castaneum*	n.a	n.a	E = 0.0 bit = 688	74

in *E. heros*, but only a partial sequence of PRP16 was identified in *P guildinii* (Cagliari et al. 2020a, Schvartzman et al. 2022).

Moreover, genes coding for Glucose dehydrogenase (GLD-1), Cytoplasmic aconitate hydratase (ACO-1), Gawky and Staufen (STAU), which acts in mRNA-associated pathways, and Elongator complex protein 1 (ELP-1), involved in siRNA pathway, are also expressed in both species. Therefore, in addition to having the complete set RNAi core machinery genes, these Pentatomidae species also possess a complete set of RISC complex-associated genes, favoring the possibility of a robust RNAi effect. In Marshall et al. (2023), using transcriptome data from salivary glands, no transcripts for STAU were identified in the species *N. viridula, C. ligata, E. conspersus* and *T. pallidovirens*, but they were found in *H. halys.* It will be interesting to search for STAU homologues in whole-body transcriptome analyses in these insects.

2.3 dsRNA Uptake

The efficiency of RNAi in pest control depends on the uptake of dsRNA (Fig. 8.3), available in the environment or plants, by the cells of the target organism (Joga et al. 2016; Wytinck et al. 2020). The internalization of dsRNA by cells usually occurs through two different pathways in insects: the transmembrane channel-mediated system and endocytosis (Joga et al. 2016). The first pathway involves two proteins identified in *Caenorhabditis elegans*, the systemic RNAi defective 1 and 2 (SID-1 and 2). SID-1 is expressed in all cells sensitive to systemic RNAi and acts cell-autonomously as a channel to transport dsRNA across cell membranes (Li et al. 2015), having orthologues described in a variety of insects (Vélez and Fishilevich 2018). SID-2 is expressed in the intestine as an endocytic receptor for ingested dsRNA and is involved in the environmental RNAi (Braukmann et al. 2021; Shih and Hunter 2011). However, no SID-2 was reported in any insect order to date (Hinas et al. 2012). Regarding the endocytic pathway, it involves scavenger receptors (SRs), a major class of pattern recognition receptors (PRRs) (Saleh et al. 2006; Joga et al. 2016). PRRs recognize molecules, usually found in pathogens or damaged cells (Li and Wu 2021), and act allowing for cellular uptake of dsRNA through receptor-mediated endocytosis (Saleh et al. 2006, Joga et al. 2016).

At least nine proteins involved in dsRNA uptake have been found to be expressed in *E. heros* and *P. guildinii* (Table 8.3) (Cagliari et al. 2020a; Schvartzman et al. 2022). For both species, no *SID* or *SID*-like genes were found, compatible with the lack of *SID* genes in *H. halys* (Sparks et al. 2014) and *M. histrionica* (Sparks et al. 2017) and corroborating the hypothesis that Pentatomidae species do not possess SID-1 channel protein-mediated uptake. The absence of SID-mediated uptake is also described in Diptera species (Tomoyasu et al. 2008).

Table 8.3 Comparison of identified genes related to dsRNA uptake in *Euschistus heros* and *Piezodorus guildinii*

Gene ID	Homologue ID	*Euschistus heros*		*Piezodorus guildinii*	
		Comparison	% Identity	Comparison	% Identity
Scavenger	XP_024218066.1 – *H.halys*	E = 0.0 – bits = 1039	96	E = 0.0 – bits = 1026	96
CG4966 = orthologous to the Hermansky Pudlak Syndrome4 (HSP4)	XP_014288755.1 – *H.halys*	E =0 .0 – bits = 1271	90	E = 0.0 – bits = 897	89
F-box protein 11 (FBX011)	XP_014287303.1 – *H.halys*	E = 0.0 – bits = 179	99	E = 0.0 – bits = 1799	99
Clathrin heavy chain (Chc)	XP_014287090.1 – *H.halys*	E = 0.0; bits = 3477	99	E = 0.0 – bits = 3485	99
AP2u (Ap50)	NP_001280510.1 – *T. castaneum*	n.a	n.a	E = 0.0 bits = 866	94
ADP-ribosylation factor-like protein 1 (ARL1)	EFA02719.2 – *T. castaneum*	n.a	n.a	E = 7.14e-115 bits = 323	85
Eater	XP_969372 – *T. castaneum*	n.a	n.a	E = 7.57e-47 – bits	37
Epsin 2 (EPN2)	XP_014270392.1 – *H. halys*	E = 0.0 – bits = 900	92	E = 7.57e-47 – bits	91
Gap Junction protein (INNEXIN2)	XP_014292574.1 – *H. halys*	E = 0.0 – bits = 736	60	E = 0.0 – bits = 1026	99

In contrast, the genes related to receptor-mediated endocytosis, such as *Scavenger*, *Eater*, *HSP4*, *Clathrin heavy chain* (*CHC*), *AP2u* (*AP50*), *F-box protein 11* (*FBX011*), *ADP-ribosylation factor-like protein 1* (*ARL1*), *Epsin 2* (*EPN2*) and *Gap Junction protein* (*INNEXIN2*) were found to be expressed in *E. heros* and *P. guildinii* (Cagliari et al. 2020a; Schvartzman et al. 2022). Except for AP50, ARL1 and Eater, not discussed in Sparks et al. (2014), all other receptor-mediated endocytosis genes (*E. heros* and *P. guildinii*) showed >90% similarity to *H. halys*. Therefore, SR-mediated endocytosis is most likely the main pathway to dsRNA uptake in Pentatomidae.

2.4 Nucleases, Antiviral and Intracellular Transport

The success of the RNAi technology in pest control also depends on cellular factors and processes capable of degrading or inactivating the RNAi mechanism in the cell (Joga et al. 2016; Cooper et al. 2019). The nuclease-related proteins exoribonuclease 1 (ERI-1), related to siRNA and miRNA pathway (Thomas et al. 2014), Nibbler, responsible for 3′-end trimming of miRNA and piRNA pathway (Xie et al. 2020) and Small RNA degrading nuclease 1 (SDN1-like), with a role in the degradation of mature miRNA (Ramachandran and Chen 2008) were found with conserved domains present in both species (Cagliari et al. 2020a, Schvartzman et al. 2022) (Table 8.4). In addition, poly(A)polymerase protein, involved in the regulation of mRNA abundance and degradation (Yamanaka et al. 2012; Hajnsdorf and Kaberdin 2018), and exosome complex exonuclease, involved in a variety of RNA degradation processes, including interaction with poly(A)polymerase (Lykke-Andersen et al. 2009; Wasmuth et al. 2014), were also found to be expressed in *P. guildinii* (Schvartzman et al. 2022). DNA/RNA non-specific endonuclease (dsRNase) was identified in *E. heros* transcriptome (seven isoforms), whereas only two isoforms were found in *P. guildinii* (Table 8.4). All dsRNase isoforms presented a DNA/RNA non-specific endonuclease (ENDONUCLEASE_NS) conserved domain, with only one isoform displaying no nuclease conserved domain, for both species (Cagliari et al. 2020a, Schvartzman et al. 2022). Based on transcriptome analysis of salivary gland samples, no RNAse homologs were found in *H. halys*, *N. viridula* and *T. pallidovirens*, while RNAse homologues were identified in *C. ligata* and *E. conspersus* (Marshall et al. 2023). It would be interesting to evaluate the success rate of the RNAi technology in these organisms correlating RNAi success with the differential expression of RNAses in these species.

Regarding antiviral RNAi response, four genes were found to be expressed in both species: *Arsenite resistance gene 2* (*ARS2*), involved in production of siRNAs and RISC complex regulation (Chen et al. 2012); *NINAC*, involved in cytoskeleton organization and vesicle transport (Saleh et al. 2006); *Beta 1,4-mannosyltransferase (EGH*), which encodes a seven-transmembrane

Table 8.4 Comparison of identified genes related to dsRNA uptake in *Euschistus heros* and *Piezodorus guildinii*

		Euschistus heros		*Piezodorus guildinii*	
Gene ID	Homologue ID	Comparison	% Identity	Comparison	% Identity
Nucleases					
Exoribonuclease 1 (ERI1)	XP_014290344.1 – *H. halys*	E = 0.0 – bits = 2701	83	E = 0.0 – bits = 1026	84
DNA/RNA non-specific endonuclease isoform 1	XP_024218583.1 – *H. halys*	E = 6e-172 – bits = 490	83	E = 1.03e-173 – bits = 494	82
DNA/RNA non-specific endonuclease isoform 3	XP_014293261.1 – *H. halys*	E = 2e-18 – bits = 83.6	56	E = 0.0 – bits = 687	74
DNA/RNA non-specific endonuclease isoform 4	XP_024218583.1 – *H. halys*	E = 8e-173 – bits = 486	85	n.a	n.a
DNA/RNA non-specific endonuclease isoform 6	XP_024218583.1 – *H. halys*	E = 4e-170 – bits = 486	83	n.a	n.a
DNA/RNA non-specific endonuclease isoform 7	XP_024218583.1 – *H. halys*	E = 4e-170 – bits = 486	83	n.a	n.a
DNA/RNA non-specific endonuclease isoform 9	XP_024218583.1 – *H. halys*	E = 5e-170 – bits = 486	83	n.a	n.a
DNA/RNA non-specific endonuclease isoform 10	XP_024218583.1 – *H. halys*	E = 7e-172 – bits = 490	83	n.a	n.a
Small RNA degrading nuclease 1 (SDN1-like)	XP_014279339.1 – *H. halys*	E = 0.0 – bits = 895	75	E = 0.0 – bits = 871	75
Exosome	XP_014288410.1 – *H. halys*	n.a	n.a	E = 0.0 – bits = 1,845	95
Poly(A)polymerase	EFA00912 – *T. castaneum*	n.a	n.a	E = 0.0 – bits = 1,026	64
Nibbler	XP_024216394.1 – *H. halys*	E = 0.0 – bits = 1402	84	E = 0.0 – bit = 1424	86
Antiviral					
ARS2	XP_014277995.1 – *H. halys*	E = 0.0 – bits = 1523	98	E = 0.0 – bits = 1,507	95
NINAC	XP_014281724.1 – *H. halys*	E = 0.0 – bits = 525	95	E = 0.0 – bits = 1,097	95
Beta 1,4-mannosyltransferase (EGH)	XP_014283435.1 – *H. halys*	E = 0.0 – bits = 914	97	E = 0.0 – bits = 918	97
CG4572	XP_014280828.1 – *H. halys*	E = 0.0 – bits = 857	89	E = 0.0 – bits = 870	90
Intracellular Transport					
Vacuolar H+ ATPase sub unit A (VHA68)	XP_014272529.1 – *H. halys*	E = 0.0 – bits = 1250	99	E = 0.0 – bits = 1256	99
Vacuolar H+ ATPase sub unit C (VHA16)	XP_014275063.1 – *H. halys*	E = 4e-100 – bits = 289	99	1.13e-102 bits = 298	98
Small Rab GTPases	XP_014286452.1 – *H. halys*	E = 3e-152 – bits = 425	99	4.24e-154 bits = 425	99

domains glycosyltransferase with role in dsRNA uptake and antiviral immunity (Saleh et al. 2009; Kemp et al. 2013); and *CG4572*, which codes for a protein with role in RNAi signal spread (Saleh et al. 2009; Karlikow et al. 2016).

Finally, three genes coding for proteins involved in intracellular vesicle transport, endosome acidification and dsRNA release by the endocytic vesicles were found to be expressed in both species: Vacuolar H + ATPase subunit A (VHA68), *Vacuolar H + ATPase subunit C (VHA16)* and *Small Rab GTPases*, respectively.

It is well described that the presence of nucleases responsible for dsRNA degradation can negatively influence RNAi efficiency (Spit et al. 2017; Guan et al. 2018; Fan et al. 2021; Zhang et al. 2022). Therefore, protection against dsRNA-degrading nuclease (dsRNase) and dsRNA degradation by environmental factors are key factors in achieving efficient gene silencing through RNAi in stink bugs (Sharma et al. 2021b).

3 Target Genes and Biological Processes

RNAi, as a tool for functional genomics, has significantly advanced studies establishing correlations between phenotype and gene function. As the potential of RNAi in agricultural pest management became evident, studies have focused on inducing mortality by knocking down genes playing essential roles in pest insects and pathogens. Therefore, the choice of the target genes has become the initial and essential step of research and development of RNAi projects and demands a thorough understanding of the biology of the target organism.

The selection of target genes can be performed by a candidate gene approach, selecting genes that are already known to play key roles in essential biological processes in other organisms, RNAi screening, usually performed in model organisms, in which genes with unknown function are silenced in large-content RNAi screenings (Dönitz et al. 2015; Schmitt-Engel et al. 2015) or by genome-wide *in silico* analyzes, in which bioinformatic tools allow for the prediction of gene function by their transcript abundance in a specific stage, tissue or condition (de Camargo et al. 2015). In all cases, understanding the target organism's biology and interaction with its environment becomes very important for target gene selection. For example, a favorable target gene may not necessarily lead to rapid and high levels of mortality but can significantly impede the insect's ability to survive in the environment or to reproduce.

Numerous genes, either directly or indirectly, can contribute to this effect. These candidates are involved with different biological processes, such as behavior, physiology, morphology, reproduction, development, microbiome, and immunity, as discussed below (Table 8.5).

Table 8.5 Summary of all studies cited on the manuscript using RNAi in Pentatomidae species cited on the manuscript

Gene	Process	Specie	Phenotype	Stage	Delivery method	Concentration	Citation
ACT-1	**Physiology and survival**	*Euschistus heros*	High mortality rate	2nd instar nymphs	Microinjection	55.2 nL (500 ng/µl)	Fishilevich et al. (2016)
ACT-2		*Euschistus heros*	High mortality rate	2nd instar nymphs	Microinjection	55.2 nL (500 ng/µl)	Fishilevich et al. (2016)
			High mortality rate	2nd instar nymphs	Microinjection	54.75 nL (500 ng/µl)	Castellanos et al. (2018)
			Increase in mortality	2nd instar nymphs	Oral	40 µL (320 ng/µL^{-1})	Castellanos et al. (2018)
			Increase in mortality	2nd instar nymphs	Oral (dsRNA + liposome)	30 µL (300 ng/µL^{-1})	Castellanos et al. (2018)
			No significant effect	2nd instar nymphs	Oral (dsRNA + EDTA)	30 µL (300 ng/µL^{-1})	Castellanos et al. (2018)
ACT-5C		*Halyomorpha halys*	Increase in mortality	Adults	Microinjection	5 µl (0,2 µg /µl)	Mogilicherla et al. (2018)
PROSA2		*Euschistus heros*	High mortality rate	2nd instar nymphs	Microinjection	54.75 nL (500 ng/µl)	Castellanos et al. (2018)
SPR54K		*Euschistus heros*	High mortality rate	2nd instar nymphs	Microinjection	54.75 nL (500 ng/µl)	Castellanos et al. (2018)
TAF-1		*Euschistus heros*	High mortality rate	2nd instar nymphs	Microinjection	54.75 nL (500 ng/µl)	Castellanos et al. (2018)
RPN7		*Euschistus heros*	High mortality rate	2nd instar nymphs	Microinjection	54.75 nL (500 ng/µL)	Castellanos et al. (2018)
SYB		*Euschistus heros*	High mortality rate	2nd instar nymphs	Microinjection	54.75 nL (500 ng/µL)	Castellanos et al. (2018)
GW		*Euschistus heros*	Increase in mortality	2nd instar nymphs	Microinjection	54.75 nL (500 ng/µL)	Castellanos et al. (2018)
CACT		*Euschistus heros*	Increase in mortality	2nd instar nymphs	Microinjection	54.75 nL (500 ng/µL)	Castellanos et al. (2018)
INR-A		*Euschistus heros*	No significantly effect	2nd instar nymphs	Microinjection	54.75 nL (500 ng/µL)	Castellanos et al. (2018)

PDF		*Plautia stali*	Impact in circadian clock	Adult females	Microinjection	1 μL (1 μg/μL)	Hasebe et al. (2022)
ATPSYNB		*Euschistus heros*	Increase in mortality	2nd instar nymphs	Microinjection	54.75 nL (500 ng/μL)	Castellanos et al. (2018)
ATPASE 1		*Nezara viridula*	No significant effect	Adults	Microinjection	2 μL (0,5 μg/μl)	Gurusamy et al. (2021)
VATPASE A		*Euschistus heros*	High mortality rate	2nd instar nymphs	Microinjection	54.75 nL (500 ng/μL)	Castellanos et al. (2018)
			Increase in mortality	2nd instar nymphs	Oral	40 μL ($320\ ng/\mu L^{-1}$)	Castellanos et al. (2018)
			Increase in mortality	2nd instar nymphs	Oral (dsRNA + liposome)	30 μL ($300\ ng/\mu L^{-1}$)	Castellanos et al. (2018)
			Increase in mortality	2nd instar nymphs	Oral (dsRNA + EDTA)	30 μL ($300\ ng/\mu L^{-1}$)	Castellanos et al. (2018)
			Increase in mortality	Adults	Microinjection	0.5 μl (3350 ng/μL)	Cagliari et al. (2020)
		Nezara viridula	Increase in mortality and immobility of legs and antennae	2nd instar nymphs	Microinjection	67 nL (1 μg/μL)	Sharma et al. (2021)[a]
			Increase in mortality and reduction in nymphal weight	2nd instar nymphs	Oral	30 μL (300 ng/μL)	Sharma et al. (2021)[a]
		Piezodorus guildinii	Increase in mortality	Adults	Microinjection	0,5 Ml (2.4 μg/μL)	Schvartzman et al. (2022)
VATPASE D		*Plautia stali*	No significant effect	1st instar nymphs	Oral	800 μL (100, 1000 or 5000 ng/μL)	Nishide et al. (2021)
			No significant effect	5th instar nymphs	Oral (plant-mediated)	1.6 mL (67 ng/μL)[2]	Nishide et al. (2021)
			High mortality rate	5th instar nymphs	Microinjection	3 μL (30-300 ng)	Nishide et al. (2021)
VATPASE H		*Plautia stali*	No significant effect	1st instar nymphs	Oral	800 μL	Nishide et al. (2021)

(continued)

Tab. 8.5 (continued)

						(100, 1000 or 5000 ng/μL)	
			High mortality rate	5th instar nymphs	Microinjection	3 μL (30-300 ng)	Nishide et al. (2021)
VATPASE E		*Plautia stali*	No significant effect	1st instar nymphs	Oral	800 μL (100, 1000 or 5000 ng/μL)	Nishide et al. (2021)
			High mortality rate	5th instar nymphs	Microinjection	3 μL (30-300 ng)	Nishide et al. (2021)
ATPASE N2B		*Murgantia histrionica*	Increase in mortality	2nd instar nymphs	Oral (plant-mediated)	0.3 mL (0.067 μg/μL)	Howell et al. (2020)
			Increase in mortality	Adults	Microinjection	1 μg	Howell et al. (2020)
		Halyomorpha halys	Increase in mortality	2nd instar nymphs	Topically	1 μL (0,5, 1, 2 or 3 μg/μL	Finetti et al. (2023)
			High mortality rate	Adults	Microinjection	5 μl (0,2 μg /μl)	Mogilicherla et al. (2018)
			Slightly increase in mortality	2nd instar nymphs	Oral (Plant-mediated)	300 μl (0.067 μg/μL)	Mogilicherla et al. (2018)
SNF7		*Plautia stali*	High mortality rate	5th instar nymphs	Microinjection	3 μL (30-300 ng)	Nishide et al. (2021)
			No significant effect	1st instar nymphs	Oral	800 μL (100, 1000 or 5000 ng/μL)	Nishide et al. (2021)
		Nezara viridula	High mortality rate	Adults	Microinjection	2 μL (0.5 μg/ μL)	Gurusamy et al. (2021)
			Increase in mortality	Adults	Oral	5 μL (1 μg/ μL)	Gurusamy et al. (2021)
			Increase in mortality	3rd instar nymphs	Oral	5 μL (1 μg/ μL)	Gurusamy et al. (2021)
		Halyomorpha halys	High mortality rate	Adults	Microinjection	5 μl (0,2 μg /μl)	Mogilicherla et al. (2018)

			High mortality rate	2nd instar nymphs	Oral (Plant-mediated)	300 µl (0.067 µg/µL)	Mogilicherla et al. (2018)
PP1A96A		*Euschistus heros*	High mortality rate	2nd instar nymphs	Microinjection	54.75 nL (500 ng/µl)	Castellanos et al. (2018)
PP1		*Halyomorpha halys*	High mortality rate	Adults	Microinjection	5 µl (0,2 µg /µl)	Mogilicherla et al. (2018)
			High mortality rate	2nd instar nymphs	Oral (Plant-mediated)	300 µl (0.067 µg/µL)	Mogilicherla et al. (2018)
			Non-significant increase in mortality	2nd instar nymphs	Topically	1 µL (0,5, 1, 2 or 3 µg/µL)	Finetti et al. (2023)
		Murgantia histrionica	High mortality rate	Adults	Microinjection	1 µg	Howell et al. (2020)
			Increase in mortality	2nd instar nymphs	Oral (plant-mediated)	0.3 mL (0.067 µg/µL)	Howell et al. (2020)
		Nezara viridula	High mortality rate	2nd instar nymphs	Microinjection	67 nL (1 µg/µL)	Sharma et al. (2021)[a]
			Slight increase in mortality	2nd instar nymphs	Oral	30 µL (300 ng/µL)	Sharma et al. (2021)[a]
			High mortality rate	Adults	Microinjection	2 µL (0,5 µg/µl)	Gurusamy et al. (2021)
SRP		*Halyomorpha halys*	Increase in mortality	Adults	Microinjection	5 µl (0,2 µg /µl)	Mogilicherla et al. (2018)
		Murgantia histrionica	Increase in mortality	Adults	Microinjection	1 µg	Howell et al. (2020)
RPS13		*Euschistus heros*	Increase in mortality	2nd instar nymphs	Microinjection	54.75 nL (500 ng/µL)	Castellanos et al. (2018)
40SRPS13		*Nezara viridula*	High mortality rate	2nd instar nymphs	Microinjection	67 nL (1 µg/µL)	Sharma et al. (2021)[a]
			Slight increase in mortality	2nd instar nymphs	Oral	30 µL (300 ng/µL)	Sharma et al. (2021)[a]
60SRPL19		*Nezara viridula*	High mortality rate	2nd instar nymphs	Microinjection	67 nL	Sharma et al.

(continued)

Tab. 8.5 (continued)

						(1 μg/μL)	(2021)[a]
			Slight increase in mortality	2nd instar nymphs	Oral	30 μL (300 ng/μL)	Sharma et al. (2021)[a]
MCO2		*Plautia stalii*	High mortality rate. Abnormal development.	5th instar nymphs	Microinjection	3 μL (30-300 ng)	Nishide et al. (2021)
			No significant effect	5th instar nymphs	Oral	6 mL (100 or 1000 ng/μL)	Nishide et al. (2021)
			No significant effect	1st instar nymphs	Oral	800 μL (100, 1000 or 5000 ng/μL)	Nishide et al. (2021)
26S		*Halyomorpha halys*	No significant effect	Adults	Microinjection	5 μl (0,2 μg /μl)	Mogilicherla et al. (2018)
26SPPRS6B		*Nezara viridula*	High mortality rate	2nd instar nymphs	Microinjection	67 nL (1 μg/μL)	Sharma et al. (2021)[a]
GPCR		*Nezara viridula*	Slight increase in mortality	Adults	Microinjection	2 μL (0,5 μg/μl)	Gurusamy et al. (2021)
		Halyomorpha halys	High mortality rate	Adults	Microinjection	5 μl (0,2 μg /μl)	Mogilicherla et al. (2018)
			Increase in mortality	2nd instar nymphs	Oral (Plant-mediated)	300 μl (0.067 μg/μL)	Mogilicherla et al. (2018)
		Murgantia histrionica	No significant effect	Adults	Microinjection	1 μg	Howell et al. (2020)
			Increase in mortality	2nd instar nymphs	Oral (plant-mediated)	0.3 mL (0.067 μg/μL)	Howell et al. (2020)
αCOP		*Nezara viridula*	High mortality rate	2nd instar nymphs	Microinjection	67 nL (1 μg/μL)	Sharma et al. (2021)[a]
			Increase in mortality. Reduction in nymphal weight.	2nd instar nymphs	Oral	30 μL (300 ng/μL)	Sharma et al. (2021)[a]
			Increase in mortality when dsRNA was	2nd instar nymphs	Oral	300 μg/μL	Sharma et al. (2021)[b]

			administered after silence of dsRNAse				
βCOP		*Euschistus heros*	High mortality rate	2nd instar nymphs	Microinjection	54.75 nL (500 ng/μL)	Castellanos et al. (2018)
γCOP		*Euschistus heros*	High mortality rate	2nd instar nymphs	Microinjection	54.75 nL (500 ng/μl)	Castellanos et al. (2018)
NSF		*Halyomorpha halys*	Increase in mortality	Adults	Microinjection	5 μl (0,2 μg /μl)	Mogilicherla et al. (2018)
KENESIN		*Halyomorpha halys*	Slight increase in mortality	Adults	Microinjection	5 μl (0,2 μg /μl)	Mogilicherla et al. (2018)
DYNAMIN		*Halyomorpha halys*	Slight increase in mortality	Adults	Microinjection	5 μl (0,2 μg /μl)	Mogilicherla et al. (2018)
ROP		*Halyomorpha halys*	No significant effect	Adults	Microinjection	5 μl (0,2 μg /μl)	Mogilicherla et al. (2018)
HSP70		*Halyomorpha halys*	No significant effect	Adults	Microinjection	5 μl (0,2 μg /μl)	Mogilicherla et al. (2018)
TRCTFC		*Nezara viridula*	No significant effect	2nd instar nymphs	Microinjection	67 nL (1 μg/μL)	Sharma et al. (2021)[a]
ABCA1		*Nezara viridula*	Increase in mortality	2nd instar nymphs	Microinjection	67 nL (1 μg/μL)	Sharma et al. (2021)[a]
ABCG4		*Nezara viridula*	Increase in mortality	2nd instar nymphs	Microinjection	67 nL (1 μg/μL)	Sharma et al. (2021)[a]
USP		*Nezara viridula*	High mortality rate	2nd instar nymphs	Microinjection	67 nL (1 μg/μL)	Sharma et al. (2021)[a]
TAR1		*Halyomorpha halys*	Decreased sensitivity towards a repellence hormone ((E)-2-decenal)	2nd instar nymphs	Topically	1 μL (100 ng/μl)	Finetti et al. (2021)
TPS		*Murgantia histrionica*	Reduction in murgantiol emission	Male adults	Microinjection	400 ng	Lancaster et al. (2018)
TROPONIN		*Euschistus heros*	Increase in mortality	2nd instar nymphs	Oral	250 ng/μL (shRNA)	Laisney et al. (2021)

(continued)

Tab. 8.5 (continued)

						or 1 ug/µL (dsRNA)	
CAT		*Halyomorpha halys*	No significant effect	Adults	Microinjection	414 nL (500 ng)	Bansal et al. (2016)
IAP		*Murgantia histrionica*	Increase in mortality	Adults	Microinjection	1 µg	Howell et al. (2020)
			Increase in mortality	2nd instar nymphs	Oral (plant-mediated)	0.3 mL (0.067 µg/µL)	Howell et al. (2020)
			Increase in mortality	2nd-3rd instar nymphs	Oral	5 µl (1 µg /µl)	Howell et al. (2020)
		Halyomorpha halys	High mortality rate	Adults	Microinjection	5 µl (0,2 µg /µl)	Mogilicherla et al. (2018)
			High mortality rate	2nd instar nymphs	Oral (Plant-mediated)	300 µl (0.067 µg/µL)	Mogilicherla et al. (2018)
			Increase in mortality	2nd instar nymphs	Topically	1 µL (0,5, 1, 2 or 3 µg/µL)	Finetti et al. (2023)
		Plautia stali	High mortality rate	5th instar nymphs	Microinjection	3 µL (30-300 ng)	Nishide et al. (2021)
			Slight increase in mortality rate	1st instar nymphs	Oral	800 µL (100, 1000 or 5000 ng/µL)	Nishide et al. (2021)
			No significant effect	5th instar nymphs	Oral (Plant-mediated)	1.6 mL (67 ng/µL)	Nishide et al. (2021)
IAP1		*Nezara viridula*	Slight increase in mortality	Adults	Microinjection	2 µL (0,5 µg/µl)	Gurusamy et al. (2021)
IAP2		*Nezara viridula*	Increase in mortality	Adults	Microinjection	2 µL (0,5 µg/µl)	Gurusamy et al. (2021)
IAP5		*Nezara viridula*	High mortality rate	Adults	Microinjection	2 µL (0,5 µg/µl)	Gurusamy et al. (2021)
PENTATOMICIN	**Reproduction**	*Plautia stali*	No significant effect in embryonic	Adult females	Microinjection	3 µL (100 ng/µL))	Nishide et al. (2022)

			development and hatching rates				
CYC		*Plautia stali*	Absence of rhythmic activity and suppression of ovarian development under long-day conditions	Adult females	microinjection	1 µL (1 µg/µL)	Tamai et al. (2018)
MIP		*Plautia stali*	Suppression of ovarian development under short-day conditions	Adult females	microinjection	1 µL (1 µg/µL)	Tamai et al. (2018)
PDF		*Plautia stali*	Inhibition of oviposition arrest or delayed oviposition, induced by the photoperiod	Adult females	microinjection	1 µL (1 µg/µL)	Hasebe et al. (2022)
DFD		*Euschistus heros*	Reduction in the number of fertilized eggs laid; High reduction in egg hatching rate; Abnormal development	Adult females	Microinjection	2.5 µL (4 µg/µl)	Cagliari et al. (2021)
LAB		*Euschistus heros*	High reduction in egg hatching rate	Adult females	Microinjection	2.5 µL (4 µg/µl)	Cagliari et al. (2021)
EXD		*Euschistus heros*	High reduction in egg hatching rate and abnormal development	Adult females	Microinjection	2.5 µL (4 µg/µl)	Cagliari et al. (2021)
MI-2		*Euschistus heros*	High mortality rate; High reduction in egg oviposition rate	Adult females	Microinjection	138 nL (1 µg/µl)	Fishilevich et al. (2016)

(continued)

Tab. 8.5 (continued)

ISWI-1		*Euschistus heros*	Slight increase in mortality rate; Reduction in egg oviposition and hatching rates	Adult females	Microinjection	138 nL (1 µg/µl)	Fishilevich et al. (2016)
ISWI-2		*Euschistus heros*	Increase in mortality rate; Slight reduction in egg oviposition rate; High reduction in egg hatching rate	Adult females	Microinjection	138 nL (1 µg/µl)	Fishilevich et al. (2016)
CHD1		*Euschistus heros*	Increase in mortality rate; Reduction in egg oviposition rate High reduction in egg hatching rate	Adult females	Microinjection	138 nL (1 µg/µl)	Fishilevich et al. (2016)
BRAHMA		*Euschistus heros*	No significant effect	2nd instar nymphs	Microinjection	55.2 nL (500 µg/µl)	Fishilevich et al. (2016)
			High mortality rate; Total impairment of egg oviposition	Adult females	Microinjection	138 nL (1 µg/µl)	Fishilevich et al. (2016)
VG		*Halyomorpha halys*	No significant effect	4th instar nymphs	Oral (Plant-mediated)	0.017 µg/µL	Ghosh et al. (2017)
			No significant effect	4th instar nymphs	Oral (Plant-mediated)	0.067 µg/µl	Ghosh et al. (2017)
MCO2		*Plautia stalii*	High mortality rate and abnormal development	5th instar nymphs	Microinjection	3 µL (30-300 ng)	Nishide et al. (2021)
			No significant effect	5th instar nymphs	Oral	6 mL (100 or 1000 ng/µL)	Nishide et al. (2021)
			No significant effect	1st instar nymphs	Oral	800 µL	Nishide et al. (2021)

						(100, 1000 or 5000 ng/μL)	
SCR		*Euschistus heros*	High reduction in egg hatching rate and abnormal development	Adult females	Microinjection	2.5 μL (4 μg/μl)	Cagliari et al. (2021)
		Halyomorpha halys	Death prior to hatching. Abnormal development	Adult females	Microinjection	2-6 μL (3 μg/μl)	Lu et al. (2017)
YEL	**Development, Morphology and Physical characteristics**	*Euschistus heros*	Increase in mortality	3rd instar nymphs	Microinjection	1 μL (1μg/μL)	Cagliari et al. (2020)
		Diceraeus melacanthus	No visual effects discussed	3rd instar nymphs	Oral	84 ug	Pinheiro et al. (2020)
WHITE		*Nezara viridula*	Alteration in coloring tone and translucent integuments. Alteration in eyes and ocelli color; Alteration in intern organs color. Paler color in eggs and nymphs from treated females	Adults	Microinjection	3 μl (50 ng/μl)	Souza et al. (2022)
			Distinct body colors, translucent integuments and alteration in eyes and ocelli color in newly emerged adults; Alteration in intern organs color	5th instar nymphs	Microinjection	2 μl (75 ng/μl)	Souza et al. (2022)
JHAMT		*Halyomorpha halys*	No visual effect	4th instar nymphs	Oral (Plant-mediated)	0.017 μg/μL	Ghosh et al. (2017)

(continued)

Tab. 8.5 (continued)

			No visual effect	4th instar nymphs	Oral (Plant-mediated)	0.067 μg/μl	Ghosh et al. (2017)
EXD		*Euschistus heros*	Impaired development in embryos, with absence of appendage structures	Adult females	Microinjection	2.5 μL (4 μg/μl)	Cagliari et al. (2021)
PB		*Euschistus heros*	Malformed rostrum, with presence of a leg-like structure. High mortality due to feeding impediment	Adult females	Microinjection	2.5 μL (4 μg/μl)	Cagliari et al. (2021)
DFD		*Euschistus heros*	Abnormal structures in head* *Only one of 13 nymphs	Adult females	Microinjection	2.5 μL (4 μg/μl)	Cagliari et al. (2021)
MCO1		*Plautia stali*	No significant effect	3rd- 5th instar nymphs	Microinjection	1-3 μL (100 ng/μL)	Nishide et al. (2020)
			No significant effect	Adult females	Microinjection	5 μL (100 ng/μL)	Nishide et al. (2020)
MCO2		*Plautia stali*	Defect in cuticle sclerotization and pigmentation	5th instar nymphs	Microinjection	3 μL (30-300 ng)	Nishide et al. (2021)
			Increase in mortality of 3rd instar nymphs and abnormal development. High mortality rate in 5th instar nymphs	3rd- 5th instar nymphs	Microinjection	1-3 μL (100 ng/μL)	Nishide et al. (2020)
			High hatching failure rates	Adult females	Microinjection	5 μL (100 ng/μL)	Nishide et al. (2020)
MCO3		*Plautia stali*	No significant effect	3rd- 5th instar nymphs	Microinjection	1-3 μL (100 ng/μL)	Nishide et al. (2020)
			No significant effect	Adult females	Microinjection	5 μL	Nishide et al.

						(100 ng/μL)	(2020)
MCO4		*Plautia stali*	No significant effect	3rd- 5th instar nymphs	Microinjection	1-3 μL (100 ng/μL)	Nishide et al. (2020)
			No significant effect	Adult females	Microinjection	5 μL (100 ng/μL)	Nishide et al. (2020)
MCO5		*Plautia stali*	No significant effect	3rd- 5th instar nymphs	Microinjection	1-3 μL (100 ng/μL)	Nishide et al. (2020)
			No significant effect	Adult females	Microinjection	5 μL (100 ng/μL)	Nishide et al. (2020)
MCO6		*Plautia stali*	No significant effect	3rd- 5th instar nymphs	Microinjection	1-3 μL (100 ng/μL)	Nishide et al. (2020)
			No significant effect	Adult females	Microinjection	5 μL (100 ng/μL)	Nishide et al. (2020)
MCO7		*Plautia stali*	No significant effect	3rd- 5th instar nymphs	Microinjection	1-3 μL (100 ng/μL)	Nishide et al. (2020)
			No significant effect	Adult females	Microinjection	5 μL (100 ng/μL)	Nishide et al. (2020)
TYR1		*Plautia stali*	No significant effect	Adult females	Microinjection	5 μL (100 ng/μL)	Nishide et al. (2020)
TYR2		*Plautia stali*	No significant effect	Adult females	Microinjection	5 μL (100 ng/μL)	Nishide et al. (2020)
TYR3		*Plautia stali*	No significant effect	Adult females	Microinjection	5 μL (100 ng/μL)	Nishide et al. (2020)
SC MUCIN		*Plautia stali*	Reduction in terminal crypts	Adult females	Microinjection	1 μL (200 ng/μL)	Moriyama et al. (2022)
ANT1		*Plautia stali*	Increase in mortality rate; Failure in ecdysis	3rd instar nymphs	Microinjection	0.3 μL (2.5 μg/μL)	Sugahara et al. (2016)
ANT2		*Plautia stali*	No significant effect	3rd instar nymphs	Microinjection	0.3 μL (2.5 μg/μL)	Sugahara et al. (2016)
PIC1		*Plautia stali*	Increase in nymphal death during ecdysis;	3rd instar nymphs	Microinjection	0.3 μL (2.5 μg/μL)	Sugahara et al. (2017)

(continued)

Tab. 8.5 (continued)

			Decreased eclosion rates				
PIC2		*Plautia stali*	Reduction of eclosion rates	3rd instar nymphs	Microinjection	0.3 µL (2.5 µg/µL)	Sugahara et al. (2017)
AWD		*Euschistus heros*	Increase in mortality. Abnormalities in wing of adults	3rd instar nymphs	Microinjection	1 µL (1µg/µL)	Cagliari et al. (2020)
TH		*Euschistus heros*	High mortality rate; High rates of defects in cuticle sclerotization and color, abdomen, antenna, and legs	3rd instar nymphs	Microinjection	1 µL (1µg/µL)	Cagliari et al. (2020)
EBONY		*Diceraeus melacanthus*	No visual effect discussed	3rd instar nymphs	Oral	84 ug	Pinheiro et al. (2020)
SCR		*Euschistus heros*	Malformed rostrum, with presence of a leg-like structure. High mortality due to feeding impediment	Adult females	Microinjection	2.5 µL (4 µg/µl)	Cagliari et al. (2021)
		Halyomorpha halys	Lack or malformation of mouthparts, with presence of a leg-like structure. Smaller bodies	Adult females	Microinjection	2-6 µL (3 µg/µl)	Lu et al. (2017)
Catalase	**Immunity and symbiont microbiota**	*Halyomorpha halys*	No significant effect described	Adults	Microinjection	414 nL (500 ng)	Bansal et al. (2016)
IMD		*Plautia stali*	No significant effect	Adults	Microinjection	3 µL (100 ng/µL)	Nishide et al. (2022)
RELISH		*Plautia stali*	No significant effect	Adults	Microinjection	3 µL (100 ng/µL)	Nishide et al. (2022)
MYD88		*Plautia stali*	Inhibition of Pentatomicin and	Adults	Microinjection	(3 µL (100 ng/µL)	Nishide et al. (2022)

			Defensin2 expression under infection				
DORSAL		*Plautia stali*	Inhibition of Pentatomicin expression under infection	Adults	Microinjection	3 μL (100 ng/μL)	Nishide et al. (2022)
PENTATOMICIN		*Plautia stali*	Increase in mortality upon bacterial infection. No effect on vertical transmission of symbiotic bacteria	Adult males	Microinjection	3 μL (100 ng/μL)	Nishide et al. (2022)
SC MUCIN		*Plautia stali*	No effect on vertical transmission of symbiotic bacteria	Adult females	Microinjection	1 μL (200 ng/μL)	Moriyama et al. (2022)
DSRNASE	**RNAi**	*Nezara viridula*	Increase in RNAi effect of target-genes	2nd instar nymphs	Microinjection	67 nL (1 μg/μL)	Sharma et al. (2021)[b]
DSRNASE		*Plautia stali*	No significant effect	1st instar nymphs	Oral	800 μL (1000 ng/μL)	Nishide et al. (2021)
FASN		*Plautia stali*	No significant effect	1st instar nymphs	Oral	800 μL (1000 ng/μL)	Nishide et al. (2021)

3.1 Behavior, Physiology and Survival

In general, insects exhibit a diverse array of behaviors and physiological traits that can be species-specific and pivotal for their achievements or setbacks in nature. Modifications in these processes can wield a detrimental influence on their behavior and body functionality, while also potentially impacting their overall reproductive success and/or survival (Chown and Terblanche 2006). One example is the inhibition of the olfactory gene *Tyramine receptor 1* (*TAR1*), known for its involvement in social and behavioral adaptability among insects. Second instar nymphs of *H. halys* treated with dsRNA for *TAR1* demonstrated reduced susceptibility to the alarm pheromone component (E)-2 decenal. The lack of the alarm pheromone in these insects resulted in normal feeding behavior on green beans in the presence of (E)-2-decenal (Finetti et al. 2021). In *M. histrionica*, silencing of the *terpene synthase (TPS)* gene, which codes for an enzyme involved in the production of molecules associated with insect communication and defense, significantly curtailed (~7-fold) the emission of murgantiol, an aggregation hormone, in males (Lancaster et al. 2018). Silencing of the gene encoding the G-protein coupled receptors (GPCR), a receptor predominantly involved in sensory, behavioral, immunological, and insecticide resistance processes, was performed through both adult microinjection and oral delivery to second instar nymphs of *H. halys* and caused approximately 95% and 50% of mortality, respectively (Mogilicherla et al. 2018). In *N. viridula*, silencing of *GPCR* through adult microinjection only led to a 17% mortality rate (Gurusamy et al. 2021), while microinjection into adults of *M. histrionica* caused no significant effect in mortality rates and feeding to second instar nymphs caused nearly 50% of mortality (Howell et al. 2020).

Reproductive behavior can also be affected using RNAi. In *Plautia stali* Scott females, silencing of the *pigment dispersing factor* (PDF) gene, crucial for the circadian clock's adaptation to the photoperiod, led to insects' inability to adjust to changes in the photoperiod. Affected individuals exhibited a delay of over 20 days in initiating oviposition during the transition from short to long days, or an incapacity to suppress oviposition when transitioning from long to short days (Hasebe et al. 2022).

Disruption of essential cellular processes strongly impacts insect's physiology and amplifies their mortality rates. One of the most targeted groups of genes encodes ATPases, enzymes critical for an array of functions, including active transportation of various compounds, cargo transfers, energy conservation, and pH homeostasis (Schumacher and Krebs 2010; Pamarthy et al. 2018). In *H. halys,* injection into adults of dsRNA against putative *ATPASE N2B* resulted in 100% mortality (Mogilicherla et al. 2018). Administering the same dsRNA molecules through oral delivery to second instar nymphs induced substantial mortality, albeit at a significantly lower level of less than 30% (Mogilicherla et al. 2018). In contrast, when topically delivered to second instar nymphs of *H. halys*, ds*ATPASE N2B* was capable to induce an increase in mortality of nearly 50% (Finetti et al. 2023). In *M. histrionica*, silencing of *ATPase N2B* resulted only in approximately 35% mortality via microinjection and 40% mortality via feeding assay (Howell et al. 2020). In *N.*

viridula, microinjection of dsRNA targeting the same gene into adults did not cause significant mortality, despite causing a 90% reduction in gene expression (Gurusamy et al. 2021). These results suggest that variations in RNAi effect could reflect species-specific factors such as dsRNA degradation rate in the gut or gene redundancy.

The gene *vATPase A* (*V-type proton ATPase catalytic subunit A*) was also used as an RNAi target in second instar nymphs of *N. viridula* through microinjection, resulting in a mean mortality higher than 90% after 14 days, accompanied by abnormal phenotypic effects that resulted in reduced mobility (Sharma et al. 2021a). When *vATPase A* dsRNA molecules were delivered orally, 43% of insects died after 14 days of treatment, accompanied by a 57% reduction in weight in the surviving population (Sharma et al. 2021a). In *E. heros*, *vATPase A* and *ATPsynβ* were silenced in second instar nymphs, yielding approximately 80% and 60% mortality rates after 14 days, respectively. These outcomes were coupled with reduced mobility, diminished weight, delayed development, and molting defects compared to the control group (Castellanos et al. 2019). Despite these high mortality rates in nymphs, silencing *V-ATPase A* in *E. heros* adults resulted in a mere 35% mortality after 96 h, along with the customary reduction in mobility (Cagliari et al. 2020a). In *P. guildinii,* microinjection of ds*V-ATPase A* into adults led to a mortality of 51.6% after 14 days (Schvartzman et al. 2022). Microinjection of ds*V-ATPase subunits D*, *H* and *E* in *P. stalli* fifth instar nymphs resulted in mortality rates ranging from 80% to 90%. However, no discernible effect was observed when delivering ds*V-ATPase D* by feeding to first instar nymphs (Nishide et al. 2020). In contrast to the previously discussed results, for *N. viridula*, the delivery of 1 μg of dsATPase1 to adults through microinjection did not induce significant mortality (Gurusamy et al. 2021).

Another class of genes frequently selected as RNAi targets in Pentatomidae are genes coding for membrane trafficking proteins. In *N. viridula*, adult microinjection of dsRNA for *SNF7*, a pivotal player in the sorting of transmembrane proteins for lysosomal degradation, resulted in 100% mortality. Feeding these molecules to adults and third instar nymphs of *N. viridula* also caused mortality of 59% and 60%, respectively, 13 days after treatment (Gurusamy et al. 2021). In *P. stali*, RNAi for *SNF7* in fifth instar nymphs resulted in mortality rates of up to 80%, 4 days post-microinjections (Nishide et al. 2021). In *H. halys*, RNAi for *SNF7* through microinjection in adults and oral delivery to second instar nymphs resulted in approximately 70% mortality in both cases, 7 days after treatment (Mogilicherla et al. 2018).

The *Dynamin* gene, involved in vesicular trafficking and endocytosis, was also used as an RNAi target in adults of *H. halys*, resulting in just 15% of mortality after microinjections (Mogilicherla et al. 2018). The *coatomer* (*αCop*) gene, which participates in transporting biosynthetic proteins from the endoplasmic reticulum to the Golgi-complex apparatus, was targeted in second instar nymphs of *N. viridula* and RNAi via microinjection, resulting in mortality rates of up to 95%, 14 days post-injection. *N. viridula* microinjected with dsRNA for *αCop* also displayed a reduction of mobility. When the same dsRNA molecules were delivered by feeding, a mortality rate of 45% and a reduction in nymphal weight by 60% were observed 14 days after treatment (Sharma et al. 2021a, b). In *E. heros,* RNAi for *coatomer subunit beta* (*βcop*) and *coatomer subunit gamma* (*γcop*) through microinjection into second instar nymphs resulted in mortality rates exceeding 80% (Castellanos et al. 2019). These results indicate that, although trafficking proteins can be considered good candidates for mortality induction via RNAi, mortality rates vary among species and for the chosen delivery methods.

Genes involved in fundamental cellular processes and homeostatic mechanisms, such as protein synthesis, apoptosis, and the cell cycle, are also frequently selected as RNAi targets. The *inhibitor of apoptosis* (*IAP*) genes, a negative regulator of cell death process (Silke and Meier 2013), were used as RNAi targets in *N. viridula, H. halys* and *P. stali*, resulting in varying mortality rates. In *N. viridula,* RNAi against *IAP* genes via microinjection into adults resulted in mortality rates of 70%, 40%, and 22%, for *IAP5*, *IAP2*, and *IAP1* genes, respectively, 13 days after treatment (Gurusamy et al. 2021). Similarly, RNAi for *IAP* in *H. halys* resulted in comparable mortality rates, with 70% mortality observed 7 days post-treatment for adult microinjection and oral delivery to second instar nymphs (Mogilicherla et al. 2018). When delivered through topical application to second instar nymphs, ds*IAP* was capable to increase mortality rates in about 33% (Finetti et al. 2023). In *P. stali*, RNAi for *IAP* caused fifth instar nymphs to fail to reach adulthood, exhibiting high mortality rates after microinjection even at lower dsRNA doses (30 ng/μL). However, feeding *IAP* dsRNA produced negligible phenotypic effects, even at doses as high as 5000 ng/μL (Nishide et al. 2021), In *M. histrionica*, adult microinjection and oral delivery of ds*IAP* to second instar nymphs resulted in a mortality increase of more than 55% compared to the control group (Howell et al. 2020).

The *Serine/threonine-protein phosphate 1* (*PP1*) gene, pivotal in functions like the cell cycle, differentiation, and protein synthesis, has also been a prime target of RNAi in stink bugs. Silencing the *PP1-beta catalytic subunit* (*PP1-β*) in *N. viridula* through adult microinjection led to 67% mortality, 13 days post-treatment (Gurusamy et al. 2021), and nearly 90% mortality when microinjected into second instar nymphs (Sharma et al. 2021a). However, oral delivery of *dsPP1-β* to second instar nymphs resulted in only 25% mortality (Sharma et al. 2021a). In *H. halys, dsPP1-β* microinjection into adults caused 100% mortality within 7 days, while oral delivery to second instar nymphs induced approximately 75% mortality at the same time (Mogilicherla et al. 2018). However, when topically applied, dsPP1 did not significantly increase mortality rates in second instar nymphs of *H. halys* (Finetti et al. 2023). In *M. histrionica,* adult microinjection with ds*PP1*was caused an increase in mortality rate of approximately 75%, while feeding ds*PP1* to second instar nymphs caused increase in mortality of nearly 40%, compared to control group (Howell et al. 2020). Lastly, RNAi for *PP1 2A catalytic subunit beta isoform* (*PP1A96A*) through microinjection into second instar nymphs resulted in mortality rates of up to 70% and 95% at 7- and 14-days post-treatment, respectively (Castellanos et al. 2019).

In *N. viridula*, silencing a diverse array of genes associated with essential cellular processes via microinjection in adults yielded mortality rates exceeding 90%. The most successful RNAi targets in *N. viridula* were *26SPRS6B* (*protease regulatory subunit 6B*), *40RPS13* (*40S ribosomal protein S13*), *60SRPL19* (*60S ribosomal protein L19*), and *USP* (*Ultraspiracle nuclear receptor*). Conversely, when the same dsRNA for genes *40SRPS13* and *60SrpL19* was orally delivered to second instar nymphs, mortality rates reached a maximum of only 17.5% (Sharma et al. 2021a). In *E. heros*, microinjection of dsRNA targeting genes *PROSA2* (*proteasome subunit alpha type-2*), *SPR54K* (*signal recognition particle 54 kDa protein*), *ACT* (*Muscle actin*), *PP1A96A* (*serine/threonine-protein phosphatase 2A catalytic subunit beta isoform*), *TAF-1* (*transcription initiation factor TFIID subunit 1*), and *RPN7* (*26S proteasome non-ATPase regulatory subunit 6*) into second instar nymphs resulted in mortality rates above 95% (Fishilevich et al. 2016; Castellanos et al. 2019).

3.2 Reproduction

For sexual reproduction to occur, a sequence of cellular and molecular mechanisms must unfold, giving rise to alterations in the morphology, physiology, and behavior of insects (Nanfack-Minkeu and Sirot 2022). The use of RNAi technology to interfere in these physiological aspects has been important not only to elucidate gene function but also to identify target genes for implementation in sterilization-based pest control strategies (Alphey 2002; Darrington et al. 2017).

Disrupting the development or physiology of male or female reproductive apparatus by RNAi can result in a reduction in the number of offspring, reduction of neonate survival rates or even fully progeny impairment due to infertility. For instance, RNAi for two genes involved in the circadian cycle, *circadian clock gene cycle* (*CYC*) and *myoinhibitory peptide* (*MIP*), in *P. stali* females, induced the suppression of ovarian development along with alterations in rhythmic activities (Tamai et al. 2019).

Subtle environmental cues can also impact or disrupt the fine-tuning of reproduction, through neurohormonal responses. Silencing the *pigment dispersing factor* (*PDF*) neuropeptide, essential for maintaining the circadian clock, led to the failure of *P. stali* females to adapt to changes in photoperiod. This resulted in a delay of over 20 days in initiating oviposition during transitions from short to long days, or an inability to suppress oviposition during transitions from long to short days (Hasebe et al. 2022).

Another conventional approach to impact fertility involves targeting genes within the oviposition pathway. In *E. heros*, silencing of various genes affects oviposition rates, particularly those involved in chromatin remodeling, crucial for gametogenesis and early embryonic development. For example, RNAi for *brahma* (*BRM*) completely inhibited oviposition, while silencing *MI-2*, *ISWI-1*, and *CHD1* resulted in a range of impacts, from high to less pronounced yet significant effects. RNAi for these genes also resulted in a reduction in the number of hatched eggs. In the case of *ISWI-2*, RNAi resulted in a modest (32%) decrease in oviposition, but in a remarkable (76%) reduction in egg hatching rate. RNAi for and *MI-2*, on the other hand, resulted in a reduction in oviposition, attributed to impairments in ovary development and ovariole elongation caused by silencing these genes (Fishilevich et al. 2016).

Another approach to decreasing pest populations involves impacting egg viability by interfering with embryonic development. For instance, in the study conducted by Nishide et al. (2020), *seven multicopper oxidase genes* (*MCOs 1-7*) and *PSTYR1* to *PSTYR3*, tyrosinases involved in cuticle formation, were evaluated for maternal RNAi in *P. stali*. The results revealed no deformations in the eggshell, nor alterations in egg number and mass, except for the gene *MCO2*. Silencing of *MCO2* led to reduced egg hatching and induced lethal effects on embryos ($P < 0.01$). The ds*MCO2* treatment caused a 33.3% lethality rate in third instar nymphs and 100% in fifth instar nymphs during molting. In a study by Cagliari et al. (2021) the silencing of *homeotic genes extradenticle* (*EXD*), *sex comb reduced* (*SCR*), *labial* (*LAB*), and *deformed* (*DFD*), crucial for embryonic body segmentation, resulted in mere 1–12% egg hatching after administration to *E. heros* females. Similar outcomes were observed when the *SCR* gene was silenced in *H. halys* (Lu et al. 2017). Abnormal embryo development was identified as the cause of mortality in all these cases.

These findings underscore the potential of targeting reproduction-related genes for stink bug control and emphasize the intricate relationship between fertility and development. One concern with this approach is the fact that oviposition itself already causes damage to crop, even if the laid eggs are not viable (Guillén et al. 2022). However, even in those cases, RNAi in these species will have a positive effect on controlling the insect population by reducing its generation success rate.

3.3 Development, Morphology and Physical Characteristics

Targeting developmental processes focusing on disrupting normal body development, leading to malformations, directly results in an increased mortality rate and is a common approach for RNAi assays (Vogel et al. 2018). One recurring method to achieve this is through the silencing of genes associated with cuticle pigmentation. For instance, when *E. heros* third instar nymphs were treated with dsRNA targeting the *tyrosine hydroxylase* (*TH*) gene, which is involved in the melanin pathway, numerous defects in cuticle sclerotization, as well as abdomen, antennae, and leg morphology, were observed. These abnormalities consequently caused mortality rates of up to 97% before reaching adulthood (Cagliari et al. 2020b). Conversely, the silencing of the *yellow* (*YEL*) gene, also part of the melanin pathway, led to a significant reduction of 45% in the success of achieving adulthood, yet no distinct phenotype was evident (Cagliari et al. 2020b). In *P. stalli* third instar nymphs, the silencing of the *MCO2* gene, implicated in cuticle formation, resulted in soft and unpigmented cuticles, preventing 100% of treated nymphs from completing the molt to the fifth instar. A similar effect was observed upon silencing dsv*ATPase* (Nishide et al. 2021). In *N. viridula*, microinjection of dsRNA targeting the *WHITE* gene induced significant changes in external body color and minor alterations in the color compounds of the eyes and ocelli. Additionally, an abnormal orange-yellow coloration was observed in the Malpighian tubule and discoloration in the gonads and metathoracic scent reservoir, despite no recorded increase in mortality. These coloration changes also affected embryos in parental RNAi applications (Souza et al. 2022).

Finally, another strategy involves targeting genes directly involved in the development of insect body parts and molting processes. As previously noted, homeotic genes are critical for embryonic development through the regulation of body segmentation. Downregulation of these genes can lead to developmental complications, malformations and, subsequently, an increase in mortality. For instance, when *E. heros* third instar nymphs were treated with dsRNA for the abnormal wing disc (AWD) gene, involved in wing development, a 45% mortality rate was observed during the molting period, with only 43% reaching adulthood, and 15% of those displaying wing abnormalities (Cagliari et al. 2020a, b). Silencing the *SCR* gene in both *H.* halys and *E. heros* resulted in mouthpart deformations, leading to the development of leg-like structures on the rostrum, impeding feeding and ultimately causing death (Lu et al. 2017; Cagliari et al. 2021). Additionally, *H. halys* nymphs subjected to *SCR* gene silencing displayed smaller bodies (Lu et al. 2017). The silencing of the *proboscipedia* (*PB*) gene in *E. heros* produced a similar leg-like structure phenotype in 91% of individuals, along with bifurcated rostra and a high mortality rate due to feeding obstruction. Silencing the *EXD* gene resulted in a complete developmental failure, with embryos lacking all appendages, while ds*DFD* treatment led to abnormal structures on one side of each head, affecting only 8% of nymphs. Interestingly, nymphs treated with ds*LAB* maintained full development and all body parts, despite perishing before hatching (Cagliari et al. 2021). In *P. stalli,* the delivery of dsRNA for *mitochondrial phosphate carrier 1* (*PIC1*) and 2 (*PIC2*) genes impacted the molting capacity of third instar nymphs, resulting in a reduction of up to 58% and 55%, respectively, in eclosion rates during ecdysis when compared to controls (Sugahara et al. 2017). In the case of silencing the *adenine nucleotide translocase 1* (*ANT1*), a nuclear-coded mitochondrial protein involved in ATP-ADP exchange, 100% of treated nymphs failed to molt (Sugahara et al. 2016).

In conclusion, it is evident that disrupting developmental processes in insects through RNAi assays, particularly by targeting genes associated with body development and molting, presents a compelling approach with significant implications for mortality rates and phenotypic outcomes. From severe malformations to subtle coloration changes, these experiments shed light on the intricate interplay of genes in shaping insect physiology, highlighting the potential of RNAi for pest management, as well as its importance as a tool for understanding developmental biology.

3.4 Microbiome and Immune System

The microbiome and immune system work in tandem to ensure the optimal fitness of the insect organism. Symbiotic microorganisms, in fact, can complement the host's diet and influence both survival and fecundity (Ferrari and Vavre 2011). Targeting this interconnected system, albeit challenging, holds the potential to disrupt the target species' homeostasis effectively and decrease its population.

For example, knockdown of Toll pathway genes *MYD88* and *Dorsal* in *P. stali* females suppressed the upregulation of the antimicrobial peptide PENTATOMICIN (Nishide et al. 2022). Conversely, silencing the IMD pathway genes *IMD* and *RELISH* had no impact, indicating that the Toll pathway controls PENTATOMICIN expression. Silencing *PENTATOMICIN* in adults heightened susceptibility to Gram-positive bacterial infections, leading to increased mortality rates. However, no effects on vertical symbiont transmission or embryonic development were observed (Nishide et al. 2022). In the context of targeting the *P. stali* microbiome, silencing *SC* mucin, a glycoprotein found in a female-specific symbiotic organ, was carried out. Nonetheless, no influence on symbiont population, vertical symbiont transmission, or egg quality was detected (Moriyama et al. 2022).

Finally, the use of dsRNA targeting *H. halys catalase*, a gene known to elevate its expression upon immune stimulation, was efficient in reducing transcript levels by 98%. However, no significant phenotypes were described (Bansal et al. 2016).

These findings underscore that while a physiological pathway may hold great promise due to its role in maintaining homeostasis, the identification of effective target genes remains a crucial step in achieving significant phenotypes and successful RNAi effects. The studies presented here also highlight the considerable potential of RNAi-mediated gene silencing to emerge as a tool for sustainable pest control in the near future. However, numerous aspects must be refined to facilitate the transition of RNAi from laboratory settings to real-world field applications, as further discussed below.

4 Challenges in Application of RNAi as a Tool to Control Stink Bugs

RNA interference is considered a very promising tool for the future of sustainable pest management (Christiaens et al. 2020). However, for its successful application in the field, several challenges need to be overcome. Together with target gene selection and a functional RNAi machinery in the target organism, successful RNAi also depends on dsRNA stability under environmental conditions, dsRNA delivery and dsRNA uptake by the target organism's cells, dsRNA degradation by the target organism, the large amount demanded to control pests in the crop field, risks of compromising off-target organisms and target resistance and/or refractoriness (Joga et al. 2016; Niu et al. 2018; Zhu and Palli 2020) (Fig. 8.2).

4.1 Environmental Factors

To achieve gene silencing, dsRNA molecules must maintain their integrity before and after reaching the target organism. DsRNA offers a variety of adversities, which can negatively affect RNAi's success. For example, dsRNA molecules tend to rapidly dissipate in soil (14 h–30 h) and waterways (up to 96 h), due to biotic degradation by microbiological communities present in both environments (Dubelman et al. 2014; Albright et al. 2017; Parker et al. 2019). When directly exposed to UV, dsRNA degradation can occur even faster, disappearing completely within 2 h (San Miguel and Scott 2016). In field trials in Puerto Rico, dsRNA foliar treatment of soybean caused a dramatic decline in detectable dsRNA, with an efficiency reduction of ~95% in the first 3 days after treatment and almost 99% reduction after 7 days (Bachman et al. 2020). However, when applied on leaf surface in a greenhouse-controlled environment, it was possible to maintain dsRNA stability and efficiency for at least 28 days, including its presence on foliar surface after watering it (San Miguel and Scott 2016). The contrasting results obtained in both cases are probably due to a combination of rainfall, UV and biodegradation existing in non-controlled environmental conditions in the field. These results indicate the high potential of feasibility of spray-induced gene silencing (SIGS) application, but also highlight the importance of formulating RNA molecules, promoting their persistence and accumulation in different environments.

In this respect, a variety of formulations have been developed aiming to preserve dsRNA integrity. The use of liposome and EDTA in dsRNA formulation for Pentatomidae assay was already described here (Castellanos et al. 2019). Another promising technology to improve dsRNA stability, uptake and/or integrity is the use of nanoparticles for both ingestion and topical application. In this process, dsRNA molecules are carried by the nanoparticles through the electrostatic interaction between the positively charged nanoparticle and the negatively charged phosphate group of nucleic acid molecules (Zhu and Palli 2020). A variety of nanocarrier-based dsRNA/siRNA has been studied aiming to improve RNAi efficiency, such as branched and guanidinium-containing polymers, conjugated polymer nanoparticles, dendritic nanocarriers, clay nanosheets, chitosan and gold nanoparticles, polymer-coated inorganic nanoparticles, and nanocarrier/dsRNA/detergent formulations (Li-Byarlay et al. 2013; Zhu and Palli 2020; Pugsley et al. 2021).

However, in addition to maintaining dsRNA integrity, it is necessary that nanoparticles also present low toxicity, high stability in formulations, be short-time biodegradable, and allow a low-cost and large-scale production, enabling its viable application in association with RNAi technology (Pugsley et al. 2021).

4.2 Delivery Methods and dsRNA Cellular Uptake

In addition to guaranteeing dsRNA integrity in the environment, RNA molecules must remain intact once in contact with, and after ingested by, the target organism. For this purpose, dsRNA delivery method needs to be compatible with insect habits, support cellular uptake and escape degradation by nucleases (Joga et al. 2016; Santos-Ortega and Flynt 2022). Usually, the main delivery methods used in RNAi technology are microinjection, oral and topical.

During the early steps in RNAi development of gene screening, the delivery method by microinjection is the most successful approach as it allows the precise control of dsRNA volume (and mass) being delivered to the target organism. Also, microinjection circumvents physiological barriers, such as hardened cuticle and degradation by nucleases in the insect digestive tract (Joga et al. 2016; Alamalakala et al. 2018). Microinjection has been successfully performed in several Pentatomidae species, including *P. stali* (Sugahara et al. 2016, 2017; Tamai et al. 2019; Hasebe et al. 2022; Moriyama et al. 2022; Nishide et al. 2020, 2022;), *E. heros* (Fishilevich et al. 2016; Cagliari et al. 2020b, 2021; Castellanos et al. 2019), *H. halys* (Bansal et al. 2016; Lu et al. 2017; Mogilicherla et al. 2018), *N. viridula* (Sharma et al. 2021a, b; Gurusamy et al. 2021; Souza et al. 2022), *M. histrionica* (Lancaster et al. 2018; Howell et al. 2020) and *P. guildinii* (Schvartzman et al. 2022) (Fig. 8.4). Overall, the experiments are performed in stages varying from second instar nymphs to adults, in concentrations ranging from 30 to 5000 ng/uL and volumes ranging from 54,75 nL to 6 uL, per insect. The results show efficient gene knockdown in all species and phenotypes of increased mortality, morphological alterations and/or decrease in fertility (Table 8.5).

Despite its high efficiency, microinjection is a laboratory-restricted application and not feasible for RNAi-based pest control in the field. For a gene screening approach, microinjections have the disadvantages of being time-consuming, demanding expertise on insect microinjection and causing mechanical damage due to needle insertion in the insects.

Oral delivery, on the other hand, allows for a less invasive method, rapidly administering dsRNA to large numbers of individuals in an easy-to-perform way,

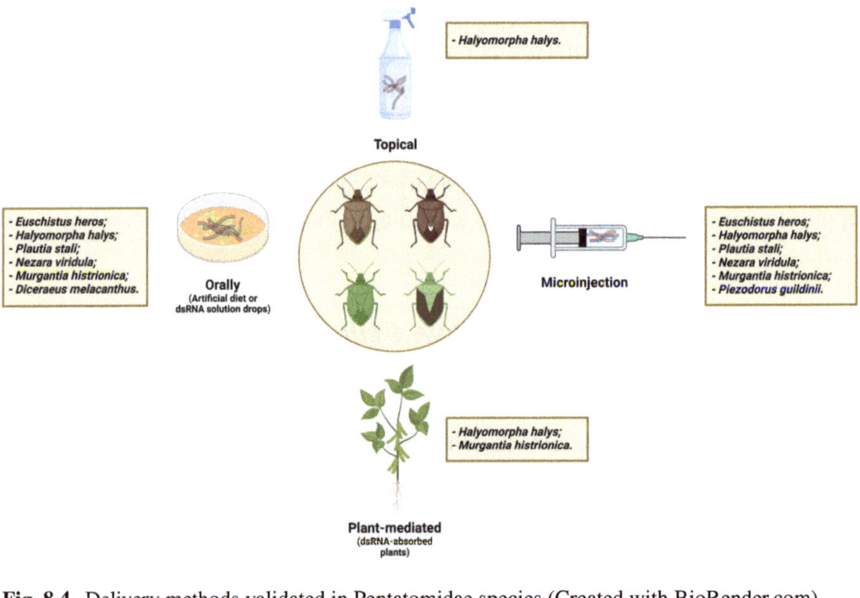

Fig. 8.4 Delivery methods validated in Pentatomidae species (Created with BioRender.com)

with the potential for high-throughput gene screening and field application (Alamalakala et al. 2018). Oral delivery through artificial diet, drop feeding or plant soaked (foliar spray, root drench, trunk injections) with dsRNA were delivered to Pentatomidae species *P. stali* (Nishide et al. 2021), *E. heros* (Castellanos et al. 2019; Laisney et al. 2021), *H. halys* (Ghosh et al. 2017; Ghosh et al. 2018; Mogilicherla et al. 2018), *N. viridula* (Gurusamy et al. 2021; Sharma et al. 2021a, b), *M. histrionica* (Howell et al. 2020) and *Diceraeus melacanthus* (Dallas) (Pinheiro et al. 2020) (Fig. 8.4). No bacterial-expressed dsRNA delivery was found. In stink bugs, published works until now target mainly first to fifth instar nymphs with concentrations ranging from 17 to 5000 ng/uL, except for *N. viridula*, in which adults were also targeted. For those, gene silencing was achieved but at a lower level compared to microinjection. Phenotypical effects were observed in almost all species, except for *P. stalli,* in which the reduction of gene expression caused no visual effects, and *D. melacanthus,* as the focus of the work was not on this scope (Table 8.5).

The lower gene silencing rates obtained through orally delivered dsRNA, compared to microinjection, reflect challenges to be overcome in agricultural application of RNAi. In the field, physical and chemical environmental conditions may lead to dsRNA degradation (Santos-Ortega and Flynt 2022). Additionally, insect biological characteristics, such as feeding behavior, may difficult molecule intake. Stink bugs feed by inserting their mouthparts into the plants, injecting saliva, and sucking the fluids (Peiffer and Felton 2014). This process not only makes oral delivery difficult but also may cause degradation of the molecule (Castellanos et al. 2019). One possible alternative for this issue is the use of transformed plants expressing dsRNA, which was already been applied in coleopteran Western corn rootworm *Diabrotica virgifera virgifera* LeConte (Baum et al. 2007). However, complications in transgenic plants regulation make this alternative less explored, also degradation by RNases in saliva remains a problem that can affect the efficiency of RNAi by this delivery method.

Finally, topical delivery relies on the application, penetration, and adsorption of dsRNA by insect cuticle, mainly by SIGS (Alamalakala et al. 2018). In the stink bug, only second instar nymphs of *H. halys* were found to be submitted to topical delivery of dsRNA until the moment (Finetti et al. 2021, 2023) (Fig. 8.4). Despite not inducing knockdown in all the tested genes, the application of a drop of dsRNA solution on the insect's abdomen cuticle showed an efficient impact on behavior (Finetti et al. 2021) and mortality (Finetti et al. 2023) using dsRNA masses ranging from 100 to 3000 ng (Table 8.5).

The future of field application of RNAi relies especially on SIGS and orally delivered dsRNA. However, dsRNA degradation by both the environment and target organism can negatively affect its success. Determining how dsRNA molecules are affected in both scenarios, as well as how formulations could improve its integrity, is a crucial step to achieve successful delivery methods (Lomate and Bonning 2016; Castellanos et al. 2019; Cantón and Bonning 2019).

4.3 Degradation by dsRNases

As discussed, efficient dsRNA delivery depends on its constant integrity from application to cellular uptake (Santos-Ortega and Flynt 2022). In fact, degradation of dsRNA by nucleases (dsRNases) in the hemolymph, midgut and/or saliva is the major responsible for the highly variable efficiency among different insect orders and/or species, being one of the most targeted challenges to be overcome in RNAi-based control (Wang et al. 2016; Yao et al. 2022). DsRNases interfere in the gene silencing in a wide variety of insects, such as Lepidoptera (Peng et al. 2020), Diptera (Tayler et al. 2019), Hymenoptera (Lei et al. 2022), Coleoptera (Spit et al. 2017) and Hemiptera (Singh et al. 2017). In some species, the use of specific dsRNAs targeting dsRNases followed by silencing of essential genes caused improvement of RNAi effect (Spit et al. 2017; Tayler et al. 2019), showing a high potential to overcome dsRNA degradation by insects using gene silencing.

DsRNases were already identified in stink bugs transcriptome (Sparks et al. 2014; Cagliari et al. 2020a; Schvartzman et al. 2022) and have proven to cause degradation of dsRNA. In *N. viridula,* > 90% mortality was achieved in second instar nymphs by Sharma et al. (2021b) via microinjection. However, oral delivery of the same dsRNAs caused ≤45% mortality due to the presence of dsRNases in the saliva of the insect. Also, it was observed that dsRNAs incubated for 30 min in saliva, but not in midgut juice, of adults were completely degraded (Sharma et al. 2021a), indicating that less effective gene silence through oral delivery is caused by the presence of RNases in saliva. Very similar results concerning dsRNA degradation rates in different body fluids of *N. viridula* adults were reported by other studies (Lomate and Bonning 2016; Singh et al. 2017). When dsRNases were silenced in second instar nymphs through microinjection, followed by oral administration of dsα*Cop*, the pest mortality rate was significantly enhanced. In addition, the presence of partial amounts of dsRNA in *ex vivo* assays using the saliva of adults was observed even after 120 min, in contrast with the complete degradation previously observed in 30 min (Sharma et al. 2021b), confirming that silencing of ds*RNase* genes can improve RNAi efficiency in oral delivery in this species. Interestingly, *ex vivo* degradation assays performed by Gurusamy et al. (2021) in hemolymph, lumen content and watery saliva of third to fourth instar *N. viridula* nymphs showed an intact dsRNA even after 12 h exposure, despite a slight degradation observed in 24 h (Gurusamy et al. 2021), although oral administration of ds*SNF7* to adults and fourth instar nymphs also caused lower mortality rate (59–60%) when compared to microinjection (100%). The variations among this *ex vivo* degradation data and the other assays here discussed can be caused due to different dsRNases expression levels and activities present in different phases of development.

In *H. halys*, dsRNA exposure to hemolymph and watery saliva for 1 h showed incomplete degradation even at higher concentrations of fluids (1.2 mg/ml) (Mogilicherla et al. 2018). In Singh et al. (2017) the degradation assay verified that a concentration of 3.07 mg/ml of body fluid for 1 h was necessary to degrade 50%

(CB_{50}) of dsRNA molecules (Singh et al. 2017). Additionally, microinjection of ds*IAP* into adults and its oral delivery to second instar nymphs caused ~75% mortality (Mogilicherla et al. 2018), indicating that oral delivery to *H. halys* nymphs is not significantly affected by dsRNA salivary or gut degradation and, although dsRNases are present in *H.* halys transcriptome and cause degradation of dsRNA molecules, their effects are relatively milder when compared to other insects.

In *M. histrionica* even lower degradation rates were observed. Howell et al. (2020) demonstrated that dsRNA was not degraded after salivary exposure for 1 h even at the highest concentration (4.1 mg/ml) and Singh et al. (2017) determined a CB_{50} of 6.57 mg/ml for this species (Singh et al. 2017; Howell et al. 2020). Microinjection of ds*PP1* and ds*IAP* into adults caused ~75% and ~65% mortality respectively, whereas plant-mediated oral delivery to second instar nymphs caused ~40% and ~57% mortality, respectively. Additionally, intact dsRNA was obtained from *M. histrionica* 48 h after feeding on plants treated with dsRNA (Howell et al. 2020). Finally, ds*IAP* delivery to second-third instar nymphs resulted in 55% mortality (Howell et al. 2020). Together, these results indicate a significant RNAi efficiency in *M. histrionica* with a low impact of dsRNases.

A contrasting scenario was observed in *P. stali*. In this species, microinjection of six genes into fifth instar nymphs caused >75% mortality, with ds*MCO2* and ds*IAP* reaching 100%. However, when low or high doses of dsRNA were orally administered (plant-mediated and/or in solution) to both first and fifth instar nymphs, no significant mortality was observed, even when dsRNA molecules targeting dsRNases or Fatty acid synthase (FASN), an RNAi blocker associated to RNAi refractoriness (Dong et al. 2017), were previously orally administered (Nishide et al. 2021). This result is very contrasting when compared to the improvement obtained by Sharma et al. (2021a, b) in *N. viridula,* after ds*RNase* silencing. The major hypothesis is that the delivery method performed is the cause as in *N. viridula* the ds*dsRNAse* was administered through microinjection, and its hemolymph has proven to have a lower activity of dsRNases, whereas in *P. stali* the ds*dsRNAse* and target-gene molecules were orally delivered, probably being degraded by saliva or gut contents before complete ingestion and cellular uptake. Until the moment, no data on effective RNAi through oral delivery was recorded for Pentatomidae (Table 8.5).

In *E. heros*, microinjection of ds*VATPase* and ds*ACT-2* in second instar nymphs resulted in ~80% and ~100% mortality respectively, whereas oral delivery of both genes caused only 30% and 33%. Exposition of dsRNAs to non-diluted adult saliva showed complete degradation at 10 min of exposure, indicating that a lower effect of RNAi through oral delivery could be caused due to its degradation by salivary RNases. Interestingly, when dsRNA formulated with liposomes was delivered, it was observed an increase of 15% and 9% in mortality for ds*VATPase* and ds*ACT-2* respectively, compared to administration of naked dsRNA. However, formulation with EDTA, which presented the higher protection of dsRNA molecules in *ex vivo* assays, increased in 21% the mortality for ds*VATPase*, but decreased the mortality for ds*ACT-2* by 11%.

Lower or higher rates of silencing and/or mortality here described in different species are proven to be in most cases related to dsRNA degradation, especially when orally delivered. However, it is worth remembering that RNAi success depends also on target gene efficiency and on the design of dsRNA molecules. In this way, different genes can cause distinct rates of mortality or phenotypes when compared to each other, as well as the same target gene with dsRNA designed differently can result in divergent efficiency in the same species. Also, dsRNA can be degraded by both host and environmental factors. Preventing dsRNA degradation is an essential step to RNAi success in the field, and new protection methods are constantly being developed and validated, as discussed below.

4.4 dsRNA Cost and Production

The future of RNAi application in agriculture depends on its possibility to be topically applied, once genetically modified (GM) crops are negatively viewed by many countries and a high variety of important agricultural species are not able to be transformed (Fletcher et al. 2020). For this purpose, target genes need to be highly efficient and high amounts of dsRNA are required to obtain acceptable rates of control in large crop areas (Zhu and Palli 2020). Furthermore, RNAi-based control price also needs to be competitive compared to the methods used nowadays, such as chemical control.

In this scenario, lower-cost and large-scale dsRNA production systems are being developed and constantly improved. The two major methods currently existing are cell-free and microorganism-based production (Guan et al. 2021; Nwokeoji et al. 2022). These approaches already reduced the cost of dsRNA gram from $12,500 (2008) to $0,50 (2021) (Rank and Koch 2021). At present, major companies known to produce and commercialize large amounts of dsRNA are GreenLight Bioscience (USA), via cell-free bioprocess, RNAgri (APSE—USA) and AgroRNA (Genolution—Korea), both using microorganism-based fermentation platform (Taning et al. 2020).

Cell-free large-scale production relies on *in vitro* transcription from DNA templates to dsRNA. This method allows the application of different polymerases using both linear or plasmid DNA templates, with highly scalable, yield and low cost for different target genes, in addition to a high purity of produced molecules (He et al. 2022; Hough et al. 2022).

On the other hand, microorganism-based large-scale production consists of using *in vivo* engineered biological systems, such as yeast, fungi, and bacteria, to generate dsRNA (Hough et al. 2022). The most described species to be genetically modified to produce dsRNA are *Escherichia coli*, *Saccharomyces cerevisiae*, *Pseudomonas syringae*, *Corynebacterium glutamicum*, *Chlamydomonas reinhardtii*, *Bacillus* sp., *Yarrowia lipolytica*, *Pichia pastoris*, *Lactobacillus plantarum* and symbiotic bacteria (*Rhodococcus rhodnii* and *BFo2*) (Suzuki et al. 2011; Van Ekert et al. 2014; Whitten et al. 2016; Thammasorn et al. 2017; Álvarez-Sánchez et al. 2018; Guan et al. 2021; Hashiro and Yasueda 2022). In the case of insect-symbiotic bacteria, they have the benefit of not triggering intense immunity response as exogenous bacteria and having additional species-specific protection against off-target effect, in addition to overcoming problems concerning dsRNA degradation due to salivary, foregut or environmental nucleases (Whitten et al. 2016; Guan et al. 2021). The microorganism-produced dsRNA molecules can be used through both live and inactivated microbes. However, the use of microorganisms to achieve RNAi response can induce an undesired immune response in target organisms, affect off-targets and transfer plasmids into the environment. Nevertheless, a strategy has been currently adopted to deal with these issues which involve the purification of the dsRNA produced before its delivery (Guan et al. 2021).

4.5 Off-Targets

Despite its high specificity for the target species (Fig. 8.5), RNAi can cause gene silencing in undesired groups. DNA sequences with more than 80% identity and siRNA and mRNA sequences sharing 10–12 pb identity are highly propitious to cause off-target effects (Chen et al. 2021; Finetti et al. 2023). In Pentatomidae, the off-target effect of specific dsRNAs for *H. halys* genes *ATPase N2B*, *PP1* and *IAP*

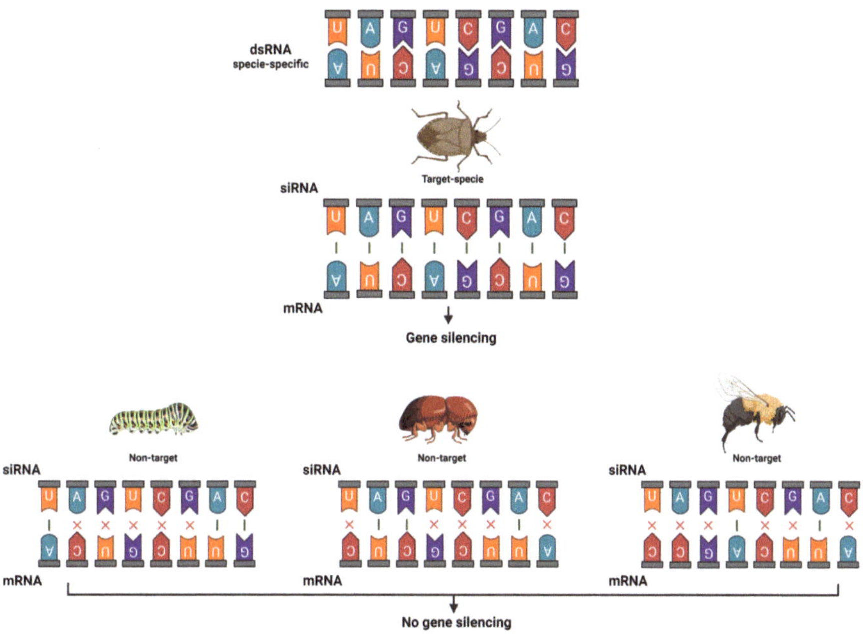

Fig. 8.5 RNAi specificity based on sequence alignment. dsRNA molecules are designed to match the target gene in the target organism while presenting mismatches to the sequence of off-target ortholog genes in off-target species (Created with BioRender.com)

were tested through microinjection in second and third instar nymphs of *Rhodnius prolixus* Stål (Heteroptera: Reduviidae). The results showed evident phenotypic effects in *H. halys,* while no gene silencing or mortality was noted in *R. prolixus,* indicating no off-target effects (Finetti et al. 2023).

To avoid non-target effects, especially in pollinators and species of importance in biological control (BC), simple steps can be taken in dsRNA molecule design. Choosing less conserved target genes, as well as carefully selecting dsRNA length and sequence using gene alignment tools, can minimize the chances of RNAi molecules affecting non-target organisms. Additionally, performing tests in natural enemies, BC-related species, pollinators, and species closely related to target organism before the dsRNA application in the field can minimize the undesired effects, guaranteeing the safety of RNAi and allowing its use in IPM alongside BC.

4.6 Resistance and Refractoriness

As commonly observed for chemical insecticides, there is a concern regarding the development of resistance to RNAi-based products in insects. Some possible mechanisms where resistance could occur can be proposed, such as target gene and RNAi machinery mutation, increased dsRNA degradation and impaired dsRNA uptake, intracellular transport, and endosomal escape (Zhu and Palli 2020). In the case of a mutation in the target gene sequence, it would hardly affect resistance to dsRNA, because this molecule generates several siRNAs complementary to different regions of the target sequence. Also, another solution would be simply to use dsRNA that targets a different region of the same gene or another gene. For mutations in the core RNAi machinery, it is also unlikely to promote resistance, once these genes are also essential for basic cellular biological processes (Cagliari et al. 2020a).

To our knowledge, there are only two studies reporting resistance to RNAi in insects and none of them involve Pentatomidae species. Khajuria et al. (2018) established dsRNA resistant *D. virgifera virgifera* colonies through maize plants expressing Dv*Snf7* dsRNA, followed by laboratory selection. These resistant populations presented impaired dsRNA uptake and this trait proved to have recessive and autosomal inheritance. In *Leptinotarsa decemlineata* (Say), Yoon et al. (2018) induced the selection of RNAi-resistant cells using dsRNA against the gene *IAP*. The lack of RNAi response was attributed to the reduced expression levels of *STAUFENC*, which is required to process dsRNA into siRNAs. In both cases, it was observed cross-resistance to other dsRNAs, indicating that resistance acquisition was not specific to the dsRNA sequence. Therefore, the characterization and understanding of insect RNAi resistance mechanisms will help to design efficient pest management strategies and assure the successful use of RNAi technology in agriculture.

Despite RNAi resistance being rare, agrochemical resistance is one of the main challenges for pest control. However, chemical insecticide resistance can be solved with the help of the gene silencing approach (Fig. 8.6). Insects have developed biological mechanisms to combat toxic compounds in the environment. One of them is

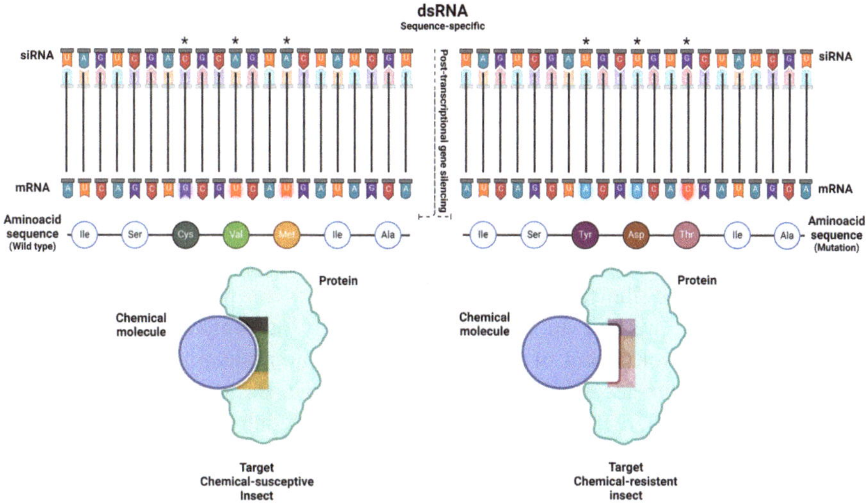

Fig. 8.6 Schematic drawing of a chemical insecticide resistance acquisition by point mutations and RNAi approach to circumvent the phenomenon via mRNA degradation of the resistance gene (Created with BioRender.com)

the cuticular barrier, which reduces or delays chemical penetration into the body of the insect (Saha and Saha 2016). The silencing of cuticle genes involved in this process decreased the resistance to neonicotinoids and pyrethroids in *L. decemlineata* (Naqqash et al. 2020) and *Culex pipiens pallens* (L.) (Fang et al. 2015), respectively. Another mechanism of insecticide resistance is the detoxification system that presents enzymes, such as cytochrome P450 monooxygenases (CYPs), glutathione S-transferases (GSTs) and esterases (ESTs), able to metabolize the toxins inside an insect's body. *CYP* genes were upregulated in insecticide resistant *Nilaparvata lugens* (Stål) compared to the sensitive strains and the knockdown of these elements increased planthopper susceptibility to imidacloprid (Zhang et al. 2016). In accordance, studies with the Asian citrus psyllid, *Diaphorina citri* Kuwayama, showed that RNAi feeding experiments of *CYP*, *GST* and *EST* genes significantly increased adult mortality. Moreover, these genes were overexpressed in imidacloprid-resistant insects in relation to responsive ones and induced under insecticide treatment (Tian et al. 2019). These observations can be helpful in the use of RNAi as a tool to improve pest susceptibility to insecticides, applying an integrated pest management approach.

5 Concluding Remarks

The RNAi technology arises as a potential tool to control agricultural pests. However, several factors affect its efficiency, including insect species, target genes and tissues, presence of RNAi-related genes, delivery strategy, nucleases, resistance to dsRNA, among others. Regarding Pentatomidae insects, many reports demonstrate the effective silencing of essential genes and transcriptome analysis of the stink bugs *E. heros* and *P. guildinii* identified the main RNAi machinery genes and suggests a dsRNA uptake through endocytosis pathway instead of transmembrane channels.

A suitable delivery method for RNAi field application is dsRNA feeding. However, studies registered low RNAi efficiency in Pentatomidae insects when using oral delivery. The action of dsRNases highlights as a key factor responsible for these effects. Nevertheless, the use of formulations can improve dsRNA integrity, enhancing gene silencing in stink bugs.

Although there are still no RNAi-based products for Pentatomidae control in agriculture, the development of large-scale dsRNA bioproduction systems makes this technology more accessible, once the costs decline. Besides that, the potential of RNAi to bypass agrochemical resistance positively contributes to an integrated approach for pest management, paving the way for a new agricultural scenario.

References

Alamalakala L, Parimi S, Patel N, Char B (2018) Insect RNAi: integrating a new tool in the crop protection toolkit. In: Trends in Insect Molecular Biology and Biotechnology, Springer, Cham. pp 193–232. https://doi.org/10.1007/978-3-319-6143-7_10

Albright VC, Wong CR, Hellmich RL, Coats JR (2017) Dissipation of double-stranded RNA in aquatic microcosms. Environ Toxicol Chem 36:1249–1253

Alphey L (2002) Re-engineering the sterile insect technique. Insect Biochem Mol Biol 32:1243–1247

Álvarez-Sánchez AR, Romo-Quinones C, Rosas-Quijano R, Reyes AG, Barraza A, Magallón-Barajas F, Angulo C, Mejía-Ruíz CH (2018) Production of specific dsRNA against white spot syndrome virus in the yeast *Yarrowia lipolytica*. Aquac Res 49:480–491

Bachman P, Fischer J, Song Z, Urbanczyk-Wochniak E, Watson G (2020) Environmental fate and dissipation of applied dsRNA in soil, aquatic systems, and plants. Front Plant Sci 11:21

Bansal R, Mittapelly P, Chen Y, Mamidala P, Zhao C, Michel A (2016) Quantitative RT-PCR gene evaluation and RNA interference in the brown marmorated stink bug. PLoS One 11:e0152730

Baum JA, Bogaert T, Clinton W, Heck GR, Feldmann P, Ilagan O, Johnson S, Plaetinck G, Munyikwa T, Pleau M, Vaughn T, Roberts J (2007) Control of coleopteran insect pests through RNA interference. Nat Biotechnol 25:1322–1326

Braukmann F, Jordan D, Jenkins B, Koulman A, Miska EA (2021) SID-2 negatively regulates development likely independent of nutritional dsRNA uptake. RNA Biol 18:888

Bueno AF, Panizzi AR, Hunt TE, Dourado PM, Pitta RM, Gonçalves J (2020) Challenges for adoption of integrated pest management (IPM): the soybean example. Neotrop Entomol 50:5–20

Cagliari D, Dias NP, dos Santos EÁ, Rickes LN, Kremer FS, Farias JR, Lenz G, Galdeano DM, Garcia FRM, Smagghe G, Zotti MJ (2020a) First transcriptome of the Neotropical pest *Euschistus heros* (Hemiptera: Pentatomidae) with dissection of its siRNA machinery. Sci Rep 10:4856

Cagliari D, Smagghe G, Zotti M, Taning CNT (2020b) Rnai and crispr/cas9 as functional genomics tools in the neotropical stink bug, *Euschistus heros*. Insects 11:1–13

Cagliari D, Taning CNT, Christiaens O, De Schutter K, Lewille B, Dewettinck K, Zotti M, Smagghe G (2021) Parental RNA interference as a tool to study genes involved in rostrum development in the Neotropical brown stink bug, *Euschistus heros*. J Insect Physiol 128:104161

Cantón PE, Bonning BC (2019) Proteases and nucleases across midgut tissues of *Nezara viridula* (Hemiptera:Pentatomidae) display distinct activity profiles that are conserved through life stages. J Insect Physiol 119:103965
Castellanos NL, Smagghe G, Sharma R, Oliveira EE, Christiaens O (2019) Liposome encapsulation and EDTA formulation of dsRNA targeting essential genes increase oral RNAi-caused mortality in the Neotropical stink bug *Euschistus heros*. Pest Manag Sci 75:537–548
Chen YH, Zhao L, Jia XT, Li XY, Li CZ, Yan H, Weng SP, He JG (2012) Isolation and characterization of cDNAs encoding Ars2 and Pasha homologues, two components of the RNA interference pathway in *Litopenaeus vannamei*. Fish Shellfish Immunol 32:373–380
Chen J, Peng Y, Zhang H, Wang K, Zhao C, Zhu G, Reddy Palli S, Han Z (2021) Off-target effects of RNAi correlate with the mismatch rate between dsRNA and non-target mRNA. RNA Biol 18:1747
Chown SL, Terblanche JS (2006) Physiological diversity in insects: ecological and evolutionary contexts. Adv In Insect Phys 33:50–152
Christiaens O, Whyard S, Vélez AM, Smagghe G (2020) Double-stranded RNA technology to control insect pests: current status and challenges. Front Plant Sci 11:451
Clark J, Rager JE (2020) Epigenetics: an overview of CpG methylation, chromatin remodeling, and regulatory/noncoding RNAs. In: Environmental epigenetics in toxicology and public health, pp 3–32
Cooper AMW, Silver K, Zhang J, Park Y, Zhu KY (2019) Molecular mechanisms influencing efficiency of RNA interference in insects. Pest Manag Sci 75:18–28
Darrington M, Dalmay T, Morrison NI, Chapman T (2017) Implementing the sterile insect technique with RNA interference – a review. Entomol Exp Appl 164:155–175
de Camargo RA, Herai RH, Santos LN, Bento FMM, Lima JE, Marques-Souza H, Figueira A (2015) De novo transcriptome assembly and analysis to identify potential gene targets for RNAi-mediated control of the tomato leafminer (*Tuta absoluta*). BMC Genomics 16:635
Dias NP, Cagliari D, Dos SE, Smagghe G, Jurat-Fuentes JL, Mishra S, Nava DE, Zotti MJ (2020) Insecticidal gene silencing by RNAi in the neotropical region. Neotrop Entomol 49:1–11
Dong X, Li X, Li Q, Jia H, Zhang H (2017) The inducible blockage of RNAi reveals a role for polyunsaturated fatty acids in the regulation of dsRNA-endocytic capacity in *Bactrocera dorsalis*. Sci Rep 7:1–11
Dönitz J, Schmitt-Engel C, Grossmann D, Gerischer L, Tech M, Schoppmeier M, Klingler M, Bucher G (2015) iBeetle-base: a database for RNAi phenotypes in the red flour beetle *Tribolium castaneum*. Nucleic Acids Res 43:D720–D725
Dubelman S, Fischer J, Zapata F, Huizinga K, Jiang C, Uffman J, Levine S, Carson D (2014) Environmental fate of double-stranded RNA in agricultural soils. PLoS One 9:e93155
Fan YH, Song HF, Abbas M, Wang YL, Li T, Ma EB, Cooper AMW, Silver K, Zhu KY, Zhang JZ (2021) A dsRNA-degrading nuclease (dsRNase2) limits RNAi efficiency in the Asian corn borer (*Ostrinia furnacalis*). Insect Sci 28:1677–1689
Fang F, Wang W, Zhang D, Lv Y, Zhou D, Ma L, Shen B, Sun Y, Zhu C (2015) The cuticle proteins: a putative role for deltamethrin resistance in *Culex pipiens pallens*. Parasitol Res 114:4421–4429
Ferrari J, Vavre F (2011) Bacterial symbionts in insects or the story of communities affecting communities. Philos Trans R Soc Lond Ser B Biol Sci 366:1389–1400
Finetti L, Pezzi M, Civolani S, Calo G, Scapoli C, Bernacchia G (2021) Characterization of *Halyomorpha halys* TAR1 reveals its involvement in (E)-2-decenal pheromone perception. J Exp Biol 224:jeb238816
Finetti L, Benetti L, Leyria J, Civolani S, Bernacchia G (2023) Topical delivery of dsRNA in two hemipteran species: evaluation of RNAi specificity and non-target effects. Pestic Biochem Physiol 189:105295
Fishilevich E, Vélez AM, Khajuria C, Frey MLF, Hamm RL, Wang H, Schulenberg GA, Bowling AJ, Pence HE, Gandra P, Arora K, Storer NP, Narva KE, Siegfried BD (2016) Use of chromatin remodeling ATPases as RNAi targets for parental control of western corn rootworm (*Diabrotica virgifera virgifera*) and Neotropical brown stink bug (*Euschistus heros*). Insect Biochem Mol Biol 71:58–71

Fletcher SJ, Reeves PT, Hoang BT, Mitter N (2020) A perspective on RNAi-based biopesticides. Front Plant Sci 11:51
Ghosh SKB, Hunter WB, Park AL, Gundersen-Rindal DE (2017) Double strand RNA delivery system for plantsap- feeding insects. PLoS One 12:e017186
Ghosh SKB, Hunter WB, Park AL, Gundersen-Rindal DE (2018) Double-stranded RNA oral delivery methods to induce RNA interference in phloem and plant-sap-feeding hemipteran insects. J Vis Exp 4:57390
Guan RB, Li HC, Fan YJ, Hu SR, Christiaens O, Smagghe G, Miao XX (2018) A nuclease specific to lepidopteran insects suppresses RNAi. J Biol Chem 293:6011–6021
Guan R, Chu D, Han X, Miao X, Li H (2021) Advances in the development of microbial double-stranded RNA production systems for application of RNA interference in agricultural pest control. Front Bioeng Biotechnol 9:842
Guillén L, Monribot-Villanueva JL, Guerrero-Analco JA, Ortega R, Altúzar-Molina A, Mena V, Ruiz-May E, Aluja M (2022) Influence of sunlight incidence and fruit chemical features on oviposition site selection in mango by *Anastrepha obliqua*: implications for management. Insects 13:141
Gurusamy D, Howell JL, Chereddy SCRR, Mogilicherla K, Palli SR (2021) Improving RNA interference in the southern green stink bug, *Nezara viridula*. J Pest Sci 94:1461–1472
Hajnsdorf E, Kaberdin VR (2018) RNA polyadenylation and its consequences in prokaryotes. Philos Trans R Soc Lond Ser B Biol Sci 373:20180166
Hasebe M, Kotaki T, Shiga S (2022) Pigment-dispersing factor is involved in photoperiodic control of reproduction in the brown-winged green bug, *Plautia stali*. J Insect Physiol 137:104359
Hashiro S, Yasueda H (2022) RNA interference-based pesticides and antiviral agents: microbial overproduction systems for double-stranded RNA for applications in agriculture and aquaculture. Appl Sci 12:2954
He L, Huang Y, Tang X, River SY (2022) RNAi-based pest control: production, application and the fate of dsRNA. Front Bioeng Biotechnol 10:1080576
Hinas A, Wright AJ, Hunter CP (2012) SID-5 is an endosome-associated protein required for efficient systemic RNAi in *C. elegans*? Curr Biol 22:1938–1943
Hough J, Howard JD, Brown S, Portwood DE, Kilby PM, Dickman MJ (2022) Strategies for the production of dsRNA biocontrols as alternatives to chemical pesticides. Front Bioeng Biotechnol 10:980592
Howell JL, Mogilicherla K, Gurusamy D, Palli SR (2020) Development of RNAi methods to control the harlequin bug, *Murgantia histrionica*. Arch Insect Biochem Physiol 104:e21690
Huang Y, Bai JY, Ren HT (2014) PiRNA biogenesis and its functions. Russ J Bioorg Chem 40:293–299
Jain RG, Robinson KE, Fletcher SJ, Mitter N (2020) RNAi-based functional genomics in Hemiptera. Insects 11:1–22
Jain RG, Robinson KE, Asgari S, Mitter N (2021) Current scenario of RNAi-based hemipteran control. Pest Manag Sci 77:2188–2196
Joga MR, Zotti MJ, Smagghe G, Christiaens O (2016) RNAi efficiency, systemic properties, and novel delivery methods for pest insect control: what we know so far. Front Physiol 7:553
Karlikow M, Goic B, Mongelli V, Salles A, Schmitt C, Bonne I, Zurzolo C, Saleh MC (2016) *Drosophila* cells use nanotube-like structures to transfer dsRNA and RNAi machinery between cells. Sci Rep 6:27085
Kemp C, Mueller S, Goto A, Barbier V, Paro S, Bonnay F, Dostert C, Troxler L, Hetru C, Meignin C, Pfeffer S, Hoffmann JA, Imler J-L (2013) Broad RNA interference-mediated antiviral immunity and virus-specific inducible responses in *Drosophila*. J Immunol 190:650
Khajuria C, Ivashuta S, Wiggins E, Flagel L, Moar W, Pleau M, Miller K, Zhang Y, Ramaseshadri P, Jiang C, Hodge T, Jensen P, Chen M, Gowda A, McNulty B, Vazquez C, Bolognesi R, Haas J, Head G, Clark T (2018) Development and characterization of the first dsRNA-resistant insect population from western corn rootworm, *Diabrotica virgifera virgifera* LeConte. PLoS One 13:e0197059

Kim DH, Rossi JJ (2008) RNAi mechanisms and applications. BioTechniques 44:613–616
Kim K, Lee YS, Carthew RW (2007) Conversion of pre-RISC to holo-RISC by Ago2 during assembly of RNAi complexes. RNA 13:22–29
Kutter C, Svoboda P (2008) miRNA, siRNA, piRNA: knowns of the unknown. RNA Biol 5:181–188
Laisney J, Loczenski Rose V, Watters K, Donohue KV, Unrine JM (2021) Delivery of short hairpin RNA in the neotropical brown stink bug, *Euschistus heros*, using a composite nanomaterial. Pestic Biochem Physiol 177:104906
Lancaster J, Khrimian A, Young S, Lehner B, Luck K, Wallingford A, Ghosh SKB, Zerbe P, Muchlinski A, Marek PE, Sparks ME, Tokuhisa JG, Tittiger C, Köllner TG, Weber DC, Gundersen-Rindal DE, Kuhar TP, Tholl D (2018) De novo formation of an aggregation pheromone precursor by an isoprenyl diphosphate synthase-related terpene synthase in the harlequin bug. Proc Natl Acad Sci USA 115:E8634–E8641
Lei J, Tan Y, List F, Puckett R, Tarone AM, Vargo EL, Zhu-Salzman K (2022) Cloning and functional characterization of a double-stranded RNA-degrading nuclease in the tawny crazy ant (*Nylanderia fulva*). Front Physiol 13:50
Li D, Wu M (2021) Pattern recognition receptors in health and diseases. Signal Transduct Target Ther 6:291
Li Z, Tariq A, Rana M (2011) Molecular mechanisms of RNA-triggered gene silencing machineries. Acc Chem Res 45:1122–1131
Li W, Koutmou KS, Leahy DJ, Li M (2015) Systemic RNA interference deficiency-1 (SID-1) extracellular domain selectively binds long double-stranded RNA and is required for RNA transport by SID-1. J Biol Chem 290:18904–18913
Liang C, Wang Y, Murota Y, Smith D, Siomi MC, Liu CQ (2015) TAF11 assembles the RISC loading complex to enhance RNAi efficiency. Mol Cell 59:807–818
Li-Byarlay H, Li Y, Stroud H, Feng S, Newman TC, Kaneda M, Hou KK, Worley KC, Elsik CG, Wickline SA, Jacobsen SE, Ma J, Robinson GE (2013) RNA interference knockdown of DNA methyltransferase 3 affects gene alternative splicing in the honey bee. Proc Natl Acad Sci USA 110:12750–12755
Lomate PR, Bonning BC (2016) Distinct properties of proteases and nucleases in the gut, salivary gland and saliva of southern green stink bug, *Nezara viridula*. Sci Rep 6:2758
Lu Y, Chen M, Reding K, Pick L (2017) Establishment of molecular genetic approaches to study gene expression and function in an invasive hemipteran, *Halyomorpha halys*. EvoDevo 8:15
Lykke-Andersen S, Brodersen DE, Jensen TH (2009) Origins and activities of the eukaryotic exosome. J Cell Sci 122:1487–1494
Maktura GC, Paranhos BAG, Marques-Souza H (2021) RNAi in fruit flies (Diptera: Tephritidae): successes and challenges. J Appl Entomol 145:740–756
Mamta B, Rajam MV (2017) RNAi technology: a new platform for crop pest control. Physiol Mol Biol Plants 23:487–501
Marshall AT, Cooper WR, Walker WB, Wildung MR, Northfield TD, Krey KL, Beers EH (2023) Salivary protein expression profiles of five species of Pentatomidae (Hemiptera). Ann Entomol Soc Am 116:6
Mello CC, Conte D (2004) Revealing the world of RNA interference. Nature 431:338–342
Mogilicherla K, Howell JL, Palli SR (2018) Improving RNAi in the brown marmorated stink Bug: identification of target genes and reference genes for RT-qPCR. Sci Rep 8:3720
Moriyama M, Hayashi T, Fukatsu T (2022) A mucin protein predominantly expressed in the female-specific symbiotic organ of the stinkbug *Plautia stali*. Sci Rep 12:7782
Nanfack-Minkeu F, Sirot LK (2022) Effects of mating on gene expression in female insects: unifying the field. Insects 13:69
Naqqash MN, Gökçe A, Aksoy E, Bakhsh A (2020) Downregulation of imidacloprid resistant genes alters the biological parameters in Colorado potato beetle, *Leptinotarsa decemlineata* Say (Chrysomelidae: Coleoptera). Chemosphere 240:124857

Nishide Y, Kageyama D, Hatakeyama M, Yokoi K, Jouraku A, Tanaka H, Koga R, Futahashi R, Fukatsu T (2020) Diversity and function of multicopper oxidase genes in the stinkbug *Plautia stali*. Sci Rep 10:3464

Nishide Y, Kageyama D, Tanaka Y, Yokoi K, Jouraku A, Futahashi R, Fukatsu T (2021) Effectiveness of orally-delivered double-stranded RNA on gene silencing in the stinkbug *Plautia stali*. PLoS One 161:e0245081

Nishide Y, Nagamine K, Kageyama D, Moriyama M, Futahashi R, Fukatsu T (2022) A new antimicrobial peptide, Pentatomicin, from the stinkbug *Plautia stali*. Sci Rep 12:16503

Niu J, Taning CNT, Christiaens O, Smagghe G, Wang JJ (2018) Rethink RNAi in insect pest control: challenges and perspectives. Adv In Insect Phys 55:1–17

Nwokeoji AO, Nwokeoji EA, Chou T, Togola A (2022) A novel sustainable platform for scaled manufacturing of double-stranded RNA biopesticides. Bioresour Bioprocess 9:1–16

Obbard DJ, Gordon KHJ, Buck AH, Jiggins FM (2009) The evolution of RNAi as a defence against viruses and transposable elements. Philos Trans R Soc Lond Ser B Biol Sci 364:99–115

Pamarthy S, Kulshrestha A, Katara GK, Beaman KD (2018) The curious case of vacuolar ATPase: regulation of signaling pathways. Mol Cancer 17:1–9

Panizzi AR (2013) History and contemporary perspectives of the integrated pest management of soybean in Brazil. Neotrop Entomol 42:119–127

Parker KM, Barragán Borrero V, Van Leeuwen DM, Lever MA, Mateescu B, Sander M (2019) Environmental fate of RNA interference pesticides: adsorption and degradation of double-stranded RNA molecules in agricultural soils. Environ Sci Technol 53:3027–3036

Peiffer M, Felton GW (2014) Insights into the saliva of the brown marmorated stink bug *Halyomorpha halys* (Hemiptera: Pentatomidae). PLoS One 9:e88483

Peng Y, Wang K, Zhu G, Han Q, Chen J, Elzaki MEA, Sheng C, Zhao C, Palli SR, Han Z (2020) Identification and characterization of multiple dsRNases from a lepidopteran insect, the tobacco cutworm, *Spodoptera litura* (Lepidoptera: Noctuidae). Pestic Biochem Physiol 162:86–95

Pinheiro DH, Moreira RO, Leite NA, Redoan AC, da Xavier AS, de Barros BA, Carneiro NP (2020) Suitable reference genes for RT-qPCR analysis in *Dichelops melacanthus* (Hemiptera: Pentatomidae). Mol Biol Rep 47:4989–5000

Price DRG, Gatehouse JA (2008) RNAi-mediated crop protection against insects. Trends Biotechnol 26:393–400

Pugsley CE, Isaac RE, Warren NJ, Cayre OJ (2021) Recent advances in engineered nanoparticles for RNAi-mediated crop protection against insect pests. Front Agron 3:9

Ramachandran V, Chen X (2008) Degradation of microRNAs by a family of exoribonucleases in *Arabidopsis*. Science 321:1490–1492

Rank AP, Koch A (2021) Lab-to-field transition of RNA spray applications—how far are we? Front Plant Sci 12:2243

Saha D (2016) Biochemical insecticide resistance in tea pests. In: Trdan S (ed) Insecticides Resistance. Intech Open, University of Ljubljana, Slovenia, pp 347–390

Saleh MC, van Rij RP, Hekele A, Gillis A, Foley E, O'Farrell PH, Andino R (2006) The endocytic pathway mediates cell entry of dsRNA to induce RNAi silencing. Nat Cell Biol 8:793–802

Saleh MC, Tassetto M, Van Rij RP, Goic B, Gausson V, Berry B, Jacquier C, Antoniewski C, Andino R (2009) Antiviral immunity in *Drosophila* requires systemic RNA interference spread. Nature 458:346–350

Samir M, Vaas LAI, Pessler F (2016) MicroRNAs in the host response to viral infections of veterinary importance. Front Vet Sci 3:212851

San Miguel K, Scott JG (2016) The next generation of insecticides: dsRNA is stable as a foliar-applied insecticide. Pest Manag Sci 72:801–809

Santos-Ortega Y, Flynt A (2022) Double-strand RNA (dsRNA) delivery methods in insects: *Diaphorina citri*. Methods Mol Biol 2360:253

Schmitt-Engel C, Schultheis D, Schwirz J, Ströhlein N, Troelenberg N, Majumdar U, Dao VA, Grossmann D, Richter T, Tech M, Dönitz J, Gerischer L, Theis M, Schild I, Trauner J, Koniszewski NDB, Küster E, Kittelmann S, Hu Y, Lehmann S, Siemanowski J, Ulrich J,

Panfilio KA, Schröder R, Morgenstern B, Stanke M, Buchhholz F, Frasch M, Roth S, Wimmer EA, Schoppmeier M, Klingler M, Bucher G (2015) The iBeetle large-scale RNAi screen reveals gene functions for insect development and physiology. Nat Commun 6:7822

Schumacher K, Krebs M (2010) The V-ATPase: small cargo, large effects. Curr Opin Plant Biol 13:724–730

Schvartzman C, Fresia P, Murchio S, Mujica MV, Dalla-Rizza M (2022) RNAi in *Piezodorus guildinii* (Hemiptera: Pentatomidae): transcriptome assembly for the development of pest control strategies. Front Plant Sci 13:804839

Sharma R, Christiaens O, Taning CNT, Smagghe G (2021a) RNAi-mediated mortality in southern green stinkbug *Nezara viridula* by oral delivery of dsRNA. Pest Manag Sci 77:77–84

Sharma R, Nji C, Taning T, Smagghe G, Christiaens O (2021b) Silencing of double-stranded ribonuclease improves oral RNAi efficacy in southern green stinkbug *Nezara viridula*. Insects 12:115

Shih JD, Hunter CP (2011) SID-1 is a dsRNA-selective dsRNA-gated channel. RNA 17:1057–1065

Silke J, Meier P (2013) Inhibitor of apoptosis (IAP) proteins–modulators of cell death and inflammation. Cold Spring Harb Perspect Biol 5:a008730

Singh IK, Singh S, Mogilicherla K, Shukla JN, Palli SR (2017) Comparative analysis of double-stranded RNA degradation and processing in insects. Sci Rep 7:1–12

Souza D, Christensen SA, Wu K, Buss L, Kleckner K, Darrisaw C, Shirk PD, Siegfried BD (2022) RNAi-induced knockdown of white gene in the southern green stink bug (*Nezara viridula* L.). Sci Rep 12:10396

Sparks ME, Shelby KS, Kuhar D, Gundersen-Rindal DE (2014) Transcriptome of the invasive brown marmorated stink bug, *Halyomorpha halys* (Stål) (Heteroptera: Pentatomidae). PLoS One 9:e111646

Sparks ME, Rhoades JH, Nelson DR, Kuhar D, Lancaster J, Lehner B, Tholl D, Weber DC, Gundersen-Rindal DE (2017) A transcriptome survey spanning life stages and sexes of the harlequin bug, *Murgantia histrionica*. Insects 8:55

Spit J, Philips A, Wynant N, Santos D, Plaetinck G, Vanden Broeck J (2017) Knockdown of nuclease activity in the gut enhances RNAi efficiency in the Colorado potato beetle, *Leptinotarsa decemlineata*, but not in the desert locust, *Schistocerca gregaria*. Insect Biochem Mol Biol 81:103–116

Sugahara R, Minaba M, Jouraku A, Kotaki T, Yamamoto T, Shinohara Y, Miyoshi H, Shiotsuki T (2016) Characterization of two adenine nucleotide translocase paralogues in the stink bug, *Plautia stali*. J Pestic Sci 41:44–48

Sugahara R, Jouraku A, Nakakura T, Minaba M, Yamamoto T, Shinohara Y, Miyoshi H, Shiotsuki T (2017) Tissue-specific expression and silencing phenotypes of mitochondrial phosphate carrier paralogues in several insect species. Insect Mol Biol 26:332–342

Suzuki H, Umekage S, Tanaka T, Kikuchi Y (2011) Artificial RNA aptamer production by the marine bacterium *Rhodovulum sulfidophilum*: improvement of the aptamer yield using a mutated transcriptional promoter. J Biosci Bioeng 112:458–461

Takeuchi H, Endo N (2012) Insecticide susceptibility of *Nezara viridula* (Heteroptera: Pentatomidae) and three other stink bug species composing a soybean pest complex in Japan. J Econ Entomol 105:1024–1033

Tamai T, Shiga S, Goto SG (2019) Roles of the circadian clock and endocrine regulator in the photoperiodic response of the brown-winged green bug *Plautia stali*. Physiol Entomol 44:43–52

Tang G (2005) siRNA and miRNA: an insight into RISCs. Trends Biochem Sci 30:106–114

Taning CNT, Arpaia S, Christiaens O, Dietz-Pfeilstetter A, Jones H, Mezzetti B, Sabbadini S, Sorteberg HG, Sweet J, Ventura V, Smagghe G (2020) RNA-based biocontrol compounds: current status and perspectives to reach the market. Pest Manag Sci 76:841–845

Tayler A, Heschuk D, Giesbrecht D, Park JY, Whyard S (2019) Efficiency of RNA interference is improved by knockdown of dsRNA nucleases in tephritid fruit flies. Open Biol 9:190198

Terenius O, Papanicolaou A, Garbutt JS, Eleftherianos I, Huvenne H, Kanginakudru S, Albrechtsen M, An C, Aymeric JL, Barthel A, Bebas P, Bitra K, Bravo A, Chevalier F, Collinge DP, Crava CM, de Maagd RA, Duvic B, Erlandson M, Faye I, Felföldi G, Fujiwara H, Futahashi R, Gandhe AS, Gatehouse HS, Gatehouse LN, Giebultowicz JM, Gómez I, Grimmelikhuijzen CJP, Groot AT, Hauser F, Heckel DG, Hegedus DD, Hrycaj S, Huang L, Hull JJ, Iatrou K, Iga M, Kanost MR, Kotwica J, Li C, Li J, Liu J, Lundmark M, Matsumoto S, Meyering-Vos M, Millichap PJ, Monteiro A, Mrinal N, Niimi T, Nowara D, Ohnishi A, Oostra V, Ozaki K, Papakonstantinou M, Popadic A, Rajam MV, Saenko S, Simpson RM, Soberón M, Strand MR, Tomita S, Toprak U, Wang P, Wee CW, Whyard S, Zhang W, Nagaraju J, Ffrench-Constant RH, Herrero S, Gordon K, Swevers L, Smagghe G (2011) RNA interference in Lepidoptera: An overview of successful and unsuccessful studies and implications for experimental design. J Insect Physiol 57:231–245

Thammasorn T, Jitrakorn S, Charoonnart P, Sirimanakul S, Rattanarojpong T, Chaturongakul S, Saksmerprome V (2017) Probiotic bacteria (*Lactobacillus plantarum*) expressing specific double-stranded RNA and its potential for controlling shrimp viral and bacterial diseases. Aquac Int 25:1679–1692

Thomas MF, L'Etoile ND, Ansel KM (2014) Eri1: a conserved enzyme at the crossroads of multiple RNA-processing pathways. Trends Genet 30:298–307

Tian F, Li C, Wang Z, Liu J, Zeng X (2019) Identification of detoxification genes in imidacloprid-resistant Asian citrus psyllid (Hemiptera: Lividae) and their expression patterns under stress of eight insecticides. Pest Manag Sci 75:1400–1410

Tibola CM, Silva L, Sgubin F, Omoto C (2021) Monitoring resistance of *Euschistus heros* (Fabricius) (Hemiptera: Pentatomidae) to insecticides by using encapsulated artificial diet bioassay. Insects 12:599

Tomoyasu Y, Miller SC, Tomita S, Schoppmeier M, Grossmann D, Bucher G (2008) Exploring systemic RNA interference in insects: a genome-wide survey for RNAi genes in Tribolium. Genome Biol 9:1–22

Van Ekert E, Powell CA, Shatters RG, Borovsky D (2014) Control of larval and egg development in *Aedes aegypti* with RNA interference against juvenile hormone acid methyl transferase. J Insect Physiol 70:143–150

Vélez AM, Fishilevich E (2018) The mysteries of insect RNAi: a focus on dsRNA uptake and transport. Pestic Biochem Physiol 151:25–31

Vogel E, Santos D, Mingels L, Verdonckt TW, Vanden BJ (2018) RNA interference in insects: protecting beneficials and controlling pests. Front Physiol 9:1912

Wang K, Peng Y, Pu J, Fu W, Wang J, Han Z (2016) Variation in RNAi efficacy among insect species is attributable to dsRNA degradation in vivo. Insect Biochem Mol Biol 77:1–9

Wasmuth EV, Januszyk K, Lima CD (2014) Structure of an Rrp6-RNA exosome complex bound to polyA RNA. Nature 511:435–439

Whitten MMA, Facey PD, del Sol R, Fernández-Martínez LT, Evans MC, Mitchell JJ, Bodger OG, Dyson PJ (2016) Symbiont-mediated RNA interference in insects. Proc Biol Sci 283:2016004

Willow J, Veromann E (2021) Highly variable dietary RNAi sensitivity among Coleoptera. Front Plant Sci 12:790816

Wytinck N, Manchur CL, Li VH, Whyard S, Belmonte MF (2020) DsRNA uptake in plant pests and pathogens: insights into RNAi-based insect and fungal control technology. Plan Theory 9:1–17

Xie W, Sowemimo I, Hayashi R, Wang J, Burkard TR, Brennecke J, Ameres SL, Patel DJ (2020) Structure-function analysis of microRNA 3′-end trimming by Nibbler. Proc Natl Acad Sci USA 117:30370–30379

Xu W, Jiang X, Huang L (2019) RNA interference technology. In: Comprehensive biotechnology, pp 560–575

Yamanaka S, Mehta S, Reyes-Turcu FE, Zhuang F, Fuchs RT, Rong Y, Robb GB, Grewal SIS (2012) RNAi triggered by specialized machinery silences developmental genes and retrotransposons. Nature 493:557–560

Yao Y, Lin DJ, Cai XY, Wang R, Hou YM, Hu CH, Gao SJ, Da WJ (2022) Multiple dsRNases involved in exogenous dsRNA degradation of fall armyworm *Spodoptera frugiperda*. Front Physiol 13:814
Ye C, An X, Di JY, Ding BY, Shang F, Christiaens O, Taning CNT, Smagghe G, Niu J, Wang JJ (2019) Induction of RNAi core machinery's gene expression by exogenous dsRNA and the effects of pre-exposure to dsRNA on the gene silencing efficiency in the pea aphid (*Acyrthosiphon pisum*). Front Physiol 10:1–10
Yoon JS, Mogilicherla K, Gurusamy D, Chen X, Chereddy SCRR, Palli SR (2018) Double-stranded RNA binding protein, Staufen, is required for the initiation of RNAi in coleopteran insects. Proc Natl Acad Sci USA 115:8334–8339
Zerbino MS, Panizzi AR (2019) The underestimated role of pest pentatomid parasitoids in Southern South America. Arthropod Plant Interact 13:703–718
Zhang Y, Yang Y, Sun H, Liu Z (2016) Metabolic imidacloprid resistance in the brown planthopper, *Nilaparvata lugens*, relies on multiple P450 enzymes. Insect Biochem Mol Biol 79:50–56
Zhang X, Fan Z, Wang Q, Kong X, Liu F, Fang J, Zhang S, Zhang Z (2022) RNAi efficiency through dsRNA injection is enhanced by knockdown of dsRNA nucleases in the fall webworm, *Hyphantria cunea* (Lepidoptera: Arctiidae). Int J Mol Sci 23:6182
Zhao JH, Guo HS (2022) RNA silencing: from discovery and elucidation to application and perspectives. J Integr Plant Biol 64:476–498
Zhou R, Rana TM (2013) RNA-based mechanisms regulating host-virus interactions. Immunol Rev 253:97
Zhu KY, Palli SR (2020) Mechanisms, applications, and challenges of insect RNA interference. Annu Rev Entomol 65:293–311

Chapter 9
The CRISPR Technology to Manage Stink Bugs: Potential, Challenges and Major Benefits

Renata Fuganti Pagliarini, Mayla Daiane Correa Molinari, Alexandre Lima Nepomuceno, and Adeney Freitas Bueno

1 Introduction

As a result of climate change, plant pests are probable to have considerable effects on crop density, distribution, yield, and crop. According to FAO (Food and Agriculture Organization) plant diseases and pests are accountable for a 20–40% reduction in global crop yields per year and this destruction can lead to the starvation of millions of people (FAO 2017; Singh et al. 2022).

Among pests, stink bugs (Hemiptera: Pentatomidae) are a big concern in many important economic crops worldwide (McPherson and McPherson 2000; Schafer and Panizzi 2000). They get their name from the strong-smelling substance they spray when threatened, disturbed, or crushed. The composition of this substance may vary between species, and even by sex or age, but usually includes aldehydes and alkanes. Descriptions of the smells vary widely and include oily, dusty, woody, earthy, and coriander-like. In some species, it contains cyanide compounds and a rancid almond scent, used to protect themselves and discourage predators (Capinera 2001).

In general, first instars feed little or not at all, but nymphs from third instar and adults feed on tender shoot tissues, buds, and fruit causing significant damage when not properly managed. They feed on plant fluids by inserting their needlelike mouthparts into stems, leaves, or seed pods. While feeding, they inject materials into the

R. F. Pagliarini (✉)
Piccola Scientific Consulting, Saskatoon, SK, Canada

M. D. C. Molinari
SEMPRE Ag Tech/WIN, São Paulo, SP, Brazil

A. L. Nepomuceno · A. F. Bueno
Embrapa Soja, Londrina, PR, Brazil

A. F. Bueno, A. R. Panizzi (eds.), *Stink Bugs (Hemiptera: Pentatomidae) Research and Management*, Entomology in Focus 9,
https://doi.org/10.1007/978-3-031-69742-5_9

plant to aid in digestion and sap removal. The more common and harmful species are *Nezara viridula* (L.), *Piezodorus guildinii* (Westwood) *Euschistus heros* (F.), *Diceraeus* spp. and *Halyomorpha halys* (Stål), among others (Schaefer and Panizzi 2000).

Nezara viridula is commonly known as the southern green stink bug (USA), southern green shield bug (UK) or green vegetable bug (Australia and New Zealand). This polyphagous species usually prefers certain species of legumes, such as beans and soybean, being an economically important pest on such crops, as well as in maize, cotton, and various fruits and vegetables (Todd 1989). All plant parts are likely to be fed but growing shoots and developing fruit are preferred. The red-banded stink bug, *P. guildinii*, causes significant damage to crops, such as soybean, alfalfa, and other legumes. In soybean, it can cause damage to the leaves, stems, flowers and pods, reducing yields, and delaying plant maturation. Pods are marked and the internal grains can shrivel (Panizzi et al. 2012).

Commonly known as the Neotropical brown stink bug, *E. heros*, feeds on several crop plants, however, is considered a major soybean pest in Brazil. Crop losses are mainly caused by the injection of saliva containing tissue-destroying enzymes resulting in damage to the fruits and seeds, and secondarily by physical damage. Popularly known as green-belly stink bugs, two species have economic importance in the Neotropics: *Diceraeus furcatus* (F.) and *D. melacanthus* (Dallas), occurring mainly in soybean, maize, and wheat. Like *E. heros*, *D. melacanthus* is considered a key pest in soybean crops, mainly in southern Brazil (Panizzi et al. 2012).

The brown marmorated stink bug, *Halyomorpha halys* is a very serious pest that feeds on more than 100 different plant species. Native to Asia, is spread throughout most of the United States, Europe, and many other parts of the world. Both adults and nymphs feed by inserting their mouthparts into the flesh tissue of fruit or vegetables, resulting in small dead areas. Besides fruits and vegetables, this bug also causes significant damage to grain crops. It is a very serious pest that feeds on more than 100 different plant species (see references in Chaps. 15 and 16).

In maize, stink bugs feeding symptoms include chlorotic lesions, tightly rolled leaves, wilting, stunting, increased tillering, delayed silk production, and smaller grain weight (Ávila and Panizzi 1995; Capinera 2001). The most dramatic symptom is the tillering of damaged plants, which appears about 10 days after the damage was caused. Tillered plants may develop misshapen ears in place of the tassel and produce little if any grain, significantly reducing the final yield (Bessin 2023).

Amongst the management strategies, the main and the one considered the highest level of effectiveness is chemical control. Insecticides are effective when applied to foliage and young fruits, with special care warranted to protect the blossoms and fruit. For some species, it's important to obtain good penetration of canopy foliage down to the soil level, because these stink bugs move to lower portions of the plant, in response to sunlight incidence and temperature variation during the time of day (Silva et al. 2021). Border treatments are especially important once bugs enter from the crop margins, making this portion of the field injured most severely (Capinera 2001). In some cases, trap plants can be used for management, such as weed control and the adoption of borders with early cycle varieties for initial management, reducing the pressure for other areas (Hokkanen 1991). In addition, among alternative control methods, natural enemies are also an option. For *Nezara viridula* for example, an insect with worldwide distribution, numerous parasitoids and predators are known. The most common parasitoids are the egg parasitoid *Trissolcus basalis* (Hymenoptera: Scelionidae) and the adult parasitoid *Trichopoda pennipes* (Diptera: Tachinidae). Common egg predators are red fire ants, *Solenopsis invicta* (Hymenoptera: Formicidae). Common nymphal predators are big-eyed bug *Geocoris* spp. (Hemiptera: Lygaeidae), damsel bug, *Reduviolus roseipennis* (Hemiptera: Nabidae), and other insects from Hemiptera and Coleoptera orders. Spiders sometimes cause significant predation (Panizzi et al. 2012).

Another example of biological control is the management of *H. halys* which includes the use of *Trissolcus japonicus* to parasitise the eggs. Predators such as spiders, ladybugs, red velvet mites and lacewings will also feed on brown marmorated stink bug eggs (see references on Chaps. 15 and 16). In general terms, the success of stink bug management depends on monitoring populations and optimizing management opportunities for this pest in the crop cultivation system, following integrated management practices. It is essential to identify the insect species and monitor the areas to define control.

Despite the availability of control methods, new possibilities on how to control/decrease stink bug populations and minimize damage in crops, are always a focus in research. As known, in the last decades, biotechnology has been an important partner of other strategies in the control and management of pests and diseases and gene editing is among these new methodologies, bringing new possibilities to researchers as well as to producers around the world.

This chapter will address the potential, basic and applied research on stink bug control using gene editing tools. The bibliography used in this chapter was selected using the automation software Harzing's Publish or Perish on the Google Scholar platform (Harzing 2010). The eligibility criterion consisted of selecting materials published between the years 2017 and 2023 from keywords contextualized to the proposed theme (Cogo 2020). The search terms used were: CRISPR or gene edit and Pentatomidae; CRISPR or gene edit and stink bug; CRISPR or gene edit and Hemiptera, CRISPR or gene edit and *Euschistus heros*; CRISPR or gene edit and *Diceraeus melacanthus*; CRISPR or gene edit and *Nezara viridula*; CRISPR or gene edit and *Vatiga illudens*; CRISPR or gene edit and *Vatiga manihotae*; CRISPR or gene edit and *Halyomorpha halys* and CRISPR or gene edit and *Piezodorus guildinii*. Results from these searches did not return many articles revealing the lack of basic and applied research using gene editing tools in stink bug management and control (Table 9.1).

Although there are not a lot of research groups applying editing tools and developing edited stink bugs, the potential of techniques such as CRISPR-Cas and its variants to the study of gene function, the importance of them in the insect's development, among other characteristics, and ultimately in the development of pest management tools is tremendous and can revolutionize insect control, reducing losses and facilitating general crop administration.

Table 9.1 Summary of research articles published on CRISPR/Cas9 gene editing in Hemiptera

Title	Species	Common name	Author
Efficient crispr/Cas9-mediated genome modification of the glassy-winged sharpshooter *Homalodisca vitripennis* (Germar)	*Homalodisca vitripennis*	Glassy-winged sharpshooter	Pacheco et al. (2022)
Genome editing of the vermilion locus generates a visible eye color marker for *Oncopeltus fasciatus*	*Oncopeltus fasciatus*	Milkweed bug	Reding et al. (2023)
High-efficiency crispr/Cas9 mutagenesis of the white gene in the milkweed bug *Oncopeltus fasciatus*	*Oncopeltus fasciatus*	Milkweed bug	Reding and Pick (2020)
An integrated protocol for targeted mutagenesis with crispr-Cas9 system in the pea aphid	*Acyrthosiphon pisum*	Pea aphid	Le Trionnaire et al. (2019)
Crispr-Cas9-based genome editing in the silverleaf whitefly (*Bemisia tabaci*)	*Bemisia tabaci*	Whitefly	Heu et al. (2020)
Functional analysis of a NPC1 gene from the whitefly, *Bemisia tabaci* (Hemiptera: Aleyrodidae)	*Bemisia tabaci*	Whitefly	Yu et al. (2023)
Crispr/Cas9-mediated mutagenesis of the Asian citrus psyllid, *Diaphorina citri*	*Diaphorina citri*	Asian citrus psyllid	Chaverra-Rodriguez et al. (2023)
BAPC-assisted crispr/Cas9 system: Targeted delivery into adult ovaries for heritable germline gene editing (Arthropoda: Hemiptera)	*Diaphorina citri*	Asian citrus psyllid	Hunter et al. (2018)
BAPC-assisted-crispr-Cas9 delivery into adult psyllid ovaries and pupae for heritable gene editing (Hemiptera). 2019.	*Diaphorina citri*	Asian citrus psyllid	Hunter and Tomich (2019)
BAPC-assisted-crispr-Cas9 delivery into nymphs and adults for heritable gene editing (Hemiptera)	*Diaphorina citri, Homalodisca vitripennis, Bemisia argentifolii*	Asian citrus psyllid, Glassy-winged sharpshooter, Whitefly	Hunter et al. (2019)
Transcriptome analysis and exploring RNAi and CRISPR in the Neotropical stink bug, *Euschistus heros*	*Euschistus heros*	Stink bug	Cagliari (2021)
RNAi and CRISPR/Cas9 as functional genomics tools in the Neotropical stink bug, *Euschistus heros*	*Euschistus heros*	Stink bug	Cagliari et al. (2020a)
First transcriptome of the Neotropical pest *Euschistus heros* (Hemiptera: Pentatomidae) with dissection of its siRNA machinery	*Euschistus heros*	Stink bug	Cagliari et al. (2020b)
Crispr-mediated knockout of cardinal and cinnabar eye pigmentation genes in the western tarnished plant bug	*Lygus hesperus*	Western tarnished plant bug	Heu et al. (2022)

(continued)

Table 9.1 (continued)

Title	Species	Common name	Author
Spirotetramat resistance in *Myzus persicae* (Sulzer) (Hemiptera: Aphididae) and its association with the presence of the A2666V mutation	*Myzus persicae*	Green peach aphid	Umina et al. (2022)
Orco mutagenesis causes deficiencies in olfactory sensitivity and fertility in the migratory brown planthopper, *Nilaparvata lugens*	*Nilaparvata lugens*	Brown planthopper	Liu et al. (2023)
A chitin synthase mutation confers widespread resistance to buprofezin, a chitin synthesis inhibitor, in the brown planthopper, *Nilaparvata lugens*	*Nilaparvata lugens*	Brown planthopper	Zeng et al. (2022)
Crispr/Cas9-mediated genome editing of gustatory receptor NlugGr23a causes male sterility in the brown planthopper *Nilaparvata lugens*	*Nilaparvata lugens*	Brown planthopper	Zhang et al. (2023a)
Crispr/Cas9-mediated knockout of NlCYP6CS1 gene reveals its role in detoxification of insecticides in *Nilaparvata lugens* (Hemiptera: Delphacidae)	*Nilaparvata lugens*	Brown planthopper	Zhang et al. (2023b)
Crispr/Cas9-mediated knockout of the NlCSAD gene results in darker cuticle pigmentation and a reduction in female fecundity in *Nilaparvata lugens* (Hemiptera: Delphacidae)	*Nilaparvata lugens*	Brown planthopper	Chen et al. (2021)
Crispr/Cas9-mediated knockout of two eye pigmentation genes in the brown planthopper, *Nilaparvata lugens* (Hemiptera: Delphacidae)	*Nilaparvata lugens*	Brown planthopper	Xue et al. (2018)
FoxO is required for optimal fitness of the migratory brown planthopper, *Nilaparvata lugens* (Hemiptera: Delphacidae)	*Nilaparvata lugens*	Brown planthopper	Zhang et al. (2022)
Importance of CYP6ER1 was different among neonicotinoids in their susceptibility in *Nilaparvata lugens*	*Nilaparvata lugens*	Brown planthopper	Zhang et al. (2023c)
Mutations in NlInR1 affect normal growth and lifespan in the brown planthopper *Nilaparvata lugens*	*Nilaparvata lugens*	Brown planthopper	Zhao et al. (2019)
Developing RNAi-and CRISPR-based control methods for the corn planthopper, *Peregrinus maidis*	*Peregrinus maidis*	Corn planthopper	Wang (2022)
Microinjection of corn planthopper, *Peregrinus maidis*, embryos for CRISPR/Cas9 genome editing	*Peregrinus maidis*	Corn planthopper	Klobasa et al. (2021)
Crispr/Cas9 genome editing introduction and optimization in the non-model insect *Pyrrhocoris apterus*	*Pyrrhocoris apterus*	Linden bug	Kotwica-Rolinska et al. (2019)

(continued)

Table 9.1 (continued)

Title	Species	Common name	Author
Functional analysis and localization of a thyrotropin-releasing hormone-type neuropeptide (EFLa) in Hemipteran insects	*Pyrrhocoris apterus*	Linden bug	Kotwica-Rolinska et al. (2020)
Efficient DIPA-CRISPR-mediated knockout of an eye pigment gene in the white-backed planthopper, *Sogatella furcifera*	*Sogatella furcifera*	White-backed planthopper	Zhang et al. (2023d)
Efficient nanoparticle-based crispr-Cas13d induced mRNA disruption of an eye pigmentation gene in the white-backed planthopper, *Sogatella furcifera*	*Sogatella furcifera*	White-backed planthopper	Ma et al. (2023)

The table lists authors, article titles, species studied, common names, the order of the insects, the year of publication, and URLs for the articles. The table serves as a compilation of recent advancements in the field of gene editing for insect pest management, with a focus on various species of Hemiptera

For information purposes, Fig. 9.1 shows in detail a schematic representation of the genome editing mechanism facilitated by the CRISPR-Cas9 technology.

2 Example of Applied Research on Stink Bug Control Using CRISPR/Cas9 Editing Toolbox

The first workflow for CRISPR/Cas9 gene knockout for a stink bug species was established in 2020 by Cagliari and co-workers in *Euschistus heros*. Initially, these authors applied RNAi to study the function of three genes, abnormal wing disc (*awd*), tyrosine hydroxylase (*th*) and yellow (*yel*). The *awd* and the *th* genes knockdown resulted, respectively, in deformed wings or lighter cuticle pigmentation/defects in the cuticle sclerotization. No distinct phenotype was observed for the *yel* gene. Further, to understand the function of this gene, the CRISPR/Cas9-mediated knockout technique was developed. The *yel* gene has two isoforms in *E. heros*, so to ensure successful expression knockout of both isoforms, a single-guide RNA was designed within the coding sequence. A total of 719 eggs were microinjected with single-guide (sgRNA) and Cas9 and six insects hatched to first-instar nymphs (~1% hatching rate). As a control, 276 eggs were injected with water and 28 successfully hatched into first-instar nymphs (10% hatching rate). Out of the 6 nymphs from the test group, three were sequenced and one insect showed a mutation (deletion) in the target region of *yel* gene, but no clear phenotype was observed. This study provided, for the first time, CRISPR/Cas9 gene editing methodology to complement RNAi for functional genomic studies in one of the most important and economically relevant stink bug species (Cagliari et al. 2020a).

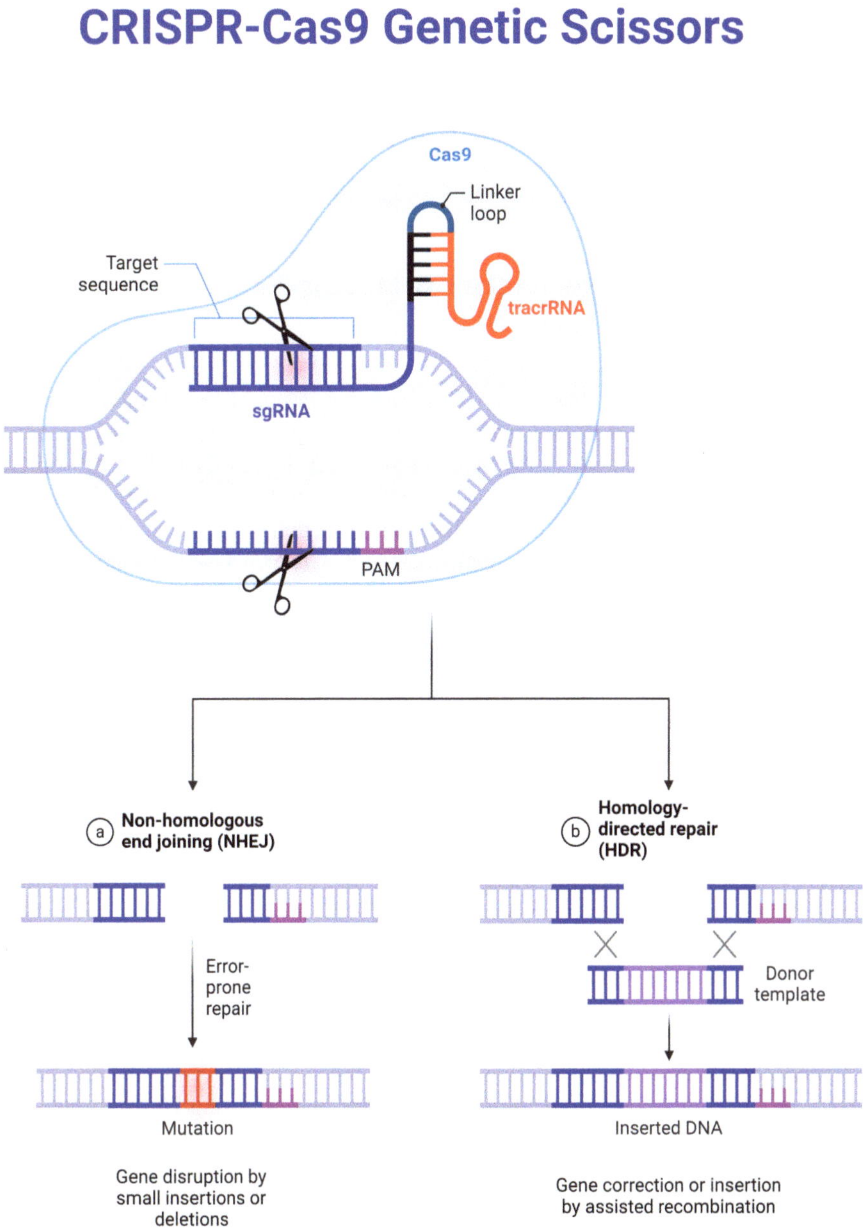

Fig. 9.1 Schematic representation of the genome editing mechanism facilitated by the CRISPR-Cas9 technology. The schema illustrates the assembly of the ribonucleoprotein complex comprising the Cas9 enzyme and the single-guide RNA (sgRNA), which targets the complex to the specific genomic site via base pairing with the target sequence and adjacency to the protospacer adjacent motif (PAM). The subsequent induction of double-stranded breaks in the DNA initiates cellular repair mechanisms, wherein non-homologous end joining (NHEJ) may result in insertions or deletions (indels) leading to mutagenesis through gene disruption, while homology-directed repair (HDR) allows for precise insertion or correction of sequences using a homologous DNA template. (This diagram was generated by the authors using the BioRender graphic design platform)

It is important to highlight that in the context of genome editing tools, the generation of a genome database for *E. heros* would greatly facilitate the use of CRISPR/Cas9 as a molecular instrument for functional genomics in this species. Many other studies are ongoing but the one described above developed by Cagliari and co-workers is the only one that effectively applied CRISPR/Cas9 tool as a control method for a stink bug species.

3 Future Perspectives on the Use of Gene-Editing Techniques in Insect Pest Management

The CRISPR/Cas9 system has arisen, in the past decades, as a classy and reasonable genetic technique, and it is expected to be broadly applied in pest control applications for crop improvement in the near future. Even though CRISPR/Cas9-based genome editing methods have made fast progress, researchers still look for improvement in some areas such as off-target effects, efficient delivery methods and genome editing efficiency. Despite these challenges, the CRISPR/Cas9 system has the potential to revolutionize the field of agriculture.

Contributing to the use of gene-editing methods, in the last decades, an increase in genome and transcriptome sequence databases for non-model insects, together with the development of high-throughput techniques for gene expression profiling and functional characterization provided tools to study the biology of non-model insects, such as stink bugs (Cagliari et al. 2020a). The transcriptome of *E. heros* recently published provides a good basis to explore the biology of this important pest species (Cagliari et al. 2020b).

Besides the use of CRISPR/Cas9 technology to directly edit genes, with the advent of this method, new life was given to the idea of using gene drives to control pests. The principle of this approach is to use CRISPR to create a "gene driver" to spread a gene rapidly through generations. In a few words, gene drives are a genetic system with the ability to 'drive' themselves and nearby genes through populations of organisms over many generations. In sexual reproduction, one set of chromosomes is provided by each parent and combined in their offspring but if one set of chromosomes contains a 'gene drive' it will cut the partner chromosome that lacks the gene drive and copy itself onto this chromosome, ensuring for example, the spread of a gene that suppresses fertility in females in a pest species population (Dearden et al. 2017). Some challenges are associated with the practical use of gene drivers in insects such as the availability of the species' genome, delivery methodology, social organization of the insect species, and the generation of resistant versions of the gene. Besides that, the risk, and consequences of extinction of an insect species must be considered worldwide and not only locally. In addition, the regulatory framework of genetically modified organisms is a complex subject in many countries and risk assessment of produced organisms through the gene drive system, among other studies, would need to be carried out and insect releases monitored following proper legislation. Another reflection to be considered is the public acceptance of the technology, which always demands the right information, time and significant and impactful results.

4 Final Considerations

In an increased food-demanding modern agriculture system, excessive and worldwide use of pesticides to control agricultural pests harms nontargeted organisms, the environment as well as on human beings. Besides that, prolonged use of chemicals may decrease crop yield gradually, followed by an accumulation in the food chain and posing a danger to human health with eroded profits. Therefore, efficient control alternative methods, as well as the development of new integrated pest management strategies, economically viable, must be developed in a way to protect plants against harmful pathogens and at the same time, show zero side effects on the non-target insect population, preserving animal and human health, ensuring crop improvement, and potential increases in yield to fulfill and guarantee food safety.

With the advance of plant biotechnology, the clustered regularly interspaced short palindromic repeats (CRISPR)/CRISPR-associated gene (Cas9) and its variants appear as a promising tool for pest control. Although much progress has been made, there are still many challenges to overcome but researchers are focused on developing and improving the technique to be applied to insect management and promising results are to come in a near future.

References

Ávila CJ, Panizzi AR (1995) Occurrence and damage by *Dichelops (Neodichelops) melacanthus* (Dallas) (Heteroptera: Pentatomidae) on corn. An Soc Entomol Brasil 24:193–194

Bessin R (2023) Stink bug damage to crop. University of Kentucky. Available in https://entomology.ca.uky.edu/ef305

Cagliari D (2021) Transcriptome analysis and exploring RNAi and CRISPR in the Neotropical stink bug, *Euschistus heros*. Doctoral dissertation, Post Graduation Program in Crop Protection, Universidade Federal de Pelotas, and in Bioscience Engineering, Universiteit Gent, 219 p

Cagliari D, Smagghe G, Zotti M, Taning CNT (2020a) RNAi and CRISPR/Cas9 as functional genomics tools in the Neotropical stink bug, *Euschistus heros*. Insects 11:838

Cagliari D, Dias NP, dos Santos EÁ, Rickes LN, Kremer FS, Farias JR, Lenz G, Galdeano DM, Garcia FRM, Smagghe G, Zotti MJ (2020b) First transcriptome of the Neotropical pest *Euschistus heros* (Hemiptera: Pentatomidae) with dissection of its siRNA machinery. Sci Rep 10:1–16

Capinera J (2001) Order Hemiptera: bugs. In: Handbook of vegetable pests. Available in https://doi.org/10.1016/B978-0-12-158861-8.X5000-5

Chaverra-Rodriguez D, Bui M, Gilleland CL, Rasgon JL, Akbari OS (2023) CRISPR-Cas9-mediated mutagenesis of the Asian citrus psyllid, *Diaphorina citri*. GEN Biotechnol 2:317–329

Chen JX, Li WX, Lyu J, Hu YT, Huang G, Zhang WQ (2021) CRISPR/Cas9-mediated knockout of the NlCSAD gene results in darker cuticle pigmentation and a reduction in female fecundity

in *Nilaparvata lugens* (Hemiptera: Delphacidae). Comp Biochem Physiol A Mol Integr Physiol 256:110921
Cogo FD (2020). Introdução à revisão sistemática e meta-análise aplicadas à agricultura. Editora UFMG. ISBN: 978-65-86832-06-8. 65 páginas. 1ª ed. https://doi.org/10.36704/9786586832068
Corrêa-Ferreira B, De Azevedo J (2002) Soybean seed damage by different species of stink bugs. Agric For Entomol 4:145–150. The Royal Entomological Society. https://doi.org/10.1046/j.1461-9563.2002.00136.x
Dearden PK, Gemmel NJ, Mercier OR, Lester PJ, Scoot MJ, Newcomb RD, Buckley TR, Jacobs JME, Goldson SG, Penman DR (2017) The potential for the use of gene drives for pest control in New Zealand: a perspective. J R Soc N Z 48:225–244
FAOSTAT (2017) The state of food security and nutrition in the world. Rome, FAO
Harzing AW (2010) The publish or perish book. Tarma Software Research Pty Limited, Melbourne
Heu CC, McCullough FM, Luan J, Rasgon JL (2020) CRISPR-Cas9-based genome editing in the silverleaf whitefly (*Bemisia tabaci*). CRISPR J 3:89–96
Heu CC, Gross RJ, Le KP, LeRoy DM, Fan B, Hull JJ, Brent CS, Fabrick JA (2022) CRISPR-mediated knockout of cardinal and cinnabar eye pigmentation genes in the western tarnished plant bug. Sci Rep 12:4917
Hokkanen HMT (1991) Trap cropping in pest management. Annu Rev Entomol 36:119–138
Hunter WB, Tomich J (2019) BAPC-assisted-CRISPR-Cas9 delivery into adult psyllid ovaries and pupae for heritable gene editing (Hemiptera). Citrus Greening Solutions–Boyce Thompson Institute, pp 1–4
Hunter WB, Gonzalez MT, Tomich J (2018) BAPC-assisted CRISPR/Cas9 system: targeted delivery into adult ovaries for heritable germline gene editing (Arthropoda: Hemiptera). BioRxiv:478743
Hunter WB, Gonzalez MT, Tomich J (2019) BAPC-assisted-CRISPR-Cas9 delivery into nymphs and adults for heritable gene editing (Hemiptera). FASEB J 33:626–622
Klobasa W, Chu FC, Huot O, Grubbs N, Rotenberg D, Whitfield AE, Lorenzen MD (2021) Microinjection of corn planthopper, *Peregrinus maidis*, embryos for CRISPR/Cas9 genome editing. J Vis Exp 169:e62417
Kotwica-Rolinska J, Chodakova L, Chvalová D, Kristofova L, Fenclova I, Provaznik J, Bertolutti M, Wu BCH, Dolezel D (2019) CRISPR/Cas9 genome editing introduction and optimization in the non-model insect *Pyrrhocoris apterus*. Front Physiol 10:455072
Kotwica-Rolinska J, Krištofová L, Chvalová D, Pauchová L, Provazník J, Hejníková M, Sehadová H, Lichý M, Vaněčková M, Doležel D (2020) Functional analysis and localisation of a thyrotropin-releasing hormone-type neuropeptide (EFLa) in hemipteran insects. Insect Biochem Mol Biol 122:103376
Le Trionnaire G, Tanguy S, Hudaverdian S, Gléonnec F, Richard G, Cayrol B, Monsion B, Pichon E, Deshoux M, Webster C, Uzest M, Herpin A, Tagu D (2019) An integrated protocol for targeted mutagenesis with CRISPR-Cas9 system in the pea aphid. Insect Biochem Mol Biol 110:34–44
Liu XY, Yan R, Chen SJ, Zhang JL, Xu HJ (2023) Orco mutagenesis causes deficiencies in olfactory sensitivity and fertility in the migratory brown planthopper, *Nilaparvata lugens*. Pest Manag Sci 79:1030–1039
Ma YF, Zhang MQ, Gong LL, Liu XZ, Long GJ, Guo H, Hull JJ, Dewer Y, He M, He P (2023) Efficient nanoparticle-based CRISPR-Cas13d induced mRNA disruption of an eye pigmentation gene in the white-backed planthopper, *Sogatella furcifera*. Insect Sci 30:1552–1564
McPherson JE, McPherson RM (2000) Stink bugs of economic importance in America north of Mexico. CRC Press, Boca Raton
Pacheco IS, Doss ALA, Vindiola BG, Brown DJ, Ettinger CL, Stajich JE, Redak RA, Walling LL, Atkinson PW (2022) Efficient CRISPR/Cas9-mediated genome modification of the glassy-winged sharpshooter *Homalodisca vitripennis* (Germar). Sci Rep 12:6428

Panizzi AR, Bueno AF, Silva FAC (2012) Insetos que atacam vagens e grãos. In: Hoffmann-Campo CB, Corrêa-Ferreira BS, Moscardi F (eds) Soja: Manejo Integrado de Insetos e outros Artrópodes-Praga. Embrapa Editora, Brasília, pp 335–398

Reding K, Pick L (2020) High-efficiency CRISPR/Cas9 mutagenesis of the white gene in the milkweed bug *Oncopeltus fasciatus*. Genetics 215:1027–1037

Reding K, Lê M, Pick L (2023) Genome editing of the vermilion locus generates a visible eye color marker for *Oncopeltus fasciatus*. Sci Rep 13:4188

Schaefer CW, Panizzi AR (2000) Heteroptera of Economic Importance. CRC Press, Boca Raton 856p

Silva PG, Vieira MCS, Vieira ECS, Silva IF, Ávila CJ (2021) Vertical distribution of *Euschistus heros* in the leaf canopy of soybean plants. J Agric Sci 13:147–154

Singh S, Rahangdale S, Pandita S, Saxena G, Upadhyay SK, Mishra G, Verma PC (2022) CRISPR/Cas9 for insect pests management: a comprehensive review of advances and applications. Agriculture 12. Available in. https://doi.org/10.3390/agriculture12111896

Todd JW (1989) Ecology and behavior of *Nezara viridula*. Annu Rev Entomol 4:273–292

Umina PA, Bass C, van Rooyen A, Chirgwin E, Arthur AL, Pym A, Mackisack J, Mathews A, Kirkland L (2022) Spirotetramat resistance in *Myzus persicae* (Sulzer) (Hemiptera: Aphididae) and its association with the presence of the A2666V mutation. Pest Manag Sci 78:4822–4831

Wang YH (2022) Developing RNAi-and CRISPR-based control methods for the corn planthopper, *Peregrinus maidis*. Ph D Thesis, North Carolina State University

Xue WH, Xu N, Yuan XB, Chen HH, Zhang JL, Fu SJ, Zhang CX, Xu HJ (2018) CRISPR/Cas9-mediated knockout of two eye pigmentation genes in the brown planthopper, *Nilaparvata lugens* (Hemiptera: Delphacidae). Insect Biochem Mol Biol 93:19–26

Yu H, Wang L, Liu S, Chen X, Wei X, Niu C, Volodymyr V, Song Q, Zhang H (2023) Functional analysis of a NPC1 gene from the whitefly, *Bemisia tabaci* (Hemiptera: Aleyrodidae). Arch Insect Biochem Physiol 114:e22048

Zeng B, Chen FR, Liu YT, Guo D, Zhang YJ, Feng ZR, Wang LX, Vontas J, Wu SF, Zhu KY, Gao CF, Gao CF (2022) A chitin synthase mutation confers widespread resistance to buprofezin, a chitin synthesis inhibitor, in the brown planthopper, *Nilaparvata lugens*. J Pest Sci 96:819–832

Zhang JL, Liu KL, Cai XY, Liu XY, Xu HJ (2022) FoxO is required for optimal fitness of the migratory brown planthopper, *Nilaparvata lugens* (Hemiptera: Delphacidae). Insect Sci 30:1352–1362

Zhang M, Hu Y, Liu J, Guan Z, Zhang W (2023a) CRISPR/Cas9-mediated genome editing of gustatory receptor NlugGr23a causes male sterility in the brown planthopper *Nilaparvata lugens*. Int J Biol Macromol 241:124612

Zhang YC, Gao Y, Ye WN, Peng YX, Zhu KY, Gao CF (2023b) CRISPR/Cas9-mediated knockout of NlCYP6CS1 gene reveals its role in detoxification of insecticides in *Nilaparvata lugens* (Hemiptera: Delphacidae). Pest Manag Sci 79:2239–2246

Zhang H, Zou J, Yang B, Zhang Y, Liu Z (2023c) Importance of CYP6ER1 was different among neonicotinoids in their susceptibility in *Nilaparvata lugens*. J Agri Food Chem 71:4163–4171

Zhang MQ, Gong LL, Zhao YQ, Ma YF, Long GJ, Guo H, He P (2023d) Efficient DIPA-CRISPR-mediated knockout of an eye pigment gene in the white-backed planthopper. Insect Sci, Sogatella furcifera. https://doi.org/10.1111/1744-7917.13286

Zhao Y, Huang G, Zhang W (2019) Mutations in NlInR1 affect normal growth and lifespan in the brown planthopper *Nilaparvata lugens*. Insect Biochem Mol Biol 115:103246

Chapter 10
Host Plant Resistance to Manage Pest Stink Bugs: The Block Technology on Soybean

Tiago Lucini and Antônio Ricardo Panizzi

1 Introduction

Stink bugs, known as notorious pests of several crops worldwide, pose a significant threat to soybean production (Panizzi et al. 2000; McPherson and McPherson 2000). In the Neotropics, the Neotropical brown stink bug, *Euschistus heros* (F.), stands out as the most abundant pest species infesting soybean crops for a long time, particularly in Brazil, one of the largest global soybean producers (Panizzi et al. 2012). Traditionally, chemical insecticides have been heavily used for stink bug control in soybean fields (Basso et al. 2016; Marques et al. 2019), leading to adverse effects on natural biological control, increased production costs, emergence of resistant stink bug populations, and environmental contamination (Sosa-Gómez and Silva 2010; Bueno et al. 2011).

To address these challenges and pursue more sustainable pest management strategies, researchers have turned their attention to alternative approaches. One such approach is the utilization of cultivars exhibiting resistance to insect pests. The concept of plant resistance has undergone refinement over time. Initially, resistance was categorized into three main types: antibiosis, tolerance, and non-preference (later renamed antixenosis) according to Painter's classification (1951). However, Stout (2013) reclassified into two categories: constitutive/inducible and direct/indirect subcategories, while tolerance refers to plant traits that reduce yield loss per unit of injury.

Previous attempts to develop soybean cultivars resistant to stink bugs has been done, such as cv. IAC 100, but their commercial success was limited due to

T. Lucini (✉) · A. R. Panizzi
Embrapa Trigo, Passo Fundo, RS, Brazil

A. F. Bueno, A. R. Panizzi (eds.), *Stink Bugs (Hemiptera: Pentatomidae) Research and Management*, Entomology in Focus 9,
https://doi.org/10.1007/978-3-031-69742-5_10

inadequate agronomic characteristics and low productivity (Rossetto et al. 1995). Nevertheless, recent breakthroughs have resulted in the development of the soybean cultivar BRS 1003 IPRO, which employs a novel approach known as the 'Block technology' that displays promising tolerance to stink bug infestation combined with high yield and good seed quality (low amount of abnormal or damaged seeds).

This chapter provides an exploration of soybean Block technology resistant to stink bugs, with a specific focus on *E. heros*. The chapter emphasizes the application of the electropenetrography (EPG) technique in studying the feeding behavior of *E. heros* on soybean cultivars bearing the Block technology. The utilization of EPG allows for a precise analysis of feeding patterns, shedding light on the underlying resistance mechanisms and their implications for stink bug population dynamics. By combining existing knowledge with new research findings, this chapter aims to contribute to the development of sustainable stink bug management strategies and the advancement of soybean breeding programs.

2 Resistance of Soybean Cultivars Bearing the Block Technology

Varietal control involves the use of cultivars that possess some form of genetic resistance and is considered an ideal method, as it allows the maintenance of pest levels below the economic threshold, without causing environmental damage. It also has the advantage of being compatible with any pest management program, providing even greater flexibility in the adoption of control measures.

A multitude of factors can exert influence over the feeding activities of insects and ultimately determine the suitability of a particular genotype for consumption. These factors encompass both physical and chemical aspects (Stout 2013). Research has suggested that specific soybean genotypes possessing higher densities of trichomes may demonstrate reduced attractiveness to stink bugs (Souza et al. 2013). This phenomenon may be attributed to the trichomes serving as physical barriers, rendering it challenging for the insects to access the plant's epidermis (War et al. 2012). In addition to these physical factors, chemical compounds also play a pivotal role in influencing insect behavior.

The Block technology involved a meticulous selection process conducted by scientists at the Embrapa Soybean Research Center in Brazil. By subjecting soybean lines to high levels of stink bug infestation over several years, researchers identified plants that exhibited greater tolerance to stink bug attack, high yield, and superior seed quality. Currently, five soybean cultivars—BRS 391, BRS 523, BRS 539, BRS 543 RR, and BRS 1003 IPRO—have been characterized as Block soybean cultivars and available for cultivating by growers (Embrapa 2023). Block cultivars merge conventional and transgenic cultivars, the main characteristics of them are presented in Table 10.1.

Compared to susceptible cultivars, these cultivars exhibit a remarkable tolerance to stink bugs, enduring over two times the economic threshold (ET) previously

Table 10.1 Main characteristics of the soybean cultivars bearing the Block technology cultivated in the Neotropics

Block soybean cultivar	Maturity group	Conventional (C) or Transgenic (T)	Growth habit	Other technologies
BRS 391	6.4	C	Determinate	–
BRS 523	5.8	C	Indeterminate	–
BRS 539	6.1	C	Indeterminate	Shield[2]
BRS 543 RR	6.0	T	Indeterminate	RR[3]
BRS 1003 IPRO	6.3 or 7.0[1]	T	Indeterminate	Intacta[4] and RR

[1]Variable according to cultivate region; [2]Resistance to Asian soybean rust (*Phakopsora pachyrhizi* Sydow & Sydow.); [3]Resistance to glyphosate; [4]Resistance to caterpillars

stablished of 2 bugs/m without any significant reduction in yield and seed quality (Arias et al. 2020). Field experiments with Block cultivars demonstrated the efficiency of the technology with seeds less damaged and absence of foliar retention compared to susceptible cultivars (TL, unpublished data). Although considered tolerant to the stink bug *E. heros*, no clear evidence has been found to elucidate what is the main factor involved in such resistance, i.e., if it is physical or chemical. According to the results obtained by Lucini et al. (2021), it was suggested that soybean Block cultivars exhibit an antixenotic effect linked to the seed endosperm, likely influenced by chemical factors. Clearly, more studies need to be conducted to fully clarify the factors involved in such resistance presented by soybean Block cultivars.

3 Feeding Behavior of *Euschistus heros* on Soybean

Recently the feeding behavior of *E. heros* on immature soybean pods has been unveiled using the technique known as electropenetrography or electrical penetration graph (both abbreviated as EPG) (Lucini and Panizzi 2018a). EPG has emerged as the most rigorous, accurate and non-invasive method for evaluating the stylet activities of piercing-sucking insects, including stink bugs, on plants in real-time (Lucini and Panizzi 2018b). By establishing an electrical circuit between the insect and its food source, EPG enables the generation of waveforms that represent distinct stylet activities such as stylet penetration, salivation, and ingestion (Walker 2000).

The EPG waveform library for *E. heros* consists of 13 distinct waveforms, encompassing four non-feeding and nine feeding activities. These waveforms are categorized into three main phases: pathway, ingestion (from xylem and seed contents), and salivation (in seed). A comprehensive overview of the major waveforms, including a brief description and proposed biological behavior, can be found in Table 10.2 (for complete details see Lucini and Panizzi 2018a).

Details of the electrical characteristics, waveform appearance and biological meanings of each major feeding waveforms are presented as follows.

Table 10.2 Major EPG waveforms captured from female adult *Euschistus heros* feeding on immature soybean pods during the R5 (pod-filling) stage, along with their corresponding suggested biological behaviors

Phase	Type or subtype	Biological activities
Non-feeding	Z	Insect standing still on pod surface
	Np	Insect moving, walking
Pathway	Eh1a	Initial stylet penetration and secretion of salivary sheath
	Eh1b	Stylet penetration toward ingestion site
	Eh1c	Stylets reach a rigid cell layer (sclerenchyma)
Ingestion	Eh2	Xylem sap ingestion
Salivation	Eh3a	Cell rupture activities on seed tissue
Ingestion	Eh3b	Ingestion of the seed contents

Source: Adapted from Lucini and Panizzi (2018a)

3.1 Pathway

The pathway phase was primarily constituted by three distinct waveform subtypes (Eh1a, Eh1b, and Eh1c), each of which possessed a distinct and specific biological significance. Eh1a was correlated with the rostrum slightly touching the plant surface as it initiated the insertion of its stylet (mouthparts) into the pod tissue. Concomitant with this action, the stink bug initiated the secretion of gelling saliva to create a salivary sheath within the pod tissue, occasionally resulting in the formation of a flange extruded from the plant tissue. Eh1a means the initial stylet activity, albeit of a brief duration, spending a mere few seconds per event (approximately 13 s), characterized by irregular peak occurrences randomly dispersed across time. Furthermore, Eh1a displayed an elevated voltage level at the initiation of stylet insertion, which gradually diminished as the stylets penetrated deeper into the plant tissue (as illustrated in Figs. 10.1a–c) (Lucini and Panizzi 2018a).

The waveform denoted as Eh1b was observed subsequently to the occurrence of Eh1a and exhibited a significantly extended event duration, taking approximately 1.4 min, in contrast to the relatively brief duration of Eh1a. In its entirety, Eh1b presented itself as an underlying waveform characterized by a continuous increase in voltage levels. Notably, it was consistently recorded in advance of Eh1c or Eh2 waves, as depicted in Figs. 10.1a–c. On certain occasions, it exhibited a distinctive pattern typified by a sudden voltage peak followed by gradual decreases, as illustrated in Fig. 10.1a. Histological analysis revealed a correlation between Eh1b and the profound insertion of the stylet into the pod tissue, accompanied by the secretion of gelling saliva, thereby forming a continuous salivary sheath that enveloped and trailed the stylets during their penetration (Fig. 10.2a).

The waveform designated as Eh1c was observed in sequence to Eh1b and presented the longest event duration, extending for approximately 5 min, surpassing all other waveforms within the pathway phase. In contrast to the preceding pathway

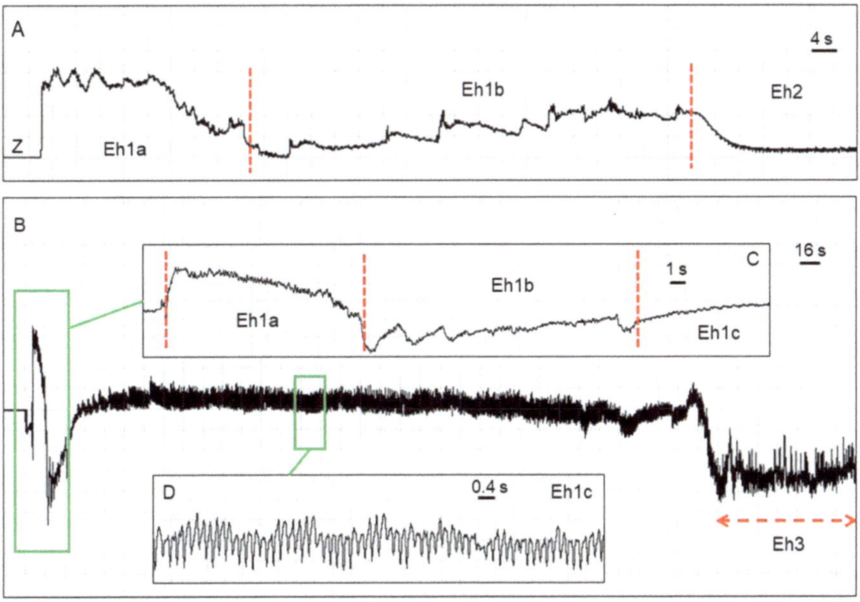

Fig. 10.1 Compressed overview of pathway waveforms (Eh1) produced by *Euschistus heros* adults on immature soybean pods at the R5 (pod-filling) stage at Ri 10^7 Ohms and 50 mV AC applied signal. Expanded view of initial (Eh1a), after a non-feeding event (wave Z), and deep stylet penetration (Eh1b) into pod tissue preceding the xylem ingestion (Eh2) (**a**). Compressed overview of pathway waveforms (Eh1a, Eh1b and Eh1c) preceding seed activities (Eh3) (**b**). Expanded view of pathway waveforms (**c**) and details of that represents a rigid cell layer (sclerenchyma) before reach seed tissue (**d**). (Source: Adapted from Lucini and Panizzi 2018a)

waveforms, Eh1c exhibited a highly regular pattern, as evidenced in Figs. 10.1b–d. Furthermore, it consistently preceded the occurrence of Eh3 (seed endosperm activities). During this waveform recording, stink bugs engaged in a distinctive behavior characterized by the downward movement of their heads, followed by the forceful insertion of the stylets into the pod tissue, with this pattern persistently repeated. Histological sections during the Eh1c revealed that the tips of the stylets were positioned within a rigid layer of cells known as sclerenchyma (Fig. 10.2b), which they overcame in order to reach the seed endosperm (Lucini and Panizzi 2018a).

3.2 *Xylem*

The xylem phase is characterized by a single waveform, named Eh2, which frequently followed either the Eh1b or, occasionally, the Eh1c waveform. Eh2 exhibits a consistent wave pattern marked by regularly distributed peaks, primarily oriented in a downward direction, as illustrated in Fig. 10.3. This waveform Eh2, represents a sustained process of sap ingestion from xylem vessels, as confirmed by analysis of

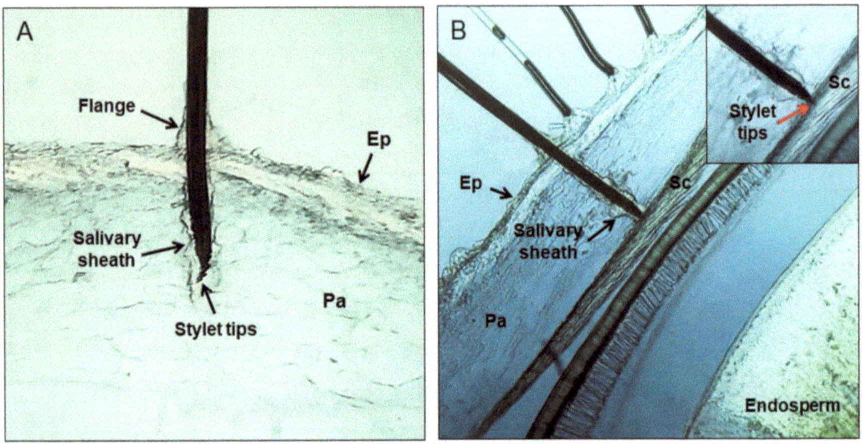

Fig. 10.2 Histological cuts of immature soybean pods (R5 stage—pod filling) bearing severed stylets and salivary sheath of adults of *Euschistus heros*. Stylet tips and salivary sheath located in the parenchyma cell layer during waveform Eh1b (**a**) and in the sclerenchyma cell layer during observed waveform Eh1c (pointed out by the red arrow in the small box in the upper corner) (**b**). Ep epidermis, Pa parenchyma, Sc sclerenchyma. (Source: Adapted from Lucini and Panizzi 2018a)

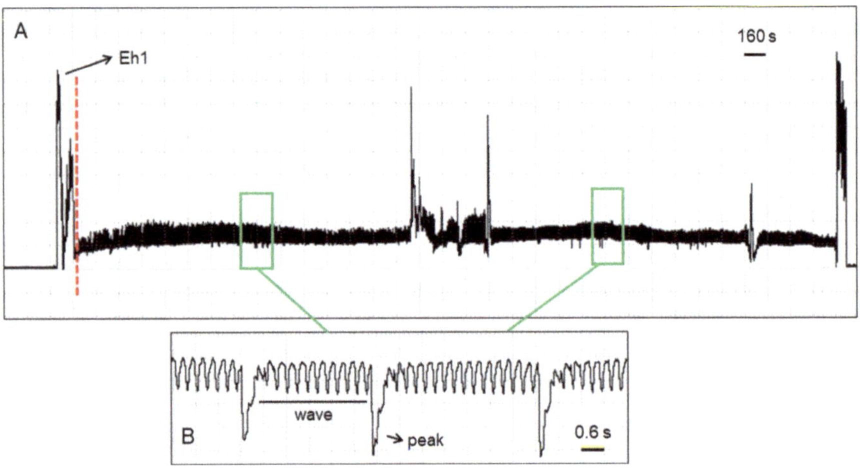

Fig. 10.3 Overview of xylem waveform (Eh2) produced by *Euschistus heros* adults on immature soybean pods at the R5 (pod-filling) stage at Ri 10^7 Ohms and 50 mV AC applied signal. Coarse structure of waveform Eh2 showing a complete feeding section (**a**). Expanded view of xylem waveform showing peak and wave portions (**b**). (Source: Adapted from Lucini and Panizzi 2018a)

histological images (Fig. 10.4). During the feeding behavior of *E. heros* on soybean pods, approximately 19% of the total recording time, which spanned for 8 h, was dedicated to xylem ingestion. Each event of ingestion was repeated approximately three times per insect and persisted for approximately 40 min per event. In aggregate, each insect expended approximately 108 min in the act of sap ingestion from xylem vessels on soybean pods (Lucini and Panizzi 2018a).

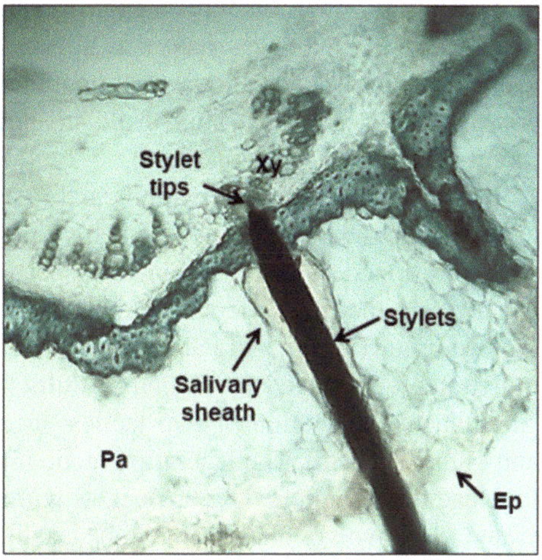

Fig. 10.4 Histological cut of an immature soybean pods (R5 stage—pod filling) bearing severed stylets and salivary sheath of adult of *Euschistus heros*. Complete salivary sheath and stylet tips positioned in the xylem vessels observed during waveform Eh2. Ep epidermis, Pa parenchyma, Xy xylem. (Source: Adapted from Lucini and Panizzi 2018a)

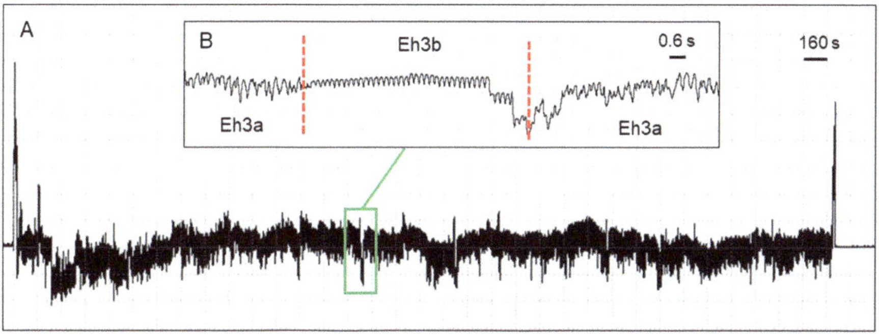

Fig. 10.5 Overview of seed activities waveforms (Eh3) produced by *Euschistus heros* adults on immature soybean pods at the R5 (pod-filling) stage at Ri 10^7 Ohms and 50 mV AC applied signal. Coarse structure of waveform Eh3 showing a complete feeding section (**a**). Expanded view of seed waveform (Eh3) showing details of waveform Eh3a (laceration/maceration activities) and Eh3b (ingestion of seed contents) (**b**). (Source: Adapted from Lucini and Panizzi 2018a)

3.3 Seed Endosperm

Seed-related activities were consistently observed subsequent to the occurrence of the Eh1c waveform. These activities encompassed two distinct waveform subtypes, namely Eh3a and Eh3b, which occurred intermittently throughout the entire recording period. Eh3a is characterized by an exceedingly irregular waveform exhibiting random peaks and spikes distributed across the waveform (Fig. 10.5). It stands out as the third-longest waveform, accounting for approximately 17% of the total recording time. Each event of Eh3a occurred about 17 times per insect, with each event lasting approximately 4 min, resulting in a cumulative duration of 80 min per insect.

In contrast, Eh3b displays a highly regular pattern, as illustrated in Fig. 10.5. Each event of Eh3b was recorded approximately 14 times per insect, but these events were of shorter duration compared to those of the Eh3a waveform, lasting approximately 13 s each. This results in a total duration of approximately 2 min per insect (Lucini and Panizzi 2018a).

Based upon visual observations and histological images, the biological significance of each waveform subtype was proposed. During the Eh3a wave, the stylets were positioned within the seed endosperm, as depicted in Fig. 10.6a. Visually, it was evident that the stylets executed rapid, continuous, and deep movements into the pod tissue (Lucini and Panizzi 2018a). This behavior is related to the cell rupturing strategy employed by *E. heros* adults for food preparation, encompassing the simultaneous execution of two tactics: laceration and macerate-and-flush (Lucini and Panizzi 2018b). The laceration tactic involves mechanical actions of the stylets that lead to the rupture of cells pockets within the plant tissue. Conversely, the macerate tactic involves chemical actions, as digestive enzymes are secreted via watery saliva to further degrade the ingested cells, as described by Backus et al. (2005). This combined approach enables *E. heros* to effectively ingest the degraded cell contents during its feeding process on seeds.

The continuous activities of laceration and maceration during Eh3 events resulted in observable damage to the seed endosperm, as evidenced in soybean pods where fresh cuts were made following such events (Fig. 10.6b). These findings provided

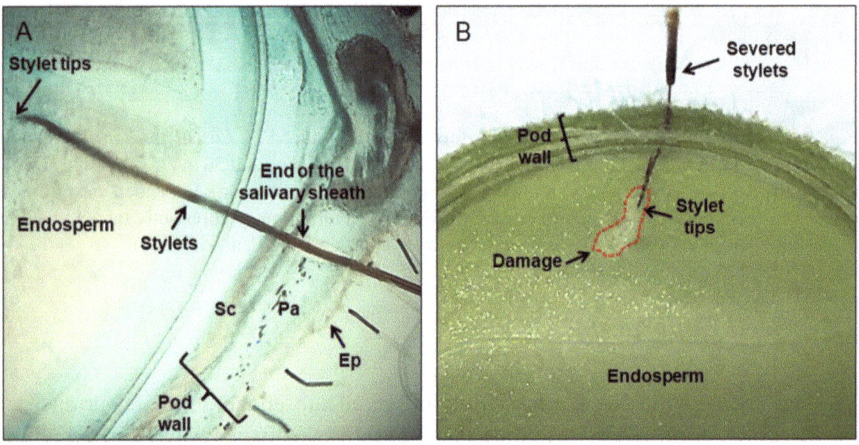

Fig. 10.6 Histological and fresh cuts of immature soybean pods (R5 stage—pod filling) bearing severed stylets and salivary sheath of adults of *Euschistus heros*. Incomplete salivary sheath secreted in the parenchyma tissue and stylet tips ending in the endosperm tissue of the seeds observed during waveform Eh3 (**a**). Fresh soybean pod bearing severed stylets and damaged area caused by stylet activities (white region delimited by the red dashed line) observed after a Eh3 event (**b**). Ep epidermis, Pa parenchyma, Sc sclerenchyma. (Source: Adapted from Lucini and Panizzi 2018a)

substantive evidence supporting the notion that the Eh3 waveform is indeed associated with feeding activities on the seeds, specifically involving the laceration and maceration processes.

The waveform subtype Eh3b exhibited a distinct behavior wherein the insect's stylets were observed to remain briefly motionless within the tissue for a few seconds, after which they moved them again (Eh3a). Based on this observation and insights from other ingestion waveforms, it is postulated that during Eh3b, ingestion of degraded cell contents, often referred to as a "slurry", occur from seeds that were previously broken down during Eh3a events. This feeding behavior was consistently reiterated throughout the Eh3 waveform (Lucini and Panizzi 2018a).

4 Feeding Frequency and Duration on Resistant Block Plants

The use of the EPG allows to quantify and compare counts and durations of each stylet activity performed by insects while feeding on resistant cultivars in comparison with a susceptible cultivar. The EPG has been extensively utilized to investigate the impact of resistant plants and genotypes on the feeding behavior of various piercing-sucking insect species (e.g., Rangasamy et al. 2015; Todd et al. 2016; Koch et al. 2018).

Previous studies have successfully employed EPG to examine the feeding responses of insects to different plant resistance mechanisms, providing valuable insights for developing effective pest management strategies. For stink bugs, the EPG has been only recently applied to assess insect-induced plant defense against the brown marmorated stink bug *Halyomorpha halys* (Stal) (Serteyn et al. 2020) and the effect of resistant soybean cultivars, bearing the Block technology, on adults of *E. heros* (Lucini et al. 2021).

Below is a summary of how three selected soybean Block cultivars (BRS 1003 IPRO, BRS 543 RR and BRS 391) may affect quantitatively the major feeding activities of *E. heros* adults (Lucini et al. (2021).

4.1 Pathway

The pathway analysis involves various stages of stylet insertion and penetration into the plant tissue. These stages are depicted for waveforms Eh1a,b,c, as previously described. However, when analyzed in separate, slight differences were observed in the number of events and duration. When combined, no significant differences were found to the number of stylet insertions per insect among cultivars, which ranged from 4.4 to 5.2 times (Fig. 10.7a). In contrast, there was a marginal difference in the total duration of pathway activities on BRS 1003 (ca. 38 min). In comparison, BRS

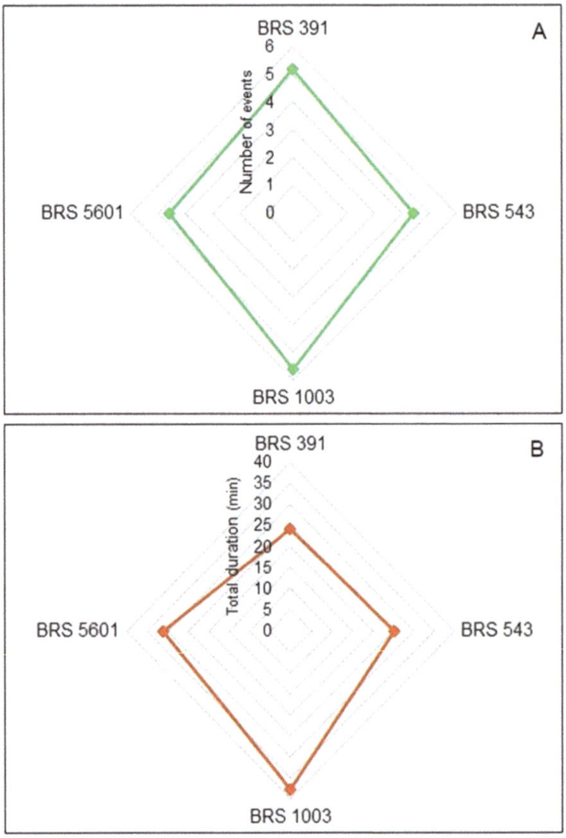

Fig. 10.7 Number of stylet insertion per insect (**a**) and total duration per insect - min (**b**) produced by *Euschistus heros* adult females on soybean pods of resistant (BRS 391, BRS 543 RR, BRS 1003 IPRO) and susceptible (BRS 5601 RR) cultivars. (Source: Adapted from Lucini et al. 2021)

391 took approximately 24 min, BRS 543 ca. of 25 min, and BRS 5601 exhibited an intermediate value, ca. 31 min (Fig. 10.7b) (Lucini et al. (2021).

The analysis of the pathway phase provide insights into the potential presence of plant resistance resulting from physical and/or chemical traits. Longer pathway activities, for instance, could indicate that stink bugs encounter challenges in locating and/or accessing specific feeding sites. In soybean pods, stink bugs must deal with a rigid layer composed of sclerenchyma cells, as represented by the Eh1c waveform, before they can reach the seed endosperm (Lucini and Panizzi 2018a, b). Despite the existence of this physical barrier, the authors discovered that the total duration of pathway waveforms, particularly Eh1c, did not exhibit significant differences among the various soybean cultivars tested. Consequently, they concluded that the resistance observed in the Block soybean cultivars is not directly associated with the presence of a physical barrier within the pod wall that would prevent or difficult stylet penetration towards the seed (Lucini et al. 2021). In summary, it can be concluded that Block soybean cultivars do not appear to substantially prevent the pathway activities of *E. heros*.

4.2 Xylem

Similar to the findings for pathway waveforms, no significant differences were reported in the number of events and the total duration of ingestion from xylem vessels (Eh2) among Block cultivars and the susceptible cultivar. The number of events ranged from 2.2 to 3.5 times per insect, with BRS 5601 exhibiting the highest value (Fig. 10.8a). Meanwhile, the total duration of ingestion varied from 46 to 58 min, with BRS 5601 and BRS 1003 displaying slightly longer durations of Eh2 (Fig. 10.8b) (Lucini et al. 2021).

This lack of differences is likely to be associated with the preferred feeding behavior of seed-feeding by stink bugs. It is well-established that these insects utilize xylem sap ingestion primarily for maintaining body hydration, rather than as a primary nutrient source (Rivera and Mitchell 2020). In addition, this behavior also allows them to achieve nutrient balance after feeding on nutrient-concentrated seeds, as suggested by Lucini et al. (2016).

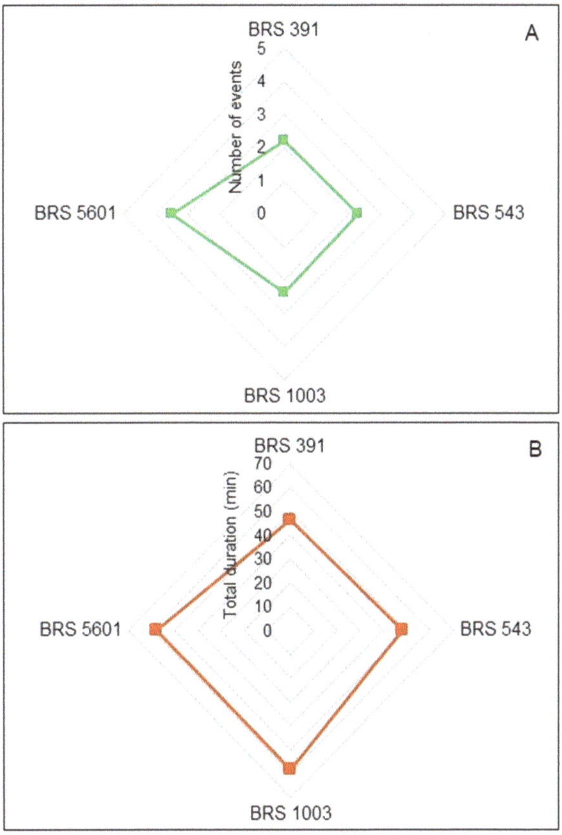

Fig. 10.8 Number of xylem events per insect (**a**) and total duration per insect - min (**b**) produced by *Euschistus heros* adult females on soybean pods of resistant (BRS 391, BRS 543 RR, BRS 1003 IPRO) and susceptible (BRS 5601 RR) cultivars. (Source: Adapted from Lucini et al. 2021)

Despite the absence of observable differences, on resistant Block plants, the insects exhibited a tendency to allocate a greater proportion of their overall feeding activities towards the ingestion of xylem sap when compared to the susceptible cultivar. This observation led to speculate that this behavior might serve as a compensatory mechanism to counterbalance the substantial reduction in stylet activities directed towards seeds, as discussed in greater detail below (Lucini et al. 2021).

4.3 Seed Endosperm

The primary distinctions in the feeding behavior of *E. heros* between resistant and susceptible soybean cultivars were directly linked to stylet activities within the seed endosperm. This encompassed both cell laceration/maceration (Eh3a waveform), and seed tissue ingestion (Eh3b waveform). When these two activities were analyzed together (i.e., sum of Eh3a and Eh3b), the total duration of these activities was significantly reduced, approximately three times, in the Block-resistant cultivars when compared to the susceptible BRS 5601. However, it's noteworthy that the number of times the bugs reached the seed exhibited no discernible differences among the various soybean cultivars, which means that stink bugs reached the seed without restrictions (Lucini et al. 2021).

Through individual analysis of each Eh3 subtype, notable differences were noticed. On resistant plants, both the number of laceration/maceration and ingestion events per insect were significantly smaller, approximately 3–5 times, when compared to the susceptible cultivar, as depicted in Fig. 10.9a, c, respectively.

Consequently, the reduced number of events on resistant plants led to a decrease in activity, which directly contributed to the 3 times shorter total duration of both waveforms Eh3a and Eh3b, than that observed in BRS 5601 (Fig. 10.9b, d, respectively) (Lucini et al. 2021). This distinctive characteristic enables the resistant cultivars to tolerate a greater number of stink bugs without reduction in yield, in contrast to the susceptible cultivar (Lucini et al. 2021).

5 Seed Damage

The Block technology was meticulously engineered with the aim of mitigating production losses in cultivars under intense stink bug infestations, exhibiting a notable advantage over susceptible cultivars in similar conditions. A series of both field and laboratory experiments has effectively showed the distinctive attributes of Block cultivars. These include a reduced occurrence of seeds displaying various forms of damage, such as greenish, fermented, and deformities caused by stink bug attack. Moreover, Block cultivars have exhibited low propensity for green stem and prolonged foliar retention syndrome (CA Arias, unpublished).

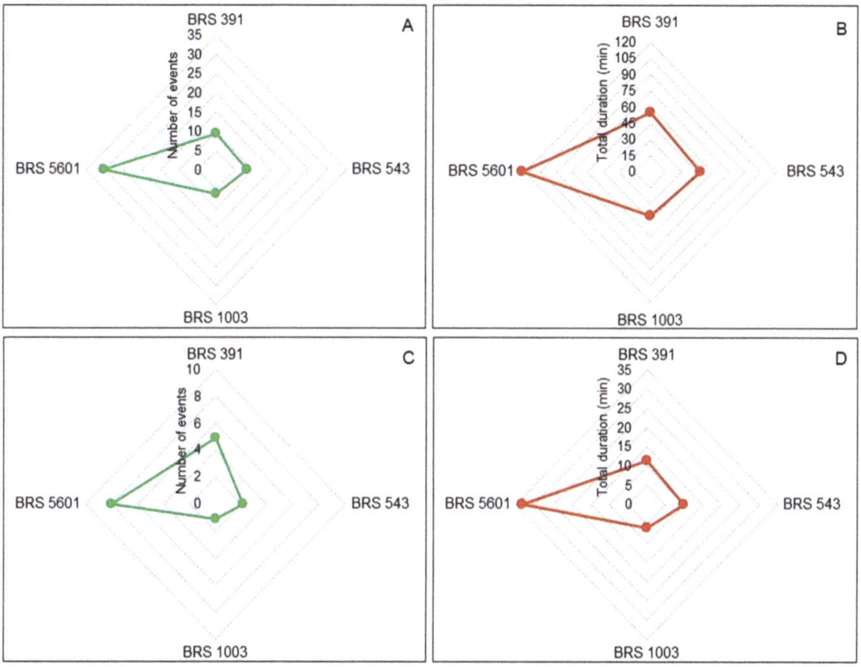

Fig. 10.9 Number of laceration/maceration events per insect (**a**) and total duration per insect - min (**b**); number of seed ingestion events per insect (**c**) and total ingestion duration per insect (**d**) produced by *Euschistus heros* adult females on soybean pods of resistant (BRS 391, BRS 543 RR, BRS 1003 IPRO) and susceptible (BRS 5601 RR) cultivars. (Source: Adapted from Lucini et al. 2021)

In their evaluation of damaged area caused by *E. heros* adults on seeds of Block and susceptible cultivars, subjected to two distinct infestation levels, Oliveira et al. (2022) uncovered significant disparities. While subtle differences were observed during the low infestation level (2 stink bugs per plant), the distinctions became considerably pronounced when confronted with the high infestation level (4 stink bugs per plant), as illustrated in Fig. 10.10.

Comparative analysis of the total proportion of damaged area per seed and the specific area damaged per seed highlighted the superior performance of the resistant cultivars. They exhibit a significant reducing in the total damaged area (ca. 3 times) compared to the susceptible cultivar (Fig. 10.10). Oliveira et al. (2022) also reported that one third of the seeds with damage from BRS 1003 IPRO exhibited no discernible damage on the inner surface of the analyzed cotyledons. Instead, the damage was confined to the seed's external surface. A similar occurrence, albeit less pronounced, was observed for the other two Block cultivars. In contrast, the susceptible cultivar showed damage at both external and internal cotyledon surfaces (Fig. 10.11).

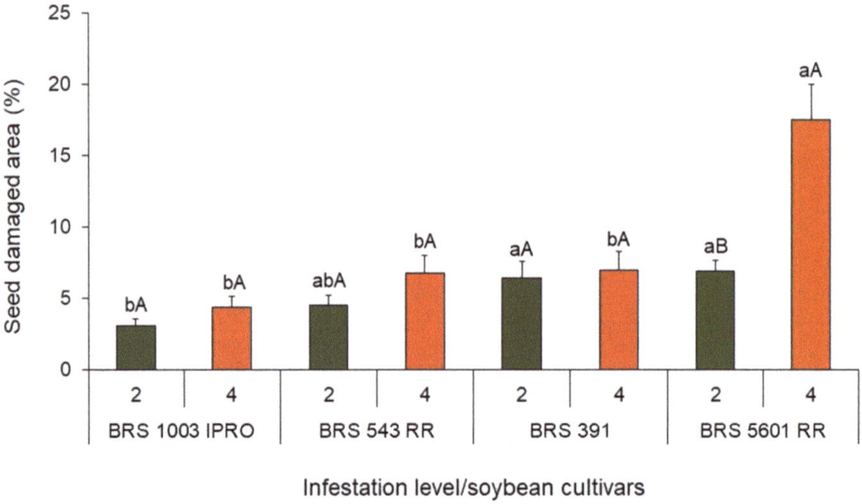

Fig. 10.10 Total seed area damaged (in %) caused by *Euschistus heros* adults on resistant soybean Block cultivars (BRS 1003 IPRO, BRS 543 RR, and BRS 391) and a susceptible cultivar (BRS 5601 RR) after a 14-day feeding period with 2 and 4 stink bugs per plant. Columns sharing the same lowercase letter (among cultivars within each infestation level) or the same uppercase letter (within each cultivar across infestation levels) indicate no significant differences using the Tukey test and Student's t-test ($P < 0.05$), respectively. (Source: Adapted from Oliveira et al. 2022)

Using the EPG technique, it was noted that the number of seed events carried out per insect was practically equivalent between Block and susceptible cultivars. Nonetheless, insects allocated significantly less time to consume seeds contents from Block cultivars in comparison to the susceptible one (Lucini et al. (2021). These findings demonstrate that stink bugs are able to feed on seeds of both, resistant soybean Block cultivars and susceptible one; however, their feeding activities are adversely impacted in the case of resistant cultivars.

Lucini et al. (2021) reported that the hardness of the pod wall, a physical factor previously considered as a potential main driver of resistance, was not the main factor involved in the resistance. This conclusion was drawn based on the analysis of pathway waveforms, which revealed that the frequency and duration of pathway activities, which are indicative of the challenge posed by stylet penetration into plant tissue, did not exhibit differences between the Block and susceptible cultivars.

In light of these results, the observed resistance might be intricately linked to the presence of certain chemical compounds in the seeds. Under pest attack, plants can promptly synthetize and released a complex of secondary metabolites (e.g., terpenoids, alkaloids, and specific proteins) as an induced defensive response (Habib and Fazili 2007). However, the factor(s) that makes soybean Block cultivar resistance to attack of *E. heros* still remains to be fully demonstrated.

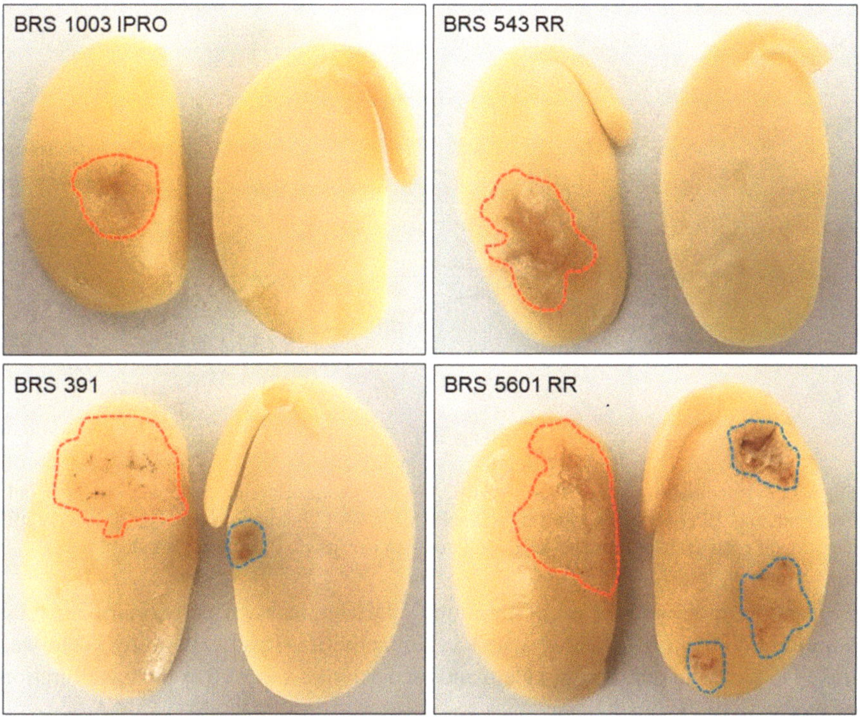

Fig. 10.11 Seed damage caused by *Euschistus heros* adults on resistant Block cultivars (BRS 1003 IPRO, BRS 543 RR, and BRS 391) and a susceptible cultivar (BRS 5601 RR) under high infestation level (4 stink bugs/plant). External and internal damaged areas in the cotyledons circled with red and blue dashed lines, respectively. (Source: Oliveira et al. 2022)

6 Multivariate Exploratory Analysis

Multivariate analysis provides insights that may not be apparent when examining variables individually. Applying this analysis, via Principal Components Analysis (PCA), valuable insights were found into the dispersion patterns among soybean Block cultivars and a susceptible cultivar regarding the quantitative EPG variables. It revealed that the majority of the variance within the dataset, totaling 85%, could be effectively explained by the first two PCA axes, namely Dim 1 and Dim 2 (Fig. 10.12). These two principal components allowed for meaningful correlations between the various factors and variables, as their values consistently exceeded the 70% threshold.

The analysis segregated the EPG variables into three main groups based on their results and relationships with soybean cultivars tested. The first group exhibited a strong association with the number of events (NWEI) in pathway (P), xylem (X), and seed (S) activities (Fig. 10.12), i.e., these group share common traits related to the frequency of events in all stylet activities. The second group was characterized

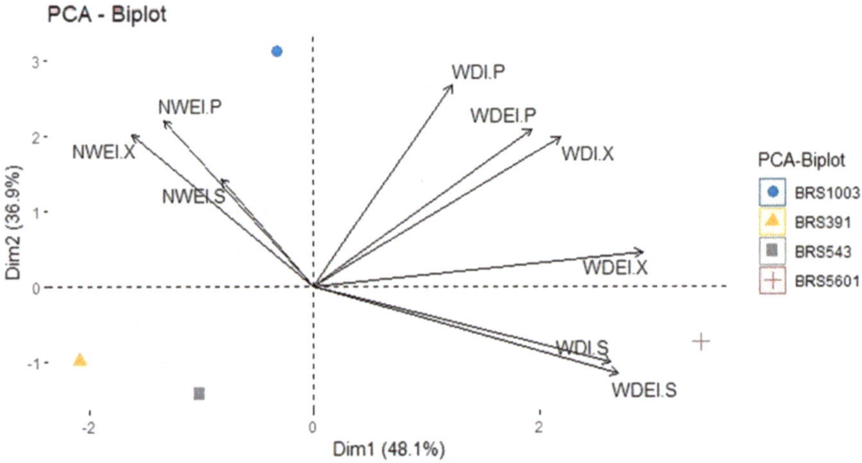

Fig. 10.12 Principal components analysis (PCA) showing the relationship between soybean cultivars and nine EPG variables (n = 6 insect/cultivar). NWEI Number of events per insect, WDEI Duration of each event per insect, WDI Total duration per insect, P pathway, X xylem, S seed

by its correlation with the event duration (WDEI) in pathway activity and the total duration (WDI) in both pathway and xylem activities. Lastly, the third group was linked to the event duration in xylem and seed activities, as well as the total duration of seed activity (Fig. 10.12).

The susceptible cultivar BRS 5601RR exhibited a positive correlation with the variables belonging to the third group. This indicates that on this particular cultivar, stink bugs displayed the longest event durations in the xylem and seed tissues, as well as the longest overall duration of feeding activities on the seeds. Conversely, these variables displayed a negative correlation with the Block cultivars, i.e., the event durations within each phase, in the xylem and especially in the seed tissues, were notably reduced, implying in a lower level of acceptance of these cultivars by the stink bugs.

Particularly on the Block cultivar BRS 1003 IPRO, there was a tendency for insects to participate in a greater number of events associated with pathway, xylem, and seed activities (group 1). This observation may imply that on this specific Block cultivar, the insects encountered greater challenges in exploiting the plant tissue, leading to increased efforts in engaging with these activities. Block cultivars BRS 391 and BRS 543RR, however, displayed contrasting results in this regard.

7 Concluding Remarks

As stink bugs are major pests of commodities such as soybean, the development and implementation of the resistant Block cultivars is a very important step toward their management. In summary, the findings herein presented and discussed allow to conclude that soybean Block cultivars are characterized by markedly reduced proportions of seed-damaged areas, as well as reduced stink bug activities related to seed consumption compared to the susceptible 'BRS 5601 RR' cultivar. Therefore, this emerges as a promising and innovative strategy for the effective management of stink bugs in soybean crops. This strategy not only holds the potential to augment crop productivity and quality but also offers the prospect of minimizing the ecological footprint associated with pest control interventions, thereby advancing sustainable agricultural practices.

References

Arias CAA, Hoffmann-Campo CB, Corrêa-Ferreira BS, Lopes ION (2020) Auxílio da genética Rev Cultivar 229:12–14

Backus EA, Serrano MS, Ranger CM (2005) Mechanisms of hopperburn: an overview of insect taxonomy, behavior, and physiology. Annu Rev Entomol 50:125–151

Basso CJ, Kuss CC, Pias OHC, Muraro DS, Cutti L (2016) Neonicotinoid insecticide systemicity in soybean plants and its effect on brown stink bugs. Pesq Agropec Trop 46:96–101

Bueno AF, Batistela MJ, Bueno RCOF, França-Neto JB, Nishikawa MAN, Libério Fo A (2011) Effects of integrated pest management, biological control and prophylactic use of insecticides on the management and sustainability of soybean. Crop Prot 30:937–945

Empresa Brasileira de Pesquisa Agropecuária – EMBRAPA (2023) Tecnologia block. Available in https://www.embrapa.br/soja/block. Accessed in June 2023

Habib H, Fazili KM (2007) Plant protease inhibitors: a defense strategy in plants. Biotechnol Mol Biol Rev 2:68–85

Koch KG, Donze-Reiner T, Baird LM, Louis J, Amundsen K, Sarath G, Bradshaw JD, Heng-Moss T (2018) Evaluation of greenbug and yellow sugarcane aphid feeding behavior on resistant and susceptible switchgrass cultivars. Bioenergy Res 11:480–490

Lucini T, Panizzi AR (2018a) EPG monitoring of the Neotropical brown-stink bug, *Euschistus heros* (F.), on soybean pods: an electrical penetration graph-histology analysis. J Insect Sci 18(5):1–14

Lucini T, Panizzi AR (2018b) Electropenetrography (EPG): a breakthrough tool unveiling stink bug (Pentatomidae) feeding on plants. Neotrop Entomol 47:6–18

Lucini T, Panizzi AR, Backus EA (2016) Characterization of an EPG waveform library for red-banded stink bug, *Piezodorus guildinii* (Hemiptera: Pentatomidae), on soybean plants. Ann Entomol Soc Am 109:198–210

Lucini T, Panizzi AR, Bueno AF (2021) Evaluating resistance of the soybean block technology cultivars to the Neotropical brown stink bug, *Euschistus heros* (F.). J Ins Physiol 131:104228

Marques RP, Cargnelutti Filho A, Melo AA, Guedes JVC, DeCarli C, Rohrig A, Pozebon H, Perini CR, Ferreira DR, Bevilaqua JCG, Patias LS, Forgiarini SE, Padilha G, Leitão JV, Moro D, Hahn L, Arnemann JÁ (2019) Managing stink bugs on soybean fields: Insights on chemical management. J Agric Sci 11:225–234

McPherson JE, McPherson RM (2000) Stink Bugs of Economic Importance in America North of Mexico. CRC Press, Boca Raton

Oliveira WP, Lucini T, Panizzi AR (2022) Seed damage by the neotropical brown stink bug, *Euschistus heros* (F.) to soybean cultivars with the block technology versus susceptible cultivar. Environ Entomol 51:451–459

Painter RH (1951) Insect resistance in crop plants. Macmillan, New York, p 520

Panizzi AR, McPherson JE, James DG, Javahery M, McPherson RM (2000) Stink bugs (Pentatomidae). In: Schaefer CW, Panizzi AR (eds) Heteroptera of Economic Importance. CRC Press, Boca Raton, pp 421–474

Panizzi AR, Bueno AF, Silva FAC (2012) Insetos que atacam vagens e grãos. In: Hoffmann-Campo CB, Corrêa-Ferreira BS, Moscardi F (eds) Soja: manejo integrado de insetos e outros artrópodes-praga. Embrapa Editora, Brasília, pp 335–398

Rangasamy M, McAuslane HJ, Backus EA, Cherry AH (2015) Differential probing behavior of *Blissus insularis* (Hemiptera: Blissidae) on resistant and susceptible St. Augustine grasses. J Econ Entomol 108:780–788

Rivera KL, Mitchell PL (2020) Probing behavior of *Nezara viridula* first instars: EPG analysis and effect of food availability on subsequent development. J Agric Urban Entomol 36:47–63

Rossetto CJ, Gallo PB, Razera LF, Bortoletto N, Igue T, Medina PF, Tisseli Fo O, Aquilera V, Veiga RFA, Pinheiro JB (1995) Mechanisms of resistance to stink bug complex in the soybean cultivar IAC-100. An Soc Entomol Brasil 24:517–522

Serteyn L, Ponnet L, Saive M, Fauconnier ML, Francis F (2020) Changes of feeding behavior and salivary proteome of brown marmorated stink bug when exposed to insect-induced plant defenses. Arthropod Plant Interact 14:101–112

Sosa-Gómez DR, Silva JJD (2010) Neotropical brown stink bug (*Euschistus heros*) resistance to methamidophos in Paraná, Brazil. Pesq Agropec Bras 45:767–769

Souza SE, Baldin ELL, Silva FGPJ, Lourenção AL (2013) Feeding preference of *Nezara viridula* (Hemiptera: Pentatomidae) and attractiveness of soybean genotypes. Chil J Agric Res 73:351–357

Stout MJ (2013) Reevaluating the conceptual framework for applied research on hostplant resistance. Insect Sci 20:263–272

Todd JC, Rouf Mian MA, Backus EA, Finer JJ, Redinbaugh MG (2016) Feeding behavior of soybean aphid (Hemiptera: Aphididae) biotype 2 on resistant and susceptible soybean. J Econ Entomol 109:426–433

Walker GP (2000) A beginner's guide to electronic monitoring of homopteran probing behavior. In: Walker GP, Backus EA (eds) Principles and Applications of Electronic Monitoring and other Techniques in the Study of Homopteran Feeding Behavior. Entomological Society of America, Lanham, pp 14–40

War AR, Paulraj MG, Ahmad T, Buhroo AA, Hussain B, Ignacimuthus S (2012) Mechanisms of plant defense against insect herbivores. Plant Signal Behav 7:1306–1320

Chapter 11
The Present and Future of Chemical Control to Manage Stink Bugs in Brazil

Renato Carvalho, Daniela Okuma, Oderlei Bernardi, and Ralf Nauen

1 Introduction

The intensification of agriculture in some regions which includes the increasing cultivated area with consecutive crop seasons and no tillage systems have contributed to the population growth of some insect pests such as stink bugs (Hemiptera: Pentatomidae). Those insects can be considered as pests of the production system since they are feeding and causing damage to different crops, including soybean and maize. In Brazil, the neotropical brown stink bug, *Euschistus heros* (F.), and the green belly stink bug, *Diceraeus (Dichelops) melacanthus* (Dallas), represent the main stink bug species attacking the production system of soybean and maize crops (Corrêa-Ferreira and Sosa-Gómez 2017).

In soybean fields, these species can feed directly on the pods by piercing the grains, leading to yield loss, and impacting the quality of grains by injecting toxins. Although causing severe impacts to the soybean production, the monitoring of stink bugs is challenging since they often prefer to stay in the lower part of the canopy, particularly the nymphs, and feeding directly on pods and not causing a significant visual damage to the plants. Therefore, the damage will be only perceived during the harvest and commercialization of the grains. The insect monitoring along the crop development and its proper management becomes a critical procedure to protect yield and quality of the grains (Panizzi et al. 2012).

R. Carvalho (✉) · D. Okuma
Bayer Crop Science, São Paulo, SP, Brazil
e-mail: renato.carvalho@bayer.com

O. Bernardi
Universidade Federal de Santa Maria, Santa Maria, RS, Brazil

R. Nauen
Bayer Crop Science, Monheim am Rhein, Germany

A. F. Bueno, A. R. Panizzi (eds.), *Stink Bugs (Hemiptera: Pentatomidae) Research and Management*, Entomology in Focus 9,
https://doi.org/10.1007/978-3-031-69742-5_11

According to the soybean IPM strategy, the trigger for insecticide management must be based on the number of insects (economic damage threshold) and crop development stage. Insect scouting should be conducted on a weekly basis in several different plots of the field. Although stink bugs are present in the field early in the vegetative stage of soybean plants, the key moment to initiate management tactics is during the beginning of pod development (R3). One or two stink bugs (nymphs beyond third instar and adults) per meter row of soybean has been established as the threshold to spray insecticides for seed and grain production, respectively (Panizzi et al. 1977; Hayashida et al. 2023). Data from soybean crop season 2022/23 confirmed most insecticide sprays (76%) to manage stink bugs occurred during the reproductive stage of soybean (Fig. 11.1).

In maize fields, stink bugs can cause severe damage, particularly during the initial plant development stages (Ávila and Panizzi 1995). In general, the critical period of stink bug attack ranges from plant emergence to vegetative stage V5. Stink bug infestations during the initial establishment of the crop can lead to significant impact on yield due to reduced plant stand and injured plants. To manage stink bugs, growers tend to treat early stages of maize plant development and even before the plant emergence. For other sucking pests such as aphids the sprays are targeted during the reproductive stage (Fig. 11.2). One stink bug per 10 maize plants or 0.5 per meter row has been established as the threshold for its management (Bianco 2004; Chiaradia et al. 2016), although higher thresholds have been suggested (Gomes et al. 2020).

For sustainable stink bug management, it is fundamental that insecticides are used only when such damage thresholds are exceeded to avoid frequent applications, and thus increased selection pressure facilitating the evolution of resistance. Furthermore, rotating insecticides with different modes of action is a critical component of IRM programs to delay the evolution of resistance. However, considering the few modes of action classes available for stink bug management it seems increasingly challenging to sustain their efficacy against the background of increasing population densities. Additionally, in many cases growers have coupled the use of insecticides with fungicide or herbicide sprays to increase the efficiency in the

Target	Growth Stage		
	Pre	Vegetative	Reproductive
Stink Bugs	2%	22%	76%

Fig. 11.1 Soybean development stages and application of insecticides targeting stink bugs during 2022/23 Brazilian crop season. (Adapted from Spark and Farmtrack 2022/23)

Target	Growth Stage		
	Pre	Vegetative	Reproductive
Sucking pests	12%	70%	18%

Fig. 11.2 Maize development stages and application of insecticides targeting sucking pests during 2022/23 Brazilian crop season. (Adapted from Spark and Farmtrack 2022/23)

operational management, by reducing the number of field visits. Some studies revealed that insecticide sprays in soybean are often done on a calendar basis, and not based in insect scouting and thresholds (Corrêa-Ferreira et al. 2013). As consequence, there is a higher risk of resistance evolution due to unnecessary applications at low pest pressure, in addition to higher costs for growers.

2 Adoption of Chemical Insecticide for Stink Bug management in Soybean and Maize Fields in Brazil

As previously mentioned, IPM practices usually consider the use of insecticides for stink bug management only after exceeding damage thresholds. However, many growers are not following the technical IPM guidelines and its operational plan for pesticide applications in the field as a key factor to spray insecticides. An important consequence of largely ignoring IPM principles are frequent insecticide treatments and thus increased amounts of insecticides applied to soybean and maize fields. A recent study by Vasconcellos et al. (2023) in Mato Grosso state in Brazil demonstrated that in soybean fields managed following IPM guidelines, growers sprayed about half, when compared to a spray regime based on a calendar (as preventive).

In addition to the increased number of insecticide sprays per hectare, the production areas for soybean and maize in Brazil have grown significantly in the last 10 years. In case of soybean, in the crop season 2022/23 44 million hectares were cultivated, compared to 27 million hectares in 2012/13. As the adoption and average number of sprays increased over the years, the Potential Area Treated (PAT = cultivated area x number of applications) increased from 43 to 120 million hectares only for stink bug management in soybean crop season 2022/23. The increased cultivated area associated with the increased number of sprays per hectare to control stink bugs have directly impacted the market value of insecticides in these past years. An impressive 1.2 billion USD was spent by growers for insecticide products to manage stink bugs in Brazilian soybean fields in the crop season 2022/23. This means a Constant Annual Growth Rate (CAGR) of 12% along the past 10 years (Fig. 11.3).

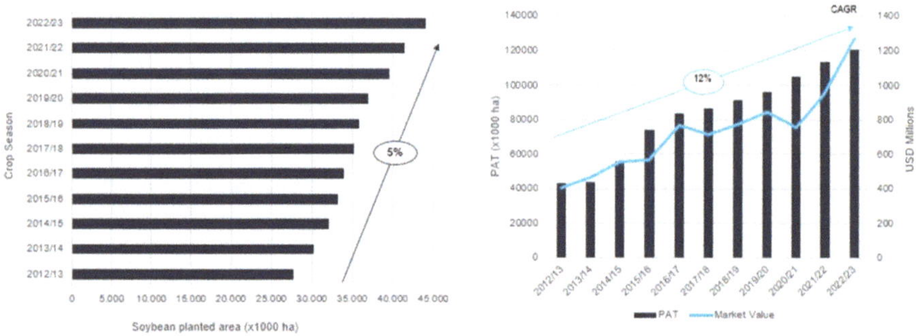

Fig. 11.3 Soybean planted area growth, Potential Treated Area (PAT) for stink bug control and market value growth over the last 10 years in Brazil. (Adapted from Spark and Farmtrack 2022/23)

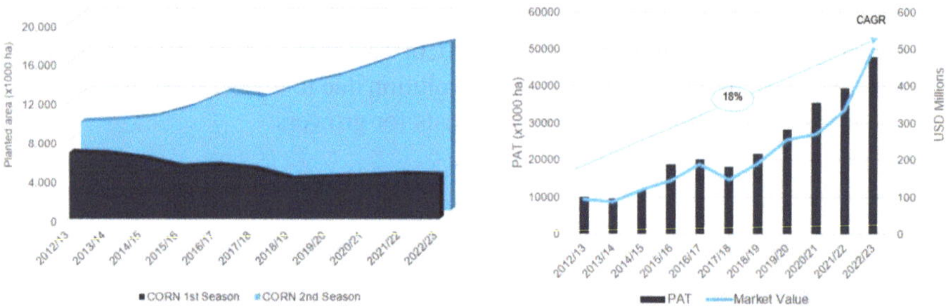

Fig. 11.4 Maize 1st and 2nd season planted area development, Potential Treated Area (PAT) for stink bug control and market value growth over the last 10 years. (Adapted from Spark and Farmtrack 2022/23)

In the Brazilian maize fields, the growth of insecticides to manage stink bugs was even more aggressive, probably associated with the significant growth of maize area planted just after the soybean harvest, the 2nd crop season known as 'safrinha'. In the crop season 2022/23, 21 million hectares were cultivated with maize in Brazil, compared to 15.8 million hectares in 2012/13. With the increase in maize acreage and number of sprays, the Potential Area Treated grew from 10 to 48 million hectares for stink bug management in maize crop season 2022/23, a CAGR of 18% in PAT along the past 10 years (Fig. 11.4). It is worth to mention the increasing importance of maize leafhopper *Dalbulus maidis* (DeLong & Wolcott), a hemipteran sucking pest and disease vector. Over the past 5 years this pest required similar insecticide applications coinciding crop phenology-wise with stink bug control measures, potentially increasing selection pressure and exacerbating its numbers.

As already shown for soybean, the increased maize cultivated area associated with the increased number of sprays per hectare to control stink bugs have drastically impacted the market value of insecticides in maize in these past years. Only in the crop season 2022/23, 563 million USD have been spent by growers on insecticide products to manage stink bugs in Brazilian maize fields, compared to 93 million 10 years before (Fig. 11.4).

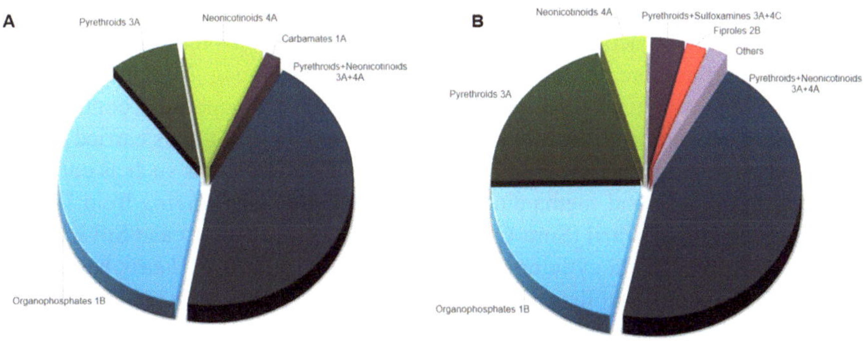

Fig. 11.5 Distribution of total insecticide used for stink bug management based on Potential Treated Area by IRAC MoA Group or Subgroup for season 2014/15 (**a**) and 2022/23 (**b**). (Adapted from Spark and Farmtrack 2022/23)

Besides this progressive growth of the insecticide market for stink bug control, the available number of insecticide classes and active ingredients for its management remained almost unchanged. Therefore, stink bug management still basically relies on old compounds launched >40 years ago such as IRAC (Insecticide Resistance Action Committee) MoA (mode of action) Group 1B Organophosphates (acephate), Group 4A Neonicotinoids (imidacloprid, thiamethoxan, acetamiprid) and Group 3A Pyrethroids (lambda-cyhalothrin, beta-cyfluthrin, bifenthrin) (Fig. 11.5). More recently additional IRAC MoA Group 4 compounds (Group 4A—dinotefuran and Group 4C—sulfoxaflor) came to the market, however, as competitive modulators of insect nicotinic acetylcholine receptors (nAChRs) they share the same target site as imidacloprid, thiamethoxam and acetamiprid, thus not expanding the options for IRM strategies. In the crop season 2021/22 a new active ingredient, ethiprole, belonging to the class of phenylpyrazoles (IRAC Group 2B) was released for commercial use in Brazil. Ethiprole antagonistically targets the GABA-gated chloride channel, a different MoA and novel chemical class for stink bug control. However, it must be mentioned that endosulfan (IRAC 2A) has been used to control stink bugs in the past and shared the same mode of action (Sosa-Gómez et al. 2001). Another new active, isocycloseram, recently approved for stink bug control, represents a new mode of action against stink bug species, targeting a different site at GABA-gated chloride channels than ethiprole (Group 30).

3 Challenges on Resistance Evolution: Current Resistance Landscape

As already stated, the management of stink bugs in soybean and maize in Brazil and other South American countries is largely carried out through chemical control. Insecticides are used in seed treatment and/or foliar applications to prevent yield losses (Panizzi 2013; Bueno et al. 2015). The limited number and the repeated use of the same mode of action for managing stink bugs have favored the evolution of resistance. Furthermore, the low dispersal of stink bugs among geographically distinct areas, due to limited flight activity, results in reduced gene flow, contributing to the rapid adaptation to insecticides (Zucchi et al. 2019). Studies carried out in 2000s reported moderate levels of resistance to older compounds such as endosulfan, monocrotophos and methamidophos in populations of *E. heros* from Paraná and São Paulo state, Brazil (Sosa-Gómez et al. 2001; Sosa-Gómez and Silva 2010). Control failures of beta-cyfluthrin and imidacloprid against *E. heros* have been described in samples collected in Goiás state, Brazil (Guedes 2017; Tuelher et al. 2018).

From 2018 onwards, studies also reported increased resistance ratios to lambda-cyhalothrin in Brazilian populations of *E. heros* (resistance ratios up to 63-fold) and *D. furcatus* (resistance ratios up to 13-fold) (Somavilla et al. 2020a, b) (Fig. 11.6). Later studies, using bean dip bioassays or encapsulated diet containing insecticides, also revealed that populations of *E. heros* collected between 2018 and 2020 in soybean also exhibited low mortality when exposed to recommended field rates or diagnostic concentrations of lambda-cyhalothrin and imidacloprid (Somavilla et al. 2020a; Tibola et al. 2021). A wide variation in control efficacy of the binary mixtures thiamethoxam + lambda-cyhalothrin and imidacloprid + bifenthrin against *E. heros* has been reported in Tocantins state, Brazil (Ribeiro et al. 2016). Metabolic detoxification has been suggested as the major mechanism associated with the resistance of stink bugs to organophosphates, neonicotinoids and pyrethroids in Brazil (Sosa-Gómez et al. 2019; Boff et al. 2022, 2023).

The levels of resistance of *E. heros* and *D. furcatus* to pyrethroids and neonicotinoids negatively affect the effectiveness of several commercial products. Studies carried out during the soybean crop seasons 2021/2022 and 2022/2023 in Rio Grande do Sul state, Brazil reported low mortality in several populations of *E. heros* and *D. furcatus* when exposed to the field dose of various pyrethroids and neonicotinoids isolated or in formulated mixtures, but not acephate and ethiprole (Moreira 2024) (Fig. 11.7). This study also reported that sulfoxaflor + lambda-cyhalothrin, acetamiprid + fenpropathrin and dinotefuran + lambda-cyhalothrin showed low efficacy against *E. heros*, whereas sulfoxaflor + lambda-cyhalothrin only showed low biological activity against *D. furcatus* (Fig. 11.8). In contrast, the mixtures of pyrethroids + neonicotinoids caused high mortality to secondary stink bugs of soybean, such as *P. guildinii* and *N. viridula* (Steinhaus et al. 2022).

The increase in resistance to pyrethroids and neonicotinoids in *E. heros* and *D. furcatus* can be explained by the widespread adoption of these chemicals. These active ingredients, alone or combined with other chemistries, represent nearly 50% of the insecticides commercially available and applied to control stink bugs in Brazil, some of them for many years. The resistance to these chemicals affects the concept of insecticide mixture for IRM—each insecticide acts individually in a way that would kill all insects susceptible to that insecticide, including those insects that are resistant to the other insecticide. From an IRM perspective, it is necessary to reduce pyrethroid and neonicotinoid use against *E. heros* and green belly stink bugs

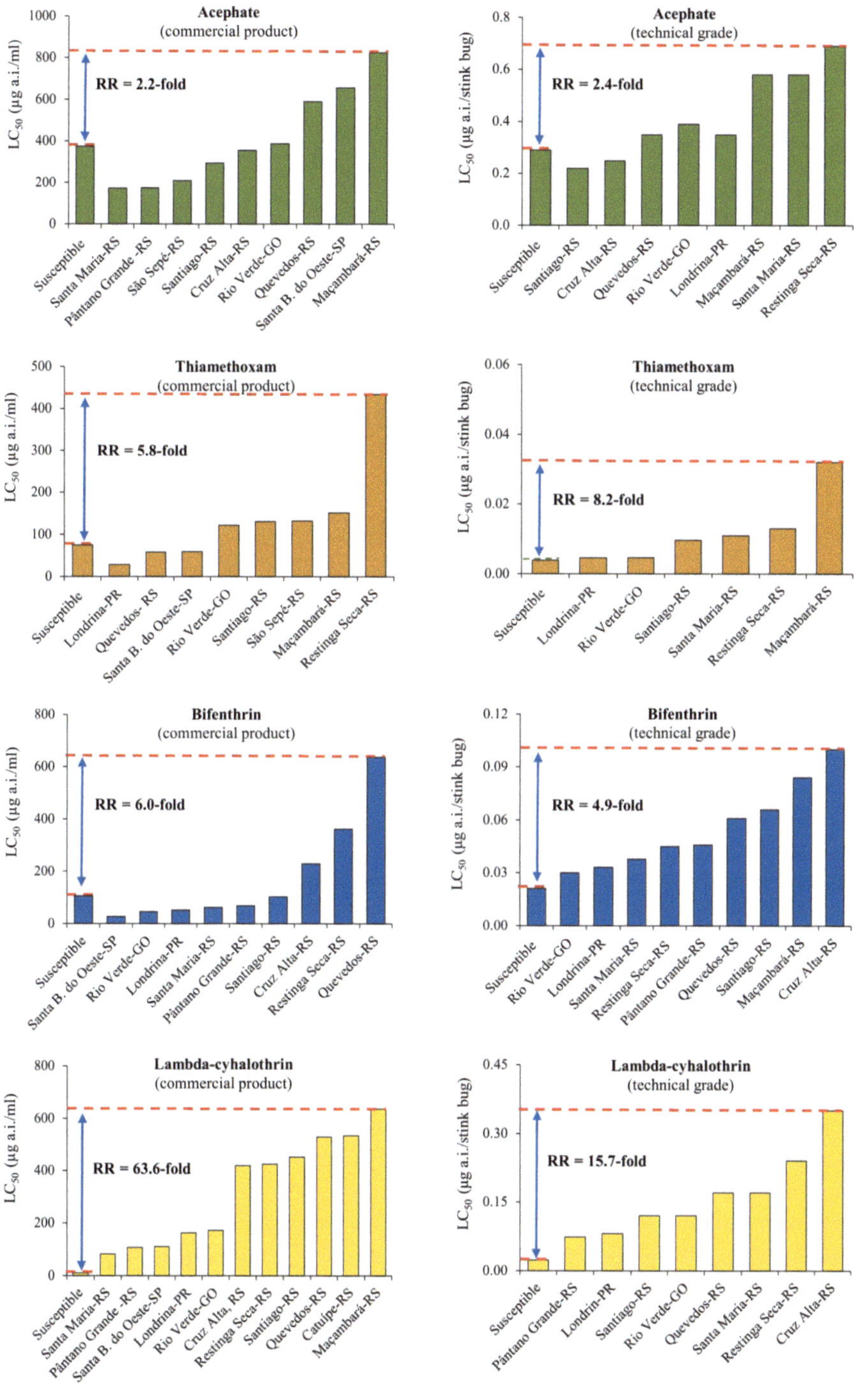

Fig. 11.6 Resistance ratios of Brazilian populations of *Euschistus heros* to insecticides in bean dip and topical bioassays. (Adapted from Somavilla et al. 2020a, b)

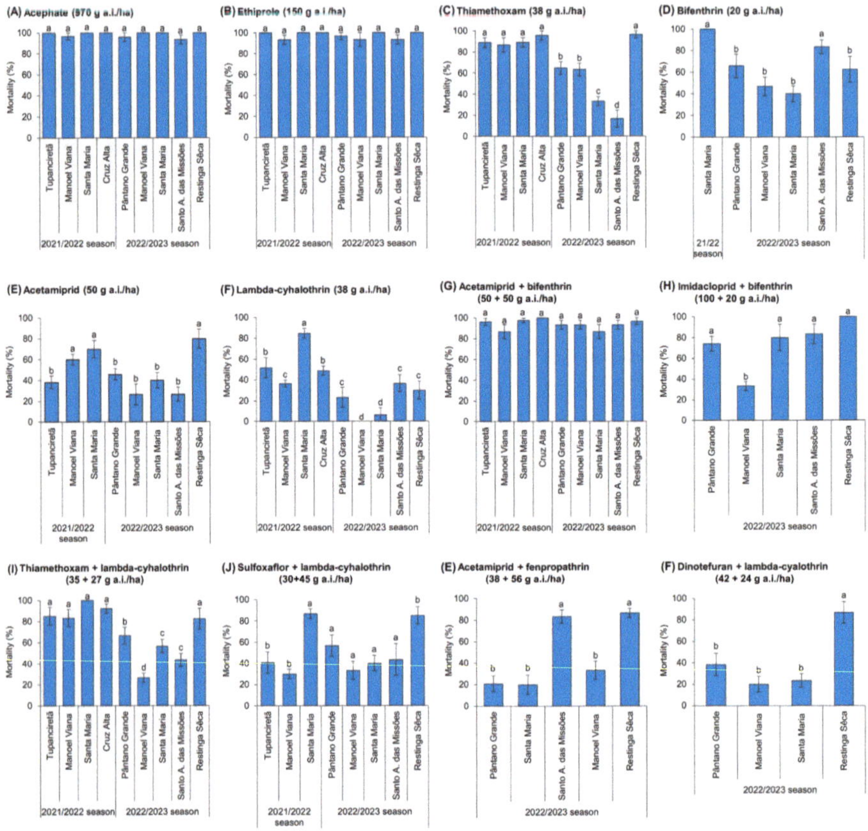

Fig. 11.7 Mortality of populations of *Euschistus heros* from Rio Grande do Sul state, Brazil exposed to the field doses of commercial insecticides in bean dip bioassays. (Adapted from Moreira 2024)

and give preference to insecticides with lower resistance ratios, thus contributing to a rotation of modes of action with those compounds showing a greater control effectiveness.

Conversely, studies reported that the majority of *E. heros* populations from Brazilian soybean fields exhibited relatively low resistance to acephate, bifenthrin, thiamethoxam and ethiprole (Somavilla et al. 2020a, b; Moreira 2024). Brazilian populations of *D. furcatus* were also tested susceptible to acephate, bifenthrin, thiamethoxam and ethiprole (Somavilla et al. 2020a, b). This was confirmed by recent studies with multiple populations of *E. heros* and *D. furcatus* collected during the 2021/2022 and 2022/2023 soybean seasons in Rio Grande do Sul state, Brazil (Moreira 2024) (Fig. 11.8). Similarly, acephate and the pre-formulated binary mixtures of thiamethoxam + lambda-cyhalothrin and imidacloprid + bifenthrin also cause high mortality of *D. melacanthus* from soybean and maize crops in Paraná state, Brazil (Paz et al. 2021). However, it is worth highlighting that significant

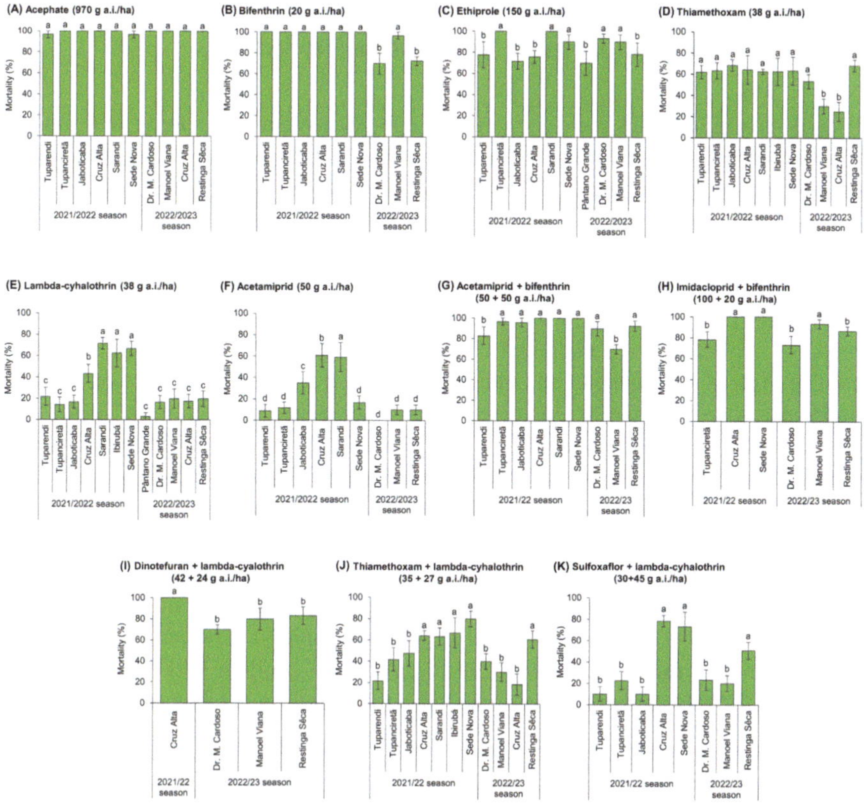

Fig. 11.8 Mortality of populations of *Diceraeus furcatus* from Rio Grande do Sul state, Brazil exposed to the field doses of insecticides in bean dip bioassays. (Adapted from Moreira 2024)

variations in susceptibility indicated the potential for resistance evolution, if IRM practices are not adopted.

Based on these recent findings, the major challenge for an effective and sustainable stink bug management in Brazil is to preserve their susceptibility to current insecticides. For that reason, insecticides should only be applied when the damage thresholds are reached, and even more important, their application should follow basic IRM principles such as rotation of modes of action by implementing a "MoA treatment window" approach as advocated by IRAC (Sparks et al. 2021), a strategy useful to delay the resistance evolution in stink bugs. Other control tactics, such as biological control, contribute to reduce the selection pressure by chemical insecticides, thus acting in concert and delaying the evolution of resistance. Nonetheless, the discovery and development of new active ingredients targeting different modes of action is essential to ensure a successful and long-term strategy for controlling stink bugs in large production areas of soybean and maize in Brazil.

4 Development of New Molecules: Timeline, Investment and Challenges to Find New Active Ingredients

Global insecticide sales are dominated by compounds acting on nerve and muscle function (85%), followed by products addressing targets interfering with insect growth and development (8%), whereas other physiological functions such as energy metabolism (mostly relevant for mite control) are underrepresented, at least in terms of global market share (Sparks and Nauen 2015) (Fig. 11.9a). Insecticides binding to neuronal targets such as pyrethroids and neonicotinoids usually act quickly and exert full efficacy on targeted pests within a few hours depending on the mode of application. All chemical classes of insecticides registered for stink bug control in Brazilian soybean and maize fields act on nerve targets representing five different modes of action, including acetylcholinesterase inhibitors (IRAC 1), GABA-gated chloride channel blockers (IRAC 2), voltage-gated sodium channel modulators (IRAC 3), nicotinic acetylcholine receptor agonists (IRAC 4) and allosteric modulators of GABA-gated chloride channels (IRAC 30) (Fig. 11.9a, inset), the latest MoA introduced for stink bug control.

The development of insecticides targeting a novel molecular site constituting a MoA not yet exploited by any other chemical class is demanding. Particularly considering the ever-increasing costs of finding and developing a new insecticide, which raised from 152 million USD in 1995 to almost 290 million USD 20 years later (McDougall 2016), which has been calculated to be equivalent to approx. 350 million

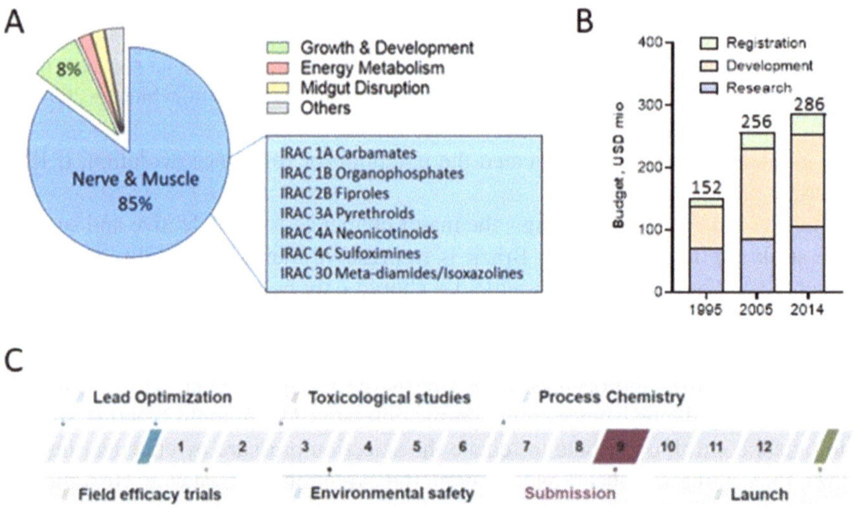

Fig. 11.9 (**a**) Insecticide market share by targeted physiology. Products acting on nerve and muscle function (blue) represented 85% of insecticide sales in 2018. The inset lists all chemical classes (n = 7) registered for stink bug control in Brazil, and which belong to five different IRAC mode of action (MoA) classes (IRAC 1, 2, 3, 4, 30). (**b**) Steady increase of costs for the development of a new active ingredient (Philips McDougall 2016). (**c**) The mean time from discovery to launch of a novel insecticide is approximately 11–12 years

USD in 2022 when corrected for inflation (Sparks and Lorsbach 2023). It has been estimated that approximately 37% of agrochemical development costs is associated with discovery, i.e., chemical synthesis, biological screens and early (eco)toxicology. Most of the costs goes towards development activities such as human and environmental safety (Fig. 11.9b) (Sparks and Lorsbach 2017). Such high discovery and development costs require a strong commitment to innovation which is these days almost exclusively restricted to larger R&D-based companies that have the resources to discover, develop and globally register novel compounds (Sparks et al. 2018).

Especially the regulatory requirements addressing environmental and ecotoxicological topics are increasing and resulted in some regions such as Europe in the loss of many older chemical insecticides and those with adverse effects on beneficial insects and non-target arthropods. Finding compounds highly selective to pest insects without harming beneficial insects such as pollinators is an intrinsically difficult task and requires sophisticated screening approaches addressing potential environmental safety issues as early as possible, for example by high throughput molecular approaches (Haas and Nauen 2021).

As mentioned above increasingly stringent regulatory standards in many regions makes it difficult to discover novel compounds that can provide the improved field efficacy, selectivity, a favorable environmental and toxicological profile as well as affordable production costs. The development time from discovery to launch for a new insecticide with a modern profile that meet the requirements mentioned above is gradually increasing and on average approx. 12 years (Fig. 11.9c) (McDougall 2016), in contrast to 5–8 years some 40 years ago (Braunholtz et al. 1981). The number of chemicals to be screened for a single new product going to the market has increased considerably over the past decades (Sparks 2013). An analysis by Philips-McDougall (2016) revealed that it requires approx.160,000 chemicals to be screened for a single product to be developed, whereas it was about 4000 to one some 40 years ago (Braunholtz et al. 1981).

Insecticide discovery requires companies to have a long staying power to explore chemical libraries encompassing millions of compounds by high-throughput screening systems, including in vivo bioassays on insect pests, representative for a broader range of taxa, in vitro testing by various forms of cell-based biochemical and/or target-site screens, recruiting omics-related research and computational approaches exploiting virtual chemical spaces for improved library design (Drewes et al. 2012; Shelton and Lahm 2015; Loso et al. 2017; Sattelle 2022; Sparks and Lorsbach 2023). However, new approaches including virtual screens and molecule optimization using computational chemistry tools may help to screen less compounds physically and thus reduce the costs and laboratory efforts linked to classical screening approaches. Indeed, stink bug species such as *E. heros* have not been included in the past as a priority in biological screening cascades across industry. This is reflected by the fact that the market is dominated by older chemistry and rather generic compounds or mixtures thereof. Specific stink bug products are still lacking, and newer compounds such as isocycloseram are broad spectrum in activity and providing stink bug activity as an "add-on". However, considering the increased (albeit regional) importance of stinkbugs in agricultural settings such as soybean and corn this may change in the future.

Providing a dedicated overview on historic and novel approaches to insecticide discovery is out of scope of this book chapter, but recent reviews and perspectives papers provided detailed insights in how the agrochemical industry is tackling the increasing challenges in their research attempts to discover and develop new insecticides with novel modes of action (Sparks and Lorsbach 2017; Sparks et al. 2018; Swale 2019; Sparks and Lorsbach 2023).

5 Final Considerations

Chemical control remains a cornerstone in stink bug control particularly in Brazilian soybean and maize fields dominated by transgenic crops expressing *Bacillus thuringiensis* insecticidal proteins to control lepidopteran pests. One of the main drivers to discover novel modes of action and chemical classes addressing them is the evolution and spread of insecticide resistance. However, insecticide resistance in stink bugs evolved rather slowly over the past decades, but recently reached levels likely to compromise recommended label rates, particularly for those chemical classes employed for stink bug control since a long time such as pyrethroids and organophosphates. Even newer compounds such as sulfoxaflor are at risk because it shares the same target-site with neonicotinoids which are employed for stink bug control for many years, either solo or in combination. The evolution of target-site mutations is likely to affect both neonicotinoids and sulfoxaflor. Ethiprole, another compound launched only very recently for stink bug control, targets a ligand-gated ion channel which has been formerly addressed by cyclodiene organochlorines such as endosulfan, a former class of insecticides utilized for stink bug control. Sustainable stink bug control is only possible by following IRM principles and alternating the MoAs currently available on the market by defining MoA treatment windows during the crop cycle avoiding treating consecutive stink bug generations with the same MoA. The implementation of IRM strategies is particularly important because farmers and other stakeholders cannot rely on a constant flow of novel MoAs launched to replace older chemistry compromised by the evolution of resistance due to frequent use. Especially the increasingly stringent regulatory requirements for chemical insecticides will make it more difficult to discover and develop new insecticides with the desired selectivity profile. Other technologies such as transgenic plants expressing traits suppressing stink bug populations or RNA interference (RNAi) based products may complement purely chemical options of stink bug control in the future.

References

Ávila CJ, Panizzi AR (1995) Occurrence and damage by *Dichelops* (*Neodichelops*) (Dallas) (Heteroptera: Pentatomidae) on corn. An Soc Entomol Brasil 24:193–194

Bianco R (2004) Nível de dano e período crítico do milho ao ataque do percevejo barriga verde (*Dichelops melacanthus*). Congresso Nacional de Milho e Sorgo 25, Resumos, Sete Lagoas, Embrapa Milho e Sorgo, p 172

Boff JS, Reis AC, Gubiani PS, Pretto VE, Garlet CG, Melo AA, Bernardi O (2022) The effect of synergistic compounds on the susceptibility of *Euschistus heros* (Hemiptera: Pentatomidae) and *Chrysodeixis includens* (Lepidoptera: Noctuidae) to pyrethroids. Environ Entomol 51:421–429
Boff JS, Reis AC, Oliveira JL, Gross RB, Fraceto LF, Melo AA, Bernardi O (2023) Development and biological evaluation of nanoencapsulated-based pyrethroids with synergists for resistance management of two soybean pests: insights for new insecticide formulations. Pest Manag Sci 79:1204–1212
Braunholtz JT, Wall C, Cremlyn RJ (1981) Crop protection: the role of the chemical industry in an uncertain future. Philos Trans R Soc B 295:19–34
Bueno AF, Bortolotto OC, Pomari-Fernandes A, França-Neto JB (2015) Assessment of a more conservative stink bug economic threshold for managing stink bugs in Brazilian soybean production. Crop Prot 71:132–137
Chiaradia LA, Nesi CN, Ribeiro LP (2016) Nível de dano econômico do percevejo barriga verde, *Dichelops furcatus* (Fabr.) (Hemiptera: Pentatomidae), em milho. Agropec Catarinense 29:63–67
Corrêa-Ferreira BS, Sosa-Gómez DR (2017) Percevejos e o sistema de produção soja-milho. Documentos Embrapa Soja 397
Corrêa-Ferreira BS, Castro LC, Roggia S, Cesconetto NL, Costa JM, Oliveira MCN (2013) MIP-Soja: resultados de uma tecnologia eficiente e sustentável no manejo de percevejos no atual sistema produtivo da soja. Documentos Embrapa Soja 341
Drewes M, Tietjen K, Sparks TC (2012) High-throughput screening in agrochemical research. In: Jeschke P, Kramer W, Schirmer U, Witschel M (eds) Modern Methods in Crop Protection Research. Wiley-VCH, Weinheim, pp 3–20
Gomes EC, Hayashida R, Bueno AF (2020) *Dichelops melacanthus* and *Euschistus heros* injury on maize: basis for re-evaluating stink bug thresholds for IPM decisions. Crop Prot 130:105050
Guedes RNC (2017) Insecticide resistance, control failure likelihood and the first law of geography. Pest Manag Sci 73:479–484
Haas J, Nauen R (2021) Pesticide risk assessment at the molecular level using honey bee cytochrome P450 enzymes: a complementary approach. Environ Int 147:106372
Hayashida R, Hoback WW, Bueno AF (2023) A test of economic thresholds for soybeans exposed to stink bugs and defoliation. Crop Prot 164:106128
Loso MR, Garizi N, Hegde VB, Hunter JE, Sparks TC (2017) Lead generation in crop protection: a portfolio approach to agrochemical discovery. Pest Manag Sci 73:678–685
McDougall P (2016) The cost of new agrochemical product discovery, development and registration, 1995, 2000, 2005–8 and 2010 to 2014. Report, CropLife America, Washington, DC
Moreira RP (2024) Suscetibilidade de *Euschistus heros* e *Diceraeus furcatus* (Hemiptera: Pentatomidae) a inseticidas no Rio Grande do Sul. M Sc dissertation, Univ Fed Santa Maria, Rio Grande do Sul, Brasil
Panizzi AR (2013) History and contemporary perspectives of the integrated pest management of soybean in Brazil. Neotrop Entomol 42:119–127
Panizzi AR, Corrêa BS, Gazzoni DL, Oliveira EB, Newman GC, Turnipseed SG (1977) Insetos da soja no Brasil. Centr Nac Pesq Soja, Londrina, PR, Brasil, Bol Tec 1, 20pp
Panizzi AR, Bueno AF, Silva FAC (2012) Insetos que atacam vagens e grãos. In: Hoffmann-Campo CB, Corrêa-Ferreira BS, Moscardi F (eds) Soja: manejo integrado de insetos e outros artrópodes-praga. Embrapa Editora, Brasília, pp 335–398
Paz MF, Diel YR, Bellon PP, Recalcatti JF (2021) Efficiency of insecticides on nymphs and adults of *Dichelops melacanthus* (Dallas, 1851) (Hemiptera: Pentatomidae). Colloquium Agrariae 17:16–29
Ribeiro FC, Rocha FS, Erasmo EAL, Matos EP, Costa SJ (2016) Management with insecticides targeting the brown stink bug control in intact soybean crop. Rev Agric Neotrop 3:48–53
Sattelle DB (2022) Invertebrate neurones, genomes, phenotypic and target-based screening; their contributions to the search for new chemical leads and new molecular targets for the control of pests, parasites and disease vectors. Pestic Biochem Physiol 187:105175

Shelton KA, Lahm GP (2015) Building a successful crop protection pipeline: molecular starting points for discovery. In: Maienfisch P, Stevenson TM (eds) Discovery and synthesis of crop protection products. Am Chem Soc, Washington, DC, pp 15–23

Somavilla JC, Reis AC, Gubiani PS, Godoy DN, Stürmer GR, Bernardi O (2020a) Susceptibility of *Euschistus heros* and *Dichelops furcatus* (Hemiptera: Pentatomidae) to selected insecticides in Brazil. J Econ Entomol 113:924–931

Somavilla JC, Gubiani PS, Reis AC, Führ FM, Machado EP, Bernardi O (2020b) Susceptibility of *Euschistus heros* and *Dichelops furcatus* (Hemiptera: Pentatomidae) to insecticides determined from topical bioassays and diagnostic doses for resistance monitoring of *E. heros* in Brazil. Crop Prot 138:105319

Sosa-Gómez DR, Silva JJ (2010) Neotropical brown stink bug (*Euschistus heros*) resistance to methamidophos in Paraná, Brazil. Pesq Agropec Brasil 45:767–769

Sosa-Gómez DR, Corso IC, Morales L (2001) Insecticide resistance to endosulfan, monocrotophos and ethamidophos in the neotropical brown stink bug, *Euschistus heros* (F.). Neotrop Entomol 30:317–320

Sosa-Gómez DR, Corrêa-Ferreira BS, Kraemer B, Pasini A, Husch PE, Vieira CED, Martinez CBR, Lopes ION (2019) Prevalence, damage, management and insecticide resistance of stink bug populations (Hemiptera: Pentatomidae) in commodity crops. Agric For Entomol 22:99–118

Spark and Farmtrack (2022/23) Pesquisas de soja e milho. Valinhos, SP

Sparks TC (2013) Insecticide discovery: an evaluation and analysis. Pestic Biochem Physiol 107:8–17

Sparks TC, Lorsbach BA (2017) Perspectives on the agrochemical industry and agrochemical discovery. Pest Manag Sci 73:672–677

Sparks TC, Lorsbach BA (2023) Insecticide discovery – "Chance favors the prepared mind". Pestic Biochem Physiol 192:105412

Sparks TC, Nauen R (2015) IRAC: mode of action classification and insecticide resistance management. Pestic Biochem Physiol 121:122–128

Sparks TC, Hunter JE, Lorsbach BA, Hanger G, Gast RE, Kemmitt G, Bryant RJ (2018) Crop protection discovery: is being the first best? J Agric Food Chem 66:10337–10346

Sparks TC, Storer N, Porter A, Slater R, Nauen R (2021) Insecticide resistance management and industry: the origins and evolution of the Insecticide Resistance Action Committee (IRAC) and the mode of action classification scheme. Pest Manag Sci 77:2609–2619

Steinhaus EA, Warpechowski LF, Braga LE, Morin M, Tenório C, Boff JS, Bernardi O, Farias JR (2022) Intra- and interspecific variation in the susceptibility to insecticides of stink bugs (Hemiptera: Pentatomidae) that attack soybean and maize in southern Brazil. J Econ Entomol 115:631–636

Swale DR (2019) Perspectives on new strategies for the identification and development of insecticide targets. Pestic Biochem Physiol 161:23–32

Tibola CM, Silva L, Sgubin F, Omoto C (2021) Monitoring resistance of *Euschistus heros* (Fabricius) (Hemiptera: Pentatomidae) to insecticides by using encapsulated artificial diet bioassay. Insects 12:599

Tuelher ES, Silva EH, Rodrigues HS, Hirose E, Guedes RNC, Oliveira EE (2018) Area-wide spatial survey of the likelihood of insecticide control failure in the neotropical brown stink bug *Euschistus heros*. J Pest Sci 91:849–859

Vasconcellos MC, Corassa JN, Pitta RM, Rolim GG (2023) Estratégias de controle de pragas em soja e suas implicações na comunidade de artrópodes e na rentabilidade da cultura. Pesq Agr Amb Nativa 11:28–43

Zucchi MI, Cordeiro EM, Wu X, Lamana LM, Brown PJ, Manjunatha S, Viana JPG, Omoto C, Pinheiro BJ, Clough SJ (2019) Population genomics of the Neotropical brown stink bug, *Euschistus heros*: the most important emerging insect pest to soybean in Brazil. Front Genet 10:1035

Chapter 12
Application Technologies for Stink Bug Management in Different Cropping Systems

Ulisses Rocha Antuniassi, Walter Boller, Clerison Regis Perini, Glauber Renato Sturmer, Juliano de Bastos Pazini, Vinadio Lucas Béga, Ivair Valmorbida, and Orcial Ceolin Bortolotto

1 Introduction

Phytophagous stink bugs (Hemiptera: Pentatomidae) are among the most important insect pests of cultivated crops. Stink bugs prefer to feed directly on reproductive structures such as flowers, pods, ears, and seeds. Feeding injuries can lead to the abortion of these structures, yield losses, and a decrease in seed quality. During feeding, stink bugs may inject toxic saliva into plant tissues, thereby affecting plant growth and development (Grigolli et al. 2016). In addition, stink bugs are difficult to control and often reach population densities that exceed economic thresholds in crop fields. For example, in the Brazilian Cerrado, an analysis of 400,000 ha of

U. R. Antuniassi
Faculdade de Ciências Agronômicas, Botucatu, SP, Brazil

W. Boller
Fundação Universidade de Passo Fundo, Passo Fundo, RS, Brazil

C. R. Perini
Proteplan, Sorriso, MT, Brazil

G. R. Sturmer
Cooperativa Central Gaúcha Ltda (CCGL), Cruz Alta, RS, Brazil

J. de Bastos Pazini
Instituto Federal Goiano, Urutai, GO, Brazil

V. L. Béga
Koppert Brasil, Piracicaba, SP, Brazil

I. Valmorbida
University of Missouri, Columbia, MO, USA

O. C. Bortolotto (✉)
Universidade Estadual de Ponta Grossa, Ponta Grossa, PR, Brazil

A. F. Bueno, A. R. Panizzi (eds.), *Stink Bugs (Hemiptera: Pentatomidae) Research and Management*, Entomology in Focus 9,
https://doi.org/10.1007/978-3-031-69742-5_12

soybean revealed that the stink-bug species complex requires, on average, 4.1 insecticidal applications during the crop cycle, with this number increasing in the last three growing seasons (Juliano de Bastos Pazini, personal information). High stink bug population densities, poor adoption of pest population monitoring, insecticide resistance evolution, reduced number of active ingredients available, dependence on chemical control, and inadequate use of application technology are factors associated with control failures and increased concerns among growers, crop consultants, agronomists, and stakeholders (Prado et al. 2010a; Corrêa-Ferreira et al. 2013; Corrêa-Ferreira and Sosa-Gómez 2017; Pazini et al. 2019; Sosa-Gómez et al. 2020).

One of the most important factors associated with the success of foliar insecticides (chemical or biological) in the management of insect pests is the proper application technology. Specifically for phytophagous stink bugs, such as the Neotropical brown stink bug *Euschistus heros* (F.) and green belly stink bugs *Diceraeus* spp. on soybean-maize-cotton cropping systems, red-banded stink bug *Piezodorus guildinii* (Westwood), southern green stink bug *Nezara viridula* (L.), and *Edessa meditabunda* (F.) on soybean, beans, sunflower, maize, and cotton, and the rice stem stink bug *Tibraca limbativentris* Stal on rice, application technology must provide thorough and uniform coverage of the target and droplet deposition into the canopy because these insects are generally found in the middle and lower parts of the plant; the available systemic insecticides do not present efficient basipetal translocation (Prado et al. 2010a; Roggia et al. 2019; Seidel et al. 2020).

Application technology comprises knowledge that integrates information on pesticides, formulations, adjuvants, spraying, information technology, targets, human resources, and the environment, aiming at safe and efficacious applications while minimizing unintended effects (Antuniassi et al. 2021a, b). When insecticide application aims to protect plants against arthropod pests, it is necessary to understand the relationship between pests, hosts (crops), and the environment (Guedes and Perini 2019). In case of stink bugs, this includes pest species, insecticide mode of action, route of entry, mechanisms of spread into the insect body, crop characteristics, application equipment and technology, and environmental variables (Antuniassi 2004).

Because phytophagous stink bugs attack several cultivated crops and at different crop growth stages, and these plants may coexist in agroecosystems, concomitantly or not (e.g., “green bridges”), special attention should be given to the unique characteristics of each target species. The differences in stink bug feeding characteristics and injuries, such as feeding on reproductive structures (e.g., soybean, cotton, beans, and sunflower, etc.) and stems (e.g., maize and rice), vertical distribution within the plant canopy, presence on the soil surface or under crop residues, pest life stage (nymphs and adults), phenology, movement, and diurnal and nocturnal activity affect the choice of insecticides (chemical and/or biological), spraying times, and application technology and equipment (e.g., foliar or seed treatment; Maziero et al. 2009, Fiorin et al. 2011).

Insecticide efficacy can be enhanced by combining knowledge of how stink bug exposure occurs after application (de Vieira 2021). These include direct contact with parts of the integument (dorsal or tarsal) and oral ingestion. Contamination by dorsal contact usually occurs with spray deposition on a target or tarsal as the insect walks over plant structures contaminated with insecticides. Systemic insecticides are absorbed and transported through plant structures, and can control stink bugs through direct contact during application and ingestion when stink bugs feed on plant tissues. In this case, the targets of insecticide application should not only be insects but also plants (Guedes and Perini 2019), ensuring adequate spray coverage in the canopy and overcoming control failures caused by the behavior of the stink bugs that are often hidden in the lower parts of canopies where droplet deposition is more difficult (Guedes et al. 2012; Bueno et al. 2021).

In this chapter, we discuss the current knowledge on the application technology and methods for the chemical and biological control of phytophagous stink bugs (Hemiptera: Pentatomidae) in different agricultural production systems. We delve into the intrinsic characteristics between pest × wild host or cultivated plant × environment and their relationships, which can impact the application technology, its performance, and pest control. We discuss the routes of exposure of stink bugs to insecticides, pest biology and behavior, the influence of their interactions on the application technology for chemical and biological insecticides, and the challenges faced by farmers in managing stink bugs. Practical examples of stink bug control in soybean-maize cropping systems are presented and discussed.

2 Routes of Contamination of Phytophagous Stink Bugs by Insecticides

The effectiveness of insecticides on targeted insects is determined by several factors, including the dosage of the active ingredient, chemical groups, physical state of the product, and mode of action. However, the most critical factor determining control effectiveness is the entry of insecticides into the insect body, which can occur via penetration through the integument of the body wall, oral intake, and/or inhalation through the spiracles (Yu 2014). This can vary for each insecticide, product formulation (Sumita et al. 2016), and insecticide application method (Hayashi 1983; Yasutomi 1987). After crossing the integument barrier, the active ingredient undergoes several physicochemical processes between stages such that only a small fraction reaches the site of action and causes a biological response (Hadaway 1971).

Insects are small, have high surface-area-to-volume ratios, and short pathways to the nervous system. This makes them prone to contact insecticides that can easily reach the insect nervous system after passing through their integument. The first stage for a contact insecticide to reach the target site involves the superficial accumulation of the toxicant on the insect body, a process determined by the insect behavior and the physical properties of the formulated product (e.g., particle or droplet size). The second stage involves the penetration of toxic substances through the integument. Even small differences in penetration rates can have a significant impact on toxicity, particularly when internal defense mechanisms, such as detoxification enzymes, are present.

Penetration through the insect cuticle is considered one of the main entry routes for contact insecticides (Ebeling 1974), such as malathion (Matsumura 1963), parathion (Ball and Beck 1951), and dichlorvos (Galley 1967). Studies have suggested that after crossing the integument, the insecticide is distributed throughout the rest of the insect body via the hemolymph until it reaches the central nervous system (Winteringham 1957; Lewis 1965; Grissom et al. 1989). However, this hypothesis was challenged by Gerolt (1965) who demonstrated that dieldrin spreads laterally over the cuticle and reaches the central nervous system through the integument of the tracheal system, a route with fewer barriers to the spread of the active ingredient.

The time taken for an insecticide to reach its target site is directly proportional to the distance between the point of application on the insect body and the site of action and is not significantly influenced by the thickness of the integument (Gerolt 1969). This finding supports the hypothesis that the penetration occurs through the tracheal system. If the hemolymph is the main route for spreading the active ingredient throughout the insect body, the speed of the circulatory system would make the distance to the site of action irrelevant. The respiratory system is also an entry route for insecticides, such as phosphine and aluminum phosphide (Lorini et al. 2015), and volatile compounds that spread in the insect body facilitated by carriers such as carbon dioxide (CO_2) (Jones 1938; Bond 1961; Terriere 1982).

The epicuticle is the outermost layer of the cuticle in insects, and acts as a barrier between insect tissues and foreign matter. It is well known that the layered structure of lipids and proteins, which is one to two microns thick, prevents tissue desiccation by restricting water transpiration. The study of factors that influence the penetration of foreign molecules through the insect cuticle is not only of academic interest but also of practical importance. Most studies on this topic have focused either directly or indirectly on the chemical control of vectors and pests of public health importance.

Spiracles found in the mesothorax of insects are an important entry route for vapor-phase pyrethroids (e.g., applied in aerosol form), as demonstrated in *Blattella germanica* (L.) (Blattodea: Blattellidae) (Sugiura et al. 2018) and *Musca domestica* (L.) (Diptera: Muscidae) (Sumita et al. 2016). This explains why this group of insecticides has a rapid knockdown effect, because entering through the spiracles provides the shortest path to the central nervous system (Gerolt 1969; Huber 1974; Burrows 1980). Furthermore, the thoracic ganglia, located in the ventral region of the thorax and connected to the central nervous cord facilitate the effects of neurotoxic insecticides that penetrate this region of the insect body (Gullan and Cranston 2017).

The third and final entry route for insecticides occurs through oral ingestion, which is commonly explored for insecticides used to control chewing insect pests, such as coleopterans and lepidopterans. This entry route is essential for biological insecticides containing *Bacillus thuringiensis* and *Baculovirus*es (Moscardi 2003). Systemic insecticides such as neonicotinoids exert a contact action when applied to target pests. They can also be incorporated into plant tissues making them toxic to sap-sucking insects (Salgado 2013). Systemic insecticides are effective in controlling stink bugs because they kill insects through contact and oral ingestion, thereby increasing the chances of achieving satisfactory control.

Contact exposure can occur topically when the insecticide is directly deposited on insects, or via the tarsi when the product is applied to plants and the pest walks or crawls over it (Yu 2014). Therefore, the target of insecticide application is not always an insect but can also be the plant, soil surface, or seeds (Guedes and Perini 2019). For stink bugs, contamination through tarsal contact is the most efficient (Cavallin 2020). Insect tarsi are considered vulnerable sites for the entry of insecticides (Matsumura 1985), especially in the openings between the tarsal segments (Balabanidou et al. 2019), as demonstrated in *M. domestica, Sitophilus oryzae* L. (Coleoptera: Curculionidae), *Rhyzopertha dominica* (F.) (Coleoptera: Bostrichidae), *Callosobruchus chinensis* (L.) (Coleoptera: Bruchidae), *Tribolium castaneum* (Herbst) (Coleoptera: Tenebrionidae) (Rajashekar et al. 2012), *Anopheles gambiae* Giles (Diptera: Culicidae) (Spielmeyer et al. 2019), and *E. heros* (F.) (Hemiptera: Pentatomidae) (Cavallin 2020).

Neurotoxic insecticides, such as organophosphates, pyrethroids, and neonicotinoids, are more effective when they enter the insect body close to the central nervous system (Gerolt 1969). These insecticides are the most commercialized worldwide (Sparks and Nauen 2015) and are widely used for stink bug control in Brazil (Agrofit 2023). The legs of insects have nerve connections with the thoracic ganglia located in the thorax, making it easy for neurotoxic insecticides to reach their sites of action (Balabanidou et al. 2019). Furthermore, the tarsi of *E. heros* have many hairs and sensilla that respond to physicochemical stimuli and regulate the tactile sensation of the legs (Chapman 1998). Sensilla are connected to sensory neurons and the muscular system, which may accelerate the entry and action of neurotoxic insecticides. Therefore, the tarsal morphology plays an important role in the absorption of the active ingredient and in the speed of killing an insect pest (Hadaway 1971).

Topical (direct) contact, although also important in stink bug control, demonstrated lower efficacy than tarsal contact. According to Cavallin (2020), this occurs because only a small fraction of the active ingredient reaches the hemolymph and is carried to the site of action. The rest is either retained on the back of the insect, metabolized, and/or sequestered in the cuticular tissue. In addition, it can be difficult to achieve direct contact when the target insects are located on the abaxial surface of the leaves or in parts of the plant canopy where spray droplets do not penetrate or have a very fine diameter or are present in smaller quantities (Guedes and Perini 2019). To avoid control failures owing to a lack of direct contact, it is important to provide uniform spray coverage, which increases the period of exposure of the target insect to the active ingredient. This increases the probability of contamination during insect movement (Matsumura 1985).

The feeding behavior of stink bugs makes them vulnerable to contamination through ingestion, as they suck sap from the reproductive structures of plants sprayed with systemic insecticides, such as neonicotinoids. Pyrethroids, on the other hand, are non-polar compounds with high molecular weight that remain on the plant surface until they degrade and practically do not cause mortality through ingestion. However, the knockdown effect justifies their combination with neonicotinoids in formulated mixtures, combining different entry routes and modes of action for the same product (Salgado 2013). According to Marques et al. (2019), this is one of the most effective strategies for controlling stink bugs on soybean.

Therefore, insecticide entry routes through topical contact, tarsal contact, and ingestion are efficient in controlling stink bugs. However, these routes may vary depending on the type of insecticide used. Understanding how stink bugs are exposed to insecticides helps to develop and implement better management strategies, which in turn can enhance insecticide efficacy and lead to more economical and sustainable pest control.

3 Decision-Making for Insecticide Application Technology Considering Phytophagous Stink Bug Behavior

Insects rely on their senses, which are determined by the relationship between the environment and receptors present in their bodies. These receptors are found on the antennae, legs, compound eyes, and digestive and reproductive systems and stimulate behavioral responses. Most insect behaviors are considered innate, meaning that they are genetically programmed in response to stimuli. However, some behaviors can be learned based on the environment in which insects live. The study of insect behavior focuses on understanding how an insect receives stimuli from its environment, processes information, and responds to it (Gullan and Cranston 2014). Sensory receptors allow insects to sense their surroundings mechanically, thermally, chemically, and visually to discriminate between hosts, environmental conditions (relative humidity, temperature, and airflow), and natural enemies (Chapman 1998).

Interactions between pests (population density, pest growth stage, behavior, and biology), crops (soybean, maize, beans, rice, and cotton), plant characteristics, critical period of damage, environmental conditions (humidity, temperature, precipitation, and wind), and the presence of natural enemies can affect the efficacy of insecticides and acaricides (Guedes and Perini 2019). Insect behavior, whether innate or learned, leads to a series of actions that affect pest behavior throughout its life cycle. Stink bug injury to soybean (seeds) and maize (stalk), their location in the plant canopy (lower, middle, and upper parts) or on the soil surface, mobility in the plant structures (leaf, stem, flower, pods, and ears), phenology, and the response to stimulus during the light and dark periods affect decision-making when choosing the active ingredients (e.g., chemical or biological), spraying times, and application technology and methods.

Many stink bug species feed on cultivated crops at different times. They also have diverse feeding behaviors and injuries, which vary according to the plant growth stage, and require monitoring, control methods, and strategies targeted at each species present and crop growth stage (Corrêa-Ferreira and Peres 2003). Both seed treatments and foliar sprays require an understanding of pest behavior to ensure the greatest potential for exposure to chemical or biological insecticides. Based on this, the application technology and equipment to increase the control efficacy can be determined (Kuss-Roggia 2009).

It is important to consider stink bug behavior within a plant to develop and implement strategies to reach the target effectively. One of these is to divide the soybean plant into three parts—lower, middle, and upper–for estimating the stink bug infestation in each part. Stink bugs that inhabit the lower and middle parts of plants are better protected from foliar sprays and natural enemies (Fig. 12.1). Stink bugs typically begin to colonize soybean at the end of the vegetative period until mid-R1-R2. They reproduce quickly, and populations increase around the R3 (beginning pod) and peak at R6 (full seed) soybean growth stage (Correa-Ferreira and Panizzi 1999). During this period, soybean with flowers, pods, and grains have a high leaf area index. Stink bugs can be found in different locations on the plant and soil surface, and their location within the plant may vary depending on the time of the day (day vs. night).

Studies on the behavior of stink bugs in soybean with *P. guildinii* (Westwood) (Hemiptera: Pentatomidae) and *E. heros* (Kuss-Roggia 2009; Puntel 2022) have demonstrated a difference in their location in the plant canopy throughout the day, and their preference for plant structures varies according to the soybean growth stage. Nymphs and adults of *P. guildinii* are found in the middle third of the plant

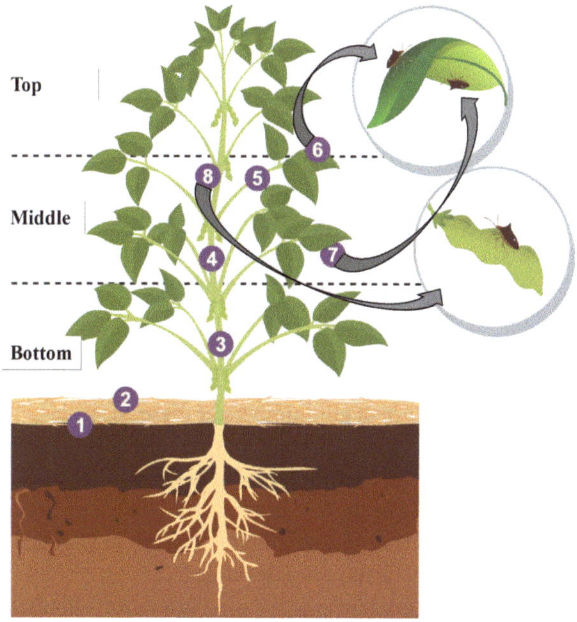

Fig. 12.1 Stink bugs are found on the lower, middle, and upper parts of soybean canopy, including: (1) on the ground under crop residue; (2) on crop residue; (3) on the main stem; (4) on the secondary stems; (5) on the petiole; (6) on the adaxial surface of the leaf; (7) on the abaxial surface of the leaf; and (8) on soybean pods, with emphasis on the leaves and pods (Perini 2024)

canopy and on pods, regardless of the time of day (Kuss-Roggia 2009). However, in case of *E. heros,* it varied depending on the pest growth stage (nymphs vs. adults), plant growth stage, and time of day. No differences in insect behavior were observed between the R1/R2 and R3/R4 soybean growth stages (before grain filling), but they differed from R5.2 and R5.3 (during grain filling). No significant differences in pest behavior were observed between the R5.2 and R5.3 soybean growth stages (Puntel 2022). The same author also observed differences in behavior at different times of the day (00:00 h, 02:00 h, and 04:00 h versus 06:00 h, 08:00 h, and 10:00 h). Brown stink bug nymphs and adults have been found in the middle and upper parts of the soybean canopy (Puntel 2022) and may be in direct contact with spray droplets carrying insecticides. These stink bugs can be found on the main stem, petiole, flowers, pods, or leaves and require spraying droplets to penetrate the crop canopy and reach the parts where the stink bugs are located. Small droplets have a greater capacity to penetrate the crop canopy and provide greater coverage and density of the target (Guedes and Perini 2019).

Additionally, insects respond to external stimuli by moving from the upper to the lower parts of a plant or from the abaxial to the adaxial side of a leaf. This behavior may be related to the insect's natural movement, field operations, and detection of a predator through receptors that stimulate acoustic defense. Stink bug nymphs move short distances in the longitudinal direction in a row of up to 12 m (Panizzi et al. 1980), which can reduce exposure to insecticides via tarsal contact. Thus, direct contact with the spraying droplets is more important. On the other hand, brown stink bug nymphs start feeding on pods and seeds when they reach the second instar (Tessmer et al. 2022) and can be exposed to systemic insecticide through ingestion.

Nymphs and adult *E. heros* are found on soybean leaves during the vegetative growth stages. However, they can be found on the pods (Puntel 2022), during pod formation and grain-filling. Therefore, the insecticide and spraying technologies used are completely different between the vegetative and reproductive stages of soybean. Systemic insecticides can be effective in managing stink bugs because they feed on pods and seeds. Conversely, insecticides with low mobility in the plant may be available on the leaves and other parts of the soybean plant, thereby contaminating stink bugs via contact as they move on the plant (Huth et al. 2012; Melo et al. 2017).

Stink bugs are also found on the surface of the soil, weeds, and crop residues. Although they may be present in small numbers at these locations, they are difficult to control using foliar sprays. However, they may be more exposed to entomopathogenic fungi in humid microclimates under crop residues and natural microorganisms present in the soil. Although *E. heros* and *P. guildinii* are present in low numbers on the soil surface (Kuss-Roggia 2009; Puntel 2022), they reproduce and increase in population density as the crop develops.

The occurrence of some stink bug species soon after crop emergence, such as the green-belly stink bug on maize, requires control strategies to protect against the most critical phase of pest damage. Yield losses caused by green-belly stink bugs to maize range from 150 to 833 kg/ha for each insect m^{-2} (Rodrigues 2011; Duarte et al. 2015; Chiaradia et al. 2016; Fernandes et al. 2020), requiring control strategies before substantial economic damage occurs to the crops.

In no-till systems, the green-belly stink bug occurs soon after maize emerges, affecting the control strategies (Fig. 12.2). During the initial stages of maize growth and development, maize plants have a low leaf area index, which favors spray coverage. However, stink bugs may be located under crop residues at certain times of the day and are less exposed to insecticides via topical contact. The behavior of *D. melacanthus* during the V3 and V4 maize growth stages varies and is characterized by remaining in the soil under maize plants and crop residues during cooler times of the day. They are more active when temperatures increase (between 9 am and 5 pm) (Clérison Perini, personal information, unpublished data). Insecticide application at night may result in low control efficacy, owing to the low potential of chemical and biological insecticides to achieve the target. As maize plants have a small leaf area, better stink bug control can be accomplished through topical contact with insecticides that have a greater contact and knockdown effect. Therefore, to effectively control stink bugs, it is necessary to understand their behavior in crops

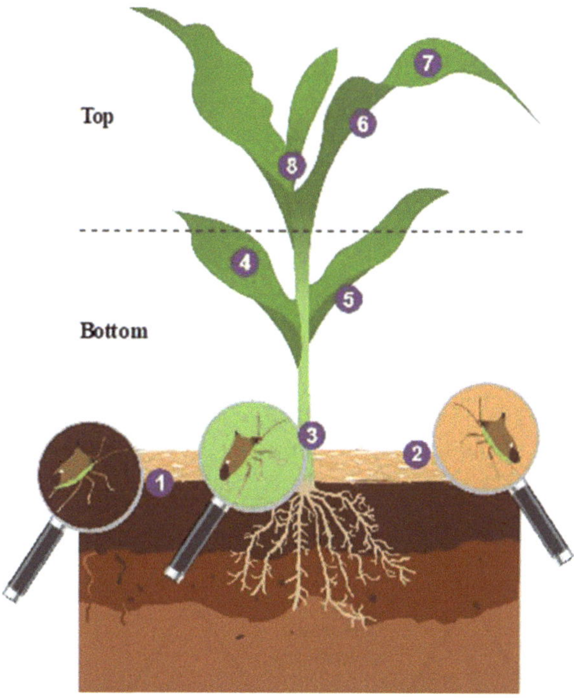

Fig. 12.2 Green-belly stink bug on maize considering the upper and bottom part of the plant: (1) on the ground under crop residue; (2) on crop residue; (3) on the base of the stalk; (4 and 5) on the adaxial and abaxial surface of leaves with visible leaf collars; (6 and 7) on the adaxial and abaxial surface of not yet fully expanded leaves and without visible leaf collars; (8) in the maize whorl (Perini 2024)

and the damage they cause during different crop growth stages. This understanding will help in selecting the most appropriate insecticide and application technology against the targeted pest species.

4 Chemical Insecticide Application Technology for Control of Phytophagous Stink Bugs

When pesticides are applied to protect cultivated plants from insect pests, all aspects that define the application technology and characteristics of the insecticides (see item 1), the behavior of the target pests (see item 2), and particularities of the cultivated plants must be taken into consideration. Based on the triangle of relationships between pests, hosts (crops), and the environment (Fig. 12.3), strategies can be implemented to maximize the performance of insecticide applications (pest control).

Because phytophagous stink bugs attack different plant species and at different stages of plant phenology, it is necessary to consider specific scenarios in which the conditions for insecticides to achieve their targets are unique. Thus, two specific approaches are discussed for managing stink bugs: on maize, these insects are important pests in the early vegetative growth stages, and on soybean, stink bugs are a concern during the reproductive stages. These two scenarios pose unique challenges for the application technology; therefore, they are discussed in detail.

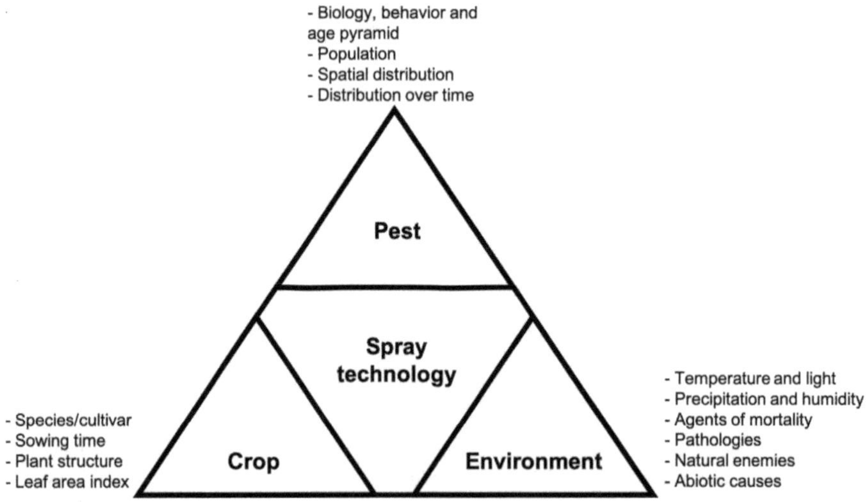

Fig. 12.3 Pest triangle maizeers: Host plant, pest, and environment (Guedes and Perini 2019)

4.1 Application Technology for Stink Bug Control on Maize

Stink bugs are typically, pests of soybean, but with the planting of maize following soybean or soybean-maize rotation, they also begin to cause damage to maize soon after plant emergence. Damage occurs during the initial phase of crop development and can cause partial or total crop loss (Viana et al. 2002).

Several species of stink bug pests on soybean, such as the green-belly stink bugs [*Diceraeus furcatus* (F.) and *D. melacanthus* (Dallas) (Hemiptera: Pentatomidae)], southern green stink bug [*Nezara viridula* (L.) (Hemiptera: Pentatomidae)] and Neotropical brown stink bug *E. heros*, can also attack maize, especially when maize follows soybean (off-season maize). Roza-Gómez et al. (2011) found that *D. furcatus* and *D. melacanthus* reduced maize plant height, the number of expanded leaves, and root dry mass and caused injuries and/or curling of the whorl. Although *E. heros* and *N. viridula* damaged the maize at different intensities, the damage caused by *E. heros* was the least intense. Furthermore, the damage caused by *D. furcatus* and *N. viridula* was similar in type and intensity to those caused by *D. melacanthus*.

The green-belly stink bug, *D. melacanthus*, is considered one of the most important stink bug pest species on maize (Ávila and Panizzi 1995) and can cause damage that are approximately three times more severe than other stink bug species (Crosariol Netto et al. 2015). *D. melacanthus* and *D. furcatus* have the greatest potential to cause damage to maize because they feed on the apical meristem of maize seedlings and inject their saliva with toxic substances when extracting sap from plants (Brustolin et al. 2011). When the apical meristem is damaged, the whorl of the seedling withers and dries, and shows a symptom called "dead heart". Damaged plants can also produce yield-robbing tillers, making them unproductive (Ávila and Panizzi 1995). The damage can be significant, especially when infestation occurs at the early growth stages of maize (V1–V3), negatively affecting the yield potential (Duarte et al. 2015). The authors also found that the economic threshold for the green-belly stink bug on maize is 0.8 stink bugs per m^2 when the pest damages maize at the V1 growth stage. Maize plants are most susceptible to *D. melacanthus* shortly after emergence, when plants are at the V1 growth stage, and the infestation lasts for at least 14 days until the V5–V7 growth stage (Silva et al. 2019).

In regions where maize is grown in succession with soybean, controlling the greenbelly stink bug on soybean to prevent migration to maize and applying insecticides before maize sowing are preventive measures to reduce infestation levels, during the early growth stages of maize (Wordell-Filho et al. 2016). Therefore, because it is an early season pest of maize, one strategy for controlling stink bugs is to use insecticide seed treatments with residual control (Viana et al. 2002).

Another way to control stink bugs is to use foliar sprays of insecticides during the first few days after maize emergence. The late adoption of control measures for this pest, even if the insects are eliminated, may not prevent damage (Cruz and Bianco 2001). A study conducted by Albuquerque et al. (2006) demonstrated satisfactory results (reduction of damaged plants between 80% and 84%) with the combination of seed treatment with a systemic insecticide (thiamethoxam) and foliar spray 8 days after seedling emergence with an insecticide (thiamethoxam + lambda-cyhalothrin) that has systemic, contact, and ingestion activities. However, the same authors concluded that green-belly stink bug control with only a foliar application of thiamethoxam + lambda-cyhalothrin 8 days after maize emergence did not prevent the insect from damaging the plants.

4.1.1 Application Technology for Protection of Crops at Early Vegetative Growth Stages

Systemic insecticides applied via seed treatment in-furrow (liquid or granular), which are absorbed by the roots and translocated to the stem and leaf tissues, allow the control of insect pests when they feed on plants (Fig. 12.4). These application methods pose less risk to natural enemies and non-target organisms because only a small part of the total area is treated with insecticides (between 55 m^2 and 500 m^2), compared to a foliar spray on a total area (10,000 m^2) (Guedes and Perini 2019; Pazini 2022).

Maize seeds treated with imidacloprid absorbed the insecticide through both seeds and roots. When absorbed, the insecticide moves upward to the stems and leaves (Tang et al. 2020). With seed coating, the active ingredient remains at high concentrations and facilitates the uptake of the systemic compound via diffusion through the pericarp (Dias et al. 2014). However, only a small amount of insecticide moves upward in plants, and most of it remains in the soil and root zones (Sur and Stork 2002). For clothianidin, a high concentration of the active ingredient was

Fig. 12.4 Seed treatment of maize with systemic insecticides, absorption and translocation into plant tissues, and effect on *Diceraeus melacanthus*. The concentration and availability of the active ingredient decreases over time (Perini 2024)

detected shortly after maize emergence, followed by an exponential decrease in the first 10 days (Alford and Krupke 2017). This might be one of the reasons why the seed treatment control period may not be sufficient to reduce stink-bug damage to maize plants because it only protects a part of the critical period to prevent crop damage. The green belly stink bug has the potential to reduce maize yield up to the mid-V7 maize growth stage (Silva et al. 2019; Perini and Machado 2023), highlighting the need to use other management strategies, such as the foliar spraying of chemical and biological insecticides.

In general, insecticides used as seed treatments are formulated as suspension concentrates (SC), with rates varying between 500 and 750 mL/100 kg of maize seeds or 100 to 150 mL per 60,000 seeds (seeds for sowing one ha). Products at a rate of 500 mL/100 kg of maize seeds can be used without dilution in water. However, at concentrations lower than 500 mL/100 kg of seeds, the formulated product can be diluted in water until it reaches a concentration of 500 mL/100 kg of maize seeds.

(a) Solution preparation: The desired amount of the product is added to a container suitable for preparing the solution. A portion of water is gradually added and mixed to form a homogeneous solution. The process is completed using the remaining water until the desired volume is reached. The solution is continuously stirred to prevent the product from settling at the bottom of the container.
(b) Application equipment: Appropriate equipment must be used to provide a uniform distribution of the desired rates over the seeds. Industrial seed treatment (IST) should be preferably performed. In this case, the seed treatment is performed in batches. The recommended steps are as follows.

    – Placing a known weight or quantity of seeds in the equipment (seed treater)

– Adding the desired volume of product (insecticide or insecticide + water) to the weight or amount of seeds.
    – Operating the equipment by moving (shaking) the seeds to obtain a uniform distribution of the product over the seeds, for 1–2 min per batch.

Other methods of seed treatment are still quite common and include manually operated mixers, rotating drums, concrete mixers, and continuous seed flow machines (portable or not). The advantage of ST with portable seed treaters is that the operation can be performed in the field, allowing the addition of insecticides, inoculants, fungicides, and other bioinputs that may be used for seed treatment just before planting. In the rotating drums and concrete mixers, a certain weight of seeds with a corresponding volume of seed treatment (not exceeding 1.0 L of product per 100 kg of maize seeds) is used, and the mixer is run for a minimum of 3 min.

For on-farm seed treatment using continuous-flow equipment, seed flow can be generated using an auger-type seed treater or gravity. To operate this equipment, the following steps should be performed:

– Measuring the flow of seeds (weight) at a given time.
– Adjusting the volume of the product to be applied based on the weight of the seeds and time.

Regardless of the equipment used for seed treatment, especially IST, when the seeds are stored for some time before sowing, it is recommended to add a dye to the solution to identify whether a treatment has been applied to the seeds and to prevent their accidental use for human or animal consumption.

Seeds treated with systemic insecticides formulated as granules and registered for in-furrow application can be used to control stink bugs in the early crop growth stages. This application method provides initial protection, and the duration of the protection varies with the product, crop, and growing conditions (Ávila 1991, 1997). However, in-furrow application of granules may require specific equipment. Alternatively, granules can be diluted in water so that the insecticide can be used as an in-furrow liquid to overcome the operational challenges associated with the in-furrow application of granular insecticides.

Currently, liquid-formulated products such as biofertilizers, inoculants, plant growth stimulants, nematicides, insecticides, and fungicides are being applied in-furrow to row crops. In some instances, the amount of pesticides and fertilizers used in seed treatment can be very high, which requires the development of new technologies or alternative methods for the in-furrow application of products that can be absorbed by the root system.

4.1.2 Application Technology for Foliar Insecticides

When stink bug populations are relatively low in the field, seed treatment is effective in controlling them and protecting the plants from damage. However, at high population densities (>1.0 stink bugs per meter of plant row), seed treatment alone is usually not enough to protect plants from damage, requiring foliar insecticide sprays (Brustolin et al. 2011; Ávila and Duarte 2012). In no-tillage systems, it is recommended to scout fields before planting maize to assess the presence of stink bugs and to determine the need for spraying an insecticide when burndown herbicides are used to kill weeds or cover crops before planting.

Insecticide sprays used to control stink bugs on maize should not just target insects and plants present in the field. Stink bugs are also found on the soil (under or on crop residue), stems, adaxial and abaxial sides of the maize leaves, and whorls (Fig. 12.2). Therefore, these are small targets that represent a small fraction of the crop surface.

Insecticide applications are recommended when the pest reaches economic threshold levels, or at the seedling or early vegetative maize growth stages (V1) followed by a second spray, 5 days later, using single or double flat jet nozzles. Spray volumes or application rates between 100 and 150 L/ha are suggested. Lower spray volumes should be paired with very fine or extremely fine droplets, whereas larger carrier volumes should be paired with medium or coarse droplets. Under favorable environmental conditions, such as temperature, relative humidity, wind speed, and the absence of air temperature inversions, hollow cone jet nozzles producing very fine droplets can be used. It is important to remember that there is an inversely proportional relationship between the target coverage (% of the surface covered) and the size of the droplets (Fig. 12.5); the smaller the droplet, the greater the coverage. Similarly, smaller spray droplets tend to be more prone to drift.

Sprayer calibration is a common practice. The nozzle flow rates are calculated based on the width of each nozzle, the spray application rate, and ground speed. When calibrating the sprayer, fine adjustments can be made by changing the working pressure and by following the recommendations for each nozzle type to reduce spray drifts and wear on the tips. When possible, it is also recommended to adopt drift reduction techniques using nozzles with higher flow rates, lower spray pressure, and application speeds, and by maintaining the height of the spray boom no more than 50 cm above the target.

The quality of the application can be improved with the use of spray additives (e.g., adjuvants) that promote less potential for drift and have the properties to improve coverage, adhesion, and penetration of insecticides into the plant cuticle. Insecticide and spray additive labels provide specific guidelines such as directions for use, precautions, and advisory statements, and are legally enforceable.

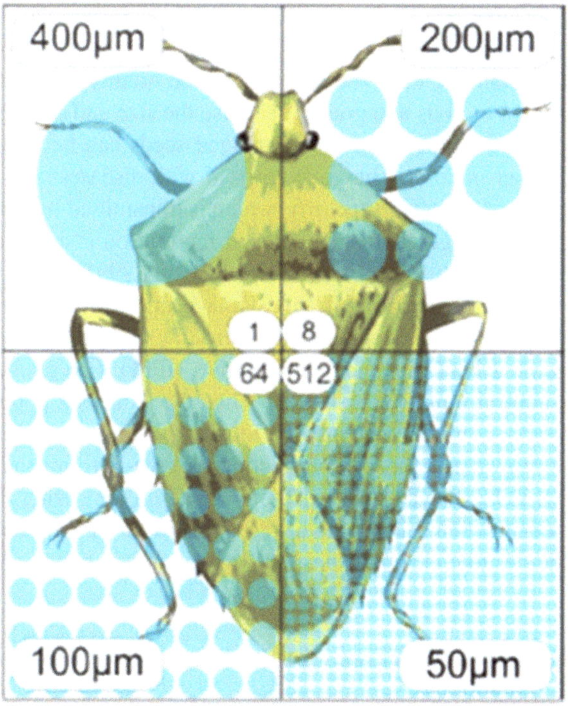

Fig. 12.5 Coverage of spray target with varying droplet sizes of pesticides (Guedes and Perini 2019)

It is important to constantly monitor and follow the recommended guidelines when applying pesticides. Insecticide labels provide specific recommendations for each product. In general, applications must be carried out with air temperature below 30 °C, relative humidity above 55%, and wind speeds between 3 and 10 km/h. In instances where wind speeds are lower than 3 km/h, farmers should pay special attention to the risks of inversion and convection. This usually occurs with cold temperatures, still air, or wind speeds below 3 km/h (a typical situation in early mornings and evenings) and without cloud cover. Convection occurs in conditions similar to those observed for inversion when wind speeds are low (less than 3 km/h), but in high-temperature conditions, normally in the afternoon of the hottest days. In both inversion and convection, a significant portion of the sprayed volume (e.g., very fine drops) remains in the suspension and drift is likely to occur.

4.2 Application Technology for Stink Bug Control on Soybean

During the reproductive phase, soybean can be attacked by stink bugs, particularly the southern green stink bug (*N. viridula*), red-banded stink bug (*P. guildinii*), and neotropical brown stink bug (*E. heros*). Other stink bug species can also cause damage to soybean, such as the green-belly stink bugs (*Diceraeus* spp.), although they are less common. Damage to soybean reduces seed quality and quantity, and can result in the abortion of pods and grains, decrease the size and oil content of grains, and affect seed germination and vigor. Damaged seeds can be smaller, wrinkled, shriveled, and darker in color. They may present with diseases such as yeast spots caused by the fungus *Nematospora corily*, which is transmitted by stink bugs during feeding. Stink bug damage can cause delayed maturation ("stay-green"), making harvesting more difficult (Correa-Ferreira and Panizzi 1999).

During the critical period of stink bug damage on soybean (beginning of pod development to grain filling), it is important to follow economic threshold levels as part of an Integrated Pest Management Program (IPM). For grain production in soybean fields, control must be carried out when the stink bug population reaches two stink bugs (adults or nymphs from the third instar onwards) per row meter, or one stink bug per row meter for seed production (Correa-Ferreira and Panizzi 1999; Bueno et al. 2013; Hayashida and de Bueno 2022). Considering that stink bugs attack soybean during their reproductive stages (R3 to R6), the preferred method of applying insecticides to control this pest on soybean is foliar application.

4.2.1 Insecticide Spraying Targets for Stink Bug Control on Soybean

To effectively reach insecticide application targets, it is important to understand pest behavior on soybean fields. Stink bugs that attack soybean can be found on the main and secondary stems, petioles, and the abaxial and adaxial sides of leaves and pods. This poses a challenge when attempting to control their infestation on soybean.

Stink bugs are also found on the ground (under and on crop residues or weeds; Fig. 12.1). Reaching the adaxial side of the leaves located in the lower third of soybean plants is challenging. Stink bugs found on the soil surface are also more difficult to reach, and their control depends on feeding on the parts of the plants that have been sprayed.

Considering the several restrictions associated with the spraying of insecticides to control stink bugs on soybean, spraying carried out with medium-to-fine and fine-to-very fine droplets is considered appropriate when the environmental conditions are suitable. Sprays with droplets greater than medium in size deposit significant amounts of insecticides only on targets located in the upper third of the plants, which is generally not sufficient to obtain satisfactory control of stink bugs. Furthermore, to ensure that droplets are deposited in the lower parts of the soybean canopy, applications should be performed when wind speeds are moderate (between 3.2 and 6.5 km/h). This condition results in a slight oscillation of the soybean leaves, facilitating the deposition of spray droplets on the lower parts (Fig. 12.6).

The use of devices attached to the sprayer boom can be an important ally in obtaining the deposit of significant amounts of insecticide in the middle and lower thirds of the soybean canopy, even reaching the insects on the soil between the soybean rows. Although the results are contradictory, the use of an adjustable air curtain to propel spray droplets is a commercial technology developed to increase droplet deposition within the soybean canopy.

Studies conducted by Bauer and Raetano (2000) demonstrated that using a spray boom with a curtain of air provided better distribution of pesticides within the crop canopy and less loss due to drift than conventional spraying. Furthermore, Bauer et al. (2008) observed a significant increase in droplet deposition on the leaves of the

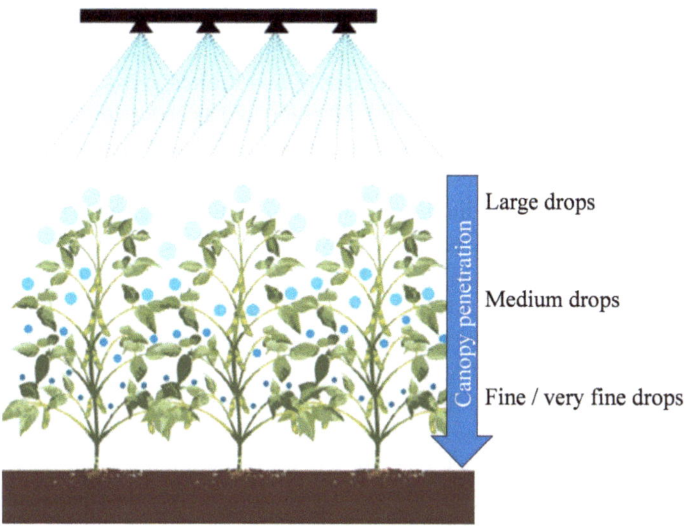

Fig. 12.6 Effect of pesticides droplet size on penetration and coverage of the soybean canopy (Guedes and Perini 2019)

lower third part of the canopy using hollow conical jet spray nozzles (JA-2) with a pressure of 600 kPa at the R5.2 soybean growth stage.

Prado et al. (2010b) evaluated the efficacy of a curtain of air with wind speeds of 0, 9, 11, and 29 km/h in a sprayer equipped with JA-2 conical jet nozzles operated at a pressure of 630 kPa, and found no significant increase in the control of *N. viridula* and *E. heros*. However, the air curtain at the maximum operating air speed (29 km h^{-1}) provided greater control of *Anticarsia gemmatalis* Hübner (Lepidoptera: Erebidae) and resulted in the highest increase in soybean yield (15%) compared to the sprayer without a curtain of air attached to the spray boom.

Prado et al. (2010a) evaluated the effects on the control of *E. heros* with flat jet (pressure of 230 kPa) and hollow conical jet (pressure of 630 kPa) nozzles, both at a rate of 200 L ha^{-1}, with and without a curtain of air at 29 km h^{-1}, the systemic insecticide thiamethoxam associated with lambda-cyhalothrin, and the contact and ingestion of the insecticide endosulfan. The results showed no significant differences between the nozzle types and the use of an air curtain for the mixture of thiamethoxam and lambda-cyhalothrin. However, the performance of endosulfan depends on its application technology. Specifically, hollow conical jet nozzles that produce smaller droplets, combined with an air curtain attached to the spray boom, may improve the distribution of the product in the soybean canopy, favoring the control of stink bugs. The combination of electrostatic spraying systems and air curtains increases droplet deposition within the crop canopy (Antuniassi et al. 2021a, b).

Fiorin et al. (2011) compared three application systems (ground with hydraulic power nozzles, aerial with hydraulic power nozzles, and aerial with rotating nozzles) and two insecticides (thiamethoxan + lambda-cyhalothrin and acephate) for controlling stink bugs in soybean. The ground sprayer was equipped with flat jet nozzles with air induction (BJ 11001 series, pressure 414 kPa), which generates coarse to very coarse droplets with a carrier volume of 120 L ha^{-1}. The aerial spray boom was equipped with conical jet nozzles (D6–45 series, pressure 262 kPa), generating medium-to-coarse droplets (200–300 kPa) and a carrier volume of 20 L ha^{-1}. The aerial spraying with rotating nozzles consisted of disc atomizers (Turboaero®) model TA-88C-5-8, with blades fixed in position three, working pressure of 207 kPa, producing very fine and medium droplets (80–200 μm), and a carrier volume of 12 L ha^{-1} (a vegetable oil-based adjuvant was added to the spray solution at a rate of 0.5 L ha^{-1}). The results showed that both insecticides had similar control efficacies, regardless of the application technology. Furthermore, ground and aerial sprays with rotating disc atomizers were more effective against stink bugs in soybean than aerial applications with a hydraulic nozzle.

Vieira (2011) conducted a field experiment to evaluate the effectiveness of flat jets and hollow cone spray nozzles at three different spray volumes (100, 150, and 200 L ha^{-1}) against brown stink bugs on soybean and demonstrated no significant differences in efficacy between the two types of spray nozzles. However, higher carrier volumes were more effective at controlling brown stink bugs, particularly nymphs.

5 Application Technology for Biological Insecticides

The incorporation of entomopathogenic agents, such as fungi and bacteria, into an Integrated Pest Management (IPM) program against stink bugs is considered an agronomic, economic, and environmental sustainability strategy. The use of biological pesticides for pest management has increased worldwide. The entomopathogenic fungi *Beauveria bassiana* and *Metarhizium anisopliae* have been used for stink bug control in large crop fields in Brazil. The use of microbes to control stink bugs and other sap-sucking species has recently included on-farm production and the use of bacteria from the genus *Chromobacterium* (Cavalcante et al. 2022).

Although several multinational and national agricultural companies are currently expanding their markets and portfolios of microbial insecticides in Brazil, their use at the field level to control insect pests remains challenging. In general, susceptibility to biological agents is age-dependent; that is, young stages (nymphs), especially in their initial instars, are more susceptible than adults (Silva et al. 2015; Portilla et al. 2022).

The adoption of biological insecticides has, as one of their main advantages, increased efficiency in cropping systems with insects resistant to chemical insecticides. However, concerns still need to be addressed to ensure the optimal performance of entomopathogenic agents when exposed to different environmental conditions (Calvin et al. 2021). Application time is a widely discussed topic, as these microorganisms are sensitive to high temperatures and exposure to sunlight (Celestino and dos Oliveira 2022; Sutanto et al. 2022). The survival of conidia and the initial stages of the fungal infection process are highly dependent on environmental conditions (Mwamburi et al. 2015). To mitigate the effects of these factors on the efficiency of biological control agents during storage, transport, and applications in the field, studies have advanced the development of new formulations capable of protecting these microorganisms, such as encapsulation (Celestino and dos Oliveira 2022).

In addition to the application technology, the selection of new biological strains with greater tolerance to the high temperatures typical of tropical agriculture should be encouraged during the development of biological pesticides. This was noted by Gava et al. (2022), who identified thermotolerant strains of *Beauveria bassiana* and *Metarhizium anisopliae* (Gava et al. 2022). In general, it is recommended that biological insecticides be applied during milder temperatures of the day, preferably at dusk or even on cloudy days, and whenever possible, target stink bug nymphs (Portilla et al. 2022).

In the field, plant protection is promoted when entomopathogenic fungi reach a target and/or are sprayed on plants. Soares (2023) observed that infection and consequent death of stink bugs by fungi can occur through ingestion and tarsal and dorsal contact. However, dorsal contact has greater control efficacy (Portilla et al. 2022; Soares 2023), demonstrating that product efficacy may vary according to the form of contamination (Soares 2023). In the case of *Chromobacterium* with pronounced action through ingestion, spraying must be carried out with a carrier volume that results in good coverage of the plant canopy; otherwise, the control efficacy may not be satisfactory. These examples demonstrate the need for studies that consider the differences between biological organisms to improve efficiency.

During spraying, biological agents are subjected to physical stress as they move through the different parts of the sprayer (agitator, pump, pipes, control valve, and spray nozzles) before being deposited onto a canopy and/or biological target. Because they are live organisms, there may be a reduction in the biological viability of these agents. Another important factor to consider when using biopesticides is the influence of droplet size and pressure used in the sprayer. The viability of entomopathogenic fungi may vary depending on the nozzle type (Kumar et al. 2023). For example, *Beauveria bassiana* exhibits greater viability with a hollow cone nozzle and reduced pressure (Kumar et al. 2023). Generally, nozzles with higher pressures and lower carrier volumes are harmful to biological organisms (Ru et al. 2020; Kumar 2021). This is even more important for bacteria, which are more vulnerable to adverse effects than fungi because of the difference in the constitution of their cell walls compared to fungi (Ru et al. 2020). Therefore, there should be specific studies on the biopesticides used in the management of stink bugs and application technologies to understand the best combinations of spray technology and biopesticides to improve pest control. Little is known about whether the current application technology is suitable for the widespread application of biopesticides to cultivated crops.

In addition, special attention should be paid to carrier volume. Higher control efficacy is obtained with a carrier volume of approximately 300 L/ha. This volume is higher than that typically used for chemical insecticides (40 and 120 L/ha, respectively). Differences in carrier volume must be considered when pest management includes chemical and biological insecticides, as they can result in varying levels of control efficacy. Pesticide manufacturers are exploring technologies that facilitate the use of biological and chemical pesticide combinations. For example, requests for new biological pesticides for aerial spraying are already being filed with the Ministry of Agriculture, Livestock, and Supply (MAPA) in Brazil. It has been demonstrated that tank mixes of chemical and biological insecticides result in greater stink bug control efficacy (Quintela et al. 2013; Soares 2023). Although promising, tank mixtures of biological and chemical insecticides may not be compatible, resulting in antagonistic effects (Celestino et al. 2020). Depending on the concentration of the spray carrier, chemical composition, and species/strains of the biological agent, chemical insecticides can inhibit and/or reduce the germination of conidia, vegetative growth, and sporulation of entomopathogenic fungi, thereby affecting the process of infection and epizootics (Ribeiro et al. 2023). To minimize risks, manufacturers' research and development departments have been conducting tests to evaluate the feasibility of mixtures and have provided technical sheets that help growers choose compatible products when performing tank mixtures with chemical and biological pesticides.

In addition to mixtures of biological and chemical insecticides, little is known about the interactions that may occur with the use of adjuvants together with microbiological products. For example, the efficiency of *Cordyceps* (=*Isaria*) *fumosorosea* in managing sap-sucking insects in citrus orchards is influenced by the type of adjuvant used. The adjuvants performed differently in terms of the adherence of the biological insecticide to the insect body and affected the viability of the conidia and spores of the entomopathogenic fungi (Arnosti et al. 2019). Gatarayiha et al. (2009) reported that the use of a silicone-based surfactant resulted in a higher survival rate of *Beauveria bassiana* conidia than that of a mineral oil-based adjuvant in the carrier solution, as well as greater efficiency against the two-spotted spider mite *Tetranychus urticae* Koch (Acari: Tetranychidae). Compatibility between fungal-based biopesticides and adjuvants/surfactants/oils used in the spray tank generally results in a significant increase in the pathogenicity of the fungal agent, conidial adhesion, durability, and efficacy against target pests (Visalakshy et al. 2006).

Beauveria bassiana may be damaged by surfactants, depending on the type of surfactant (Swathi et al. 2018; Kaiser et al. 2020) and the fungal isolate (Mwamburi et al. 2015). Studies have demonstrated challenges regarding the use of adjuvants to improve biopesticide efficacy (Holka and Kowalska 2023). This includes the formulation components and their toxicological risks (Lin et al. 2023), indicating that the use of adjuvants associated with biopesticides requires an in-depth understanding of possible interactions to develop best management practices (Holka and Kowalska 2023). It is also important to highlight that the order for tank mixing products when a biological pesticide is added may change from what is usually established as proper mixing. To ensure product efficacy, biological insecticides should be added last to the tank mix, regardless of the formulation, to minimize the effects of pH variation on the viability of biological agents. In addition, a jar test should be performed to evaluate their compatibility before adding them to the tank.

Despite recent technological advances that have allowed significant growth in the use of microbiological pesticides worldwide, challenges remain regarding the use of biopesticides, especially those related to the efficacy of controlling stink bugs under field conditions. Facing such challenges involves research and technological advances to improve and integrate biopesticides with other control strategies. More recently, "self-dissemination" systems ("lure and kill" or "attract and kill"), employing semiochemicals that attract the pest, have been gaining importance to increase contamination and dispersion, and the protection of the entomopathogenic microorganisms. Although it has been evaluated against many pests (Gonzalez et al. 2016), important advances are still required for its practical application against stink bugs (Hemiptera: Pentatomidae). Therefore, as highlighted here, additional advances are needed in product formulations and application technology, which are the determining factors for the success of insecticide sprays (Valicente 2019).

6 Concluding Remarks

Various operational, biological, and environmental factors must be considered when managing phytophagous stink bugs on soybean and on maize. These factors include operational, biological, and environmental aspects. In recent years, insecticide application technologies have advanced to achieve more effective pest control. Among the technological aspects, the formulation, timing of application, nozzle type, carrier volume, and other technologies must be considered. Effective pest management requires refining the use of techniques such as industrial seed treatment or in-furrow application early in the season, operational tools for aerial spraying, and best management practices for biopesticides, chemical insecticides, tank mixtures, and adjuvants. Combining application technologies and considering the biological aspects of the target pest species leads to successful management strategies.

References

Agrofit (2023) Sistema de Agrotóxicos Fitossanitários. https://extranet.agricultura.gov.br/agrofit_cons/principal_agrofit_cons

Albuquerque FA, Borges LM, Iacono TO, Crubelati NCS, Singer AC (2006) Eficiência de inseticidas aplicados em tratamento de sementes e em pulverização, no controle de pragas iniciais do milho. Rev Bras Milho e Sorgo 5:15–25

Alford A, Krupke CH (2017) Translocation of the neonicotinoid seed treatment clothianidin in maize. PLoS One 12:e0173836

Antuniassi UR (2004) Tecnologia de aplicação de defensivos. Fundação de Apoio a Pesquisa Agropecuária de Mato Grosso - Fundação MT, Rondonópolis, Embrapa Soja, Londrina, Boletim Técnico de Soja 8, pp165–177

Antuniassi UR, Carvalho FK, Mota AAB, Chechetto RG (2021a) Entendendo a tecnologia de aplicação 2 ed rev ampl. FEPAF, Botucatu, 64p

Antuniassi UR, Mertens TB, Carvalho FK, Mota AAB, Chechetto RG, Silva DP, Madeira RG (2021b) Assistência de ar em conjunto com a transferência de carga elétrica na deposição de fungicidas na cultura do algodão. Rev Plantio Direto 31:18–22

Arnosti A, Delalibera Júnior I, Conceschi MR, D'Alessandro CP, Travaglini RV, Camargo-Mathias MI (2019) Interactions of adjuvants on adhesion and germination of *Isaria fumosorosea* on adults of *Diaphorina citri*. Sci Agric 76:487–493

Ávila CJ (1991) Pragas e seu controle. In Milho, Informações técnicas. EMBRAPA-UEPAE, Dourados, MS

Ávila CJ (1997) Eficiência do inseticida terbufós (Counter 5 G) no controle de larvas de *Diabrotica speciosa* em milho (*Zea mays* L.). In Reunião Sul-Brasileira de Insetos de Solo, 4, EMBRAPA-CNPT, Passo Fundo, p 144

Ávila CJ, Duarte MM (2012) Eficiência de inseticidas, aplicados nas sementes e em pulverização, no controle do percevejo barriga-verde *Dichelops melacanthus* (Dallas) (Hemiptera: Pentatomidae), na cultura do milho. BioAssay 7:1–6

Ávila CJ, Panizzi AR (1995) Occurrence and damage by *Dichelops* (*Neodichelops*) *melacanthus* (Dallas) (Heteroptera: Pentatomidae) on maize. An Soc Entomol Bras 24:193–194

Balabanidou V, Kefi M, Aivaliotis M, Koidou V, Girotti JR, Mijailovsky SJ, Juárez MP, Papadogiorgaki E, Chalepakis G, Kampouraki A, Nikolaou C, Ranson H, Vontas J (2019) Mosquitoes cloak their legs to resist insecticides. Proc R Soc B Biol Sci 286:20191091

Ball HJ, Beck SD (1951) The role of the circulatory and nervous system in the toxic action of parathion. J Econ Entomol 44:558–564

Bauer FC, Raetano CG (2000) Assistência de ar na deposição e perdas de produtos fitossanitários em pulverizações na cultura da soja. Sci Agrár 57:271–276

Bauer FC, Almeida E, Marques DC, Rossi T, Pereira FAR (2008) Deposição de pontas de pulverização AXI 11002 e JA-2 em diferentes condições operacionais. Cienc Rural 38:1610–1614

Bond EJ (1961) The action of fumigants on insects. II. The effect of hydrogen cyanide on the activity and respiration of certain insects. Can J Zool 39:437–444

Brustolin C, Bianco R, Neves PMOJ (2011) Inseticidas em pré e pós emergência do milho (*Zea mays* L.), associados ao tratamento de sementes, sobre *Dichelops melacanthus* (Dallas) (Hemiptera: Pentatomidae). Rev Bras Milho Sorgo 10:215–223

Bueno AF, Bortolotto OC, Pomari A, Roggia S, Corrêa-Ferreira BS, França-Neto JB (2013) Quando se deve controlar percevejos na soja? In: Reunião de Pesquisa de Soja da Região Central do Brasil, 33, 2013. Embrapa Soja, Londrina, pp 50–53

Bueno AF, Parra JRP, Colombo FC, Colmenarez YC, Narde BVF, Pereira FF (2021) Manejo de pragas com parasitoides. In: Meyer MC, Bueno AF, Mazaro SM, Silva JC (eds) Bioinsumos na Cultura da Soja. Embrapa, Brasília, pp 417–434

Burrows M (1980) The tracheal supply to the central nervous system of the locust. Proc R Soc Lond B 207:63–78

Calvin W, Beuzelin JM, Liburd OE, Branham MA, Simon LJ (2021) Effects of biological insecticides on the sugarcane aphid, *Melanaphis sacchari* (Zehntner) (Hemiptera: Aphididae), in sorghum. Crop Prot 142:105528

Cavalcante JKG, Conte H, Daquila BV, Caleffe RRT (2022) Biotechnological potential of *Chromobacterium subtsugae*: a short review on the biological management of pest arthropods. Rev Ibero-Am Ciênc Ambient 13:200–211

Cavallin LA (2020) Eficácia e vias de transferência de inseticidas no controle de *Euschistus heros* (Fabricius, 1798) (Hemiptera: Pentatomidae). Master thesis, Universidade Federal de Santa Maria, Santa Maria, RS, Brasil

Celestino MF, dos Oliveira JAS (2022) Encapsulation methods of entomopathogenic fungi for their application in the biological control. Res Soc Dev 11:123111638031

Celestino FN, Pratissoli D, Santos Júnior HJG, Machado LC (2020) In vivo compatibility between *Beauveria bassiana* (Bals.) Vuillemin and castor oil on *Hypothenemus hampei* (Ferrari). Coffee Sci 15:151771

Chapman RF (1998) The insects: structure and function, 4th edn. Cambridge University Press, 788p

Chiaradia LA, Nesi CN, Ribeiro LP (2016) Nível de dano econômico do percevejo barriga-verde, *Dichelops furcatus* (Fabr.) (Hemiptera: Pentatomidae), em milho. Agropec Catarinense 29:63–67

Correa-Ferreira BS, Panizzi AR (1999) Percevejos da soja e seu manejo. EMBRAPA-CNPSO, Londrina, Circular Técnica 24, 45p

Corrêa-Ferreira BS, Peres WAA (2003) Comportamento da população dos percevejos-pragas e a fenologia da soja. In: Corrêa-Ferreira BS (ed) Soja orgânica: alternativas para o manejo dos insetos-pragas. Embrapa Soja, Londrina, pp 27–32

Corrêa-Ferreira BS, Sosa-Gómez DR (2017) Percevejos e o sistema de produção soja-milho. Embrapa Soja, Londrina, Documentos 397, 99p

Corrêa-Ferreira BS, de Castro LC, Roggia S, Cesconetto NL, da Costa JM, Oliveira MCN (2013) MIP Soja: resultados de uma tecnologia eficiente e sustentável no manejo de percevejos no atual sistema produtivo da soja. Embrapa Soja, Londrina, Documentos 341, 55p

Crosariol Netto J, Michelotto MD, Grigolli JFJ, Galli JA, Pirotta MZ, Busoli AC (2015) Damages caused by *Dichelops melacanthus* (Heteroptera: Pentatomidae) in conventional and transgenic maize hybrids. Biosci J 31:1092–1101

Cruz L, Bianco R (2001) Manejo de pragas na cultura de milho safrinha. In: Seminário Nacional de Milho Safrinha, 6, 2001. IAPAR, Londrina, pp 79–112

de Vieira ECS (2021) Modos de exposição de inseticidas, estratégias químicas e tecnologia de aplicação para o manejo de percevejos fitófagos nas culturas do milho e da soja. Master thesis, Universidade Federal da Grande Dourados, Dourados, MS, Brasil

Dias MAN, Taylor AG, Cicero SM (2014) Uptake of systemic seed treatments by maize evaluated with fluorescent tracers. Seed Sci Technol 42:101–107

Duarte M, Ávila CJ, Santos V (2015) Danos e nível de dano econômico do percevejo barriga verde na cultura do milho. Rev Bras Milho Sorgo 14:291–299

Ebeling W (1974) Permeability of insect cuticle. In: Rockstein M (ed) The physiology of Insecta v6. Academic, New York, pp 507–556

Fernandes PHR, Ávila CJ, Silva IF, Zulin D (2020) Damage by the green-belly stink bug to maize. Pesq Agropec Bras 55:e01131

Fiorin RA, Stürmer GR, Guedes JVC, Costa IFD, Perini CR (2011) Métodos de aplicação e inseticidas no controle de percevejos na cultura da soja. Semina: Ciências Agrárias 32:139–145

Galley DJ (1967) The effect of gas velocity on the fumigant action of nicotine, dichlorvos, and hydrogen cyanide. J Stored Prod Res 3:17–27

Gatarayiha M, Laing M, Miller R (2009) Effects of adjuvant and conidial concentration on the efficacy of *Beauveria bassiana* for the control of the two spotted spider mite, *Tetranychus urticae*. Exp Appl Acarol 50:217–229

Gava CAT, Leal CM, Vieira de Sá A, Paranhos BAJ (2022) Selecting thermal tolerant strains of entomopathogenic fungi to control *Ceratitis capitata* (Wiedeman) in tropical semi-arid conditions. Biol Control 176:105062

Gerolt P (1965) The fate of dieldrin in insects. J Econ Entomol 58:850–857

Gerolt P (1969) Mode of entry of contact insecticides. J Insect Physiol 15:563–580

Gonzalez F, Tkaczuk C, Dinu MM, Fiedler Ż, Vidal S, Zchori-Fein E, Messelink GJ (2016) New opportunities for the integration of microorganisms into biological pest control systems in greenhouse crops. J Pest Sci 89:295–311

Grigolli JFJ, Grigolli MMK, Lourenção ALF, de Gitti DC (2016) Estratégias de controle químico do percevejo barriga verde *Dichelops melacanthus* (Dallas) (Heteroptera: Pentatomidae) no sistema de sucessão soja e milho safrinha. In: Congresso Nacional de Milho Sorgo, 31. Associação Brasileira de Milho e Sorgo, Sete Lagoas, pp 248–253

Grissom RE, Konno T, Motoyama N, Dauterman WC (1989) Comparison between in vivo and in vitro cutaneous penetration of fenvalerate in tobacco budworm (Lepidoptera: Noctuidae). J Econ Entomol 82:41–44

Guedes JVC, Perini CR (2019) Tecnologia de aplicação de inseticidas em culturas anuais. In: Antuniassi UR, Boller W (eds) Tecnologia de aplicação para culturas anuais. Aldeia Norte, Passo Fundo, pp 303–318

Guedes JVC, Arnemann JA, Stürmer GR, Melo AA, Bigolin M, Perini CR, Sari BG (2012) Percevejos da soja: novos cenários, novo manejo. Rev Plantio Direto 1(2):24–30

Gullan PJ, Cranston PS (2014) Sensory systems and behaviour. In: Gullan PJ, Cranston PS (eds) The insects: an outline of entomology. Wiley-Blackwell, pp 95–124

Gullan PJ, Cranston PS (2017) Insetos: Fundamentos da Entomologia, 5th edn. Guanabara Koogan, Rio de Janeiro, 460p

Hadaway AB (1971) Some factors affecting the distribution and rate of action of insecticides. Bull World Health Organ 44:221–224

Hayashi A (1983) Practice of pest control and insecticide. Nanzan-Do Press, pp 4–8

Hayashida R, de Bueno AF (2022) Níveis de ação em soja: efeito conjunto de desfolha e ataque de percevejos fitófagos em soja. In: Congresso Brasileiro de Soja, 9. Embrapa, Brasília, p 97

Holka M, Kowalska J (2023) The potential of adjuvants used with microbiological control of insect pests with emphasis on organic farming. Agriculture 13:1659

Huber F (1974) Neural integration (central nervous system). In: Rockstein M (ed) The physiology of Insecta v4. Academic, New York, pp 12–24

Huth C, Roggia S, Cantone W, Guedes JVC (2012) Efeito sistêmico de inseticida (tiametoxam + lambda-cialotrina) em planta de soja. In: Congresso Brasileiro de Soja, 6. Embrapa, Brasília, pp 1–4

Jones RM (1938) Toxicity of fumigant-CO2 mixtures to the red flour beetle. J Econ Entomol 31:298–309

Kaiser D, Handschin S, Rohr RP, Bacher S, Grabenweger G (2020) Co-formulation of *Beauveria bassiana* with natural substances to control pollen beetles – synergy between fungal spores and colza oil. Biol Control 140:104106

Kumar M (2021) Studies on spray characteristics of selected nozzles with bio-pesticides. Dissertation, Indian Agricultural Research Institute, New Delhi, India

Kumar M, Mehta CR, Agrawal KN, Tripathi MK (2023) Optimization of operating parameters for spraying microbial (*Bacillus thuringiensis* and *Beauveria bassiana*) based bio-pesticide solutions for foliar application. Int J Pest Manag. https://doi.org/10.1080/09670874.2023.2213183

Kuss-Roggia RCR (2009) Distribuição espacial e temporal de percevejos da soja e comportamento de *Piezodorus guildinii* (Westwood, 1837) (Hemiptera: Pentatomidae) na soja (*Glycine max* (L.) Merrill) ao longo do dia. Disssertation, Universidade Federal de Santa Maria, Santa Maria, RS, Brasil

Lewis CT (1965) Influence of cuticle structure and hypodermal cells on DDT absorption by *Phormia terranovae* R-D. J Insect Physiol 11:683–694

Lin F, Mao Y, Zhao F, Idris AL, Liu Q, Zou S, Guan X, Huang T (2023) Towards sustainable green adjuvants for microbial pesticides: recent progress, upcoming challenges, and future perspectives. Microorganisms 11:364

Lorini I, Krzyzanowski FC, França-Neto JB, Henning AA, Henning FA (2015) Manejo integrado de pragas de grãos e sementes armazenadas. Embrapa, Brasília, 970p

Marques RP, Melo AA, Guedes JVC, Carli CD, Rohrig A, Pozebon H, Perini CR, Ferreira DR, Bevilaqua JG, Patias LS, Forgiarini SE, Padilha G, Leitão JV, Moro D, Hahn L, Arnemann JA (2019) Managing stink bugs on soybean fields: insights on chemical management. J Agric Sci 11:225–234

Matsumura F (1963) The permeability of the cuticle of *Periplaneta americana* (L.) to malathion. J Insect Physiol 9:207–221

Matsumura F (1985) Toxicology of Insecticides, 2nd edn. Plenum Press, New York, 598p

Maziero H, Guedes JVC, Farias JR, Rodrigues RB, Dalazen G, Prá ED (2009) Volume de calda e inseticidas no controle de *Piezodorus guildinii* (Westwood) na cultura da soja. Cienc Rural 39:1307–1312

Melo AA, Guedes JVC, Luchese EF, Hahn L, Almeida ML, Machado MC (2017) Tecnologia de aplicação para percevejos em soja. Rev Plantio Direto 160. https://plantiodireto.com.br/artigos/188

Moscardi F (2003) Uso de baculovírus e *Bacillus thuringiesis* no controle de lagarta da soja, *Anticarsia gemmatalis*. In: Corrêa-Ferreira BS (ed) Soja orgânica: alternativas para o manejo dos insetos-pragas. Embrapa Soja, Londrina, pp 15–25

Mwamburi LA, Laing MD, Miller RM, Mwamburi LA, Laing MD, Miller RM (2015) Effect of surfactants and temperature on germination and vegetative growth of *Beauveria bassiana*. Braz J Microbiol 46:67–74

Panizzi AR, Galileo MH, Gastal HA, Toledo JF, Wild CH (1980) Dispersal of *Nezara viridula* and *Piezodorus guildinii* nymphs in soybean. Environ Entomol 9:293–297

Pazini JB, Padilha AC, Cagliari D, Bueno FA, Rakes M, Zotti MJ, Martins JFS, Grützmacher AD (2019) Differential impacts of pesticides on *Euschistus heros* (Hem.: Pentatomidae) and its parasitoid *Telenomus podisi* (Hym.: Platygastridae). Sci Rep 9:6544.

Pazini (2022) Seletividade de inseticidas aos inimigos naturais: aplicações práticas para o MIP-Soja. Rev Contr Biol 4:16–20

Perini CR, Machado DN (2023) Application periods against *Diceraeus* (*Dichelops*) *melacanthus* on maize and their significant response on damage and grain yield in the Brazilian Midwest. Crop Prot 172:106344

Portilla M, Zhang M, Glover JP, Johnson CP, Reddy GV (2022) Lethal concentration and sporulation by contact and direct spray of the entomopathogenic fungus *Beauveria bassiana* on different stages of *Nezara viridula* (Heteroptera: Pentatomidae). J Fungi 8:1164

Prado EP, Raetano CG, de Christovam RS, Aguiar-Júnior HO, Dal Pogetto MHF d A, Gimenes MJ (2010a) Tecnologias de aplicação de produtos fitossanitários no controle de percevejos pragas na cultura da soja. Arq Inst Biol 77:265–274

Prado EP, Raetano CG, Aguiar-Júnior HO, de Christovam RS, Dal Pogetto MHF d A, Gimenes MJ (2010b) Velocidade do fluxo de ar em barra de pulverização no controle químico de *Anticarsia gemmatalis*, Hübner e percevejos na cultura da soja. Bragantia 69:995–1004

Puntel L (2022) Efeito de compostos modificadores de comportamento aplicados no manejo de *Euschistus heros* (Fabricius, 1798) na soja. Master thesis, Universidade Federal Santa Maria, Santa Maria, RS, Brasil

Quintela ED, Mascarin GM, Silva RA, Barrigossi JAF, Martins JFS (2013) Enhanced susceptibility of *Tibraca limbativentris* (Heteroptera: Pentatomidae) to *Metarhizium anisopliae* with sublethal doses of chemical insecticides. Biol Control 66:56–64

Rajashekar Y, Rao LJM, Shivanandappa T (2012) Decaleside: a new class of natural insecticide targeting tarsal gustatory sites. Sci Nat 99:843–852

Ribeiro LR, Amandio DTT, Rakes M, Belegante F, Moresco C, Nesi CN, Zanardi AM, Grutzmacher AD, Bernardi D, Zanardi OZ (2023) Insecticides for maize leafhopper management versus entomopathogenic fungal isolates: *In vitro* compatibility, physical-chemical interactions, and on-farm assessments. Crop Prot 174:106417

Rodrigues RB (2011) Danos do percevejo-barriga-verde *Dichelops melacanthus* (Dallas, 1851) (Hemiptera: Pentatomidae) na cultura do milho. Master thesis, Universidade Federal de Santa Maria, Santa Maria, RS, Brasil

Roggia S, Utiamada CM, Hirose E, Tomquelski GV, Jakoby GL, Guedes JVC, Roy JMT, Grigolli JFJ, Rattes JF, Vivan LM, Marchioro L, Curioletti LE, Sato LN, Peixoto MF, Goussain Júnior MM, Tamai MA, Grigolli MMK, Martins MC, Madalosso T (2019) Eficiência de inseticidas no controle do percevejo-marrom (*Euschistus heros*) em soja, na safra 2018/2019: resultados sumarizados de ensaios cooperativos. Embrapa Soja, Londrina, Circular Técnica 154, 19p

Roza-Gómez MF, Salvadori JR, Pereira PRVS, Panizzi AR (2011) Injúrias de quatro espécies de percevejos pentatomídeos em plântulas de milho. Cienc Rural 41:1115–1119

Ru Y, Liu Y, Qu R, Patel MK (2020) Experimental study on spraying performance of biological pesticides in aerial rotary cage nozzle. Int J Agric Biol Eng 13:1–6

Salgado VL (2013) BASF – Insecticide mode of action: technical training manual. https://agriculture.basf.com/global/assets/en/Crop%20Protection/innovationBASF_Insecticide_MoA_Manual_2014.pdf. Acessed 13 Sept 2023

Seidel EJ, de Pazini JB, Tomazella VLD, Vieira AMC, Silva FF, Martins JFS, Barrigossi JAF (2020) Predicting rice stem stink bug population dynamics based on GAMLSS models. Environ Entomol 49:1145–1154

Silva RA, Quintela ED, Mascarin GM, Pedrini N, Lião LM, Ferri PH (2015) Unveiling chemical defense in the rice stalk stink bug against the entomopathogenic fungus *Metarhizium anisopliae*. J Invertebr Pathol 127:93–100

Silva PR, Istchuk AN, Hunt TE, Bastos CS, Torres JB, Campos KL, Foresti J (2019) Susceptibility of maize to stink bug (*Dichelops melacanthus*) and its management through seed treatment. Aust J Crop Sci 13:2015–2021

Soares FS (2023) Manejo biológico e químico no controle de percevejo marrom da soja. Master thesis, Universidade Federal de Santa Maria, Santa Maria, RS, Brasil

Sosa-Gómez DR, Corrêa-Ferreira BS, Kraemer B, Pasini A, Husch PE, Vieira CED, Martinez CBR, Lopes ION (2020) Prevalence, damage, management and insecticide resistance of stink bug populations (Hemiptera: Pentatomidae) in commodity crops. Agric For Meteorol 22:99–118

Sparks TC, Nauen R (2015) IRAC: mode of action classification and insecticide resistance management. Pestic Biochem Physiol 121:122–128

Spielmeyer A, Schetelig MF, Etang J (2019) High-throughput analysis of insecticides on malaria vectors using liquid chromatography tandem mass spectrometry. PLoS One 14:e0211064
Sugiura M, Horibe Y, Kawada H, Takagi M (2018) Insect spiracle as the main penetration route of pyrethroids. Pestic Biochem Physiol 91:135–140
Sumita Y, Kawada H, Minakawa N (2016) Mode of entry of a vaporized pyrethroid knockdown agent into the body of the housefly, *Musca domestica* (Diptera: Muscidae). Appl Entomol Zool 51:653–659
Sur R, Stork A (2002) Uptake, translocation and metabolism of imidacloprid in plants. Bull Insectol 56:35–40
Sutanto KD, Husain M, Rasool KG, Malik AF, Al-Qahtani WH, Aldawood AS (2022) Persistency of indigenous and exotic entomopathogenic fungi isolates under ultraviolet B (UV-B) irradiation to enhance field application efficacy and obtain sustainable control of the red palm weevil. Insects 13:103
Swathi P, Visalakshy PNG, Das SB (2018) In vitro evaluation for compatibility of additives with *Beauveria bassiana* (Balsamo) Vuillemin. Egypt J Biol Pest Control 28:13
Tang X, Hao F, Yuan H, Yan X, Yang D, Taylor AG (2020) Uptake and translocation of imidacloprid via seed pathway and root pathway during early seedling growth of maize. Pest Manag Sci 76:3792–3799
Terriere LC (1982) The biochemistry and toxicology of insecticides. Oregon State University, Corvallis
Tessmer MA, Kuhn TMA, Gloria BA, Lopes JRS, Erler G, Bonani JP (2022) Histology of damage caused by *Euschistus heros* (F.) nymphs in soybean pods and seeds. Neotrop Entomol 51:112–121
Valicente FH (2019) Entomophogenic viruses. In: Souza B, Vázquez LL, Marucci RC (eds) Natural enemies of insect pests in neotropical agroecosystems: biological control and functional biodiversity. Springer Nature, pp 235–244
Viana PA, Cruz I, Waquil JM (2002) Cultivo do milho: pragas iniciais. Embrapa Milho e Sorgo, Sete Lagoas, Comunicado Técnico 59, 13p
Visalakshy PNG, Krishnamoorthy A, Kumar AM (2006) Compatibility of plant oils and additives with *Paecilomyces farinosus*, a potential entomopathogenic fungus. J Food Agric Environ 4:333–335
Winteringham FPW (1957) Comparative biochemical aspects of insecticidal action. Chem Ind 36:1195–1202
Wordell-Filho JÁ, do Ribeiro LP, Chiaradia LA, Madalóz JC, Nesi CN (2016) Pragas e doenças do milho: diagnose, danos e estratégias de manejo. Epagri, Florianópolis, Boletim Técnico 170, 82p
Yasutomi K (1987) Guidance of chemical methods for the control of insect pests. Life Environ 32:30–33
Yu SJ (2014) The Toxicology and Biochemistry of Insecticides, 2nd edn. CRC Press, Boca Raton, 380p

Chapter 13
Insecticide Resistance Management of Stink Bugs

Daniel Ricardo Sosa-Gómez and Oderlei Bernardi

1 Introduction

As they are R strategists and show polyphagy, species of stink bugs (Heteroptera: Pentatomidae) are important pests of several crops, such as soybean, cotton, green beans, vegetables, and fruit trees.

When the combinations of favorable factors converge, unusual high population densities have been reported (Nielsen and Hamilton 2009; Leskey et al. 2012b; Husch et al. 2018). Probably nowadays, the most relevant and common Pentatomidae species are *Halyomorpha halys* Stål in the United States and Europe; the Neotropical brown stink bug, *Euschistus heros* (F.) and the green belly stink bugs, *Diceraeus melacanthus* (Dallas) and *D. furcatus* (F.) in Brazil (Panizzi 2015; Sosa-Gómez et al. 2020). Despite the wide distribution of *Nezara viridula* (L.) (Knight and Gurr 2007) reports of field resistance has not been published in the last 30 years (see Sect. 2.4).

The invasive species *H. halys* was detected in USA in Allentown, PA, in 1996. Because of its capacity and plasticity to adapt to local conditions, since 2001 *H. halys* populations have increased to high densities in multiple states (Nielsen and Hamilton 2009; Musolin et al. 2022; Reznik et al. 2023). Conversely, *E. heros* and *Diceraeus* species are native of the Neotropical Region and their importance, host ranges and distribution have been revised (Panizzi 2015).

Since the use of biological control of stink bugs is limited, control strategies most of the time rely on applications of broad-spectrum insecticides such as

D. R. Sosa-Gómez (✉)
Embrapa Soja, Rodovia Carlos João Strass, Acesso Orlando Amaral, Londrina, PR, Brazil
e-mail: daniel.sosa-gomez@embrapa.br

O. Bernardi
Universidade Federal de Santa Maria, Santa Maria, RS, Brazil

A. F. Bueno, A. R. Panizzi (eds.), *Stink Bugs (Hemiptera: Pentatomidae) Research and Management*, Entomology in Focus 9,
https://doi.org/10.1007/978-3-031-69742-5_13

neonicotinoides, pyrethroids, organophosphates, and more recent group of insecticides (butenolides group, isooxazoline class and others) (Sydnor et al. 2023). As stink bugs are harmful and cause economic damage (e. g., Soria et al. 2017; Bueno et al. 2021) and have great potential to build up high densities, usually they require action measures at low densities (e.g., 1 to 2 stink bugs per meter of soybean row) (Panizzi et al. 1977; Bueno et al. 2015; Sosa-Gómez et al. 2020) or higher than that (Owens et al. 2013).

In this chapter, we consider the concept of field resistance defined by Tabashnik et al. (2014) as "a decrease in susceptibility to a pesticide in a population, caused genetic changes after the exposure to the pesticide in the field", and susceptible and tolerance concepts defined by Sosa-Gómez et al. (2020) as "the sensitivity of an insect pest that has not undergone a selection process, and can be controlled easily with lower doses of an insecticide compared with its conspecifics. Conversely, tolerance is defined as the innate capacity of an insect species to survive higher doses of an insecticide compared with other species. This means that the tolerant species has not gone through a selection process as a result of an increasing exposure to insecticide applications".

2 Laboratory Bioassays and Baseline Studies with Stink Bugs

In the insecticide market, baseline studies on the general impacts are not conducted before commercialization release of new active ingredients. Therefore, little is known about the genetic background response of unselected populations to active ingredients, especially in countries where the agriculture is intensive (multiple growing seasons per year) and demand frequent insecticide treatments. Natural genetic variability could be masked by early selection of unsusceptible phenotypes. Baseline studies are scarce and for most of the insecticides it is not known the LC_{95} or LC_{99} data to define diagnostic doses/concentrations. For better comparison among different laboratories is advisable that lethal doses should be expressed in units of weight (mg, μg or ηg) of active ingredient per bug (Baur et al. 2010; Sosa-Gómez et al. 2020; Tibola et al. 2021; Boff et al. 2022) or per unit of area (Baur et al. 2010; Tuelher et al. 2017). Unless, that the purpose of the bioassay is to determine systemic activity (Tibola et al. 2021).

2.1 *Brown Marmorated Stink Bug,* **Halyomorpha halys** *(Stål)*

The brown marmorated stink bug (BMSM) acquired special importance as a serious pest after its introduction in the USA in the late 1990s (Hoebeke and Carter 2003) and in Europe (see Chaps. 15 and 16, respectively). This species is native to east

Asia (China, Japan and Korea) has a high dispersal capacity and from its first detection in Pennsylvania had spread to most of the states of the United States, and Canada (Valentin et al. 2017). In Europe, it was detected in in 2004 in Switzerland for the first time, and after that it was detected in several countries of the central, west and east region of Europe (Maistrello et al. 2017) and Turkey in 2017 (Ak et al. 2019). The polyphagous nature of the BMSB is testified by its wide range record of more than 170 host plants (Leskey and Nielsen 2018).

Leskey et al. (2012a) studies performed with *H. halys* proposed a lethality index to evaluate overall efficacy of insecticides for this species. A more traditional approach of baseline determination for neonicotinoids and pyrethroids was done by Nielsen et al. (2008) and LD_{99} determination for bifenthrin and thiamethoxam has been provided by Alford et al. (2020).

No record of resistance to insecticide of this species have been published (Sparks et al. 2020), however, monitoring studies are of paramount importance considering its considerable genetic variation (Parvizi et al. 2023).

2.2 Neotropical Brown Stink Bug, **Euschistus heros** *(F.)*

The neotropical brown stink bug, *Euschistus heros* (F.), is one of the most important pests occurring on Brazilian soybean fields. Its distribution has expanded or increased during the last 40 years to the Southern, Central and Northern regions of Brazil. This long history as prevalent species made of him a frequent target of insecticide application being submitted their populations to a selection pressure, and selecting resistant phenotypes. Recent surveys in diverse localities of Paraná state, ranked *E. heros* as the most prevalent species (64%), followed by the green belly stink bug, *Diceraeus melacanthus* (18%) and in third place by the southern green stink bug, *Nezara viridula* (8%) (Oliveira et al. 2022a, b). In a wide sampling across soybean-growing regions of Brazil, Saldanha et al. (2024) observed that NBSB was the most abundant species, comprising 82.4% of the adults and 84.1% of the nymph samples overall.

Therefore, most of the chemical applications against stink bugs are targeting *E. heros*. Until 2004, this species was managed almost exclusively with chemicals using organophosphates (OPs) and endosulfan. The first mixture of neonicotinoids (imidacloprid) and pyrethroids (beta-cyfluthrin) to control stink bugs was recommended in 2004 (Tecnologias de produção de soja 2005) and, in the following year, a new mixture of thiamethoxan plus lambda-cyhalothrin was introduced into the official recommendation (Tecnologias de produção de soja 2006). The reduction of susceptibility to OPs is possibly due to the use of these products for over 60 years.

Since the early 1990s, control failures with endosulfan have been observed on NBSB populations. In 1999, it was observed a slight reduction of susceptibility to OPs in populations at Pedrinhas Paulista, São Paulo, by using vial bioassays

(Sosa-Gómez et al. 2001). In addition, in 2000-decade, control failures with OPs, such as methamidophos, had become frequent in Paraná state, Brazil, leading growers to increase of recommended rates.

In 2000s, studies reported resistance of *E. heros* to other long time used insecticides such as endosulfan, monocrotophos and methamidophos in Paraná and São Paulo (Sosa-Gómez et al. 2001; Sosa-Gómez and Silva 2010). Bioassays performed with acephate, methamidofos and endosulfan showed decreased susceptibility in geographical populations of *E. heros* (Sosa-Gómez et al. 2009). From 2010 onwards, relative high levels of resistance to lambda-cyhalothrin were detected in *E. heros* populations from various regions of Brazil and in *Diceraeus* species in southern Brazil (Sosa-Gómez et al. 2020; Somavilla et al. 2020a, b) (Table 13.1).

The moderate-high degree of resistance of *E. heros* and *D. furcatus* to pyrethroids and to some neonicotinoids detected in laboratory studies has also impacted their control efficacy under field conditions. From 2020s onwards, studies revealed that *E. heros* showed reduced mortality (low control efficacy) when exposed to the field-recommended doses or diagnostic concentrations of lambda-cyhalothrin and imidacloprid (Somavilla et al. 2020a; Tibola et al. 2021; Moreira et al. 2024). A large variation in control efficacy of the binary mixtures of thiamethoxam + lambda-cyhalothrin and imidacloprid + bifenthrin against *E. heros* has also been verified in populations from various regions of Brazil (Ribeiro et al. 2016; Moreira et al. 2024). Control failures of beta-cyfluthrin (pyrethroid) and imidacloprid (neonicotinoid) against *E. heros* was also reported in soybean fields in Goiás (Guedes 2017; Tuelher et al. 2018). In contrast, in Rio Grande do Sul state, the binary mixtures of pyrethroids + neonicotinoids remain effective against secondary stink bugs of soybean, such as *P. guildinii* and *N. viridula* (Steinhaus et al. 2022). Conversely, studies revealed that *E. heros* exhibited relatively low resistance levels to acephate, bifenthrin and ethiprole (Sosa-Gómez et al. 2020; Somavilla et al. 2020a, b; Moreira et al. 2024). Detoxification enzymes seems to be the major component in the insecticide resistance of *E. heros* (Sosa-Gómez et al. 2009, 2020; Boff et al. 2022; Lira et al. 2023) (Table 13.2).

In some occasions, stink bugs recover within a few days after pyrethroid/neonicotinoid applications and return to feeding activity (Sosa-Gómez, unpublished).

2.3 Green Belly Stink Bugs, **Diceraeus melacanthus** *(Dallas) and* **D. furcatus** *(F.)*

Succession of soybean and maize crops has favored the increase in GBSB populations over the years. In the southern region of Brazil, soybean is typically planted early, in mid-September to early October, and are harvested in January–February. Immediately after the soybean harvest, maize is planted to be harvested in May–June. This succession of crops allows populations that have built up during the

Table 13.1 Resistance levels of the two major stink bug pests of soybean and maize in Brazil

Species	Season: collection site	Bioassay method	Insecticide (MoA)	Resistance level	References
Euschistus heros	2000: Paraná and São Paulo	Vial	Endosulfan (2A)	<8.7-fold	Sosa-Gómez et al. (2001)
			Monocrotophos (1B)	<3.1-fold	
			Metamidophos (1B)	<2.7-fold	
	2006/07 and 2015: Paraná and São Paulo	Topical	Acephate (1B)	<10-fold	Sosa-Gómez et al. (2020)
	2010 and 2015: Paraná and São Paulo		Lambda-cyhalothrin (3A)	1.3–13.2-fold	
	2009/10 and 2015: Paraná and São Paulo		Thiamethoxam (4A)	1.2–26.1-fold	
	2018/19: Rio Grande do Sul, Paraná, São Paulo and Goiás	Dip	Acephate (1B)	<5.9-fold	Somavilla et al. (2020a)
			Thiamethoxam (4A)	2.0–15.1-fold	
			Bifenthrin (3A)	1.7–23.8-fold	
			Lambda-cyhalothrin (3A)	8.2–63.6-fold	
	2018–2020: Rio Grande do Sul, São Paulo, Paraná, Goiás, Mato Grosso and Bahia	Diet encapsulation	Thiamethoxam (4A)	2.3–22.1-fold	Tibola et al. (2021)
			Imidacloprid (4A)	1.6–21.9-fold	
	2019/20: Rio Grande do Sul, Paraná and Goiás	Topical	Acephate (1B)	<2.4-fold	Somavilla et al. (2020b)
			Thiamethoxam (4A)	<8.2-fold	
			Bifenthrin (3A)	<4.9-fold	
			Lambda-cyhalothrin (3A)	3.1–15.7-fold	
Diceraeus furcatus	2019/20: Rio Grande do Sul	Topical	Acephate (1B)	<1.4-fold	Somavilla et al. (2020b)
			Thiamethoxam (4A)	<2.1-fold	
			Lambda-cyhalothrin (3A)	<2.1-fold	
	2018/19: Rio Grande do Sul	Dip	Acephate (1B)	<2.8-fold	Somavilla et al. (2020a)
			Thiamethoxam (4A)	<8.4-fold	
			Bifenthrin (3A)	<3.1-fold	
			Lambda-cyhalothrin (3A)	1.3–13.4-fold	

Table 13.2 Insecticides recommended against stink bugs considering its mode of action (MoA). Performance of these insecticide are widely variable considering species, geographic locations, thus is not a recommendation

Mode of Action	Insecticides examples	Bibliographic source
Ache inhibitors	Acephate, chlorpirifos, carbosulfan	Fukuto (1990)
It interferes with the passage of chloride ions through the gamma-aminobutyric acid (GABA)-regulated chloride channel, thereby disrupting central nervous system (CNS) activity. This disruption in chloride ion flow affects the inhibitory neurotransmission mediated by GABA, leading to altered neuronal excitability and function	Ethiprole, Fipronil	Gant et al. (1998) Caboni et al. (2003)
Act as agonists on nicotinic acetylcholine receptors (nAChRs) opening cation channels. However, voltage-gated calcium channels are also involved in their insecticidal activity	Acetamiprid, dinotefuran imidacloprid, thiamethoxam,	Simon-Delso et al. (2015)
Pyrethroids bind to a protein that regulates the voltage-gated sodium channel. This gate opens and pyrethoids prevent it from closing normally, which results in continuous nerve stimulation, neural cells produce repetitive discharges and eventually cause paralysis	Bifenthrin, Cyfluthrin, Lambda-cyalothrin, zeta-cypermethrin	He et al. (2008) Bradbury and Coats (1989), Shafer and Meyer (2004)
Nicotinic acetylcholine receptor (nAChR) competitive modulator/agonist	Sulfoxaflor	Watson et al. (2021)
Acts as a noncompetitive antagonist of the invertebrate γ-aminobutyric acid (GABA) receptor at a site distinct to that of fiproles and cyclodienes	Isocycloseram	Jeschke (2021)

soybean growing season to find maize fields, which are more suitable hosts, for further growth. Both species were reported to fed on at least 18 host plant families in South America (Smaniotto and Panizzi 2015).

The GBSB usually is controlled on maize using systemic neonicotinoid insecticides, treating the seed at the moment of sowing. However, despite the lower efficiency than seed treatment, foliar application are done with mixtures of neonicotinoids and pyrethroids. Susceptibility line studies have been carried out by Sosa-Gómez et al. (2020) and they observed differences from 5 to 10 times more tolerance of *D. melacanthus* than *E. heros* depending on the insecticide and the population geographic origin. Similarly, differences in susceptibility were observed in *D. furcatus* populations and *E. heros* populations from Rio Grande do Sul, being *D. furcatus* more tolerant to acephate, thiametoxam and lambda-cyhalothrin

(Somavilla et al. 2020a, b). Populations of *D. furcatus* also remains susceptible to acephate, bifenthrin, thiamethoxam and ethiprole (Somavilla et al. 2020a, b; Moreira et al. 2024). There are no reports of field resistance to these species.

2.4 *Southern Green Stink Bug,* Nezara viridula *(L.)*

The first documented case of insecticide resistance in stink bugs was reported in Australia, involving *N. viridula* and DDT (Georghiou and Lagunes-Tejeda 1991). At that time, no information regarding lethal concentrations or quantitative data on field failures was available. Since then, there have been no reported cases of resistance worldwide. In Brazil, until the 1990s, this species used to reach action levels on soybean crops, but occurrences have become less consistent in last decades (Panizzi et al. 2022). However, more recently, this stink bug has reach pest status in other regions of the world (Marcu and Grozea 2020).

2.5 *Red Banded Stink Bug,* Piezodorus guildinii *(Westwood)*

Studies on susceptibility to insecticides of *P. guildinii* have been reported in Brazil and United States by Baur et al. (2010). Studies performed in Argentina found no differences in susceptibility of this species to endosulfan (Stadler et al. 2006). Bioassays performed by Temple et al. (2013) found that red banded stink bugs collected in Louisiana State localities were four to eight-fold less susceptible to pyrethroids and two to eight-fold less susceptible to organophosphates than was the southern green stink bug. However, until now, there are no reports of field resistance to these species.

2.6 *Painted Bug,* Bagrada hilaris *(Burmeister)*

The painted bug is native to eastern and southern Africa and Asia. It was detected in California in 2008, and it has spread throughout multiple southwestern USA states and Mexico (Palumbo et al. 2016; Torres-Acosta and Sanchez-Peña 2016). *Bagrada hilaris* has preference for Brassicaceae species (radish, *Raphanus sativus* L., red cabbage, *Brassica oleracea* var. *capitata* (Palumbo et al. 2016)). Median lethal concentration for this species were determined using multiple bioassay methods (leaf-spray bioassay, leaf-dip Petri dish bioassay, soil systemic bioassay) for pyrethroids, organophosphates neonicotinoids and other insecticides (Palumbo et al. 2015). Updated information about bioinsecticide-susceptibility or natural products were reported (Felipe-Victoriano et al. 2023; Palumbo (2023).

2.7 *Harlequin Bug,* Murgantia histrionica *(Hahn)*

This species of stink bug originally from Central America and Mexico, have invaded the southern part of the USA. The first report of its presence was made by Walsh in 1866 based on samples collected in Texas. Since then, the harlequin bug spread to several states in USA, but it is more common in the south where it causes economic losses on brassicaceous crops (McPherson et al. 2018). Chemical control in *Brassica oleracea* var. *capitata* is done with systemic insecticides such as flupyradifurone from the butenolides group applied as soil drench or isocycloseram from the isooxazoline class in foliar applications (Sydnor et al. 2023). No baseline is available for this species.

3 Resistance Management Strategies of Stink Bug Pests in Brazil

Strategies to circumvent resistance selection are based on rationally minimizing exposure to the same active ingredient; define baselines and standardize replicable bioassays to detect resistance; monitor pest populations to detect initial shifts in susceptibility; establish diagnostic doses; and perform cross-resistance studies. Also, provide training and support to the extension service to follow detailed records of insecticide applications (products, mode of actions, doses, application frequency, dates, treated areas, and prevalent stink bug species) are fundamental. These measures will help in planning and adopting future control measures. Nowadays, the continuous monitoring of the susceptibility to insecticide should be imperative due to rapid shifts in the susceptibility status in agricultural pests over short periods.

The major challenge for the sustainable stink bug management in Brazil is to preserve their susceptibility to current insecticides. All chemical classes of insecticides registered for managing stink bugs in Brazilian soybean and maize fields act on nerve targets representing five different modes of action, including acetylcholinesterase inhibitors (IRAC 1), GABA-gated chloride channel blockers (IRAC 2), voltage-gated sodium channel modulators (IRAC 3), nicotinic acetylcholine receptor agonists (IRAC 4) and allosteric modulators of GABA-gated chloride channels (IRAC 30) (Table 13.2).

The Insecticide Resistance Action Committee (IRAC) recommends applying insecticide with different mode of actions (MoA) against stink bugs on soybean and maize crops using a "MoA treatment window" approach (Sparks et al. 2021) (Fig. 13.1). A window is defined by the estimated duration of an insect generation (egg–adult period) or the duration of the effect of insecticide application (residual activity). On this basis, IRAC recommends a window of 30 days for most pest species of soybean, maize and cotton, including stink bugs. Therefore, insecticides with the same MoA must be applied only within a defined window (Fig. 13.1). Nevertheless, the same MoA can be used in a later window but only once a different

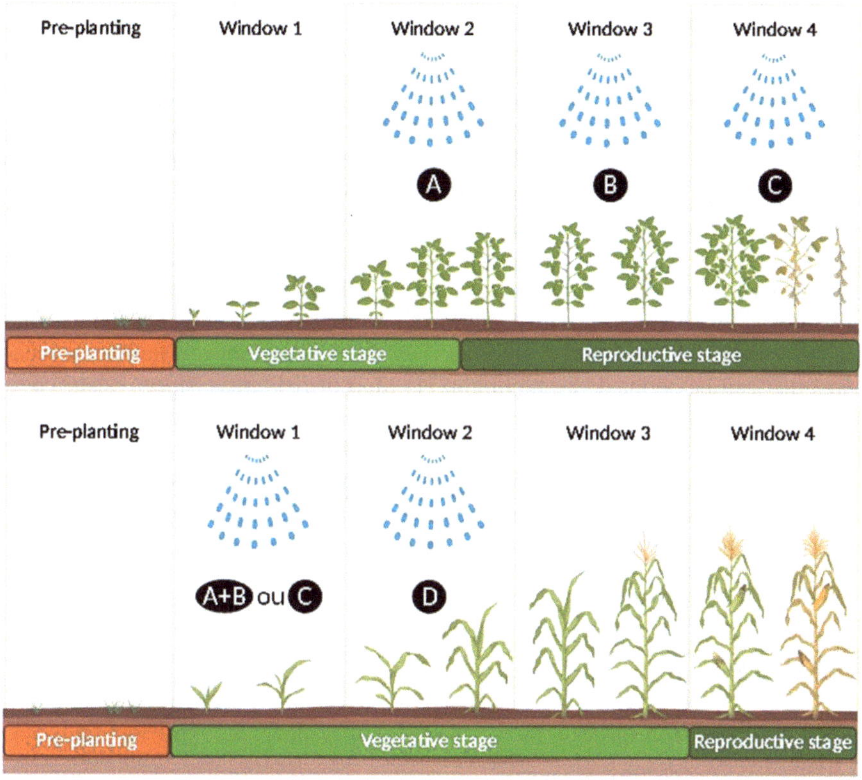

Fig. 13.1 Schematic representation of the MoA treatment window approach for resistance management of stink bugs on soybean and maize in Brazil. Distinct letters represent different MoA. (Created with BioRender.com)

MoA is previous used, and in the case of planting treated seeds, it is recommended the spray of insecticides (if necessary) with same MoA only within the 30-day window.

Additionally to the MoA treatment window approach, some other agronomic practices are recommended for improving resistance management of stink bugs on soybean and maize in Brazil, such as: (i) Improve control of weeds and volunteer plants during the off-season to reduce stink bug infestations during the regular season; (ii) monitor stink bug density and apply integrated pest management measures when the action thresholds is reached; (iii) use information about local or regional susceptibility of stink bugs to insecticides to support decision-making about insecticide choice; (iv) avoid use of insecticides (alone or in mixtures) with reduced control effectiveness (i.e., lambda-cyhalothrin–containing insecticides); (v) implement rotation of insecticides using a "MoA treatment window" approach to reduce selection pressure for resistance; (vi) do not increase the dose of insecticide when control fails but, instead, use an insecticide with a different MoA; and (vii) use other control tactics [e.g. egg parasitoids and soybean Block technology cultivars, resistant to stink bug damage (Oliveira et al. 2022a, b)].

4 Final Considerations

The chemical control will continue to be the main control tactic used for managing stink bugs on soybean and maize fields in Brazil, at least in the next few decades. The control of stink bug must be based on the current insect resistance management principles through the rotation of the MoA's by using MoA treatment windows during the regular crop season, avoiding the successive exposure of stink bug generations to the same MoA. Therefore, the adoption of IRM strategies to avoid evolution of resistance is critical, as farmers and other stakeholders cannot rely on a constant flow of new insecticides with different MoA's from those currently available on the market to replace chemicals compromised by the field resistance. Among the measures that can be adopted, we highlight: (1) Map of risk areas, since resistance to stink bugs is limited and can be confined to specific geographic areas, with abusive and inappropriate use of products; (2) maintain a historical record of the frequency and insecticides used for the proper selection of products to be used onwards; and (3) delay initial applications of broad-spectrum or non-selective insecticides, considering the action levels recommended by research institutions.

References

Ak K, Uluca M, Aydin Ö, Goktuk T (2019) Important invasive species and its pest status in Turkey: *Halyomorpha halys* (Stål) (Heteroptera: Pentatomidae). J Plant Dis Prot 126:401–408

Alford A, Kuhar TP, Hamilton GC, Jentsch P, Krawczyk G, Walgenbach JF, Welty C (2020) Baseline toxicity of the insecticides bifenthrin and thiamethoxam on *Halyomorpha halys* (Hemiptera: Pentatomidae) collected from the Eastern United States. J Econ Entomol 113:1043–1046

Baur ME, Sosa-Gómez DR, Ottea J, Leonard BR, Corso IC, Silva JJ, Temple J, Boethel DJ (2010) Susceptibility to insecticides used for control of *Piezodorus guildinii* (Heteroptera: Pentatomidae) in the United States and Brazil. J Econ Entomol 103:869–876

Boff JS, Reis AC, Patricia DSG, Pretto VE, Garlet CG, Melo AA, Bernardi O (2022) The effect of synergistic compounds on the susceptibility of *Euschistus heros* (Hemiptera: Pentatomidae) and *Chrysodeixis includens* (Lepidoptera: Noctuidae) to pyrethroids. Environ Entomol 51:421–429

Bradbury SP, Coats JR (1989) Comparative toxicology of the pyrethroid insecticides. In: Ware GW (ed) Reviews of environmental contamination and toxicology. Springer, Dordrecht, pp 133–177

Bueno AF, Bortolotto OC, Pomari-Fernandes A, França-Neto JB (2015) Assessment of a more conservative stink bug economic threshold for managing stink bugs in Brazilian soybean production. Crop Prot 71:132–137

Bueno NM, Baldin ELL, Melotto G, Silva IF, Santos TLB, Ribeiro LP, Koch RL (2021) Potential for injury from different life stages of *Diceraeus melacanthus* and *Euschistus heros* (Hemiptera: Pentatomidae) on corn seedlings. Phytoparasitica 49:883–892

Caboni P, Sammelson RE, Casida JE (2003) Phenylpyrazole insecticide photochemistry, metabolism, and GABAergic action: ethiprole compared with fipronil. J Agric Food Chem 51:7055–7061

Felipe-Victoriano M, Villegas-Luján R, Treviño-Cueto D, Sánchez-Peña SR (2023) Combined activity of natural products and the fungal entomopathogen *Cordyceps farinosa* against *Bagrada hilaris* (Hemiptera: Pentatomidae). Fla Entomol 106:154–160

Fukuto TR (1990) Mechanism of action of organophosphorus and carbamate insecticides. Environ Health Perspect 87:245–254
Gant DB, Chalmers AE, Wolff MA, Hoffman HB, Bushey DF (1998) Fipronil: action at the GABA receptor. In: Kuhr RJ, Motoyama N (eds) Pesticides and the future: minimizing chronic exposure of humans and the environment. IOS Press, pp 147–156
Georghiou GP, Lagunes-Tejeda A (1991) The occurrence of resistance to pesticides in arthropods. FAO, No. 04, QH545, P4, G4, Rome
Guedes RNC (2017) Insecticide resistance, control failure likelihood and the first law of geography. Pest Manag Sci 73:479–484
He LM, Troiano J, Wang A, Goh K (2008) Environmental chemistry, ecotoxicity, and fate of lambda-cyhalothrin. In: Whitacre DM (ed) Reviews of environmental contamination and toxicology, vol 195. Springer, pp 71–91
Hoebeke ER, Carter ME (2003) *Halyomorpha halys* (Stål) (Heteroptera: Pentatomidae): A polyphagous plant pest from Asia newly detected in North America. Proc Entomol Soc Wash 105:225–237
Husch PE, Ferreira DG, Seraphim N, Harvey N, Silva-Brandão KL, Sofia SH, Sosa-Gómez DR (2018) Structure and genetic variation among populations of *Euschistus heros* from different geographic regions in Brazil. Entomol Exp Appl 166:191–203
Jeschke P (2021) Current trends in the design of fluorine-containing agrochemicals. organofluorine chemistry: synthesis, modeling, and applications, p 363–395
Knight KM, Gurr GM (2007) Review of *Nezara viridula* (L.) management strategies and potential for IPM in field crops with emphasis on Australia. Crop Prot 26:1–10
Leskey TC, Nielsen AL (2018) Impact of the invasive brown marmorated stink bug in North America and Europe: history, biology, ecology, and management. Annu Rev Entomol 63:599–618
Leskey TC, Lee D-H, Short BD, Wright SE (2012a) Impact of insecticides on the invasive *Halyomorpha halys* (Hemiptera: Pentatomidae): analysis of insecticide lethality. J Econ Entomol 105:1726–1735
Leskey TC, Short BD, Butler BR, Wright SE (2012b) Impact of the invasive brown marmorated stink bug, *Halyomorpha halys* (Stål), in Mid-Atlantic tree fruit orchards in the United States: Case studies of commercial management. Psyche 2012:1–14
Lira EC, Nascimento AR, Bass C, Omoto C, Cônsoli FL (2023) Transcriptomic investigation of the molecular mechanisms underlying resistance to the neonicotinoid thiamethoxam and the pyrethroid lambda-cyhalothrin in *Euschistus heros* (Hemiptera: Pentatomidae). Pest Manag Sci 79:5349–5361
Maistrello L, Vaccari G, Caruso S, Costi E, Bortolini S, Macavei L, Foca G, Ulrici A, Bortolotti PP, Nannini R, Casoli L (2017) Monitoring of the invasive *Halyomorpha halys*, a new key pest of fruit orchards in northern Italy. J Pest Sci 90:1231–1244
Marcu V, Grozea I (2020) Analysis of different host plants and identifying non-polluting bioinsecticide strategies to control the invasive pest *Nezara viridula*. Rom Biotechnol Lett 25:1877–1885
McPherson JE, Bundy SC, Kuhar TP (2018) *Murgantia histrionica* (Hahn). In: McPherson JE (ed) Invasive stink bugs and related species (Pentatomoidea): biology, higher systematics, semiochemistry, and management. CRC Press, Boca Raton, pp 333–349
Moreira RP, Palharini RB, Massoli GS, Diniz LHM, Godoy DN, Warpechowski LF, Steinhaus EA, Stürmer GR, Bernardi O (2024) Geographic and interspecific variation in susceptibility of *Euschistus heros* and *Diceraeus furcatus* (Hemiptera: Pentatomidae) to selected insecticides in southern Brazil. Crop Prot 179:106625
Musolin DL, Dolgovskaya MY, Zakharchenko VY, Karpun NN, Haye T, Saulich AK, Reznik SY (2022) Flying over Eurasia: Geographic variation of photoperiodic control of nymphal development and adult diapause induction in native and invasive populations of the brown marmorated stink bug, *Halyomorpha halys* (Hemiptera: Heteroptera: Pentatomidae). Insects 13:522
Nielsen AL, Hamilton GC (2009) Life history of the invasive species *Halyomorpha halys* (Hemiptera: Pentatomidae) in northeastern United States. Ann Entomol Soc Am 102:608–616

Nielsen AL, Shearer PW, Hamilton GC (2008) Toxicity of insecticides to *Halyomorpha halys* (Hemiptera: Pentatomidae) using glass-vial bioassays. J Econ Entomol 101:1439–1442
Oliveira WP, Lucini T, Panizzi AR (2022a) Seed damage by the Neotropical brown stink bug, *Euschistus heros* (F.) to resistant soybean cultivars with the Block technology versus a susceptible cultivar. Environ Entomol 51:451–459
Oliveira AB, Gomes EC, Possamai EJ, Silva GC, Reis EA, Roggia S, Prando AM, Conte O (2022b) Resultados do manejo integrado de pragas da soja na safra 2020/2021 no Paraná. Empresa Brasileira de Pesquisa Agropecuária Embrapa Soja. Ministério da Agricultura, Pecuária e Abastecimento Embrapa Soja, Londrina, PR, Documentos 443
Owens DR, Herbert Jr DA, Dively GP, Reisig DD, Kuhar TP (2013) Does feeding by *Halyomorpha halys* (Hemiptera: Pentatomidae) reduce soybean seed quality and yield? J Econ Entomol 106:1317–1323
Palumbo JC (2023) Contact activity of biopesticides against *Bagrada hilaris* adults. Arthropod Manag Tests 48:1–2
Palumbo JC, Prabhaker N, Reed DA, Perring TM, Castle SJ, Huang TI (2015) Susceptibility of *Bagrada hilaris* (Hemiptera: Pentatomidae) to insecticides in laboratory and greenhouse bioassays. J Econ Entomol 108:672–682
Palumbo JC, Perring TM, Millar JG, Reed DA (2016) Biology, ecology, and management of an invasive stink bug, *Bagrada hilaris*, in North America. Annu Rev Entomol 61:453–473
Panizzi AR (2015) Growing problems with stink bugs (Hemiptera: Heteroptera: Pentatomidae): Species invasive to the US and potential Neotropical invaders. Am Entomol 61:223–233
Panizzi AR, Corrêa BS, Gazzoni DL, Oliveira EB, Newman GG, Turnipseed SG (1977) Insetos da soja no Brasil. EMBRAPA-CNPSo. Bol Téc 1:1–20
Panizzi AR, Lucini T, Aldrich JR (2022) Dynamics in pest status of phytophagous stink bugs in the Neotropics. Neotrop Entomol 51:18–31
Parvizi E, Dhami MK, Yan J, McGaughran A (2023) Population genomic insights into invasion success in a polyphagous agricultural pest, *Halyomorpha halys*. Mol Ecol 32:138–151
Reznik SY, Dolgovskaya MY, Karpun NN, Zakharchenko VY, Saulich AK, Musolin DL (2023) The invasive caucasian populations of the brown marmorated stink bug *Halyomorpha halys* (Hemiptera: Heteroptera: Pentatomidae) rapidly adapt their ecophysiological traits to the local environmental conditions. Insects 14:424
Ribeiro FC, Rocha FS, Erasmo EAL, Matos EP, Costa SJ (2016) Management with insecticides targeting the brown stink bug control in intact soybean crop. Rev Agric Neotrop 3:48–53
Saldanha AV, Horikoshi R, Dourado P, Overejo RVL, Berger GU, Martinelli S, Head GP, Moraes T, Correa AS, Schwertneret CF (2024) The first extensive analysis of species composition and abundance of stink bugs (Hemiptera: Pentatomidae) on soybean crops in Brazil. Pest Manag Sci 80:3945–3956
Shafer TJ, Meyer DA (2004) Effects of pyrethroids on voltage-sensitive calcium channels: a critical evaluation of strengths, weaknesses, data needs, and relationship to assessment of cumulative neurotoxicity. Toxicol Appl Pharmacol 196:303–318
Simon-Delso N, Amaral-Rogers V, Belzunces LP, Bonmatin JM, Chagnon M, Downs C, Furlan L, Gibbons DW, Giorio C, Girolami V, Goulson D, Kreutzweiser DP, Krupke CH, Liess M, Long E, McField M, Mineau P, Mitchell EAD, Morrissey CA, Noome DA, Pisa L, Settele J, Stark JD, Tapparo A, Van Dyck H, Van Praagh J, Van der Sluijs JP, Whitehorn PR, Wiemers M (2015) Systemic insecticides (neonicotinoids and fipronil): Trends, uses, mode of action and metabolites. Environ Sci Pollut Res 22:5–34
Smaniotto LF, Panizzi AR (2015) Interactions of selected species of stink bugs (Hemiptera: Heteroptera: Pentatomidae) from leguminous crops with plants in the Neotropics. Fla Entomol 98:7–17
Somavilla JC, Gubiani PS, Reis AC, Führ FM, Machado EP, Bernardi O (2020a) Susceptibility of *Euschistus heros* and *Dichelops furcatus* (Hemiptera: Pentatomidae) to insecticides determined from topical bioassays and diagnostic doses for resistance monitoring of *E. heros* in Brazil. Crop Prot 138:105319

Somavilla JC, Reis AC, Gubiani PS, Godoy DN, Stürmer GR, Bernardi O (2020b) Susceptibility of *Euschistus heros* and *Dichelops furcatus* (Hemiptera: Pentatomidae) to selected insecticides in Brazil. J Econ Entomol 113:924–931

Soria MF, Degrande PE, Panizzi AR, Toews MD (2017) Economic injury level of the neotropical brown stink bug *Euschistus heros* (F.) on cotton plants. Neotrop Entomol 46:324–335

Sosa-Gómez DR, Silva JJ (2010) Neotropical brown stink bug (*Euschistus heros*) resistance to methamidophos in Paraná, Brazil. Pesq Agrop Brasileira 45:767–769

Sosa-Gómez DR, Corso IC, Morales L (2001) Insecticide resistance to endosulfan, monocrotophos and ethamidophos in the neotropical brown stink bug, *Euschistus heros* (F.). Neotrop Entomol 30:317–320

Sosa-Gómez DR, Silva JJ, Lopes ION, Corso IC, Almeida AM, Moraes GCP, Baur ME (2009) Insecticide susceptibility of *Euschistus heros* (Heteroptera: Pentatomidae) in Brazil. J Econ Entomol 102:1209–1216

Sosa-Gómez DR, Corrêa-Ferreira BS, Kraemer B, Pasini A, Husch PE, Vieira CED, Martinez CBR, Lopes ION (2020) Prevalence, damage, management and insecticide resistance of stink bug populations (Hemiptera: Pentatomidae) in commodity crops. Agric Forest Entomol 22:99–118

Sparks ME, Bansal R, Benoit JB, Blackburn MB, Chao H, Chen M, Cheng S, Childers C, Dinh H, Doddapaneni HV, Dugan S, Elpidina EN, Farrow DW, Friedrich M, Gibbs RA, Hall B, Han Y, Hardy RW, Holmes CJ, Hughes DST, Ioannidis P, Jarvela AMC, Johnston JS, Jones JW, Kronmiller BA, Kung F, Lee SL, Martynov AG, Masterson P, Maumus F, Munoz-Torres M, Murali SC, Murphy TD, Muzny DM, Nelson DR, Oppert B, Panfilio KA, Paula DO, Pick L, Poelchau MF, Qu J, Reding K, Rhoades JH, Rhodes A, Richards S, Richter R, Robertson HM, Rosendale AJ, Tu ZJ, Velamuri AS, Waterhouse RM, Weirauch MT, Wells JT, Werren JH, Worley KC, Zdobnov EM, Gundersen-Rindal DE (2020) Brown marmorated stink bug, *Halyomorpha halys* (Stål), genome: putative underpinnings of polyphagy, insecticide resistance potential and biology of a top worldwide pest. BMC Genomics 21:1–26

Sparks TC, Storer N, Porter A, Slater R, Nauen R (2021) Insecticide resistance management and industry: the origins and evolution of the insecticide resistance action committee (IRAC) and the mode of action classification scheme. Pest Manag Sci 77:2609–2619

Stadler T, Buteler M, Ferrero AA (2006) Susceptibility to endosulphan and resistance monitoring in green stinkbug populations, *Piezodorus guildinii* (Insecta, Heteroptera: Pentatomidae), in soya crops in Argentina. Rev Soc Entomol Argent 65:109–119

Steinhaus EA, Warpechowski LF, Braga LE, Morin M, Tenório C, Boff JS, Bernardi O, Farias JR (2022) Intra-and interspecific variation in the susceptibility to insecticides of stink bugs (Hemiptera: Pentatomidae) that attack soybean and maize in southern Brazil. J Econ Entomol 115:631–636

Sydnor T, McIntyre K, Bekelja K, Kuhar TP (2023) Efficacy of selected insecticides against flea beetle and harlequin bug in cabbage in Virginia, 2022. Arthropod Manage Tests 48:1–2

Tabashnik BE, Mota-Sanchez D, Whalon ME, Hollingworth RM, Carrière Y (2014) Defining terms for proactive management of resistance to Bt crops and pesticides. J Econ Entomol 107:496–507

Tecnologias de produção de soja - Região Central do Brasil (2005) Sistemas de Produção. Outubro, 2004. Londrina: Embrapa Soja: Embrapa Cerrados: Embrapa Agropecuária Oeste: Fundação Meridional, 239 p

Tecnologias de produção de soja - Região Central do Brasil (2006) Sistemas de Produção. Outubro, 2005. Londrina: Embrapa Soja: Embrapa Cerrados: Embrapa Agropecuária Oeste: Fundação Meridional, 220 p

Temple JH, Davis JA, Hardke J, Moore J, Leonard BR (2013) Susceptibility of southern green stink bug and redbanded stink bug to insecticides in soybean field experiments and laboratory bioassays. Southwest Entomol 38:393–406

Tibola CM, Silva L, Sgubin F, Omoto C (2021) Monitoring resistance of *Euschistus heros* (Fabricius) (Hemiptera: Pentatomidae) to insecticides by using encapsulated artificial diet bioassay. Insects 12:599

Torres-Acosta RI, Sanchez-Peña SR (2016) Geographical Distribution of *Bagrada hilaris* (Hemiptera: Pentatomidae) in Mexico. J Entomol Sci 51:165–167

Tuelher ES, Silva EH, Freitas HL, Namorato FA, Serrão JE, Guedes RNC, Oliveira EE (2017) Chlorantraniliprole-mediated toxicity and changes in sexual fitness of the Neotropical brown stink bug *Euschistus heros*. J Pest Sci 90:397–405

Tuelher ES, Silva ÉH, Rodrigues HS, Hirose E, Guedes RNC, Oliveira EE (2018) Area-wide spatial survey of the likelihood of insecticide control failure in the Neotropical brown stink bug *Euschistus heros*. J Pest Sci 91:849–859

Valentin RE, Nielsen AL, Wiman NG, Lee DH, Fonseca DM (2017) Global invasion network of the brown marmorated stink bug, *Halyomorpha halys*. Sci Rep 7:9866

Watson GB, Siebert MW, Wang NX, Loso MR, Sparks TC (2021) Sulfoxaflor–A sulfoximine insecticide: Review and analysis of mode of action, resistance and cross-resistance. Pestic Biochem Phys 178:104924

Chapter 14
Case Study 1: *Euschistus heros* (F.) on Soybean in Brazil

Adeney Freitas Bueno, Antônio Ricardo Panizzi, and Weidson Plauter Sutil

1 Introduction

Among the different insects that attack soybean, not only in Brazil but in other countries as well, stink bugs are key pests responsible for up to 60% of all insecticide sprayed on soybean fields (Carnevalli et al. 2022; Bueno et al. 2023). They are noteworthy for feeding directly on pods, seriously affecting crop yields beyond the physiological and sanitary quality (Fig. 14.1) of the seeds (Panizzi et al. 1979; Corrêa-Ferreira and Azevedo 2002). These stink bugs (Hemiptera: Pentatomidae) are formed for at least 54 different species (Panizzi and Slansky Jr 1985) reported on soybean fields from South America (especially at latitudes between 0° and 23°) (Panizzi and Corrêa-Ferreira 1997) to the Southeastern USA (Ademokoya et al. 2022).

Despite the low abundance of the Neotropical brown stink bug, *Euschistus heros* (Fig. 14.2), in the 1970s, today this species is the most common stink bug species in Brazilian soybean fields (Panizzi et al. 2022), becoming a serious phytosanitary problem for the crop management (Bueno et al. 2023). Currently, the primary pest-control method adopted by soybean growers as the first line of defense against stink bugs is spraying insecticides (Bueno et al. 2021). However, the overuse of some harmful insecticides has negatively impacted human health as well as the environment (Jacquet et al. 2022). Synthetic chemical insecticides can be responsible for

A. F. Bueno (✉)
Embrapa Soja, Londrina, PR, Brazil
e-mail: adeney.bueno@embrapa.br

A. R. Panizzi
Embrapa Trigo, Passo Fundo, RS, Brazil

W. P. Sutil
Universidade Federal do Paraná, Curitiba, PR, Brazil

A. F. Bueno, A. R. Panizzi (eds.), *Stink Bugs (Hemiptera: Pentatomidae) Research and Management*, Entomology in Focus 9,
https://doi.org/10.1007/978-3-031-69742-5_14

Fig. 14.1 Injured soybean seeds by *Euschistus heros* (F.) colored using the tetrazolium test. (Photo: A.F. Bueno)

significant negative side-effects (Desneux et al. 2007). Their overuse can reduce natural biological control agents (Torres and Bueno 2018) and pollinators (Kuldna et al. 2009), select stink bug for resistance of populations on soybean fields (Sosa-Gómez et al. 2020), and trigger pest resurgence and/or outbreaks of secondary pests, among other negative side-effects (Panizzi et al. 1977a, b; Bueno et al. 2021). Consequently, reducing synthetic chemical use not only in the stink bug management but in agriculture in a broader perspective, to reach an even cheaper and more sustainable food production system (Maciel and Bueno 2022), has become a global goal (Lee et al. 2019).

Among the most sustainable stink bug management strategy available, the adoption of integrated pest management (IPM) programs stands out as the most efficient option adopted by soybean farmers (Bueno et al. 2023). The global concept of IPM was established in the late 1950s based upon plant tolerance to certain levels of injury without economically yield losses (Higley and Pedigo 1996). In Brazil, IPM on soybean was first applied in the late 1960s and early 1970s and was rapidly disseminated by extension agents and technical publications that triggered the adoption of the technology by soybean farmers (Kogan 1998; Bueno et al. 2021).

Soybean-IPM combines different stink bug control recommendations including plant tolerance (Fig. 14.3) (discussed in more details in Chap. 10), adoption of economic thresholds (ETs) as the population level at which control measure should be

Fig. 14.2 *Euschistus heros* (F.) eggs (**a**), nymph (**b**) and adult (**c**). (Photos: J.J. Silva)

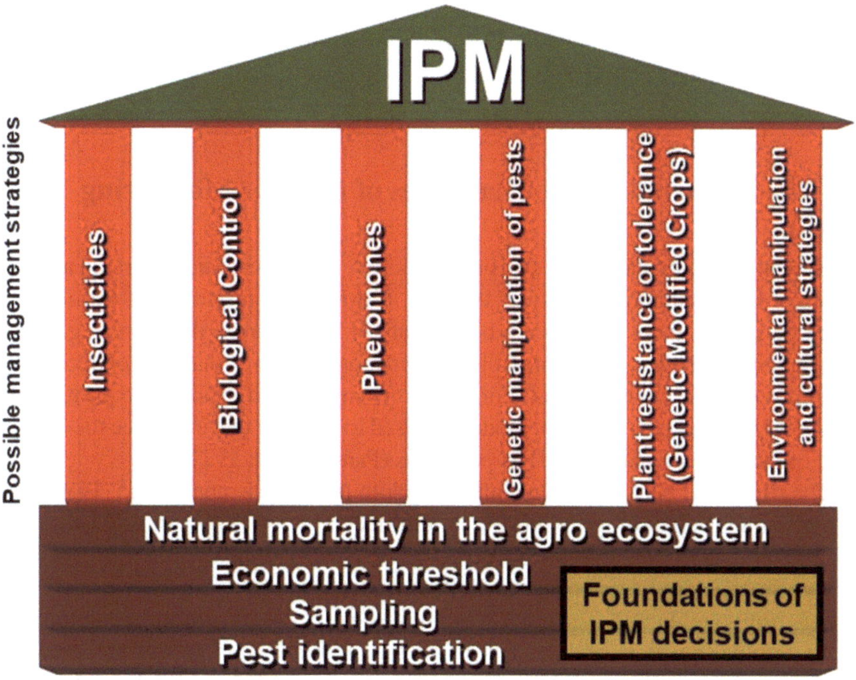

Fig. 14.3 Integrated pest management main components

initiated, scouting procedures, use of selective chemical insecticides, adoption of augmentative biological control (discussed in more details in Chap. 4), as well as other sustainable and greener pest management strategies (Bueno et al. 2021; Maciel and Bueno 2022). The benefits of adopting soybean-IPM for the management of stink bugs, spraying insecticides only when the pest population reaches or surpasses ETs, have been remarkable when extensionists provide consultation to hundreds of soybean farmers in Paraná State, Brazil (Bueno et al. 2023).

When economic thresholds (ETs) are reached or surpassed and pest control must be taken, among the different tools currently available for stink bug control in soybean-IPM, augmentative biological control (ABC) stands out as the most sustainable one (van Lenteren et al. 2018). However, the successful application of ABC programs depends on the adoption of IPM to provide a more favorable agroecosystems to the released biocontrol agents (Bueno et al. 2021). Although in some cases multiple field experiments have shown negative impacts of chemical insecticides on biological control (Janssen and van Rijn 2021), some selective insecticides exist that have lower impact on biocontrol agents due to their physiological selectivity. In addition, non-selective insecticides can be applied in a way that reduces contact with beneficial organisms, and consequently reduces impact on them, a strategy called ecological selectivity (Torres and Bueno 2018). Both physiological and ecological selectivity of insecticides are important options to be taken into consideration for soybean-IPM to succeed, besides the use of chemical control only when the ETs are reached or surpassed.

2 The Importance and Challenges of *E. heros* Monitoring

Correct stink bug management decisions require reliable, accurate, and rapid assessments of the population density of the pest. If not done precisely and within adequate frequency, there will always be a great risk of mistaken decisions regarding the use of insecticides, which might be anticipating sprayings that are not needed or dispensing the necessary ones. For pests on soybean in Brazil, the beat cloth (Fig. 14.4) has been the main methodology used to monitor the crop since the beginning of the implementation of IPM programs (Panizzi et al. 1977a, b).

The use of the beat cloth is simple. The cloth properly wrapped (Fig. 14.4, step 1) and without disturbing the plants, is introduced between two adjacent rows of soybean and extended onto the soil (Fig. 14.4, step 2). Quickly, the plants of one row of soybean are leaned over the cloth and beat vigorously in order to move the insects from the plants to the cloth (Fig. 14.4, step 3). The plants are then returned to the original position, and the insects on the cloth are counted and recorded on monitoring sheets (Fig. 14.4, step 4). Despite of its efficacy, soybean farmers started to complain about the difficulties of this process. The bottleneck to scale this sampling technology to large extension on soybean fields is related to the difficulties regarded the required amount of work to sample insects. It includes the time consumed to sample large soybean areas, low qualified workers to do the samples,

Fig. 14.4 Insect pests sampling procedures on soybean using the beat-cloth. Step 1—Choosing and inserting the beat-cloth in between two soybean rows. Step 2—Unfolding one side of the beat-cloth over one soybean row. Step 3—Bending and shaking soybean plants of one row over the beat-cloth. Step 4—Unfolding the beat-cloth and recording the number of insects dislodged over it. (Photos: A.F. Bueno)

among other issues. This fostered soybean farmers to overuse insecticides in a calendar basis, due to the easiness of this "identify and spray strategy" approach and the low prices of several chemicals, which represented an enormous drawback to IPM on soybean worldwide (Bueno et al. 2011). To trigger a higher adoption of soybean-IPM in large areas, a pest-sampling technology as precise as the beat cloth but faster and consequently cheaper is urgently needed.

Time consumed in insect sampling and the necessity of qualified workers for a precise result have always been some of the greatest constraints to be overcome in IPM, mainly considering extensive crops such as soybean. Some promising results have been obtained with the use of insect pheromones in baited traps. Especially, for some Lepidoptera and Hemiptera species, several studies indicate the attraction and capture of those pests by the traps containing sexual pheromone compounds (Borges et al. 1998; Pires et al. 2000; Schmidt et al. 2003) or even ammonia compounds to attract stink bugs (Paiva 2024). Although these results obtained with those compounds are quite promising, they did not attract nymphs that also damage soybean pods. Therefore, the number of insects capture in traps has not a clear correlation with the injuring-pest population in the field and, therefore, its use for ET decisions might not carry the required precision at this moment.

Another promise in insect sampling is the use of aerial image (satellite or drone images). With cameras going more powerful and cheaper in the market, this technology might certainly be helpful for insect monitoring with great facility, speed and low costs. However, similarly to the use of pheromones traps, the use of aerial images still needs further investigations in order to perform some calibration and adjustments of this sampling method. Then, even though with new technologies coming up, the most traditional sampling procedure (beat cloth) is still necessary to deliver the required precision in insect monitoring, for IPM decisions. Of course, an easier and faster sampling procedure than the beat cloth would be welcome and facilitate the implementation of soybean IPM. However, since the minimum of points required is of only one point sampled for each 10 hectares of soybean, the work required to perform such samples cannot be an acceptable excuse for not adopting such sampling procedure. Despite the high technology or old fashion sampling techniques, one should be aware that is not possible to be successful in IPM without having an efficient pest population monitoring.

3 Economic Thresholds (ETs) for Stink Bugs on Soybean

Both, chemical control and augmentative biological control ABC, should only be applied on soybean fields when necessary to avoid economic loss and ecological harm. Soybean plants can tolerate certain levels of stink bugs without economically significant yield reduction (Bueno et al. 2013). Therefore, no type of applied control should be adopted preventively, and both chemical insecticides and bioinsecticides should be used only when the stink bug population is equal to or greater than the established ET (Fig. 14.5).

The recommended ETs for the total number of stink bugs, including *E. heros*, can slightly differ among countries. In the USA, the ET for stink bugs is three bugs larger than 0.6 cm per row meter (Andrews et al. 2009). In Brazil, the ETs are variable depending on the production purpose (for grain or seed production). For grain

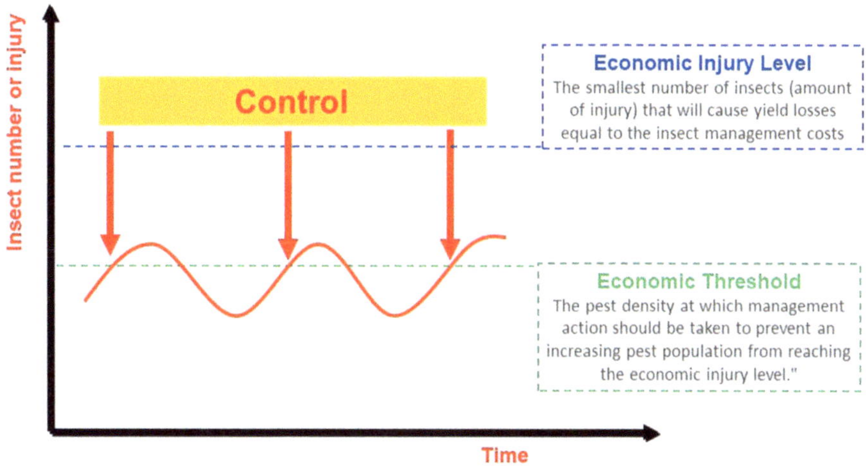

Fig. 14.5 Economic injury level (EIL) and economic threshold (ET) relation to pest control decision and avoidance of economic yield reduction

production, the ET is two stink bugs larger than 0.5 cm per row meter (including nymphs from 3rd instar to adults). In the case of seed production, the ET is only one bug larger than 0.5 cm (Panizzi et al. 1977a; Bueno et al. 2015).

It is important to emphasize that despite being less harmful than chemicals, even biological control products should be used with caution. For instance, the massive release of *Cotesia flavipes* (Cameron) in sugarcane to control *Diatraea saccharalis* (F.) in ABC carried out for years in Brazil probably displaced tachinid parasitoid flies from the environment where they used to be commonly found (Rossi and Fowler 2004). Such undesirable consequences need to be avoided on soybean fields with the adoption of a sound ET for adopting biologicals against *E. heros*, especially considering the large area cultivated with this crop (around 45 million hectares only in Brazil—crop season 2022/23). Thus, less pest control applied on soybean fields will lead to a better preservation of the greatest number of biocontrol agents as well as a higher biodiversity on general terms. Such reduced applied pest control can only be reached with the adoption of IPM and sound ETs.

Soybean-IPM (and the ETs) has been adopted in Brazil and results were recorded in several counties of the State of Paraná, in a joint effort by Embrapa Soybean and the State Government through its official institute "Paraná Rural Development Institute (IDR-Paraná)". Pests were evaluated weekly by an IDR-Paraná extensionist throughout the soybean growing season and control decisions were taken respecting the ET for stink bug's control. At the end of the soybean season, fields were individually harvested and evaluated. Simultaneously, a survey was carried out using a questionnaire for farmers that were not assisted by the soybean-IPM program. Afterwards, a comparison was made between non-assisted (Non-IPM) and assisted farmers (IPM) and the data regarding only stink bug control are presented and discussed (Table 14.1).

Table 14.1 Soybean-IPM program[a] results (mean) in Paraná State, South Brazil

Variable	Comparison	2015/16	2016/17	2017/18	2018/19	2019/20	2020/21
Number of insecticide applications during the soybean growing season to control stink bugs	IPM	1.40 (123 growers)	1.30 (141 growers)	1.06 (196 growers)	1.15 (241 growers)	1.13 (255 growers)	1.27 (191 growers)
	Non-IPM	1.90 (314 growers)	1.9 (390 growers)	2.0 (615 growers)	2.10 (773 growers)	1.94 (553 growers)	2.22 (518 growers)
Days until first insecticide application to control stink bugs	IPM	70.4 days	78.4 days	86.4 days	82.3 days	82.1 days	83.3 days
	Non-IPM	66.0 days	69.5 days	65.8 days	63.1 days	68.3 days	62.6 days
Stink bug control costs (% of total number of insecticide applications)	IPM	80.0 kg/ha (66.7%)	89.7 kg/ha (65.0%)	59.8 kg/ha (70.7%)	85.2 kg/ha (67.7%)	74.0 kg/ha (68.5%)	44.3 kg/ha (73.8%)
	Non-IPM	120 kg/ha (50.0%)	126.3 kg/ha (51.4%)	115.5 kg/ha (58.8%)	151.9 kg/ha (61.8%)	115.6 kg/ha (64.2%)	78.1 kg/ha (65.1%)
Yield (kg/ha)	IPM	3,426.0	3,870.0	3,702.0	3,006.0	3,864.0	3,654.0
	Non-IPM	3,282.0	3,828.0	3,624.0	2,916.0	3,804.0	3,618.0

Adapted from Conte et al. (2016, 2017, 2018, 2019, 2020) and Oliveira et al. (2022)

[a]Program where public consultants (from IDR—Paraná) sampled pests across the seasons and took all decisions on pest management in IPM areas for selected farmers. At the end of the season, the results of IPM areas were compared with other non-IPM areas of Paraná, Brazil

The adoption of IPM and the appropriated ETs for stink bugs in soybean resulted in an average reduction of 40.1% of insecticides applied to control stink bugs (considering the growing seasons from 2015/16 to 2020/21), ranging from 26.3% (2015/16) to 53.0% (2017/18). Not only did the adoption of IPM reduce the number of insecticide applications to control stink bugs but also increased the time (days) from sowing to the application of the first insecticide to control stink bugs by 14.8 days. This time period was also the average from 2015/16 to 2020/21 growing seasons (Table 14.1).

Each day that chemical pesticides are avoided, biocontrol is better preserved, a service that has great impact on the stink bug population. Thus, a reduction of applied chemicals is of great value for the management of stink bug outbreaks on soybean fields, especially considering the usually low selectivity of chemical insecticides available on the market to be used to control stink bugs.

It is important to mention that the recorded reduction of insecticide use, together with the longer period of time until the first insecticide application for stink bug control, benefit the preservation of biological control agents in the agroecosystem without any yield reduction. IPM fields had a 75-kg higher average yield compared with non-IPM fields. Yields in IPM fields were between 36 kg/ha and 144 kg/ha higher in 2020/21 and 2015/16, respectively because of the better pest management. In addition, areas adopting IPM had lower stink bug control costs. Control costs, which were calculated and transformed to their equivalent in value of kg soybean per ha for each crop season, were from 33.8 kg/ha (2020/21) to 55.7 kg/ha (2017/18) lower for IPM fields compared with non-IPM fields. Taking together the increase in yield and the reduction of costs, adopting IPM for the stink bug problem on soybean resulted in a higher profit of 120.7 kg/ha (45.7 kg/ha of reduction of stink bug control costs +75.0 kg/ha of increase on soybean yield) (Table 14.1). Adopting IPM in large-scale crops can therefore be considered one of the most sustainable stink bug management strategy due to the combined results of reducing pesticide use (preserving biocontrol agents) and increasing farmers' profits.

4 Biological Control of Stink Bugs

Several species of natural enemies of *E. heros* have been recorded in Brazil, one of the most important soybean producers in the world (Table 14.2). Eggs, late nymphal stages and adults of stink bugs are most commonly attacked by parasitoids, but eggs and early nymph stages can also be attacked by predators (Ademokoya et al. 2022).

There is a significant number of microhymenopteran species that are parasitoids of stink bugs, most of which are egg parasitoids. Twenty-three species of stink bug egg parasitoids have already been identified on soybean (Bueno et al. 2022), and these natural enemies are considered the most important natural agents of mortality for this group of pests (Koppel et al. 2009; Laumann et al. 2010; Bueno et al. 2012). They are often responsible for naturally keeping stink bug populations below economic injury levels (without the need to adopt any additional control strategy)

Table 14.2 *Euschistus heros* parasitoids of importance on soybean in Brazil

Species	[a]Stage(s) attacked	Parasitism	References
Hexacladia smithii (Hymenoptera: Encyrtidae)	A	0.6–90%	Corrêa-Ferreira et al. (1998), Nunes (2000), Turchen et al. (2015), and Barakat et al. (2022)
Telenomus podisi Ashmead (Hymenoptera: Scelionidae)	E	43.4%	Corrêa-Ferreira and Moscardi (1995)
	E	59.3–62.5%	Godoy et al. (2005)
Trissolcus basalis (Wollaston) (Hymenoptera: Scelionidae)	E	10.6%	Corrêa-Ferreira and Moscardi (1995) and Cingolani et al. (2014)
Trissolcus urichi (Crawford) (Hymenoptera: Scelionidae)	E	No information	Cingolani et al. (2014) and Paz-Neto et al. (2015)
	E	No information	Sousa et al. (2019)

[a]*E* egg, *A* adult

(Bueno et al. 2022). They can also be enhanced in augmentative biological control (ABC) programs what is described in more detail in Chap. 4.

Not only releasing biocontrol agents in ABC programs but also preserving them are quite important to reduce stink bug outbreaks. Conservation biological control (CBC) strategies can be adopted to preserve or even promote the natural increase of biocontrol agents in the field, which has been proved to promote good yields as well as environmental protection. CBC is crucial to the establishment of more resilient and balanced agroecosystems. In this concept, actions that consider the system as a whole, and improve agriculture through what is today known as landscape architecture (LA), are required (Martin et al. 2019). Thus, it integrates landscape management, combined with food production and with the conservation of ecosystem services, in particular biodiversity (Patterson 2004). For example, the simple maintenance of wild vegetation, especially flowering plants, on the periphery of crops, alongside roads, or along border fences, contributes to the preservation of various natural enemies, as well as parasitoids released in large quantities to control specific pests (Jahnke and Da Silva 2021).

In addition, the diversification of plant strata in windbreak barriers, which are widely used in different agricultural crops, is another example of a simple measure of CBC that could be easily adopted. Field windbreaks designed to work in CBC must include different strata of plants, varying in height and structure. In addition, flowering plants or plants with extra-floral nectary increase resources for natural enemies present in the landscape (Podhrázská et al. 2021). Those are topics that need to be better studied for soybean, including the identification of the best plant species to be used, and whether these landscape alterations could also lead to an increase of the number of pests.

5 Use of Selective Pesticides to Control *E. heros*

Despite the importance of adopting ABC strategies and ETs for stink bug management as previously discussed, the use of chemical pesticides is still necessary within the current soybean system, at least in a short-term scenario. In this context, multiple pest control tactics were included in the IPM concept (Fig. 14.3), demonstrating that the use of more selective and harmless insecticides is compatible with biological control, fitting inside IPM and providing a sustainable stink bug management (Croft 1990; Bueno et al. 2011, 2021). Thus, selective insecticides are of great value for stink bug management (Torres and Bueno 2018). A significant advantage of these products is their effectiveness with minimal side effects for natural enemies of stink bugs (Silva et al. 2022). Information obtained from research on the selectivity of pesticides for beneficial organisms is useful for the success of soybean IPM, and compounds of lower toxicity should be prioritized by soybean farmers whenever possible. Furthermore, it is important not only to consider the selectivity of the insecticides used against *E. heros* but also the selectivity of any other pesticides applied over soybean fields.

Among the most commonly used and studied pesticides to be used on soybean, biopesticides are generally the most harmless for biocontrol agents of importance for stink bugs, followed by fungicides, herbicides, and insecticides from the insect growth regulator (IGR) group. Despite some harm caused by ethiprole, this insecticide is still more selective than pyrethroids, carbamates or organophosphates (Table 14.3).

It is important to mention that despite IGR's being regarded as less harmful to beneficial insects (Costa et al. 2023) compared with other chemical groups, negative side-effects have been also reported, especially concerning their impact on immatures of different biocontrol species (Santos et al. 2006). Although not killing adults, lufenuron, for example, significantly reduced egg viability in *Chrysoperla externa* Hagen (a predator) after ingestion of the insecticide. Therefore, it was classified as harmful (class 4) for this stage of development (Bueno and Freitas 2004).

Insect growth regulators are used to control Lepidoptera on soybean but not to control stink bugs. However, it is important to emphasize that when caterpillars need to be controlled on soybean, the use of IGR's is preferable to more harmful pesticides because it helps to protect natural enemies of importance to stink bugs as well. Among the recommended insecticides against stink bugs, biological ones are the most selective products, and among the chemical products, ethiprole was noted to cause less harm to biocontrol agents compared with the ready-to-use mixture of pyrethroids and neonicotinoids (Silva et al. 2022).

Not only physiological but also ecological selectivity should be taken into consideration when applying pesticides (Torres and Bueno 2018). Ecological selectivity aims to reduce the exposure of natural enemies to the most harmful insecticides, which can be done through applications that consider the seasonality or circadian cycle of pests and natural enemies (Heimpel and Mills 2017). Another strategy

Table 14.3 Classification of insecticide selectivity to *Telenomus podisi* according to the "International Organization for Biological Control" (IOBC) on different days after spraying (DAS) or days after adult emergence (DAE). Class 1 = Harmless; Class 2 = Slightly harmful; Class 3 = Moderately harmful; Class 4 = Harmful

	Parasitoid development stage					
	Pupae			Adult		
Treatment (grams/hectare)	Sprayed pupae	1 DAE	3 DAE	1 DAS	3 DAS	References
Beta-cyfluthrin 7.5	1	1	1	1	1	Stecca et al. (2018)
Beta-cyfluthrin 12.5 + imidacloprid 100	1	1	1	4	4	Stecca et al. (2018)
Bifenthrin 5	1	1	1	4	4	Stecca et al. (2018)
Chlorantraniliprole 10/15/20/30/50	−1	1/1/1/1/1	1/1/1/1/2	-	–	Silva et al. (2018) and Stecca et al. (2018)
Chlorantraniliprole 7.5/10/20/30 + lambda-cyhalothrin 3.75/5/10/15	1/1/1/1	1/1/1/1	1/1/1/1	3/-/-/-	3/-/-/-	Silva et al. (2018) and Stecca et al. (2018)
Chlorfluazuron 37.5	1	1	1	1	1	Stecca et al. (2018)
Chlorpyrifos 480/640/960	2-1-2	3-1-2	3-2-2	0-4-2	0-4-2	Silva et al. (2018, 2022) and Stecca et al. (2018)
Deltamethrin 7.5	1	1	1	2	2	Stecca et al. (2018)
Ethiprole 100/133.3	1/1	1/1	1/1	1/1	1/1	Silva et al. (2022)
Flubendiamide 33.6	1	1	1	1	1	Stecca et al. (2018)
Lambda-cyhalothrin 7.5	1	1	1	3	3	Stecca et al. (2018)
Lufenuron 7.5	1	1	1	1	1	Stecca et al. (2018)
Methoxyfenozide 21.6	1	1	1	1	1	Stecca et al. (2018)
Novaluron 7.5	1	1	1	1	1	Stecca et al. (2018)
Spinetoran 3	1	1	1	1	1	Stecca et al. (2018)
Spinosad 24	1	1	1	1	1	Stecca et al. (2018)
Sulphoxaflor 13.3/20 + lambda-cyhalothrin 20/30	3/3	½	2/2	3/3	3/4	Silva et al. (2022)
Tebufenozide 30	1	1	1	1	1	Stecca et al. (2018)
Teflubenzuron 7.5	1	1	1	1	1	Stecca et al. (2018)

includes using precision agriculture tools, with pesticides being applied only in areas where the pest is occurring, avoiding spaces in the crop where the ET has not been reached (Moses-Gonzales and Brewer 2021). For stink bugs on soybean in particular, kitchen salt (NaCl) can be added to the insecticide in the proportion of 0.5% for prolonging stink bug feeding and therefore prolonging the insect contact with insecticides (Niva and Panizzi 1996). This mixture can be applied in stripes, and therefore reduces the area at which the insecticide is sprayed (Corso 1990). This is another important example of ecological selectivity to help preserving natural biological control while still guaranteeing an acceptable insecticide performance.

6 Final Considerations

Higher productivity was always recorded when IPM principles were adopted to manage stink bugs, including *E. heros*, with lower use of insecticides. Therefore, the intensification of soybean-IPM adoption should be incentivized in all possible ways since it is probably the only way to keep soybean production sustainable throughout the years. With a global population that is expected to reach 10 billion in 2050, the increasing demand for food will put more pressure on agriculture. Brazil has all the climate conditions of temperature, humidity and light favorable to the crop. However, the intensification of soybean crop areas will also favor stink bug outbreaks as well as other pests. Therefore, it is necessary to develop greener pest management tools such as new biological control products, RNAi, and other modern strategies that control pests with minimal impact on natural biological control. In Brazil, biological control programs with the use of bioinputs have been increasing in the last 5 years at a rate from 20% to 30% a year, mainly with applications of microorganisms (fungi, bacteria and viruses). It is important to mention that on soybean there are several species of natural enemies that contribute to the regulation of pest population growth, whether stink bugs or caterpillars. Therefore, studies about the compatibility of chemical and biological methods in management programs for these pests are of great importance. Not only the adoption of ABC and CBC strategies but also of other sustainable management will increase in the IPM areas in near future. This association of safe sustainable strategies is crucial for the success of stink bug management on soybean.

References

Ademokoya B, Athey K, Ruberson J (2022) Natural enemies and biological control of stink bugs (Hemiptera: Heteroptera) in North America. Insects 13:932

Andrews G, Daves C, Koger T, Reed J, Burdine B, Dodds D, Larson E, Robbsins J, Catchot A, Gore J, Musser F, Smith J, Cook D, Jackson R, Mckibben P, Winters S (2009) Cotton insect management. In: Catchot A (ed) Insect control guide for agronomic crops. Mississippi State University, Starkville, pp 2–21

Barakat MC, Aquino DA, Cingolani MF (2022) Potential of *Hexacladia smithii* (Hymenoptera: Encyrtidae) to parasitize *Piezodorus guildinii* (Hemiptera Pentatomidae) adults. Bull Insect 75:177–182

Borges M, Schmidt FGV, Sujii ER, Medeiros MA, Mori K, Zarbin PHG, Ferreira JTB (1998) Field responses of stink bugs to the natural and synthetic pheromone of the Neotropical brown stink bug, *Euschistus heros,* (Heteroptera: Pentatomidae). Physiol Entomol 23:202–207

Bueno AF, Freitas S (2004) Effect of the insecticides abamectin and lufenuron on eggs and larvae of *Chrysoperla externa* under laboratory conditions. BioControl 49:277–283

Bueno AF, Batistela MJ, Bueno RCOF, França-Neto JB, Nishikawa MAN, Filho AL (2011) Effects of integrated pest management, biological control and prophylactic use of insecticides on the management and sustainability of soybean. Crop Prot 30:937–945

Bueno AF, Sosa-Gómez DR, Côrrea-Ferreira BS, Moscardi F, Bueno RCOF (2012) Inimigos naturais das pragas da soja. In: Hoffmann-Campo CB, Côrrea-Ferreira BS, Moscardi F (eds) Soja: Manejo Integrado de Insetos e outros Artrópodes-Praga. Embrapa, Brasília, pp 493–630

Bueno AF, Paula-Moraes SV, Gazzoni DL, Pomari AF (2013) Economic thresholds in soybean-integrated pest management: old concepts, current adoption, and adequacy. Neotrop Entomol 42:439–447

Bueno AF, Bortolotto OC, Pomari-Fernandes A, França-Neto JB (2015) Assessment of a more conservative stink bug economic threshold for managing stink bugs in Brazilian soybean. Crop Prot 71:132–137

Bueno AF, Panizzi AR, Hunt TE, Dourado PM, Pitta RM, Gonçalves J (2021) Challenges for adoption of integrated pest management (IPM): the soybean example. Neotrop Entomol 50:5–20

Bueno AF, Parra JRP, Colombo FC, Colmenarez YC, Narde BVF, Pereira FF (2022) Manejo de pragas com parasitoides. In: Meyer MC, Bueno AF, Mazaro SMM, Silva JC (eds) Bioinsumos na Cultura da Soja. Embrapa, Brasília, pp 417–434

Bueno AF, Colmenarez YC, Carnevalli RA, Sutil WP (2023) Benefits and perspectives of adopting soybean-IPM: the success of a Brazilian programme. Plant Health Cases 1:1–16

Carnevalli RA, Oliveira AB, Gomes EC, Possamai EJ, Silva GC, Reis EA, Roggia S, Prando AM, Lima D (2022) Resultados do manejo integrado de pragas da soja na safra 2021/2022 no Paraná. Embrapa Soja, Londrina, Doc 448, p 43

Cingolani MF, Greco NM, Liljesthröm GG (2014) Egg parasitism of *Piezodorus guildinii* and *Nezara viridula* (Hemiptera: Pentatomidae) in soybean, alfalfa and red clover. Rev Fca Uncuyo 46:15–27

Conte O, Oliveira FT, Harger N, Corrêa-Ferreira BS, Roggia S, Prando AM, Seratto CD (2016) Resultados do manejo integrado de pragas da soja na safra 2015/16 no Paraná. Londrina, EMBRAPA-CNPSo, Doc 375, p 59

Conte O, Oliveira FT, Harger N, Corrêa-Ferreira BS, Roggia S, Prando AM, Seratto, CD (2017) Resultados do manejo integrado de pragas da soja na safra 2016/17 no Paraná. Londrina, EMBRAPA-CNPSo, Doc 394, p 70

Conte O, Oliveira FT, Harger N, Corrêa-Ferreira BS, Roggia S, Prando AM, Seratto CD (2018) Resultados do manejo integrado de pragas da soja na safra 2017/18 no Paraná. Londrina, EMBRAPA-CNPSo, Doc 402, p 66

Conte O, Oliveira FT, Harger N, Corrêa-Ferreira BS, Roggia S, Prando AM, Possmai EJ, Reis EA, Marx EF (2019) Resultados do manejo integrado de pragas da soja na safra 2018/19 no Paraná. Embrapa Soja, Londrina, Doc 416, p 63

Conte O, Possamai EJ, Silva GC, Reis EA, Gomes EC, Corrêa-Ferreira BS, Roggia S, Prando AM (2020) Resultados do manejo integrado de pragas da soja na safra 2019/20 no Paraná. Embrapa Soja, Londrina, Doc 431, p 66

Corrêa-Ferreira BS, Azevedo J (2002) Soybean seed damage by different species of stink bugs. Agric For Entomol 4:145–150

Corrêa-Ferreira BS, Moscardi F (1995) Seasonal occurrence and host spectrum of egg parasitoids associated with soybean stink bugs. Biol Control 5:196–202

Corso IC (1990) Uso de sal de cozinha na redução da dose de inseticida para controle de percevejos da soja. Embrapa Soja, Londrina, Com Téc 45:1–7

Corrêa-Ferreira BS, Nunes MC, Uguccioni LD (1998) Ocorrência do parasitóide *Hexacladia smithii* Ashmead em adultos de *Euschistus heros* (F.) no Brasil. An Soc Entomol Brasil 27:495–498

Costa MA, Farias ES, Andrade ED, Carvalho VC, Carvalho GA (2023) Lethal, sublethal and transgenerational effects of insecticides labeled for cotton on immature *Trichogramma pretiosum*. J Pest Sci 96:119–127

Croft BA (1990) Arthropod biological control agents and pesticides, 2nd edn. Wiley, New York, 826pp

Desneux N, Decourtye A, Delpuech JM (2007) The sublethal effects of pesticides on beneficial arthropods. Annu Rev Entomol 52:81–106

Godoy KB, Galli JC, Ávila CJ (2005) Parasitismo em ovos de percevejos da soja *Euschistus heros* (Fabricius) e *Piezodorus guildinii* (Westwood) (Hemiptera: Pentatomidae) em São Gabriel do Oeste, MS. Cien Rural 35:455–458

Heimpel GE, Mills NJ (2017) Biological Control: Ecology and Applications. Cambridge University Press, Cambridge, 386pp
Higley LG, Pedigo LP (1996) The EIL concept. In: Higley LG, Pedigo LP (eds) Economic Threshold for Integrated Pest Management. University of Nebraska Press, Lincoln, pp 9–21
Jacquet F, Jeuffroy MH, Jouan J, Le Cadre E, Litrico I, Malausa T, Reboud X, Huyghe C (2022) Pesticide-free agriculture as a new paradigm for research. Agron Sustain Dev 42:8
Jahnke SM, Da Silva GS (2021) Challenges in the applied use of parasitoids to control agricultural pests. In: Awasthi LP (ed) Biopesticides in organic farming: recent advances. CRC Press, pp 261–266
Janssen A, van Rijn P (2021) Pesticides do not significantly reduce arthropod pest densities in the presence of natural enemies. Ecol Lett 24:2010–2024
Kogan M (1998) Integrated pest management: historical perspectives and contemporary developments. Annu Rev Entomol 43:243–270
Koppel AL, Herbert DA Jr, Kuhar TP, Kamminga K (2009) Survey of stink bug (Hemiptera: Pentatomidae) egg parasitoids in wheat, soybean, and vegetable crops in Southeast Virginia. Environ Entomol 38:375–379
Kuldna PK, Peterson H, Poltimäe J (2009) An application of DPSIR framework to identify issues of pollinator loss. Ecol Econ 69:32–42
Laumann RA, Moraes MC, Silva JP, Vieira AM, Silveira SD, Borges M (2010) Egg parasitoid wasps as natural enemies of the neotropical stink bug *Dichelops melacanthus*. Pesq Agropec Bras 45:442–449
Lee R, den Uyl R, Runhaar H (2019) Assessment of policy instruments for pesticide use reduction in Europe; learning from a systematic literature review. Crop Prot 126:104929
Maciel RMA, Bueno AF (2022) The role of integrated pest management for sustainable food production: the soybean example. In: Galanakis CM (ed) Biodiversity, functional ecosystems and sustainable food production. Springer Nature, Switzerland, pp 117–139
Martin EA, Dainese M, Clough Y, Báldi A, Bommarco R, Gagic V, Garratt MPD, Holzschuh A, Kleijn D, Kovács-Hostyánszki A, Marini L, Potts SG, Smith HG, Al Hassan D, Albrecht M, Andersson GKS, Asís JD, Aviron S, Balzan MV, Baños-Picón L, Bartomeus I, Batáry P, Burel F, Caballero-López B, Concepción ED, Coundrain V, Dänhardt J, Diaz M, Diekötter T, Dormann CF, Duflot R, Entling MH, Farwig N, Fischer C, Frank T, Garibaldi LA, Hermann J, Herzog F, Inclán D, Jacot K, Jauker F, Jeanneret P, Kaiser M, Krauss J, Féon VL, Marshall J, Moonen A-C, Moreno G, Riedinger V, Rundlöf M, Rusch A, Sheper J, Schneider G, Schüepp C, Stutz S, Sutter L, Tamburini G, Thies C, Tormos J, Tscharntke T, Tshumi M, Uzman D, Wagner C, Zubair-Anjum M, Steffan-Dewenter I (2019) The interplay of landscape composition and configuration: new pathways to manage functional biodiversity and agroecosystem services across Europe. Ecol Lett 22:1083–1094
Moses-Gonzales N, Brewer MJ (2021) A special collection: drones to improve insect pest management. J Econ Entomol 114:1853–1856
Niva CC, Panizzi AR (1996) Efeitos do cloreto de sódio no comportamento de *Nezara viridula* (L.) (Heteroptera: Pentatomidae) em vagem de soja. An Soc Entomol. Brasil 25:251–257
Nunes MC (2000) Efeito do parasitismo de *Hexacladia smithii* Ashmead (Hymenoptera: Encyrtidae) na capacidade reprodutiva e no dano de *Euschistus heros* (Fabricius) (Hemiptera: Pentatomidae) causado a soja. Dissertação Mestrado, Universidade Federal do Paraná, PR, Brazil
Oliveira AB, Gomes EC, Possamai EJ, Silva GC, Reis EA, Roggia S, Prando AM, Conte O (2022) Resultados do manejo integrado de pragas da soja na safra 2020/2021 no Paraná. Embrapa Soja, Londrina, Doc 443, p 68
Paiva HC (2024) Armadilhas para amostragem e controle de percevejos na cultura da soja. Master's Thesis, State University of Londrina (UEL), Londrina, PR, Brazil, 64p
Panizzi AR, Corrêa-Ferreira BS (1997) Dynamics in the insect fauna adaptation to soybean in the tropics. Trends Entomol 1:71–88
Panizzi AR, Slansky F Jr (1985) Review of phytophagous pentatomids (Hemiptera: Pentatomidae) associated with soybean in the Americas. Fla Entomol 68:184–203

Panizzi AR, Corrêa BS, Gazzoni DL, Oliveira EB, Newman GG, Turnipseed SG (1977a) Insetos da soja no Brasil. EMBRAPA-CNPSo, Bol Téc 1:1–20
Panizzi AR, Corrêa BS, Newman GG, Turnipseed SG (1977b) Efeito de inseticidas na população das principais pragas da soja. An Soc Entomol Brasil 6:264–275
Panizzi AR, Smith JG, Pereira LAG, Yamashita J (1979) Efeitos dos danos de *Piezodorus guildinii* no rendimento e qualidade da soja. Anais Semin Nac Pesq Soja 1, Londrina, PR, 2:9–78
Panizzi AR, Lucini T, Aldrich JR (2022) Dynamics in pest status of phytophagous stink bugs in the Neotropics. Neotrop Entomol 51:18–31
Patterson EL (2004) Agriculture, landscape architecture, & ecological design: a foundation for collaboration between ecologists and landscape architects. PhD thesis, The University of Georgia, 103 pages
Paz-Neto AA, Querino RB, Margaría C (2015) Egg parasitoids of stink bugs (Hemiptera: Coreidae and Pentatomidae) on soybean and cowpea in Brazil. Fla Entomol 1:929–932
Pires CSS, Sujii ER, Schmidt FGV, Santos HM, Pais JSO, Borges M (2000) Potencial de utilização de armadilhas iscadas com o feromônio sexual do percevejo marrom, *Euschistus heros*: uma nova metodologia para o monitoramento populacional de percevejos praga da soja. Embrapa-Cenargen, Brasília, Circ Tec 7:1–24
Podhrázská J, Kucera J, Doubrava D, Doležal P (2021) Functions of windbreaks in the landscape ecological network and methods of their evaluation. Forests 12:67
Rossi MN, Fowler HG (2004) Spatial and temporal population interactions between the parasitoids *Cotesia flavipes* and Tachinidae flies: considerations on the adverse effects of biological control practice. J Appl Entomol 128:112–119
Santos AC, Bueno AF, Bueno RCOF (2006) Seletividade de defensivos agrícolas aos inimigos naturais. In: Pinto AS, Nava DE, Rossi MM, Malerbo-Souza DT (eds) Controle Biológico de Pragas na Prática, pp 221–227
Schmidt FGV, Pires CSS, Sujii ER, Borges M, Pantaleão DC, Lacerda ALM, Azevedo VCR (2003) Comportamento e captura das fêmeas de *Euschistus heros* em armadilhas iscadas com feromônio sexual. Embrapa-Cenargen, Brasília, Com Tec 93:1–4
Silva GV, de Freitas BA, Favetti BM, Neves PM (2018) Selectivity of chlorantraniliprole and lambda-cyhalothrin to the egg parasitoid *Telenomus podisi* (Hymenoptera: Platygastridae). Sem Ciênc Agrár 39:549–563
Silva DM, Carvalho GA, Souza WR, Bueno AF (2022) Toxicity of insecticides to the egg parasitoids *Telenomus podisi* and *Trissolcus teretis* (Hymenoptera: Scelionidae). Rev Brasileira Entomol 66:e2022003
Sousa KKA, Silva NNP, Querino RB, Silva PHS, Grazia J (2019) Diversity, seasonality, and egg parasitism of hemipteran (Coreidae and Pentatomidae) from a cowpea crop in northeastern Brazil. Fla Entomol 102:29–35
Sosa-Gómez DR, Corrêa-Ferreira BS, Kraemer B, Pasini A, Husch PE, Delfino Vieira CE, Reis Martinez CB, Negrao Lopes IO (2020) Prevalence, damage, management and insecticide resistance of stink bug populations (Hemiptera: Pentatomidae) in commodity crops. Agric For Entomol 22:99–118
Stecca CS, Bueno AF, Pasini A, Silva DM, Andrade K, Zirondi Filho DM (2018) Impact of insecticides used in soybean crops to the egg parasitoid *Telenomus podisi* (Hymenoptera: Platygastridae). Neotrop Entomol 47:281–291
Torres JB, Bueno AF (2018) Conservation biological control using selective insecticides – a valuable tool for IPM. BioControl 126:53–64
Turchen LM, Golin V, Favetti BM, Butnariu AR, Costa VA (2015) Natural parasitism of *Hexacladia smithii* Ashmead (Hymenoptera: Encyrtidae) on *Euschistus heros* (F.) (Hemiptera: Pentatomidae): new record from Mato Grosso State, Brazil. Arq Inst Biol 82:1–3
Van Lenteren JC, Bolckmans K, Köhl J, Ravensberg WJ, Urbaneja A (2018) Biological control using invertebrates and microorganisms: plenty of new opportunities. BioControl 63:39–59

Chapter 15
Case Study 2: *Halyomorpha halys* (Stål) in Europe

Lara Maistrello

1 Introduction

The brown marmorated stink bug (BMSB), *Halyomorpha halys* (Stål) (Hemiptera, Pentatomidae), native to Eastern Asia (Lee et al. 2013) is one of the most serious pests of fruit and seed crops globally (Haye and Weber 2017). The first detection of BMSB in Europe, believed to be in Switzerland in 2007 (Wermelinger et al. 2008), was actually in a light trap in Balzers, Liechtenstein (near the Swiss border) in 2004 (Arnold 2009). In 2008, the European Plant Protection Organization (EPPO) added *H. halys* to the "EPPO Alert List" (EPPO 2013) due to its polyphagy and potential damage to agricultural crops in Europe. However, the pest was removed from the list in November 2013 as no country requested any action. Meanwhile, BMSB had been reported in urban areas in Germany (Heckmann 2012), France (Callot and Brua 2013), Greece (Milonas and Partsinevelos 2014) and Italy (Maistrello et al. 2014, 2016a). In 2015, a short time after BMSB finding in Italy, the first serious damage occurred in orchards in Emilia Romagna (Bariselli et al. 2016; Maistrello et al. 2017), one of the most important fruit-producing regions in Europe (Regione Emilia Romagna 2022). Since then, BMSB has become the key pest of orchards throughout northern Italy, causing severe damage and completely disrupting current integrated pest management (IPM) programmes. A similar situation was observed in Georgia with hazelnut in 2016–2020. Since 2016, BMSB has been included in the unified list of quarantine pests (A1 list) of the Eurasian Economic Union (EAEU), which includes Russia, Armenia, Belarus, Kazakhstan and Kyrgyzstan,

L. Maistrello (✉)
Department of Life Sciences, Inter-Departmental Centre BIOGEST-SITEIA, University of Modena and Reggio Emilia, Reggio Emilia, Italy

NBFC, National Biodiversity Future Center, Palermo, Italy
e-mail: lara.maistrello@unimore.it

A. F. Bueno, A. R. Panizzi (eds.), *Stink Bugs (Hemiptera: Pentatomidae) Research and Management*, Entomology in Focus 9,
https://doi.org/10.1007/978-3-031-69742-5_15

and since 2019 it has also been classified as a pest of quarantine significance (A1 list) for Ukraine.

Less than 20 years after first being recorded, BMSB is now found in almost every country in continental Europe. It poses a serious threat to agricultural production, particularly fruit and hazelnuts, mainly in southern and south-eastern Europe. In all cases, it first appeared in urban areas as a domestic nuisance, then damaged vegetable gardens and, within a few years, caused significant crop losses in the countryside. This is very similar to what has happened in the United States. BMSB was first detected in Pennsylvania in 1996–2002 (Hoebeke and Carter 2003). It quickly invaded surrounding states and became a nuisance pest in the Mid-Atlantic region (2002–2009), where major losses to orchards and other crops were reported in 2010 (Leskey et al. 2012).

Originally from East Asia, BMSB has spread to almost all continents in recent years. It is currently present in North America (USDA-NIFA SCRI 2021), South America (Faúndez and Rider 2017), Europe, and North Africa (van der Heyden et al. 2021). The fast spread of BMSB worldwide is due to the hitchhiking nature of overwintering adults, which seek shelter into objects (e.g., containers, packs, vehicles, luggage, etc.) and are carried with trade and movements of people (Maistrello et al. 2018). Finally, overwintering in man-made structures also makes these insects annoying household pests. This close association with humans has been used in citizen science surveys for early detection and track their spread in real time, while raising public awareness and education.

This chapter aims to provide an update of studies carried out on BMSB in Europe and/or by European researchers. Information was gathered from scientific literature, e-mail survey, and interviewing researchers and managers of plant protection services in different European countries. The purpose of the requests was to obtain up-to-date information for each country on the following aspects: current presence and distribution of BMSB; existence of any initiative to monitor its distribution; biology and ecology, with particular reference to surveys of natural enemies; whether BMSB is considered a serious crop pest; the crops most affected and estimated damage; and management using biocontrol with native or exotic parasitoids.

2 Invasion History in Europe

2.1 BMSB in Europe: The Importance of Human-Assisted Spread

A survey of all countries in continental Europe and neighbouring regions was carried out to check for the presence of BMSB, using extensive literature search and records using biodiversity platforms INaturalist (iNaturalist contributors 2023a) and GBIF (Global Biodiversity Information Facility) (GBIF 2023a).

Records have been reported from 43 countries (Table 15.1), including South Caucasus (Abkhazia) and countries in the Mediterranean area (Algeria) and West Central Asia (Uzbekistan). BMSB is currently present in 80% of the European countries. Since 2004, when it was first found in Liechtenstein, it has rapidly colonized virtually the entire continent in less than 20 years. The only countries where it is not known to be present in Europe (as of November 2023) are Andorra, Belarus, Cyprus, Estonia, Iceland, Ireland, Kosovo, Latvia, Lithuania, Monaco and Norway. In Iceland, 36 records have been reported between 2003 and 2023 in Reykjavík and many other locations, all related to specimens found in imported goods and/or luggage, near sea ports, airports and warehouses (M. Alfreðsson, p.c.). However, BMSB has not become established in the country, probably due to low summer temperatures (T. Ísberg and M. Alfreðsson, p.c.). Reports from the Swedish Plant Health Service indicate that all findings consisted solely of adult specimens. Despite the installation of pheromone traps near finding sites and orchards in the southern region of the country, BMSB were never caught in the field. Therefore, there is no evidence indicating that BMSB has become established in Sweden (J. Jansson, p.c.). Similarly, despite reports of findings, there is no indication that BMSB is established in Denmark and Finland.

Most of the first BMSB records, considering only those for which it was possible to trace the context (61% of the total), occurred in urban settings (46%), mainly inside houses, warehouses and other buildings (Table 15.1). This percentage rises to 83%, considering all records that BMSB in man-made structures facilitates discovery by people and thus also explains the success of citizen science initiatives. Records in agricultural settings decrease to 13% to 37%, based on field monitoring using traps with aggregation pheromones, specially activated following initial reports in urban settings, gardens or vegetable gardens.

Occasionally, BMSB has been found inside cars, tractors, in railway carriages and stations, in ports and airports (Malek et al. 2018, 2019; Maistrello et al. 2018). Hitchhiking via major routes is the most likely hypothesis to explain the rapid spread of BMSB to different countries on the European continent. BMSB's invasion into northern Italy along the main communication routes is described (Maistrello et al. 2016a, 2018; Malek et al. 2018, 2019). Similarly, the spread of BMSB in Germany also occurred along major routes between major cities (Hess et al. 2022). The first findings in Russia in Sochi in 2013 seem to be associated with the arrival of ornamental plant material from Europe (especially Italy, where it was already established) for the staging of the XXII Olympic Winter Games (Musolin et al. 2018). Another example is the invasion in Serbia, where it was first recorded repeatedly in 2015 close to the Bucharest-Belgrade railway line, suggesting BMSB came from Bucharest (where it had been recorded since 2014) by railroad cars or in freight (Musolin et al. 2018). The detection along major roads and/or railway lines, together with the many occurrences in transport and transit areas, strongly supports the hypothesis that human-assisted spread plays a major role in the invasive potential of BMSB.

Table 15.1 History of the invasion of *Halyomorpha halys* in Europe (including surrounding areas) and within each country, and citizen science initiatives to track its spread. For each country, the year of the first record is given in bold, followed by subsequent records for different areas of the same country (region/department/province). Only the first finding for each region/area is reported and referenced. Asterisk (*) on the name of a country indicates that there is not yet evidence indicating establishment of BMSB. The right part of the table refers to the existence, for each country, of citizen science initiatives aimed at collecting data on the spread of the insect on the national territory. Context: *U* urban (cities, villages, buildings), *A* agricultural (crops), *G* gardens, botanical gardens, ornamental plants, *X* unknown. Citizen science initiatives type: *ACS* Active Citizen Science= active request (publicized in various ways) by the organizers to contribute to reporting cases of *H. halys* findings (pictures, dead specimens) via email, websites and/or smartphone applications; *PCS* Passive Citizen Science: retrieval of presence data (photographs with date and location) from existing online platforms. In all cases, the records had been/are being validated by experts

First detections					Citizen science initiatives		
Year	Country	Location	References	Context	Year	Type	Reference
2004	**Liechtenstein**	Balzers	Arnold (2009)	U	–	–	
2004	**Switzerland**	Zurich	Haye et al. (2014b) later discovery	U	2012–19	ACS: Multimedia survey, data collected via a webpage (www.halyomorphahalys.com) where an updated map was available	Haye et al. (2014b)
2007		Erlenbach	Wermelinger et al. (2008)	UGA			
2010		Zurich spread	Mueller et al. (2011)	U			
2011		11 cantons/26	Haye et al. (2014b)	UA			
2018		20 cantons/26	Claerebout et al. (2018)	X			
2019		All bioclimatic suitable areas	Stoeckli et al. (2020)	UGA			
2007	**Italy**	Genova (Liguria)	Maistrello and Dioli (2014) later discovery	X	2013–16	ACS: Multimedia survey, data collected via email by L. Maistrello	Maistrello et al. (2016a, 2018)
2012		Magreta (Emilia)	Maistrello et al. (2014) first official record	U			
2013		Locations in Emilia, Lombardy, Piedmont	Pansa et al. (2013), Maistrello et al. (2016a)	UGA			

2014		Central Alps, Veneto, Romagna, Friuli Venezia Giulia, Lazio Marche, Abruzzo	Maistrello and Dioli (2014), Cianferoni et al. (2018), Maistrello et al. (2018)	UGA	2018-ongoing	ACS: Smartphone application (BugMap)	Malek et al. (2018, 2019)
2015		Valtellina, Tuscany, Calabria	Limonta et al. (2016), Cianferoni et al. (2018), Maistrello et al. (2018)	UGA			
2016		Valdaosta, Trentino, Alto Adige, Puglia Campania, Sardinia	Dioli et al. (2016), Claerebout et al. (2018), Cianferoni et al. (2018), Maistrello et al. (2018), Loru et al. (2018)	U UG			
2017		Umbria, Basilicata, Sicily	Carapezza and Lo Verde (2017), Claerebout et al. (2018), Yonow et al. (2021)	UGA UGA			
2018		Molise	P. Sciarretta p.c.	X			
2011	**Germany**	Konstanz	Heckmann (2012)	U	2017-ongoing	ACS: Multimedia survey, data collected by LTZ Augustenberg or JKI, and by CS platforms: iNaturalist, naturgucker.de, GBIF.org	Hess et al. (2022)
2015		Baden Württemberg	Haye and Zimmermann (2017)	UG			
2016		Berlin	Haye and Zimmermann (2017)	UG			
2016		Rhine-Palatinate	Hanselmann (2017)	U			
2018		Hessen, North Rhine-Westphalia, Bavaria	Claerebout et al. (2018)	X			
2019		Cologne	Kott (2019)	G			

(continued)

Table 15.1 (continued)

First detections					Citizen science initiatives		
Year	Country	Location	References	Context	Year	Type	Reference
2020		Brandenburg, Sachsen Niedersachsen, Rhine-Pfalz	Hess et al. (2020)	X			
2021		Hamburg	Hartung et al. (2022)	UG			
2022		All landers	Hess et al. (2022)	UA			
2011	**Belgium**	Soignies (Hainaut)	Claerebout et al. (2018)	X	2011-ongoing	ACS-PCS; data collected from www.observation.org and www.pcfruit.be (fruit growers are encouraged to report records)	S. Claerebout, T. Belien p.c.
2017		East Flanders		X			
2023		Many sites	GBIF (2023b)	X			
2011	**Greece**	Athens	Milonas and Partsinevelos (2014)	U	2017-ongoing	PCS; data collected from the specific project on the CS platform iNaturalist (iNaturalist contributors 2023b)	S. Andreadis p.c.
2011		Thessaloniki	(Andreadis, p.c.)	U			
2015–18		Crete, Attica, Chalkidiki, Central Macedonia, Salamis, Peloponnese, Thessaly	Claerebout et al. (2018)	X			

2017		Pieria	Andreadis et al. (2018)	A			
2018–19		Imathia	Damos et al. (2020)	A			
2019		Lesbos isl.	Cianferoni et al. (2019)	X			
2020		N. Greece (Naoussa)	Koutsogeorgiou et al. (2022)	A			
2021		Rhodes isl.	Corsini-Foka et al. (2021)	U			
2023		Many sites, mainly in the North	Inaturalist, project "BMSB in Greece" (iNaturalist contributors 2023b)	X			
2012	**France**	Strasbourg (Bas-Rhin)	Callot and Brua (2013)	U	2012–2023	ACS: Multimedia survey, data collected from smartphone applications (Agiir by INRAE, http://ephytia.inrae.fr/fr/P/128/Agiir; INPN-Espèces by MNHN: https://inpn.mnhn.fr/espece/jeudonnees/21092) and by the CS platform GBIF (GBIF 2023a)	Chartois et al. (2021), Streito et al. (2021)
2013		Paris (Ile de France)	Garrouste et al. (2015)	U			
2015		Alpes-Maritimes, Haute-Garonne	Claerebout et al. (2018) Claerebout et al. (2018)	X X			
2016		Haut-Rhin	Claerebout et al. (2018)	X			
2016		Corse-du-Sud	Maistrello et al. (2018)	U			
2017		Savoie	Claerebout et al. (2018)	X			

(continued)

Table 15.1 (continued)

First detections					Citizen science initiatives		
Year	Country	Location	References	Context	Year	Type	Reference
2018		Bouches-du-Rhône, Drôme, Gironde, Gers, Haute-Corse, Hauts-de-Seine, Hérault, Isère, Languedoc-Roussillon, Meurthe-et-Moselle, Midi-Pyrénées, Provence-Alpes-Côte d'Azur, Pyrénées-Orientales, Rhône, Saône-et-Loire, Val-de-Marne, Var, Vaucluse	Claerebout et al. (2018)	X			
2018		Loire-Atlantique, Gironde	Cherpitel and Casset (2018)	U			
2019		Haute-Savoie	Claerebout et al. (2018)	X			
2022		Updated map and history	Chartois et al. (2021), Streito et al. (2021)	X			
2013	**Hungary**	Budapest	Vétek et al. (2014)	U	2016-ongoing	ACS: Multimedia survey, data collected via a webpage (izeltlabuak.hu 2023)	Vétek et al. (2018) izeltlabuak.hu (2023)
2017–19		Many locations all over the country	Claerebout et al. (2018), Vétek et al. (2018)	G			
2023		Almost everywhere	izeltlabuak.hu (2023)	X			

2013	**Russia**	Sotchi	Mityushev (2016), Gapon (2016)	UG	2014-ongoing	ACS: (1) annual requests to agronomists, biologists in border regions; data collected via email; (2) request to citizens to collect samples to study haplotype diversity via the website (Musolin 2015)	D. Musolin, N. Karpun, S. Reznik p.c.
2015–17		Krasnodar, Sotchi districts	Musolin et al. (2018)	UGA			
2019		Adygea, Crimea, Rostov oblast, Stavropol	Mityushev (2020, 2021)	UG			
2022		Stavropol, Rostov, Dagestan	Musolin et al. (2022b)	UGA			
2013	**Sweden***	Västervik (Småland)	Liljeblad (2023) interception on goods from USA	U	2016	PCS: data collected from the Swedish biodiversity portal https://www.artportalen.se/ and reported on GBIF	Liljeblad (2023)
2016		Danderyd (Uppland)	Liljeblad (2023) first official record	U			
2018		Mjölby (Östergötland), Malmö (Skåne), Piteå (Norrbotten)	Liljeblad (2023)				
2019		Blekinge (Blekinge), Huddinge (Södermanland)	Liljeblad (2023)	U			

(continued)

Table 15.1 (continued)

First detections					Citizen science initiatives		
Year	Country	Location	References	Context	Year	Type	Reference
2020		Kiruna (Torne lappmark)	Liljeblad (2023)	U			
2022		Uppvidinge (Småland)	Liljeblad (2023)	U			
2014	**Abkhazia**	Not indicated	Neimorevets (2018)	X	–		
2015		All seven districts	Musolin et al. (2018)	A			
2016		Pitsunda	Gapon (2016)	U			
2014	**Romania**	Bucarest	Macavei et al. (2015)	UG	2016–17	ACS-PCS: Multimedia survey and CS platforms: iNaturalist, GBIF.org, EASIN, (https://easin.jrc.ec.europa.eu/); data collected by R. Ciceoi et al.	Ciceoi et al. (2017a)
2016		Ilfov, Giurgiu, Buzău Dâmbovița	Ciceoi et al. (2016, 2017d)	UGA			
2017		Mures	Claerebout et al. (2018)	X			
2017–2018		Southern, southeast, western areas	Ciceoi et al. (2017b), De Michele and Grozea (2018)	X			
2021		Northeast, northwest	Muntean and Grozea (2021), Grozea, p.c.	X			
2015	**Georgia**	Samegrelo-Verkhnyaya	Gapon (2016)	UA	2017-ongoing	PCS: data collected on the platform http://parosana.ge/	N. Kharabadze p.c.
2015		Adjara	Claerebout et al. (2018), Kharabadze (2022	X			
2015		Guria, Imereti	Burjanadze et al. (2020b), Kharabadze (2022)	A			
2016		Batumi	Dumbadze et al. (2019)	A			

2017		Kakheti	Claerebout et al. (2018)	X			
2019		Achara	Japoshvili et al. (2022)	A			
2019		Agruni, Lanchkhuti, Junetseri, Shida-Kartli	Talamas and Japoshvili (2022), Kharabadze (2022) N. Kharabadze p.c.	A U			
2019		Tbilisi	M. Kereselidze p.c, N. Kharabadze p.c.	U			
2023		Almost all the country	M. Kereselidze p.c, N. Kharabadze p.c.	X			
2015	**Serbia**	Vršac, Belgrade, Vatin	Šeat (2015)	UG	2015-ongoing	PCS: data collected from the FaceBook page "Insekti Srbije"	A. Konjevic p.c.
2016		Bela Crkva	Musolin et al. (2018)	UGA			
2017		Northern Serbia	A. Konjevic p.c.	UGA			
2018		Central Serbia	Claerebout et al. (2018)	X			
2019		South and East Serbia	A. Konjevic p.c.	UA			
2023		Many regions	Kavran et al. (2023)	X			
2015	**Austria**	Vienna	Rabitsch and Friebe (2015)	U	2015-ongoing	PCS: data collected from museums and on the science platform: https://observation.org/	G. Friebe p.c.
2016		Höchst, Lochau	Friebe et al. (2019)	U			
2018		Dornbirn	Friebe et al. (2019)	U			
2018		Thüringen, Lustenau	Claerebout et al. (2018)	X			
2019		Styria, Mödling	Claerebout et al. (2018)	X			

(continued)

Table 15.1 (continued)

First detections					Citizen science initiatives		
Year	Country	Location	References	Context	Year	Type	Reference
2016	**Vatican city**	Vatican city	D. Di Domenico p.c.	U	–		
2016	**Spain**	Girona	Dioli et al. (2016)	U	2017-ongoing	ACS: multimedia survey data collected by IRTA, Un. Barcelona, Dep. Agriculture of the Catalan Government; request to upload observations on CS platform Natusfera https://natusfera.gbif.es/taxa/21937-Halyomorpha-halys	A. Escudero p.c.
2017		Barcelona	Roca-Cusachs et al. (2018)	U			
2018		Gipuzkoa	Pagola-Carte and Zabalegui (2019)	U			
2018		Seville	Claerebout et al. (2018)	X			
2019		Madrid	Claerebout et al. (2018)	X			
2020		Lleida, Tarragona	Escudero-Colomar (2020)	U			
2021		Tenerife	van der Heyden and Petrovan (2023)	U			
2022		Zaragoza	Tanco (2022)	U			
2023		Pamplona	Tanco (2023)	U			
2023		Palma de Mallorca	Tanco and Canyelles (2023)	U			

2016	**Bulgaria**	Sofia	Simov (2016)	UG	2016-ongoing	PCS: data collected from the FaceBook page Насекомите и ентомолозите [Insects and entomologists]—https://www.facebook.com/groups/634529326 ACS: BMSB included in educational programs and toolkits of the National Museum of Natural History	N. Simov p.c.
2018		Plovdiv	Hristozova and Harizanova (2022)	AG			
2023		Almost all country, especially in lowlands, up to 1000–1300 m	N. Simov, p.c.	UGA			
2016	**Slovakia**	Štúrovo (Southern Slovakia)	Hemala and Kment (2017)	U	–		
2023		Established in many sites, especially in west	P. Kment p.c.	X			
2016	**Turkey**	Artvin (Kemalpaşa)	Gokturk and Tozlu (2019)	X	–		
2017		Artvin, Hopa, Arhavi	Güncan and Gümüş (2019), Gokturk and Tozlu (2019)	G			
2017–18		Istanbul	Çerçï and Koçak (2017), Cianferoni et al. (2019)	UA			
2018		Sakarya (Hendek)	Öztemiz et al. (2019)	A			

(continued)

Table 15.1 (continued)

First detections					Citizen science initiatives		
Year	Country	Location	References	Context	Year	Type	Reference
2019-20-21		Black Sea region: many locations in the provinces Artvin, Rize, Trabzon, Giresun, Ordu, Samsun	Ak et al. (2019, 2023a)	A			
2020		Bornova, Kavacık (Izmir)	Çerçi (2021), Ak et al. (2023b)	UG			
2022		Bartın, Bursa, Düzce, Canakkale, İzmit, Kırklareli, Sakarya, Sinop, Yalova, Zonguldak	Tuncer and Ozdemir (2023), Ak et al. (2023a)	X			
2016	**Ukraine**	Odessa	Uzhevskaya (2017) UkrBIN platform	U	2017-ongoing	PCS: data collected from the CS platform UkrBIN [Halyomorpha halys. Ukrainian Biodiversity Information Network. URL: https://ukrbin.com/index.php?id=347082&action=distribution.] and GBIF (https://www.gbif.org/occurrence/search?country=UA&taxon_key=4485843&year=1000,2017)	T. Stefanovska p.c.
2018–19		Carpathian Lowland, Transcarpathia, Zakarpattia	Szanyi et al. (2022)	UGA X			

2018–19		Sevastopol, Simferopol (Crimea)	Stryukova and Stryukov (2022)	GA			
2020		Kiev	GBIF platform	X			
2020		Kherson, Dnipropetrovsk Dnipro	Meshkova (2022), UkrBIN platform	X			
2021		Zaporizhia	UkrBIN platform	X			
2022		Mykolaiv, Lviv	GBIF platform	X			
2023		Zaporizhia city	First official record by Plant Protection Authority SSUFSCP	X			
2016–18	**Kazakhstan**	Almaty, Almaty Oblast	Temreshev et al. (2018)	UA	–		
2017	**Albania**	Vlora	Claerebout et al. (2018)	X	–		
2018		Berat	Claerebout et al. (2018)				
2019		Durrës	iNaturalist contributors (2023a)	X			
2021		Fier	iNaturalist contributors (2023a)	X			
2022		Korçë	iNaturalist contributors (2023a)	X			
2023		Gjirokastër	iNaturalist contributors (2023a)	U			
2017	**Slovenia**	Šempeter pri Gorici, Nova Gorica, Bilje, Miren, Kromberk	Rot et al. (2018)	UGA	2017-ongoing	ACS: multimedia survey as part of the project LIFE Artemis. Data collected from the online form: https://www.tujerodne-vrste.info/vrste/marmorirana-smrdljivka/	M. Rot p.c.

(continued)

Table 15.1 (continued)

First detections					Citizen science initiatives		
Year	Country	Location	References	Context	Year	Type	Reference
2018		Locations in Primorska, Murska Sobota, Central and Southeast Slovenia, Drava, Savinja regions	Claerebout et al. (2018), Rot et al. (2019)	UGA			
2017	**Croatia**	Rijeka (Primorje-Gorski Kotar)	Šapina and Jelaska (2018)	UG	–		
2018–19		Zagreb, Sveta Nedelja, Vukovar, Split, Opatija	Claerebout et al. (2018), Pajač Živković et al. (2021)	UA			
2022		Istria, Osijek-Baranja	I. Pajač Živcovič p.c.	A			
2017	**Azerbaijan**	Baku, Zagatala	Nuriyeva (2018)	U	–		
2018-19	**Portugal**	Pombal (Leiria)	Grosso-Silva et al. (2020)	A	2018-ongoing	ACS: Multimedia awareness Campaign, data collected by FLOWer lab, from the platforms https://www.facebook.com/groups/BMSBPT https://www.inaturalist.org/projects/percevejo-asiatico-halyomorpha-halys-pt	H. Gaspar p.c.
2020		Locations in Braga, Coimbra, Lisbon	Grosso-Silva et al. (2020)	U			
2021		Locations in Aveiro, Bragança, Viana do Castelo, Braga	Gaspar et al. (2023)	X			

2022		Locations in Porto, Beja, Madeira island	Gaspar et al. (2023)	X			
2010	**United Kingdom**	London, **interception** on luggage from USA	Malumphy (2014)	U	2018-ongoing	ACS: Multimedia survey, data collected via email by NIAB (National Institute Agricultural Botany), RHS (Royal Horticultural Society) and NHM (National History Museum)	F. Wamonje p.c.
2013		Teesport, N-Yorkshire **interception** on stones imported from China					
2018		New Alresford (Hampshire)	Powell et al. (2021) first official record	U			
2019		Farnham (East Hampshire)	Cianferoni et al. (2019)	X			
2020		London, Rainham Marshes (Essex), Rowley Fields (Leicester)	Powell et al. (2021)	UA X			
2020		Egham (Surrey)	Terrestrial Heteroptera Recording Scheme—Shieldbugs & allied species (2023)	U			
2021		Danbury (Chelmsford), Bristol, Five Mile Drive (Oxford), Isle of Skye (Scotland), North Yorkshire	Terrestrial Heteroptera Recording Scheme—Shieldbugs & allied species (2023)	U			

(continued)

Table 15.1 (continued)

First detections					Citizen science initiatives		
Year	Country	Location	References	Context	Year	Type	Reference
2021		Suffolk	PPC (2021)	U			
2022		Warren Heath (Ipswich)	Terrestrial Heteroptera Recording Scheme—Shieldbugs & allied species (2023)	U			
2018	**Malta**	Birżebbuġa	Tassini and Mifsud (2019)	A	–		
2018	**Poland**	Dobczyce (Western Beskids)	Claerebout et al. (2018)	X	2018	PCS: data collected during entomological studies and occasional occurrences, map available at: https://www.heteroptera.us.edu.pl/mapy/halyomorpha-halys.html	W. Karnkowski p.c.
2019		Cieszyn, Boguszowice (Beskid Zachodni)	Gierlasiński et al. (2020a)	U			
2019		Warsaw, Nizina Mazowiecka	Gierlasiński et al. (2020b)	U			
2020		Przemyśl (Beskid Wschodni)	Bury (2021)	U			
2021		Kraków (Kraków-Wieluńska Upland)	Gierlasiński et al. (2021)	X			
2021		Lipnik (Beskid Wschodni), Bolestraszyce (Nizina Sandomierska)	Bury and Mazepa (2022)	UG			

2021		Kraków (Wyżyna Krakowsko-Wieluńska)	Gierlasiński et al. (2022)	X			
2022		Wrocław (Śląsk Dolny)	Smolis and Kadej (2022)	U			
2022		Kielce (Świętokrzyskie)	iNaturalist contributors (2023a)	X			
2023		Mielec (Subcarpathian)	iNaturalist contributors (2023a)	X			
2018 2019	**North Macedonia**	Skopje, Gevgelija	Claerebout et al. (2018), Konjević (2020)	U	–		
2018	**Czechia**	Praha-Holešovice (Bohemia centr.); Luková u Přerova (Moravia centr.)	Kment and Březíková (2018)	UG	2018-ongoing	ACS: request to collect information on the CS platform https://www.najdije.cz/pf/knezice-mramorovana/ Data used to update the certified map	J. Skuhrovec p.c.
2022		Updated map	Skuhrovec et al. (2022)	X			
2018	**Bosnia-Herzegovina**	Mostar, Citluk	Zovko et al. (2019)	X	–		
2020–21		Banja Luka	Vidović et al. (2023)	AG			
2021		Đurđevik (Tuzla), Jajce	iNaturalist contributors (2023a)	X			
2022		Pop Ljubina (Doboj)	iNaturalist contributors (2023a)	X			
2023		Banja Vrućica (Teslić), Brčko, Studenci (West Herzegovina)	iNaturalist contributors (2023a)	X			

(continued)

Table 15.1 (continued)

First detections					Citizen science initiatives		
Year	Country	Location	References	Context	Year	Type	Reference
2023		Southern part of the country	N. Karic p.c.	X			
2018	**Netherlands**	Amsterdam, Oldenzaal, Doodewaard, Beesel, Treebeek, Sevenum, Heerlerheide	Aukema et al. (2019)	UG	2020-ongoing	ACS; Multimedia survey, data collected via email and the webpage Waarneming.nl (largest nature platform in the Netherlands)	K. Winkler, p.c.
2023		Many sites, mainly central-southern areas	Waarneming.nl (2023)	UGA			
2019	**Finland***	Haaga (Helsinki, Uusimaa)	Finnish Biodiversity Information Facility (2023)	X	2019-ongoing	PCS: data collected from the Finnish biodiversity portal https://laji.fi and reported on GBIF	Finnish Biodiversity Information Facility (2023)
2021		Solo Sokos (Varsinais-Suomi)	Same as above	U			
2022		Pohjois-Häme (Keski-Suomi), Kahilanniementie (Etelä-Karjala)	Same as above	X			
2023		Pohjois-Parkano Kuivasjärvi (Satakunta)	Same as above	X			
2019	**Moldova**	Chişinău	Derjanschi and Chimişliu (2019)	UG	–		
2020		Bălţi	iNaturalist contributors (2023a)	X			
2020–21	**Armenia**	Yerevan district Lori district	Kalashian et al. (2022)	U	–		

2020	**Denmark***	Sjælland, Helsingør	iNaturalist contributors (2023a)	X	2020-ongoing	PCS: data collected from the Danish biodiversity portal https://arter.dk and reported on GBIF	Arter project group (2023)
2021		Rødby (Sjælland), Nordborg (Syddanmark)	Arter project group (2023) Arter project group (2023)	X X			
2020	**Montenegro**	Kotor	iNaturalist contributors (2023a)	X	–		
2021		Bijela (Herceg Novi), Budva	iNaturalist contributors (2023a)	X			
2022		Dobra voda (Bar)	iNaturalist contributors (2023a)				
2020–22	**Uzbekistan**	Urgench, Ferghana	Gandjaeva et al. (2022)	A	–		
2022		Tashkent	iNaturalist contributors (2023a)	X			
2021	**Algeria**	Skikda	van der Heyden et al. (2021)	U	–		
2021	**San Marino**	San Marino	iNaturalist contributors (2023a)	X	–		
2022	**Luxembourg**	Luxembourg, a few locations	iNaturalist contributors (2023a), confirmed by A.M. Weigand, F. Vitali, T. Se Sousa, J. Reinhers, p.c.	X	–		

2.2 Benefits of Citizen Science Initiatives

The majority of countries surveyed (58%) have or have had citizen science initiatives to monitor the presence and spread of BMSB. In 39% of cases, BMSB detection occurred from pictures via email, websites and/or smartphone applications. These initiatives are important for early detection of introduced and potentially invasive species. Also, they are effective tools for promoting scientific literacy, useful for public outreach and education to raise public awareness of local environmental and economic concerns (Malek et al. 2018, 2019).

In addition, data collected has been used to develop specific models. In northern Italy, data revealed a stratified dispersal pattern of BMSB, with both continuous and human-mediated dispersal contributing to the species' widespread distribution (Maistrello et al. 2018). The model also showed exponential growth, suggesting that populations can reach damaging levels for crops in a very short period of time. Moreover, the model, based on data collected between 2012 and 2016, estimated the beginning of the invasion in northern Italy in 2009. A later discovery showed that the first specimen had been detected in Genoa in 2007 (Maistrello and Dioli 2014). In Trentino, northern Italy, a model based on data from the smartphone application "BugMap" allowed predicting the distribution and risk areas for BMSB establishment and revealed seasonal invasion dynamics, providing insights into the phenological patterns, life cycle and management methods (Malek et al. 2018, 2019).

In Switzerland, observations by a citizen science initiative between 2004 and 2019 were used to validate a purposefully modified version of the CLIMEX model (Stoeckli et al. 2020). The results showed that 15+ years after its accidental introduction, BMSB has colonised most bioclimatically suitable areas.

In France, data collected from the crowdsourcing survey (Table 15.1), in addition to reveal the phenology and habitat preferences, allowed to develop a model to identify the potential expansion of BMSB in Europe and the impact of climate change on its distribution (Chartois et al. 2021; Streito et al. 2021).

Crowdsourcing studies targeting BMSB show that one-to-one communication (e.g., by email exchange) can raise awareness of invasive species' negative impacts, encourage recruitment, and foster a sense of importance in scientific research (Maistrello et al. 2018). However, this approach only improves participation in citizen science surveys when researchers demonstrate personal commitment and passion (Ciceoi et al. 2017a).

2.3 Unravelling Invasion Pathways Through Genetic Studies

Substantial information on the reconstruction of invasion pathways of BMSB has been obtained through the use of molecular and population genetic tools. Mitochondrial DNA (mtDNA) sequence analysis is one of the most widely used methods for investigating genetic diversity and determining the origin of an invasive species.

The first study of BMSB samples collected from Switzerland in 2012 (Gariepy et al. 2014) analysed the cytochrome C oxidase subunit 1 (COI) gene and showed the presence of 3 haplotypes, of which one was dominant (H3), consistent with Asian samples from Hebei and Beijing, and two were novel (H8 and H9). The absence of haplotypes in common with samples from North America ruled out invasion of Switzerland by this route.

Initial analyses of the COI and COII genes of BMSB samples collected in 2013 in the northern Italian regions of Lombardy and Emilia Romagna and in Canton Ticino (southern Switzerland) showed that in Emilia Romagna there was a single haplotype (H1), already found in China, Korea and North America, and that in Lombardy there were several haplotypes, two of which were shared with the Canton of Ticino (H3 and H8), suggesting the Swiss origin of the Lombardy population (Cesari et al. 2015). Further analyses on the same genes (Cesari et al. 2017), performed on samples collected in 2013–2016 in different Italian regions through the crowdsourcing study (Maistrello et al. 2018) and from Greece, Romania and Switzerland, show that for the COI marker, among the 111 haplotypes scored, H1 and H3 were the most abundant and widespread, and that seven of them (H150–156), all from Italian samples, were detected for the first time. In addition, it was shown that Italy (13 haplotypes, with differences between north eastern and north western regions), followed by Greece (7 haplotypes), has the greatest biodiversity of BMSB outside the area of origin, demonstrating the existence of multiple invasions still in progress, mainly from Asia (especially China, Korea, Japan) and possibly from other European countries (Switzerland- France). In contrast, only one haplotype (H1) was detected in Romania.

According to Gariepy et al. (2015), who analysed the COI gene in samples collected in 2013–2014 from the Alsace region (France), Athens (Greece), Budapest (Hungary) and the Canton Ticino, the most common haplotypes were H1 and H3, with H1 predominant in Hungary (followed by H3) and H3 most common in France (followed by H1 and H8). Greece showed 7 haplotypes and Canton Ticino 3. The study confirmed the occurrence of multiple invasions from Asia as well as a 'bridgehead effect', i.e. secondary invasions originating from a particularly successful invasive population within Europe.

Lee et al. (2018), who compared existing COI sequences with those from new samples from Korea, showed that Korean populations were genetically distinct from those in China, Europe and North America, suggesting that BMSB populations in Europe and North America originated from multiple Chinese invasions and that Korean populations did not spread to other countries.

Schuler et al. (2021) analysed COI from samples collected in 2016–2019 in South Tyrol and found high haplotype diversity (with H1 predominant, followed by H3, H153 and 12 less frequent), with evidence of BMSB migration from neighbouring regions, as well as a haplotype previously found only in Greece. Phylogenetic analysis showed that the majority of haplotypes clustered with those found predominantly in Japan, suggesting an additional invasion pathway not previously considered.

Martinez-Sañudo et al. (2020) analysed the correlation between the COI of BMSB and the genetic structure of its bacterial symbiont, *Candidatus (Pantoea) carbekii*, using adults collected in 2017–2019 from native (China, South Korea and Japan) and invaded (United States, Italy and Hungary) countries. The host species showed higher haplotype diversity than symbionts, suggesting a long-term association. Populations from north eastern Italy showed the highest genetic diversity, suggesting multiple ongoing introductions. Some symbiont-host haplotypes were shared only by populations from northern Italy and Japan, confirming this invasion route.

Kapantaidaki et al. (2019) used two molecular markers: the COI gene and ribosomal internal transcribed spacer 1 (ITS1) to analyse samples from eight populations in Greece, Italy, Canada and the USA. They identified 14 haplotypes for COI and 24 for the ITS1 region, showing high genetic diversity for Greece and also for samples from the USA.

In the study by Gariepy et al. (2021), COI haplotype diversity and distribution were investigated for BMSB collected between 2016 and 2019 in Serbia, Ukraine, Russia, Georgia and Kazakhstan. It revealed the successful establishment of the H1 haplotype along the eastern invasion front in Eurasia (it was the only one detected in Ukraine, Russia, Georgia and was also dominant in Kazakhstan), whereas Serbia, in the centre of the invaded range, had a higher diversity (H3 dominant, followed by H1, H8, H33 and H80), likely due to multiple secondary invasions from neighbouring countries.

According to Yan et al. (2021) analyses of COI and COII from samples from 12 countries, indicate that BMSB populations from Chile, Georgia, Hungary, Italy, Romania, Turkey and the USA are undergoing demographic expansion, while those from Austria, Serbia and Slovenia show a potential population bottleneck.

3 Biology, Ecology and Behaviour

3.1 Biology

The BMSB exhibits a facultative winter adult diapause that is regulated by day length. This enables the insect to produce varying numbers of annual generations in different regions of its native or invaded range, depending on local temperature and other environmental factors (e.g., food availability, precipitation, etc.) (Saulich and Musolin 2017). Data obtained from studies carried out in Switzerland, Italy and Slovenia (Haye et al. 2014a; Costi et al. 2017; Rot et al. 2022a) are shown in Table 15.2. In all cases, individuals collected in the previous year from overwintering shelters were kept in outdoor cages, protected from precipitation and direct sunlight, with abundant and varied food available, and were observed daily to record all parameters. Adult survival during overwintering period is highly variable. Activity starts with 13 h photoperiod and 14 °C+ average daily temperature; oviposition starts with photoperiod >14 h and temperature exceeds 15 °C in Switzerland, 17 °C in Slovenia and 21 °C in Italy.

Table 15.2 Biological parameters of *Halyomorpha halys* in European countries. *NGY* number of generations per year, *ov.* overwintering, *ovip.* oviposition, *Su.* summer, *Mort.* mortality, *F* female, *FF* females, *Dev.* development, *R0* net reproductive rate of increase; Type of study: *CO* cage outdoor, *FO* field observations. Exit ov., Start ovip. and ovip. Peaks are expressed in number of calendar week (when available) or month. Exit Ov. = indicates the first week of emergence from overwintering sites or, in the case of CO studies, the range from the first to the last exit. Oviposition peak and Development time egg-adult: in the case of two generations, the first number is related to the first generation and the second to the second generation

Country	Location	Year	NGY	Ov. surv. (%)	Exit ov.	Start ovip.	N. Eggs/F Ov. FF	N. Eggs/F Su. FF	Ovip. peak	Tot. mort. Ov. Gen. (%)	Tot. mort. Su. Gen. (%)	Dev. Time egg-adult (days)	R0 Ov. Gen	R0 Su. Gen	Type of study (CO/FO)	Reference
Switzerland	Basel	2014	1	61	14	24	79.18	–	26	86.7	–	60–131	5.7	–	CO	Haye et al. (2014a)
	Basel	2018	2	–	–	–	–	–	–	–	–	–	–	–	CO	Stoeckli et al. (2020)
		2019	2	–	–	–	–	–	–	–	–	–	–	–	CO	
Italy (Northern)	Emilia	2015	2	28	12–21	19	285	215	24; 30	56.5	97.5	44; 60	24.4	5.4	CO	Costi et al. (2017)
	Emilia	2015	2	–	14	20	–	–	–	–	–	–	–	–	FO	Maistrello et al. (2016b)
	Emilia	2016	2	63.5	11–25	22	–	–	–	–	–	–	–	–	CO	Maistrello unp. data
	Emilia	2017	2	37.9	9–21	21	–	–	–	–	–	–	–	–	CO	Maistrello unp. data
	Emilia	2018	2	52	9–21	20	–	–	–	–	–	–	–	–	CO	Tommasini et al. (2019)
	Romagna	2018–19	2	–	16	22	–	–	–	–	–	–	–	–	FO	Francati et al. (2021)
Slovenia	Nova Gorica	2019	2	39.9	13	20	174	76	24; 31	60.9	89.9	48; 59	14.8	5.6	CO	Rot et al. (2022a)
		2020	2	40.3	13	18			23; 31			59; 54			CO	Rot et al. (2022a)

(continued)

Table 15.2 (continued)

Country	Location	Year	NGY	Ov. surv. (%)	Exit ov.	Start ovip.	N. Eggs/F Ov. FF	N. Eggs/F Su. FF	Ovip. peak	Tot. mort. Ov. Gen. (%)	Tot. mort. Su. Gen. (%)	Dev. Time egg-adult (days)	R0 Ov. Gen	R0 Su. Gen	Type of study (CO/FO)	Reference
		2021	2	36.5	13	20			24; 29			49; 54			CO	Rot et al. (2022a)
	Ljubljana	2021	1	–	13	–	–	–	–	–	–	–	–	–	FO	Bohinc et al. (2022)
Serbia	Belgrade	2016	2	–	–	26	–	–	–	–	–	–	–	–	FO	Musolin et al. (2018)
	Novi Sad	2018–21	2	–	–	22	–	–	–	–	–	–	–	–	FO	Konjevic (2022)
	Irig (Fruška gora), Sombor	2022	2	–	–	20	–	–	–	–	–	–	–	–	FO	A.Konjevic p.c.
Spain	Girona	2017	2	–	–	–	–	–	–	–	–	–	–	–	FO	A.Escudero p.c.
Azerbaijan	Zagatala	2017	3	–	13	18	–	–	–	–	–	40	–	–	FO	Nuriyeva and Nadirova (2020)
Ukraine	Kherson, Dnipro	2018	3	–	–	–	–	–	–	–	–	–	–	–	FO	Yanovsky (2021)
Georgia	Batumi	2018	2	–	17	–	–	–	–	–	–	–	–	–	CO	Dumbadze et al. (2019)
Russia	Sotchi	2018–21	2	–	–	20–21	–	–	24; 32	–	–	–	–	–	FO	Reznik et al. (2022)

Greece	Imathia Pieria	2020	2	–	–	–	–	–	–	–	–	–	–	–	FO	Andreadis et al. (2021)
France	Garonne	2021	2	–	–	–	–	–	23; 32	–	–	–	–	–	FO	Bout et al. (2022)
Germany	Bodensee	2019	2	—	–	–	–	–	–	–	–	–	–	–	FO	Bauer Pilla et al. (2022)
	Constance	2021	2	–	14	22	–	–	–	–	–	–	–	–	FO	ISIP (2022)
	Karlsruhe	2021	2	–	14	24	–	–	–	–	–	–	–	–	FO	ISIP (2022)
	Constance	2022	2	–	15	21	–	–	–	–	–	–	–	–	FO	ISIP (2022)
	Constance	2023	2	–	15	23	–	–	–	–	–	–	–	–	CO	R. Bauer Pilla p.c.
	Karlsruhe	2023	2	–	15	22	–	–	–	–	–	–	–	–	FO	R. Bauer Pilla p.c.
Bosnia-Herzegovina	Banja Luka	2021	1	–		July	–	–	–	–	–	–	–	–	FO	Vidović et al. (2023)
Turkey	Sakarya	2022	2	–	–	–	–	–	–	23	90.2	48; 47	–	–	CO	Özdemir et al. (2023a)
	Artvin, Rize	2023	1/2	–	–	24	–	–	–	–	–	–	–	–	FO	Tuncer and Ozdemir (2023)

According to Koutsogeorgiou et al. (2022) in northern Greece, BMSB overwintering females undergo facultative reproductive diapause, requiring additional time for termination before reaching reproductive maturity.

Local weather conditions, particularly during the spring and summer, have a significant impact on the number of generations produced per year, even with the same photoperiod. Thus, overwintered females anticipate oviposition if the spring season is warm, and in summer season the completion of the second generation may occur. For instance, in Basel, Switzerland, BMSB was monovoltine in 2014, whereas two generations were observed in 2018 and 2019, which may be an effect of climate change (Stoeckli et al. 2020).

Two generations of BMSB occur annually in the countries surrounding the Mediterranean and Black Seas in Europe, while only one generation occurs in the colder regions of Central Europe (Table 15.2). Based on field observations, BMSB could potentially generate three generations annually in specific regions of Azerbaijan and Ukraine.

The climatic conditions in northern Italy appear to be extremely favourable for BMSB. The realised net reproductive rates (R_0) of the Italian BMSB population were 24.04 and 5.44 for the overwintering and summer generations, respectively, indicating an overall extremely high growth potential of this pest (Costi et al. 2017). These growth rates, much higher than those for Slovenia and Switzerland, may partly explain the very short time elapsed between the year of first official detection (2012) and the outbreaks observed since 2015, together with the severe economic losses in orchards in northern Italy.

The development time from egg to adult varies with temperature. In Swiss monovoltine populations, development takes between 2 and 4 months, but in populations with two generations per year, the first generation (which appears in early summer) develops faster (50.0 d) than the 2nd (56.8 d). Under controlled conditions of 20°, 25° and 30 °C, Swiss BMSB take, from egg to adult, 75.8, 42.3 and 33.2 days respectively. No development occurs below 15 or above 35 °C (Haye et al. 2014a). In northern Italy, the lifespan of BMSB is about 1 year for individuals of the overwintering generation, and 70–80 days for the summer generation (Tommasini et al. 2019).

BMSB collected in Sotchi (Russia) show faster nymphal development under a short-day photoperiod compared to a long-day photoperiod; short days induced diapause in adults, while long days promoted reproduction; and the critical day length for diapause induction was around 15.5 hours at 24 °C (Musolin et al. 2019). They investigated the photoperiodic control of nymphal development and adult diapause induction by comparing the critical day length for diapause induction among one native (Andong, South Korea) and three invasive (Torino, Italy; Basel, Switzerland; and Sochi, Russia) populations. The critical day lengths required for diapause induction in both sexes ranged from 14.5 to 15.0 hours in the Korean population and from 15.0 to 15.5 hours in the European populations (Musolin et al. 2022a). The findings indicate that microevolution may have been insufficiently rapid to keep pace with the fast expansion of the invader throughout Eurasia. In a phenological study conducted in Sotchi, reproductively active adult BMSB were recorded from the end of May to the beginning of September. The seasonal activity of BMSB was correlated with the day length; when the day length fell below the experimentally critical value (15.0–15.5 hours), the percentage of females with fully developed reproductive organs fell abruptly to zero (Reznik et al. 2022).

In Romania, BMSB showed different behaviour at different temperatures, seeking shelter at low temperatures (below 18–20 °C), especially in door and window frames of buildings, and feeding on still green plants at higher temperatures (above 20 °C) during the day (Neacsu et al. 2019). In eastern Hungary BMSB preferred artificial overwintering sites to natural ones; larger females had higher overwintering success, indicating that internal factors are more important than environmental conditions (Aletta et al. 2021). According to Bozsik et al. (2021) higher numbers of both BMSB and *Nezara viridula* (L.) were attracted to artificially heated shelters compared to unheated ones.

Papa and Negri (2020) observed that overwintering populations of BMSB exhibit cannibalistic behaviour and hypothesised that this is driven by the depletion of metabolic reserves and desiccation that occurs in winter. Cannibalism may be facilitated by aggregation of individuals and suppression of species-specific signals that prevent predation on conspecifics.

Scaccini et al. (2019) characterised the mortality response curves of adult BMSB to short exposure to high temperatures and compared field-collected individuals entering and exiting overwintering. Adults exiting overwintering were found to be more susceptible to high temperatures than those entering diapause.

Scaccini et al. (2020b) investigated the lethal and sub-lethal effects of exposure to low temperatures on BMSB. Adults exiting overwintering are more likely to face higher mortality than those entering overwintering when the temperature is low. Additionally, low temperatures reduce fecundity in females entering overwintering. Higher nutrient indices and weight increased survival rates in BMSB adults. The study highlights the increased mortality of BMSB adults during spring frost events.

Grisafi et al. (2021) found that microclimate conditions significantly affect the transpiration of BMSB, with hot and dry climates being unfavourable. In the overwintering generation water loss is critical. Therefore, upon exiting diapause, the primary nutritional requirement for individuals is hydration. The study suggests that microclimatic conditions within crops can affect the distribution and activity of BMSB, highlighting the importance of crop management practices.

3.2 Morphology, Physiology and Behaviour

The external male and female genitalia of adult BMSB specimens are described by Vétek et al. (2014). A study of the morphology of adult BMSB (Pajač-Živković et al. 2022) showed that males and females differed in all measurements with female-biased sexual size dimorphism. Larger and stronger make females the better sex for migratory flight and thus responsible for the rapid expansion of the population into new areas.

To gain insight into the sensory system of BMSB, Ibrahim et al. (2019) investigated the morphology, localization, abundance and distribution of the sensilla on the antennae of adult and fifth instars using scanning and transmission electron microscopy. Four basic types of sensilla were found on the antennae (basiconica -subtypes A, B, C, D, and E, coeloconica, trichoidea, and chaetica—subtypes A and B). The sensilla were determined to be significant for olfaction, mechanoreception, gustation, thermo-reception, and hygro-reception. The study also found that there is no sexual dimorphism in the sensilla.

A study measured the spectral sensitivity of BMSB's compound eyes using electroretinography and conducted behavioural experiments to determine its attraction to various light stimuli (Egri et al. 2023). The findings indicate that the compound eyes are highly sensitive to the green range, with a minor sensitivity peak in the UV region. The behavioural experiments revealed that the attraction of BMSB to light increased with decreasing wavelength, emphasizing the importance of short wavelength lights (368–455 nm) for attracting BMSB.

Communication among adult stink bugs is facilitated by substrate-borne vibrational signals. The vibrational signal repertoire exhibited by male and female BMSB during mating behaviour was characterised by Polajnar et al. (2016). Among the various signals, female song 2 (FS2) was identified and induces directional movements in males towards the point of emission (Mazzoni et al. 2017). Further research has identified the spectral and temporal features of the FS2 signal that prompt male responsiveness most effectively. This may help to create ideal vibrational signals to attract BMSB males and be implemented in the development of effective pest control methods (Caorsi et al. 2021).

4 Plant Damage and Crop Losses

4.1 *Polyphagy and Insect-Plant Interactions*

The polyphagy of BMSB in Europe (Haye et al. 2015b; Maistrello et al. 2016a) includes crop and non-crop plants of many families (Hamilton et al. 2017). According to a survey conducted between 2014 and 2019 in agrocenoses, ornamental areas and forest regions of the humid subtropical Black Sea coast in the Caucasus (Sochi, Abkhazia), BMSB feeds on 107 plant species belonging to 48 families. Fabaceae, Rosaceae, Rutaceae, Sapindaceae and Solanaceae were found to have the highest species diversity. Notably, *Metasequoia glyptostroboides* and *Taxodium mucronatum* are among the conifers that serve as host plants (Zakharchenko et al. 2020). In addition, adult BMSB feeding on panicles in rice-growing areas of Lombardy, Italy, provides the first evidence of a link between the pest and rice, a previously unrecorded host plant (Lupi et al. 2017).

To understand how BMSB selects and locates its hosts, Koutsogeorgiou et al. (2023) compared the volatile organic compounds (VOCs) emitted by host plants (kiwi, red delicious and golden delicious apples, cotton, and cannabis) and found that 13 out of the 68 identified compounds were shared among the hosts. This suggests that the recognition of host plants by BMSB is likely driven by a combination of attractive and repellent compounds.

According to Serteyn et al. (2020) the strong pest status of BMSB can be explained by the uniquely large plasticity of its feeding strategies, as demonstrated by their research through analysis of electropenetrography waveforms. Results demonstrate that BMSB is capable of utilising two unique feeding strategies, sheath feeding and cell rupture feeding, on the same plant structure. This is likely accomplished by exploiting three distinct cell types within plants—xylem cells, phloem sieve elements, and mesophyll/parenchyma. Such a diversity in the types of cells that can be fed on is not generally observed in other Heteroptera. Additionally, BMSB is able to produce X waves, which are considered to indicate the inoculation behaviour of the plant pathogen *Xylella fastidiosa*. Consequently, BMSB could potentially transmit similar plant diseases.

A further investigation (Serteyn et al. 2020) revealed that BMSB feeding intensified lipoxygenase activity in faba bean leaves and influenced the insect's feeding behaviour. When exposed to plant defences, BMSB probes were delayed, sustained ingestion events were shorter, and the salivary gland proteins involved in plant allelochemical detoxification changed. Overall, the findings indicate that BMSB can detect plant defences and modify its feeding strategies accordingly.

According to Rondoni et al. (2018), *Vicia faba* plants recognise the presence of BMSB eggs and pre-activate defence mechanisms against insect's offspring. The response involves activation of both the jasmonic acid (JA) and salicylic acid (SA) pathways.

A preliminary investigation showed that BMBS samples collected from the field tested positive for "stolbur" (16SrXII-A) and "aster yellow" (16Srl-B) phytoplasmas, and their capacity to transmit phytoplasmas to healthy plants was also confirmed (Paltrinieri et al. 2016).

4.2 Crop Losses

The damage caused by BMSB varies according to the plant species, variety, phenological stage, time of attack and intensity (number of bites). It causes discolouration of the fruit, spots, varying degrees of deformity, changes in texture, subcutaneous necrosis (suberification of the flesh), rendering products unmarketable. An overview of crop losses caused by BMSB in European countries is shown in Table 15.3. Damage severity varies from 1 to 4, with 4 representing the highest losses (more than 70%). The most significant losses (damage severity $\geq$ to 3) were recorded mainly in hazelnut groves in Georgia and in orchards (peach, apricot, pear, apple and kiwi) in Italy. These are the only countries in which an economic estimate of the losses was made, which was worth USD 20–24 mln in Georgia (2016) and almost EUR 600 mln for the entire large fruit-growing area of northern Italy (2019). Among vegetables, peppers are the most damaged, especially in Switzerland, Italy

Table 15.3 Overview of crop damage caused by *Halyomorpha halys* in European countries over time. Assessments based on the review of scientific literature and by researchers and managers of plant protection services in the respective countries. DS = Damage Severity: 1 = some damage detected not precisely quantified or losses less than 10%; 2 = significant damage not precisely quantified or losses between 10% and 30%; 3 = serious damage not precisely quantified or losses between 30% and 70%; 4 = very serious damage, total crop loss (>70%). * = Estimated damage includes both impact in terms of not-harvested fruits and on post-harvest and supply chain. (*N*- northern, *E*- eastern, *C*- central, *S*- southern, *W*- western)

Country	Year	Crop and (location)	DS	Quantification and/or type of damage	References
Switzerland	2012	Pepper (tunnels, German Switzerland)	1	Quality degradation	Sauer (2012)
	2016	Apple (Ticino)	1	% Fruit damage	C. Marazzi p.c.
	2017	Apple, Pear, Peach, Olive (Ticino)	2	% Fruit damage	C. Marazzi p.c.
	2017	Pepper (German Switzerland)	1–3	% crop loss	Sauer and Total (2019)
	2018	Apple, Peach, berries (different regions)	2	% Fruit damage	C. Marazzi p.c.
	2018	Pear (Aargau, Luzern, St. Gallen, Thurgau, Zug, Zurich)	1	% Fruit damage	Schweizer Obstverband SOV in Bauernzeitung (2019)
	2019	Pear (Aargau, Luzern, St. Gallen, Thurgau, Zug, Zurich)	2	% Fruit damage	Schweizer Obstverband SOV in Bauernzeitung (2019)
	2019	Pepper (German Switzerland)	1–4	% crop loss	Sauer and Total (2019)
	2021	Apple, pear, cherry, apricot (different regions)	1	% Fruit damage	Stäheli et al. (2022)
Italy	2013	Peach (Piedmont)	1	Fruit damage	Pansa et al. (2013)
	2015	Pear (Emilia)	2–3	Damage at harvest	Maistrello et al. (2017)
	2015	Apricot, plum, peach, apple, persimmon, tomato (N-Italy)	2–3	Damage at harvest	Bariselli et al. (2016)
	2015–16	Hazelnut (Piedmont)	2	Up to 19% damaged kernels	Bosco et al. (2018)
	2019	Pear (N-Italy) apple (N-Italy) peach/nectarine (N-Italy) kiwi (N-Italy)	3–4	267.4 Mln € Estimated damage* 177.3 Mln € Estimated damage* 88.9 Mln € Estimated damage* 54.7 Mln € Estimated damage*	CSO Italy—Centro Servizi Ortofrutticoli (2019)

Abkhazia	2016	Peach, mandarin orange, persimmon, other major crops	2–4	11.7–85.2% losses at harvest	Musolin et al. (2018)
Georgia	2015	Hazelnut (Samegrelo)	1	Up to 4% damaged kernels	Bosco et al. (2018)
	2016	Hazelnut (Samegrelo)	3	Up to 51% damaged kernels	Bosco et al. (2018)
	2016	Hazelnut (Samegrelo)	3–4	21–24 million USD (loss to nut growers and exporters)	OC Media (2017)
	2018	Hazelnut, kiwi, persimmon, tangerine, apple, pear, peach, grapes, small fruits, fig, walnut tree, beans, peas, corn, paprika, tomato, ornamentals (Batumi)	1–2	n. individuals, losses	Dumbadze et al. (2019)
	2019–20	Hazelnut (Samegrelo, Guria)	3–4	Up to 100% kernel damage	Kereselidze et al. (2022)
	2021	Blueberry, mandarin (Guria)	4	Up to 100% fruit damage	Kereselidze et al. (2022)
Romania	2016	Goji (Bucharest area)	1–4	Attack intensity	Ciceoi et al. (2016)
	2016	Maize (Bucharest area)	1	Quality degradation	Ciceoi et al. (2016)
	2017	Maize (Bucharest area)	3	Up to 37% damage	Ciceoi et al. (2017c)
	2019	Apricot (Timis county)	1–2	n. individuals	Grozea and Costea (2020)
	2020	Ornamental plants (Timisoara)	1–2	n. individuals	Muntean and Grozea (2021)
	2010–23	Tomato, pepper, eggplant, asparagus (Timis county)	1	n. individuals, losses	I. Grozea p.c.
Hungary	2016	Dry bean 'Etna' (Budapest area)	4	94% damaged pods	Vétek (2017)
	2016	Hybrid green hot pepper 'Daras' (Budapest area)	4	100% damaged peppers	Vétek (2017)
	2023	Peach, nectarine, apple, pear, tomato, sweet pepper, sweet corn, bean and soybean (all country)	1–2	Presence and damage	J. Fail p.c.

(continued)

Table 15.2 (continued)

Country	Year	Crop and (location)	DS	Quantification and/or type of damage	References
Bulgaria	2016–18	Corn, tomato, apple, grapes, raspberry, mulberry, hibiscus (Plovdiv)	1–2	Presence	Hristozova and Harizanova (2022)
	2023	Same crops (all country)	1–2	Presence	N. Simov p.c.
Serbia	2017	Peach (N-Serbia—Vojvodina)	2	Damage before and at harvest	S. Šestić & A. Konjević, p.c.
	2017	Hazelnut (N- and W-Serbia)	3	Damage before and at harvest	
	2018	Apple (N-Serbia—Vojvodina—Titel, Čelarevo, Šid, Inđija)	2	Damage before and at harvest	
	2019	Peach (N-Serbia—Vojvodina—Slankamen)	3	Damage before and at harvest	
	2020	Pear (N- Serbia—Vojvodina—Šid)	2	Damage before and at harvest	
	2020	Hazelnut (N-Serbia—Vojvodina—Aleksa Šantić, Irig, Novi Sad, Sombor)	1	Damage at harvest	
	2021	Blueberry (W-Serbia—Šabac)	1	Damage at harvest	
	2021	Hazelnut (N- and C-Serbia -Aleksa Šantić, Irig, Velika Plana, Ub)	1	Damage at harvest	
	2021	Hazelnut (N- and W- Serbia—Novi Sad, Šabac, Sombor)	2	Damage at harvest	
	2022	Hazelnut (W- Serbia—Šabac)	2	Damage before and at harvest 20%	
	2022	Hazelnut (N- and C-Serbia -Aleksa Šantić, Irig, Velika Plana, Ub)	2	Damage at harvest	
	2022	Hazelnut (N- Serbia—Kikinda, Novi Sad, Sombor)	3	Damage at harvest	
	2022	Hazelnut (N- and W-Serbia—Crvenka, Šabac)	1	Damage at harvest	
	2022	Apple (N-Serbia—Vojvodina Titel, Čelarevo, Šid, Inđija)	1	Damage before and at harvest 5–10%	
	2022	Pear (N-Serbia—Vojvodina—Šid)	1	Damage before and at harvest 5%	

	2022	Blueberry (W-Serbia—Šabac, and C- Serbia, Šumadija)	1	Damage at harvest 3%	
	2022	Peach (N- Serbia—Vojvodina—Fruška gora)	3	Damage before and at harvest	
	2023	Hazelnut (N-Serbia—Aleksa Šantić, Crvenka Kikinda)	1	Damage before and at harvest	
	2023	Hazelnut (N- Serbia—Irig, Novi Sad)	2	Damage before and at harvest	
	2023	Hazelnut (N- and C-Serbia—Sombor, Velika Plana, Ub)	3	Damage before and at harvest	
	2023	Hazelnut (W-Serbia—Šabac)	4	Damage before and at harvest	
Greece	2017	Kiwi (Imathia, Pieria, Arta)	2	Fruit damage	Andreadis et al. (2018)
	2018	Olive—peach (Imathia)	4	Fruit damage	Damos et al. (2020)
	2019	Olive—peach (Imathia)	3	Fruit damage	Damos et al. (2020)
	2020	Hemp (Thessaloniki)	3	Quality degradation	Andreadis et al. (2022)
	2021–22	Peaches (Imathia)	2	Fruit damage	S. Andreadis p.c.
	2023	Peaches (Imathia)	1	Fruit damage	S. Andreadis p.c.
	2018–23	Kiwi (Imathia, Pieria, Arta, Pella)	2	Fruit damage	S. Andreadis p.c.
	2022–23	Apples (Centr. Macedonia, Thessaly)	3	Fruit damage	S. Andreadis p.c.
	2021–23	Cherry (Pella)	2	Fruit damage	S. Andreadis p.c.
	2022–23	Pears (Thessaly)	2	Fruit damage	S. Andreadis p.c.
	2023	Cotton (Evros)	2	Quality degradation	S. Andreadis p.c.
Turkey	2018	Hazelnut, kiwi, maize, apple, persimmon (Artvin, Eastern Black Sea)	1	n. individuals, losses	C. Tuncer (evaulation of some studies, personal communications and observations) p.c.
	2019	Hazelnut, kiwi, maize, apple, persimmon (Artvin, Rize, Eastern Black Sea)	1	n. individuals, losses	
	2020	Hazelnut, kiwi, maize, apple, persimmon (Artvin, Rize, Eastern Black Sea)	1	n. individuals, losses	
	2021	Hazelnut, kiwi, maize, apple, persimmon (Artvin, Rize, Trabzon, Eastern Black Sea)	1	n. individuals, losses	

(continued)

Table 15.2 (continued)

Country	Year	Crop and (location)	DS	Quantification and/or type of damage	References
	2022	Hazelnut, kiwi, maize, apple, persimmon (Artvin, Rize, Trabzon, Giresun, Ordu, Samsun, Mid- and Eastern Black Sea)	1	n. individuals, losses	
	2023	Hazelnut, kiwi, maize, apple, persimmon (Artvin, Rize, Trabzon, Giresun, Ordu, Samsun, Mid and Eastern Black Sea)	1–2	n. individuals, losses	
Turkey	2021–23	Kiwi (Yalova, Eastern Marmara)	1	n. individuals, losses	I.O. Ozdemir (individual field observations) p.c.
	2021–23	Peach (Yalova, E-Marmara)	1	n. individuals	
	2021–23	Hazelnut (Sakarya, E-Marmara)	1	n. individuals, losses	
	2023	Kiwi, Persimmon (Sakarya, E-Marmara)	1	n. individuals, losses	
	2023	Maize (Sakarya, E-Marmara)	1	n. individuals	
	2021–23	Hazelnut (Düzce, West-Black Sea)	1	n. individuals, losses	
	2023	Kiwi (Düzce, W-Black Sea)	1	n. individuals, losses	
France	2018–19	Pear, Apple (Chambery, Savoie)	3–4	Damage at harvest	Alison et al. (2020), Bout et al. (2022)
	2019	Kiwi, hazelnut (Drôme)	1–2	Damage at harvest	Alison et al. (2020), Bout et al. (2022)
	2020	Kiwi, fruit trees (Rhône valley, Gard, Hérault)	2–3	Damage at harvest	Alison et al. (2020), Bout et al. (2022)
	2021	Hazelnut (Tarn-et-Garonne)	1–2	Damaged kernels	Bout et al. (2022)
	2022	Apples, pears, peaches, apricots, cherries, kiwifruit, hazelnuts (all previous regions plus Loire valley)	2–3	Damage at harvest	B. Alison p.c.
	2023	Same crops and regions as above	1–2	Damage at harvest	B. Alison p.c.

Slovenia	2019–22	Sweet cherry	1	Quality degradation	M. Rot p.c.
	2019–22	Apricot, peach apple, pear, persimmon (Goriška region)	2–4	Fruit damage, damage at harvest	Rot et al. (2022b)
	2021–22	Hazelnut (Goriška region)	3	Up to 50% damaged kernels	M. Rot p.c.
	2021	Pepper	2	20–40% damage	M. Rot p.c
	2022	Olive (Goriška region, Slovenian Istria)	2	Premature fruit drop	Hoblaj et al. (2022)
	2023	Olive (Goriška region, Slovenian Istria)	2	Premature fruit drop	Devetak et al. (2023)
Germany	2019	Apple (Constance)	2	Fruit damage, damage at harvest	Bauer Pilla et al. (2022)
	2019	Pear (Constance)	1	Fruit damage, damage at harvest	Bauer Pilla et al. (2022)
	2020	Apple and pear orchards	1	Fruit damage, damage at harvest	R. Bauer Pilla p.c.
	2021	Apple and pear orchards	1	Fruit damage, damage at harvest	R. Bauer Pilla p.c.
	2022	Apple and pear orchards	1	Fruit damage, damage at harvest	R. Bauer Pilla p.c.
	2020–23	Cucumber, peppers, eggplant (in greenhouse), beans (field), apple, peach, pear, berry crops (raspberry mostly), hazelnut	1–4	Presence, high variability of damage	C. Dieckhoff p.c.
Ukraine	2019	Apple (Kherson, Dnipro)	2	Fruit damage	Yanovsky (2021)
Croatia	2022	Hazelnut, pear, apple, cherry, apricot, fig, grapevine, plum, olive	1	Presence	I. Pajač Živcovič p.c.
Belgium	2023	Pear, apple, grapes, soft berry fruits	1	Presence	T. Belien p.c.

(Table 15.4) and Hungary. Damage by BMSB has been observed also on row crops: maize in Romania, Turkey and Italy, and soybean in Hungary and Italy (Tables 15.3 and 15.4). The intensity of attacks in different countries depends on the time elapsed since introduction, and specific weather conditions. In countries where BMSB has been detected for the longest time with favourable conditions, such as Italy, population and crop damage increase exponentially in the first few years (Maistrello et al. 2018). Therefore, there is a great concern about damage increase in recently invaded countries, particularly in areas around the Mediterranean and the Black Sea. The recent increased presence of *Pentatoma rufipes* (L.), a stink bug associated with woody plants, in orchards (apple and pears) in central and northern European countries (Peusens and Belien 2012; Powell 2020, B. Hegger, T. Haye, p.c.), cause similar fruit damage to that of BMSB, and should not be confused.

4.2.1 Crop Damage in Italy

Italy is the country in Europe where BMSB has caused the most economic damage, given the suitability of different Italian regions for various fruit and vegetable crops production and exportation. For this reason, its presence and damage to crops has been carefully monitored. The results of this survey are shown in Table 15.4. Since 2015, BMSB has become a key pest in orchards (Maistrello et al. 2017), and on vegetable and row crops. Pear, which is mainly grown in the Po Valley (Emilia Romagna accounts for 70% of Italian production), is the crop most damaged, and in 2019 almost the entire harvest was lost. BMSB is one of the reasons that have contributed to the progressive and dramatic decline in the area planted with pear trees in recent years (Regione Emilia Romagna 2022). Apple, peach/nectarine and kiwifruit trees grown in the same area, have also been affected, particularly in 2019 (CSO Italy—Centro Servizi Ortofrutticoli 2019). In a field study carried out in Romagna with kiwi fruit by Francati et al. (2021) fruit damage was found to be significantly related to the growth stage of the kiwi fruit, the BMSB trap catch and the cultivar, with yellow-fleshed cultivar being more susceptible to damage than the green-fleshed cultivar.

Hazelnuts are also one of the most affected crops (Bosco et al. 2018), and the progressive spread of BMSB in central and southern regions of Italy is of great concern; it also damages olive trees (2020) in northern Italy. Considerable damage is also reported on walnuts from 2019 in Veneto and on pepper, maize and soybean in the Po Valley. A study conducted in maize fields in Piedmont revealed that BMSB feeding activity reduces the size and quality of maize ears leading to economic losses, poor grain quality and increased susceptibility to secondary pathogens (Rancati et al. 2017). The “stay green” phenomenon was detected in the 10–15 m border area of soybean crops in Friuli Venezia Giulia. This is the result of BMSB feeding, which prevents pod maturation and harvesting (Signor et al. 2018).

Table 15.4 Overview of crop damage caused by *Halyomorpha halys* in Italian regions over eight years. Evaluations from researchers and managers of the Italian regional plant health services. For each crop, region and year damage severity is reported (1 = some damage detected not precisely quantified or losses less than 10%; 2 = significant damage not precisely quantified or losses between 10% and 30%; 3 = serious damage not precisely quantified or losses between 30% and 70%; 4 = very serious damage, total crop loss (>70%); /= data not recorded due to frost. Acronyms for the Italian regions: *ER* Emilia Romagna, *Ven* Veneto, *Pie* Piedmont, *Lom* Lombardy, *Tre* Trentino, *AA* Alto Adige, *VDA* Val d'Aosta, *FVG* Friuli Venezia Giulia, *Lig* Liguria, *Tus* Tuscany, *Mar* Marche, *Umb* Umbria, *Laz* Lazio, *Abr* Abruzzo, *Cam* Campania, *Cal* Calabria, *Bas* Basilicata, *Pug* Puglia, *Sar* Sardinia

Crop	Region	2016	2017	2018	2019	2020	2021	2022	2023
Pear	ER	3	2	3	4	2–3	3	2	2–3
	Ven	3	3	3	4	2–3	/	2	2–3
	Pie	3	3	3	4	2	2–3	3	3
	Lom	3	3	3	4	2–3	3	3	3
Apple	ER	2	2	3	3	2	3	2	2
	Ven	3	3	2	3	2	2	2	2
	Pie	2	2	3	3	2	2	2–3	3
	Lom	2	3	2	3	2	2	3	2
	Tre	1	1	1	3	2	2	1	1
	AA	1	1	1	2	2	2	2	1
	VDA	–	–	–	–	1	1	1	1
	FVG	3	3	3	3	2	1–2	1	1–2
	Umb	–	–	–	3	2	2	2	2
	Cam	–	–	–	–	2	2	2	2
Peach/nectarine	ER	–	–	3	3	2	2–3	2	2–3
	Ven	–	–	2	3	1	/	2	2
	Pie	–	–	3	3	2	2	2	2–3
	Lom	–	–	3	3	–	–	–	–
	Mar	–	–	–	3	2	2	2	2
	Umb	–	–	–	–	–	–	–	2
	Cam	–	–	–	–	1	1	1–2	1–2
	Bas	–	–	–	–	1	1	1	1
	Cal	–	–	–	–	1	1	1	1
Apricot	ER	–	–	1	1	2	2–3	1	2
	Ven	–	–	2	2	2	/	1	1
	Pie	–	–	1	1	1	1	1	1
	Cam	–	–	–	–	1	1	1	1
	Cal	–	–	–	–	1	1	1	1
Cherry	ER	–	–	1	1	1	1	1	2
	Ven	–	–	1	2	3	/	1	1
	Pie	–	–	2	2	1	2	–	–
	Tre	–	–	1	1	1	2	1	1
	Cam	–	–	–	–	1	1	1–2	1–2

(continued)

Table 15.4 (continued)

Crop	Region	2016	2017	2018	2019	2020	2021	2022	2023
Plum	ER	–	–	1	1	1	2	1	2
	Ven	–	–	1	1	1	/	1	1
	Pie	–	–	–	–	1	1	–	–
	Cal	–	–	–	–	1	1	1	1
Kiwi	ER	–	–	2	3	2	3	2	2
	Ven	–	–	2	3	2	2	2	2–3
	Pie	–	–	2	2	1–2	1–2	2	1–2
	Lom	–	–	–	2	3	2–3	–	–
	Tre	–	–	–	–	1–2	2	1	1
	FVG	–	–	3	2	2	1	1–2	2–3
	Cam	–	–	–	–	2	2–3	2	2
	Cal	–	–	–	–	1	1	1	1
Grapevine	ER	1	1	1	1	1	1	1	1
	Ven	1	1	1	1	1	1	1	1
	Pie	1	1	1	2	1	2	1	1
	Lom	1	1	1	1	1	1	1	1
	Tre	1	1	2	1	1	1	1	1
	AA	1	1	1	1	1	1	1	1
	VDA	–	–	–	–	1	1	1	1
	FVG	1	1	2	1	1	1	1	1
	Tus	1	1	1	2	1	1	1	1
	Mar	–	–	1	1	1	1	1	1
	Abr	–	–	1	1	1	1	1	1
	Pug	–	–	–	1	1	1	1	1
	Cal	–	–	–	–	1	1	1	1
	Sar	–	–	–	–	–	1	1	1
Olive	Ven	–	–	2	/	3	2–3	2	2
	Lom	–	–	2	3	3	3	–	–
	Tre	–	–	–	2	2	3	2	2
	FVG	–	–	2	2	3	3	2	2
	Lig	–	–	–	–	3	3	2	1
	Umb	–	–	–	1	2	2	2–3	2–3
	Mar	–	–	–	–	1	1	–	–
	Abr	–	–	–	–	1	1	–	–
	Pug	–	–	–	–	1	1	–	–
	Cam	–	–	–	–	1	1	–	–
	Cal	–	–	–	–	1	1	–	–
Hazelnut	Ven	–	–	–	–	2	2	2	2
	Pie	–	3	2	2	2	2	2	2
	Lom	–	–	–	–	3	–	–	–
	Umb	–	–	–	1	2	2	2	2–3
	Laz	–	–	–	–	–	1	1	2
	Cam	–	–	–	–	2	2–3	2–3	3

(continued)

Table 15.4 (continued)

Crop	Region	2016	2017	2018	2019	2020	2021	2022	2023
Walnut	Ven	–	–	–	3	2	2	2	2
Soybean	Ven	–	–	2	1	2	2	2	2
	Lom	–	–	2	1	2–3	–	–	–
	FVG	–	2	2	1	1	1	–	–
	Umb	–	–	–	–	–	–	2	2–3
Maize	Ven	–	–	2–3	2	2	2	2	2
	Lom	–	–	2	2	–	–	–	–
Wheat	ER	–	–	1	2	1	1	1	1
	Ven	–	–	2	2	2	2	1	1
	Pie	–	–	2	1	1	2	1	1
	Lom	–	–	2	1	2	2	–	–
Pepper	Ven	–	–	–	–	2	2	2	2
	Pie	–	–	3	2	2	2	2	2
Tomato	ER	–	–	–	–	1	1	–	–
	Pie	–	–	–	–	1	1	1	1
	Umb	–	–	–	2	2	2	–	–
	Pug	–	–	–	–	1	1	–	–
	Cam	–	–	–	–	1	1	–	–
Green Bean	Pie	–	–	3	2	2	2	3	2
Bean	Pie	–	–	3	2	2	1	1	1–2

Damage caused by BMSB has also been reported to plants of forest interest; on poplars it causes bark necrosis and dieback, stem swelling, and trunk malformation in both nurseries and plantations (Gennaro and Giorcelli 2019).

4.3 Impact on the Quality of Fruit and Vegetables

The effect of BMSB on crops is not limited to physical damage. BMSB-affected fruit and vegetables are unsuitable for commercial production as they undergo metabolic and organoleptic changes that cause off-flavors. Zamljen et al. (2021a) evaluated the chemical composition of apple fruits (cv. 'Red Pinova') around puncture wounds from BMSB. The affected fruits showed a higher total sugar content, lower total organic acid content, and a higher total phenolic content compared to the control apples. The puncture wound area showed a strong phenolic response, with significant increases in flavanol and hydroxycinnamic acid contents. Weber et al. (2021) demonstrated that BMSB-affected 'Clery' strawberry exhibits a decrease in total sugar and an increase in shikimic acid, catechin, epicatechin and total phenolic content (ellagic acid derivatives, hydroxycinnamic acids, and anthocyanins contents).

The study by Zamljen et al. (2021b) on BMSB damage on 'Eris F1' hot peppers and 'Lombardo tago' non-hot peppers found reduced total sugar content and increased metabolites, particularly phenolics and capsaicinoids, in both varieties. Infested hot peppers showed a 15-fold increase in capsaicinoid content, while non-hot peppers showed a four-fold increase. After 1 month of cold storage, capsaicinoids levels decreased, but remained higher than control treatments, suggesting the feeding of BMSB on pepper fruits has a lasting impact.

Ivancic et al. (2022) examined the impact of BMSB infestation on phenolic response and olive fruit quality. Results showed significant increases in total phenolic content and antioxidant capacity in damaged fruits. Oleuropein was the most responsive compound, with increased content in pierced tissue of damaged fruits. An increase in phenolic compounds, specifically oleuropein and verbascoside levels, which are vital for the quality of olive oil, were noticed in a field study (Daher et al. 2023). Feeding by BMSB resulted in increased fruit dropping during the pre-pitting stage of the olive fruit.

The impact of BMSB on hazelnut is greater on early-ripening compared to late-ripening cultivars (De Benedetta et al. 2023). This can be attributed to the timing of stink bug attacks and the susceptibility of hazelnut fruits in different stages of development. Özdemir et al. (2023b) analysed the metabolic response of hazelnut kernels to feeding damage inflicted by BMSB. Phenolics compounds were significantly elevated in damaged kernels when compared to healthy ones. The feeding of BMSB caused a change in the fatty acid composition of hazelnut kernels, resulting in a decrease in oleic acid and an increase in linoleic and stearic acids.

Although BMSB feeds on grapevines, it appears to have no significant impact on grapevines and on wine production (Nannini et al. 2018), even during years when other crops suffered severe losses (Table 15.4). According to the results of laboratory analyses and wine-tasting tests after microvinifications carried out on Lambrusco, Pinot Grigio, Ribolla Gialla and Merlot with high quantities of dead or live bugs (3 per 150 g average bunch), the wines did not show the presence of specific aldehydes produced by the bugs nor any particular unpleasant taste (Tommasini et al. 2019; Benvenuto et al. 2020). Similarly, BMSB has been monitored in Bordeaux vineyards of France (Delbac et al. 2022), shortly after its first detection in Bordeaux in 2018 (Streito et al. 2021). Currently it is widely observed in Bordeaux, Cognac and Dordogne. According to the study conducted in 2021–22 in Bordeaux, the insect's prevalence was high, but the population level was very low. For instance, in 2022 there were 92% of plots with less than 2 individuals per 100 bunches, which is below harmful thresholds (Mohekar et al. 2017; Kehrli et al. 2021). The sorting system employed at harvest totally eliminates the insects on bunches (Delbac et al. 2023).

Kehrli et al. (2021) investigated how BMSB bugs influence the aroma and flavour of 'Chardonnay' and 'Merlot' musts and wines. They artificially contaminated grapes with varying BMSB densities before crushing and tested the resulting juices, musts, and wines. The study found that while BMSB contamination can affect grape juices and musts quality, it does not significantly affect the flavor of processed wines.

5 Natural Antagonists

Studies have been carried out in several European countries to determine the presence of natural antagonists of BMSB in the native fauna. Results from the early surveys were reviewed by Conti et al. (2021).

5.1 Parasitoids

An overview of studies carried out in Europe on egg parasitoids of the BMSB is given in Table 15.5. Different types of sampling have been used (see Table 15.5 for references). The collected eggs are kept in laboratories to check for hatching or the emergence of parasitoids, and examined under a microscope for signs of predation and other features. Sabbatini Peverieri et al. (2020b) describe in detail the characteristics of the meconia (faeces excreted by the larvae prior to pupation in the host egg) produced by different species of egg parasitoids attacking BMSB in Europe, providing a visual aid for identifying species and genus within the parasitoid guild. Emerged parasitoids are then identified using specific keys (Matsuo et al. 2014; Talamas et al. 2017; Tortorici et al. 2019; Moraglio et al. 2021b).

From the 51 studies presented (Table 15.5), 71% of which were related to naturally laid eggs, the native species *Anastatus bifasciatus* (Geoffroy) (Hymenoptera: Eupelmidae) (Fig. 15.1) was detected in 67% of cases, followed by the exotic species *Trissolcus japonicus* (Ashmead) (Hymenoptera: Scelionidae) (Fig. 15.2) (43%, detected in Switzerland, Italy, Germany and Serbia); *Acroclisoides sinicus* (Huang & Liao) (Hymenoptera: Pteromalidae) (Fig. 15.3) (33%, in Switzerland, Italy, Germany); *Trissolcus mitsukurii* (Ashmead) (Hymenoptera: Scelionidae) (Fig. 15.4) (33%, in Italy, Slovenia, France, Serbia); and the native ones: *Telenomus* spp. (Hymenoptera: Scelionidae) (29%, mostly represented by *T. turesis* Walker, followed by *T. chloropus* (Thomson); *Trissolcus basalis* (Wollaston) (Fig. 15.5, left) (24%); *Trissolcus cultratus* (Talamas, Johnson & Buffington) (24%), *Ooencyrtus* spp. (Hymenoptera, Encyrtidae) (18%, mostly *O. telenomicida* (Vassiliev), followed by *O. pityocampae* (Mercet) [Fig. 15.6, right]); *Trissolcus semistriatus* (Nees) (Fig. 15.6, left) (18%); *Trissolcus belenus* (Walker) (Fig. 15.5, right) (9%); *Trissolcus kozlovi* Ryakhovskii (6%).

Among the exotic species, *T. japonicus*, known as "samurai wasp", is considered the most efficient natural antagonist for BMSB (Zhang et al. 2017), together with *T. mitsukurii* (Arakawa and Namura 2002) and they are both native to Asia, where they occur in sympatry (Yonow et al. 2021). Adventive populations of *T. japonicus* were first detected in southern Switzerland in 2017 (Stahl et al. 2019a) and northern Italy in 2018 (Sabbatini-Peverieri et al. 2018b), whereas *T. mitsukurii* was first detected in Northeastern Italy in 2016 (Scaccini et al. 2020a). *Acroclisoides sinicus* is an obligate hyperparasitoid for species of Scelionidae (Giovannini et al. 2021) which was first detected in northeastern Italy in 2017 (Sabbatini-Peverieri et al.

Table 15.5 Parasitoids emerged from *Halyomorpha halys* eggs from field surveys in Europe. *EMC* Egg mass collection (*N* naturally laid egg masses, *D* devitalised sentinel egg masses, i.e. frozen/refrigerated, *A* alive/fresh sentinel egg masses, *S* suction sampling). *Eup* Eupelmidae, *Enc* Encyrtidae, *Pte* Pteromalidae, *Ab Anastatus bifasciatus, Oo Ooencyrtus sp., OT Ooencyrtus telenomicida, OP Ooencyrtus pityocampae, TrBa Trissolcus basalis, TrSe Trissolcus semistriatus, TrBe Trissolcus belenus, TrKo Trissolcus kozlovi, TrCu Trissolcus cultratus, TrCo Trissolcus colemanii, TrJa Trissolcus japonicus, TrMi Trissolcus mitsukurii, Tr sp. Trissolcus sp., Te Telenomus sp., TeTu Telenomus turesis, TeCh Telenomus chloropus, AcS Acroclisoides sinicus, NI* species not identified. The asterisk (*) in the column of *T. japonicus* indicates that in the indicated region a classic biocontrol program using this parasitoid was ongoing

				Eup.	Enc	Scelionidae									Pte		
Country	Region	Year	EMC	*Ab*	*Oo*	*Tr Ba*	*TrSe*	*TrBe*	*TrKo*	*TrCu*	*TrCo*	*Tr Ja*	*TrMi*	*Te*	*AcS*	NI	Reference
Switzerland	Jura, Valais	2013	D	X						X							Haye et al. (2015a)
	Ticino	2017–18	D									X					Stahl et al. (2019a)
	Zurich	2019	N									X			X		Sabbatini-Peverieri et al. (2019)
	Basel-stadt, Zurich, Bern,	2019-20-21	N	X	*OT*					X		X			X		Haye et al. (2023)
	Ticino		D									X				X	

Italy	ER	2014-15-16	A	X													Costi et al. (2018)
	Tus	2016	D	X	*OT*												Roversi et al. (2016)
	Pie	2015	N	X													Abram et al. (2017)
	Lom/FVG	2018	N	X								X	X				Sabbatini-Peverieri et al. (2018b)
	ER	2018–19	N	X													Francati et al. (2021)
	Pie/Lom	2016-17-18	N	X		X	X		X	X		X	X	*TeTu*	X		Moraglio et al. (2019)
	North	2016-17-18	N	X		X			X			X	X		X		Scaccini et al. (2020a)
	North	2017–2018	N	X								X	X		X		Sabbatini-Peverieri et al. (2019)
	North	2019	N	X		X	X	X		X		X	X	X	X		Zapponi et al. (2021)
	Tre/AA	2019	N	X		X		X		X			X	X	X		Zapponi et al. (2020)
			D	X								X	X		X		
			A	X								X	X				
	Pie	2019–20	N	X	*OT*					X		X	X	*TeTu*	X		Haye et al. (2023)
			A									X				X	
	Ven, FVG	2019–20	N	X									X		X		Mele et al. (2022)
	ER	2020	D	X	*OP*		X	X				X*					Costi, unp. data
			N	X	*OP*	X	X					X*	X	X	X	X	
	ER	2020	N	X								X*	X		X		Maistrello et al. (2022)
		2021	N	X								X*	X		X		
	Tre/AA	2020-21-22	N	X					X	X		X*	X	X	X	X	Falagiarda et al. (2023)

(continued)

Table 15.2 (continued)

Country	Region	Year	EMC	Eup.	Enc	Scelionidae									Pte	NI	Reference
				Ab	*Oo*	*Tr Ba*	*TrSe*	*TrBe*	*TrKo*	*TrCu*	*TrCo*	*Tr Ja*	*TrMi*	*Te*	*AcS*		
Georgia	Samegrelo	2017	N	X													Kereselidze et al. (2018)
	Nagobilevi (Guria)	2020	N							X							Japoshvili et al. (2022)
France	Different regions	2018-19-20	NDA	X	X								X	X		*Tr sp.*	Bout et al. (2021)
Slovenia	Western Slov.	2019–20	N	X		X							X	X			Rot et al. (2021)
			A	X	*OT*								X				
	Western Slov.	2021	N	X		X							X				Rot et al. (2022b)
Germany	Upper Rhine Valley northern Wuerttemberg	2020	N	X		X	X			X		X		X			Dieckhoff et al. (2021)
			S			X	X					X		X			
			D	X								X					
	Lake Constance	2021	D							X				X			Bauer Pilla et al. (2023)
	Lake Constance	2021	N							X		X		*TeCh*	X		Bauer Pilla et al. (2023)
	Lake Constance	2022	N							X		X			X		R. Bauer Pilla unp. data
	Lake Constance	2023	N									X			X		R. Bauer Pilla unp. data

Greece	Thessaloniki	2020	N	X	*OT*												Andreadis et al. (2021)
	Thessaloniki	2021	N		*OT*									*TeTu*			S. Andreadis, p.c.
	Imathia	2021	N								X						S. Andreadis, p.c.
	Imathia	2022	N	X													S. Andreadis, p.c.
Serbia	Srpski Miletić	2021	N			X		X									A.Konjevic & S. Popovici, unpublished data
	Novi Sad	2022	N	X		X		X				X	X				Konjevic et al. (2023, submitted)
	Sombor		N	X		X											
	Bački vinogradi		N	X		X											
	Srpski Miletić		N				X										
	Ribari		N			X								*TeTu*			
	Bački Petrovac		N											*TeTu*			
Turkey	Black Sea area	2021	N	X													Altanlar et al. (2023)
	Sakarya	2022	A											*TeTu*			Özdemir et al. (2023a)

Fig. 15.1 *Anastatus bifasciatus:* female (left), male (right). (Photos M. Violi)

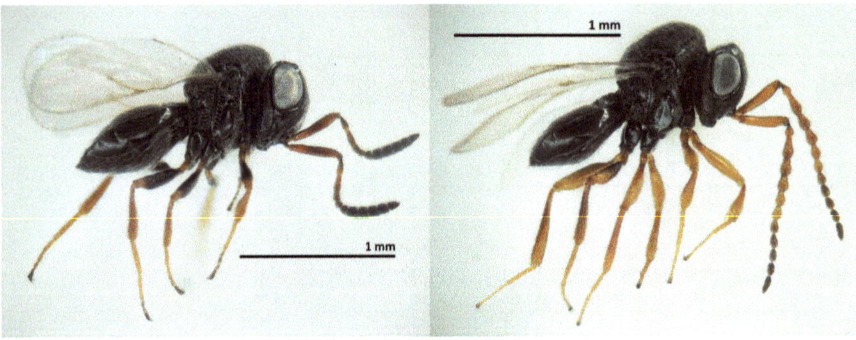

Fig. 15.2 *Trissolcus japonicus:* female (left), male (right). (Photos M. Violi)

Fig. 15.3 *Acroclisoides sinicus:* female (left), female on an eggmass of *Halyomorpha halys* (right). (Photos M. Violi)

Fig. 15.4 *Trissolcus mitsukurii:* female (left), male (right). (Photos M. Violi)

Fig. 15.5 Left: *Trissolcus basalis,* female; right: *Trissolcus belenus,* female. (Photos M. Violi)

Fig. 15.6 Left: *Trissolcus semistriatus,* female; right: *Ooencyrtus pityocampae*, female. (Photos M. Violi)

2019), and, according to Mele et al. (2021) it showed a clear preference for parasitising *T. mitsukurii*.

The actual pathway of entry of these exotic species is unknown, but it is likely to be the introduction of diapausing adults or plants with parasitised egg masses of Asian pentatomids. Molecular analyses indicate that the *T. japonicus* found in Switzerland and northern Italy originate from Japan (Sabbatini-Peverieri et al. 2018b; Stahl et al. 2019a); that the *T. mitsukurii* from northeastern Italy are genetically close to those from Japan and China (Scaccini et al. 2020a); and the European *Acroclisoides* did not originate from the USA but from an unknown region (Sabbatini-Peverieri et al. 2019).

In northeastern Italy, BMSB populations were high until 2017, with gradual decrease thereafter. This decrease is associated with a simultaneous increase in BMSB egg masses parasitised by *T. mitsukurii* (Malossini et al. 2020). According to Mele et al. (2022), adventive populations of *T. mitsukurii* are favoured by large patches of non-crop habitat, such as riparian vegetation buffers, which promote its presence in surrounding orchards, thus increasing parasitism on BMSB.

Some countries and regions have considerable egg parasitoid biodiversity, while others have very few species (Table 15.5). It is hypothesised that the reason for the severe depletion of the western Georgian parasitoid fauna recorded in 2019–20 could be an effect of extensive use of broad-spectrum insecticides in response to high populations of BMSB, which had caused significant economic losses in previous years (Japoshvili et al. 2022).

Egg parasitoids represent the vast majority of BMSB parasitoids. In Artvin (Turkey), in 2021, *Cylindromyia bicolor* (Olivier) (Diptera: Tachinidae) was found to parasitise adult of BMSB (Gokturk et al. 2023).

5.1.1 Competition, Tritrophic Interactions and Chemical Ecology

A number of studies have analysed the impact of BMSB establishment in European ecosystems to the community of natural enemies of native pentatomids. Rondoni et al. (2017) found that *A. bifasciatus* and *O. telenomicida* can parasitise BMSB and respond positively to both volatiles emitted by the stink bug and the host plants, whereas *T. basalis* only responded to female volatiles. Moreover, the co-occurrence of *Nezara viridula* and BMSB on the same plant inhibits the attraction of *T. basalis* to the volatiles emitted by *N. viridula*. This implies that the presence of BMSB disrupts the established semiochemical interactions in the tritrophic system (Martorana et al. 2017). According to Peri et al. (2021), BMSB acts as an evolutionary trap for *T. basalis*, resulting in the unsuccessful development of the majority of eggs laid by this host species.

Konopka et al. (2017) suggest that *T. japonicus* and *A. bifasciatus* engage in counterbalanced competition with *T. japonicus* being a superior extrinsic competitor (egg guarding and aggressiveness) and *A. bifasciatus* being a superior intrinsic competitor (successful development from multiparasitised eggs of all ages); they both can potentially coexist and act synergistically in the control of BMSB.

Trissolcus mitsukurii and *T. japonicus* are able to parasitise egg masses already parasitised by the other species, but the first species to oviposit had a reproductive advantage (Costi et al. 2022). When both species were released on the same egg mass, *T. mitsukurii* showed more aggressive behaviour. Overall, the study suggests that the biocontrol potential of parasitoids is maintained even in the presence of competition. Giovannini et al. (2022c) concluded that multiparasitism between *T. japonicus* and *T. mitsukurii* resulted in a general waste of search time, energy, eggs laid and reduced reproductive potential of the second arriving species on the host egg mass.

Arif et al. (2021) investigated the role of contact chemical cues in host egg-seeking behaviour and adult host sex discrimination of *T. japonicus*. Parasitoids spent more time exploring patches contaminated with chemicals associated with adult female stink bugs, and found that a non-polar chemical compound influences host sex discrimination. According to Malek et al. (2021) *T. japonicus* exhibited increased foraging behaviour in response to BMSB female footprints containing n-tridecane and (E)-2-decenal, which are likely to be key components of the kairomone used for host location. Synthetic n-tridecane was found to prolong the residence time of the parasitoid.

Scala et al. (2022) investigated the preference of the two exotic parasitoids for the chemical cues of the footprints of two stink bug species and found that *T. japonicus* preferred BMSB cues whereas *T. mitsukurii* preferred *Nezara viridula* cues, suggesting that it could act as a potential biocontrol agent for this latter species. Rondoni et al. (2022a) investigated the olfactory responses of *T. mitsukurii* to plants attacked by target and non-target stink bugs and found that it is attracted to plants exposed to feeding and egg deposition by BMSB and *N. viridula*, but is repelled by plants bearing eggs of the beneficial *Arma custos*, suggesting that this parasitoid has a limited host range and may pose little risk to non-target stink bugs.

Chierici et al. (2023) found that *T. japonicus'* olfactory responses to plant-host complex odours are influenced by its rearing host, emphasizing the need to consider host-induced phenotypic plasticity when evaluating parasitoids for biological control.

5.2 Predators

Data on egg predation averages 5–10% and generally it does not exceed 50%.

Table 15.6 summarises the results of six studies on the predation of BMSB by different arthropod species. Most of the studies were conducted in Italy (Castracani et al. 2017; Bulgarini et al. 2021a, c, 2022), one in Georgia (Kharabadze et al. 2022) and one in Azerbaijan (Nuriyeva and Nadirova 2020). The majority of the studies were carried out by testing, after a fasting period of at least 24 hours, the ability to prey on different BMSB instars by potential predator species collected in the field or obtained from a biocontrol company. These surveys identified several native and one introduced species (*Harmonia axyridis*) that are generalist insect predators and

Table 15.6 Recognized predators of *Halyomorpha halys* in Europe. Abbreviations: *A* adult, *FO* field observation, *L* Larva, *LA* laboratory, *MO* molecular analysis, *N* Nymph, *PI* predator instar, *SF* semi-field, *TS* type of study, *U* unknown

Order	Family	Genus/Species of the predator	PI	*Halyomorpha halys* instars								TS	Reference
				Eggs	N1	N2	N3	N4	N5	A	U		
Mantodea	Mantidae	*Hierodula transcaucasica* (Brunner von Wattenwyl, 1878)	A							X		LA	Kharabadze et al. (2022)
			N3		X							SF	Kharabadze et al. (2022)
Orthoptera	Mogoplistidae	*Arachnocephalus vestitus* (Costa, 1855)	U								X	MO	Bulgarini et al. (2022)
	Tettigoniidae	*Eupholidoptera chabrieri* (Charpentier, 1825)	A	X	X	X						LA	Bulgarini et al. (2021a)
		Phaneroptera falcata (Poda, 1761)	U								X	MO	Bulgarini et al. (2022)
		Tylopsis liliifolia (Fabricius, 1793)	U								X	MO	Bulgarini et al. (2022)
		Yersinella raymondi (Yersin, 1860)	U								X	MO	Bulgarini et al. (2022)
Dermaptera	Forficulidae	*Forficula auricularia* (Linnaeus, 1758)	U								X	MO	Bulgarini et al. (2022)

Heteroptera	Nabidae	*Himacerus mirmicoides* (Costa, 1834)	U								X	MO	Bulgarini et al. (2022)
			A		X							LA	Bulgarini et al. (2021a)
	Reduviidae	*Nagusta goedelii* (Kolenati, 1857)	U								X	MO	Bulgarini et al. (2022)
			A	X	X							LA	Bulgarini et al. (2021a)
		Rhynocoris iracundus (Poda, 1761)	A		X	X				X		LA	Bulgarini et al. (2021a)
Coleoptera	Coccinellidae	*Adalia bipunctata* (Linnaeus, 1758)	A	X								LA	Bulgarini et al. (2021a)
			L	X								FO	Nuriyeva and Nadirova (2020)
		Coccinella septempunctata (Linnaeus 1758)	L	X								FO	Nuriyeva and Nadirova (2020)
		Harmonia axyridis (Pallas, 1773)	U								X	MO	Bulgarini et al. (2022)
		Hippodamia variegata (Goeze, 1777)	U								X	MO	Bulgarini et al. (2022)
		Oenopia conglobata (Linnaeus, 1758)	U								X	MO	Bulgarini et al. (2022)
Neuroptera	Chrysopidae	*Chrysoperla carnea* (Stephens, 1836)	L2		X							LA	Bulgarini et al. (2021a)

(continued)

Table 15.6 (continued)

Order	Family	Genus/Species of the predator	PI	*Halyomorpha halys* instars								TS	Reference
				Eggs	N1	N2	N3	N4	N5	A	U		
Hymenoptera	Formicidae	*Crematogaster scutellaris (Olivier, 1792)*	A		X	X	X	X	X			LA	Castracani et al. (2017)
		Lasius niger (Linnaeus, 1758)	A			X	X					LA	Bulgarini et al. (2021c)
Opiliones	Phalangiidae	*Mitopus morio* (Fabricius, 1799)	U								X	MO	Bulgarini et al. (2022)
		Opilio canestrinii (Thorell, 1876)	U								X	MO	Bulgarini et al. (2022)
Araneae	Anyphaenidae	*Anyphaena* sp. (Sundevall, 1833)	U								X	MO	Bulgarini et al. (2022)
	Araneidae	*Araneus* sp. (Clerick, 1757)	U								X	MO	Bulgarini et al. (2022)
	Philodromidae	*Philodromus* sp. (Walckenaer, 1826)	U								X	MO	Bulgarini et al. (2022)
	Salticidae	*Calositticus* sp. (Lohmander, 1944)	U								X	MO	Bulgarini et al. (2022)

arachnids capable of preying on BMSB instars in Europe. The role of ants seems to be crucial. In particular, *C. scutellaris*, which is very common in Europe, both in natural and agricultural systems, arboreal and quite aggressive, proved able to prey on all juvenile instars except eggs (Castracani et al. 2017).

Schifani et al. (2023) found that ants and egg parasitoids can control BMSB without interfering with each other. Ants did not harm parasitized eggs or emerging parasitoids, and their presence slightly slowed parasitization in *T. mitsukurii*.

With regard to vertebrates, there is anecdotal evidence that some poultry, such as chickens and guinea fowl, are voracious predators of BMSB during the overwintering period when hundreds/thousands of bugs aggregate in farm buildings. Bats, specifically *Myotis* and *Nyctalus*, have been identified as predators of this invasive species in Piedmont, Aosta Valley, and Tuscany regions (Piemontese 2020).

5.3 Entomopathogens

Microsporidian pathogen *Nosema maddoxi* Becnel et al. was found as an infectious agent of BMSB in a survey conducted in Georgia (Kereselidze et al. 2019). This species is native to the United States and is known to infect three native stink bug species and BMSB (Hajek et al. 2018; Preston et al. 2020). BMSB samples of overwintering adults from the Guria region had the highest prevalence of *N. maddoxi*; laboratory bioassays indicate that *N. maddoxi*-infected BMSB display decreased fecundity and longevity (Preston et al. 2019).

Gorgadze et al. (2017) found entomopathogenic nematodes (local *Steinernema gurgistana*, *S. tbilisiensis*, *S. thesami* and introduced species *S. carpocapsae* and *Heterorhabditis bacteriophora*) and entomopathogenic fungi *Isaria fumosorosea* (Wize) to be more effective against nymphs than adults. The combined application of nematodes and fungi was more effective than application of them separately.

Burjanadze et al. (2020a) compared the efficacy of different strains of entomopathogenic nematodes from Georgia (*H. bacteriophora* (HRB, GEO), *Steinernema borjomiense* GS-BOR) and Italy (*H. bacteriophora* HRB (IT) IH-LU1, and *Steinernema apuliae* IS-MS3) on BMSB adults. The Italian strains HRB (IT) and *S. apuliae* showed higher suppressive efficacy compared to the Georgian strains. These findings suggest that entomopathogenic nematodes have potential as biocontrol agents of BMSB.

Tozlu et al. (2019) investigated the efficacy of various bacterial strains and the fungus *Beauveria bassiana*, in controlling BMSB nymphs. Mortality rates ranged from 75% to 100%. *Bacillus cereus* GC subgroup B and *Pantoea agglomerans* GC subgroup exhibited the highest potential control (100%).

Özdemír et al. (2022) assessed the effectiveness of native Turkish isolates of *Beauveria bassiana* and *B. pseudobassiana* against BMSB adult specimens. They caused mortality rates of 92% and 96% in BMSB adults.

Aşkin et al. (2022) evaluated the biological activities of commercial preparations with *Verticillium lecani, Beauveria bassiana* and *Metarhizium anisopliae*, as well as Azadiractin on adult BMSB. *V. lecani* was found to be the most effective (72% letal effect after 7 days); *B. bassiana* and *M. anisopliae* had 56% and 40% effects, respectively. Azadirachtin was the most effective (94%). However, abiotic effects, formulation and field-specific issues were identified as major limitations.

The virome of an Italian BMSB population has been studied by Giovannini et al. (2022a), leading to the identification of seven novel viral sequences, categorised into different taxonomic groups. Papa et al. (2023) investigated the virome associated with ten field populations of BMSB in Italy, using RNA-Seq data. The analysis revealed six complete viral genomes, three of which were previously unknown, classified in the orders of Reovirales, Articulavirales, Ghabrivirales, Durnavirales, and Picornavirales. Among these, *Halyomorpha halys* ifla-like virus 2 was the most widespread, detected in more than 50% of the insects investigated. Further investigation with virus-virus interactions and their relationship with insect hosts may lead to successful biocontrol of BMSB.

6 Field Monitoring

6.1 Monitoring Techniques

Immediately after the first discovery in Emilia, a three-year (2014–2016) monitoring programme was carried out to assess the abundance and impact of BMSB and other Heteroptera species in pear orchards and surrounding areas (Maistrello et al. 2017). Visual sampling was used on trees and shrubs, and sweep netting was used on grassy areas. The abundance of BMSB was highest in the late season and it was mainly found in orchard margins and hedgerows. A 2015 study (Maistrello et al. 2016b) compared different monitoring techniques, including visual inspection, tree beating, sweep netting and aggregation pheromone traps in pear orchards and adjacent uncultivated areas. The results showed that BMSB mainly target orchard borders and hedgerows, preferring the shaded upper parts of plants. Early in the season, all monitoring techniques showed similar detection levels, with traps showing better performance from mid-June onwards. BMSB populations increased progressively overtime causing significant yield losses, particularly in orchard borders.

Aggregation pheromone traps can attract stink bugs from a distance, but they are not effective at capturing individuals; few are caught, some escape and/or remain outside the trap on surrounding plants where they can cause damage ("trap spillover") (Sargent et al. 2014). In a study in pear orchards using Rescue traps, Vaccari et al. (2018) found that damage was particularly high on plants within 4–5 m of the traps, while acceptable levels of damage were found on plants more than 10 m from the traps.

A joint European (Hungary, Greece, Switzerland, various regions in Italy) and US (Maryland) research project (Morrison et al. 2017) conducted in 2016 showed that despite having different haplotypes in different locations, BMSB were equally attracted to traps baited with aggregation pheromone and synergist (MDT), concluding that pheromone-based technology can be useful for BMSB monitoring globally.

A work carried out in 2016–2018 (Vaccari et al. 2019) aimed to identify the most reliable system for monitoring BMSB in pear orchards in Emilia Romagna, evaluating the effectiveness of different trap designs. Results showed that the combination of Trécé baits with pyramid models (either Rescue or AgBio) provided better trap reliability in pear orchards. The results also indicate that the correct positioning of the Rescue trap is essential for its proper functioning, i.e. ensuring that the base is securely tied and in contact with the trunk or branch of the plant on which it is installed.

In a collaborative study between the USA and Europe, researchers evaluated different loading rates of pheromones (2-component aggregation pheromone + the synergist MDT), dispenser substrate and the inclusion of an additional compound (EDT). Results indicated that reducing the pheromone loading rate had a significant effect on trap catches, but increasing the synergist loading rate could offset this reduction; the standard bait dispenser outperformed experimental polyethylene dispensers (Leskey et al. 2021).

The use of aggregation pheromone traps is currently widespread in Europe, both for early detection in countries of recent introduction and for monitoring BMSB populations for management purposes. As of 2020, a territorial monitoring project 'CIMICE.NET' has been activated in Emilia-Romagna to provide real-time information on BMSB populations (Vaccari et al. 2023). The project uses pheromone traps, visual sampling and tree-beating to monitor the pest. Monitoring data is collected, processed and displayed on an open-access online platform (OpenStreetMap Contributors 2020), enabling stakeholders to access timely and reliable information for effective pest management strategies (Forresi et al. 2024).

6.2 Monitoring Innovations

A novel multimodal trap, using LED lights, synthetic aggregation pheromone, and odours from trapped individuals, demonstrated that by combining UV-A with blue or green visible wavelengths, attraction to BMSB increased by up to eight times compared to conventional sticky or small pyramid traps (Rondoni et al. 2022b).

To address the "trap spillover" of aggregation pheromone based traps, a bi-modal trap has been developed that incorporates an aggregation pheromone dispenser, a recording of a female vibrational signal, and an electric shock device to kill the attracted bugs (Polajnar et al. 2019). An enhanced version of the "vibrotrap" was subsequently created, employing the optimized vibrational signal of female FS2 signal, which was confirmed to be extremely attractive to male BMSB both in confined settings and field trials. Field trials showed that, despite FS2 being selected to target males, there was also an increase in the number of trapped females. The results indicate that the utilization of both semiophysicals and semiochemicals in conjunction can increase the efficacy of BMSB traps (Zapponi et al. 2023).

To improve catch rates several baited traps with different aggregation pheromone formulations and design have been developed. Among them are the "Nazgûl" and the "rotating live" traps with average improvement 3.5 to 7 times, compared to sticky panels (Suckling et al. 2019a, b).

An autonomous data collection system has been designed and implemented by an international multidisciplinary research team within the 'HALY.ID' project (HALY.ID 2023) to detect BMSB in real-time. This innovative approach eliminates the drawbacks and limited reliability of aggregation pheromone-based traps, as well as the time-consuming and labor-intensive process of BMSB scouting traditionally carried out by plant protection operators. The designed hardware architecture comprises a UAV (Unmanned Aerial Vehicle), an RGB vision chip, a novel IoT trap, as well as micro-climate stations. Furthermore, it investigates the application of deep learning models to recognize BMSB using a dataset of images (Almstedt et al. 2023). A specific application was created that gathers data on the presence of BMSB by flying a drone near the trees and capturing videos and photos (Betti-Sorbelli et al. 2022). The proposed method was executed on a DJI drone and verified using a real-world environment (a BMSB infested orchard in northern Italy). Images of the orchard were gathered with an RGB camera-equipped drone and other devices, including smartphones. Despite the drone camera's suboptimal image quality, a dedicated computer vision algorithm for BMSB detection was able to produce positive outcomes (Betti-Sorbelli et al. 2023a). Automated decision-making in BMSB monitoring was optimised by refining Machine Learning (ML) algorithms and correcting image imperfections (Betti-Sorbelli et al. 2023b). Field trials conducted in pear orchards demonstrated that adult BMSB reacted by increasing their freezing behavior in the presence of a drone and increasing their flying behavior in the presence of a walking operator. This behaviour endorses drone usage for image acquisition and the automatic recognition using ML algorithms (Giannetti et al. 2024).

A further step in the same project was to overcome the problem of BMSB mimicry by evaluating the potential of Near Infrared Hyperspectral Imaging (NIR-HSI). In a laboratory experiment, hyperspectral images of BMSB were acquired in the 980–1660 nm range, considering different vegetation backgrounds, which had been selected to mimic a real field application scene (Ferrari et al. 2023). Classification models were developed using chemometric approaches and convolutional neural networks to analyse the spectral and spatial features of the hyperspectral images. The results showed promising classification performances, and the merged strategies combining spectral and spatial based methods improved the detection capability of the models.

7 Integrated Management Approaches

BMSB is extremely difficult to manage in the field, due to its high polyphagy, high mobility of all instars, high rate of population growth and the overlap of different instars and generations during the summer. In European countries farmers have increased the use of broad-spectrum insecticides that disrupted integrated pest management (IPM) programmes and caused negative economic and environmental impacts (Maistrello et al. 2017; Japoshvili et al. 2022).

IPM guidelines were modified to cope with the increasing damage caused by BMSB in pear orchards in Emilia Romagna (Table 15.7). Prior to 2014, when

Table 15.7 Changes over time in IPM guidelines (considering insecticides only) for pears in Emilia Romagna

Year	Chemical group	Active substances (max n° allowed treatments)	Max n° allowed treatments per crop cycle
2014–15	Organophosphates	Clorpirifos methyl (max 1) Clorpirifos (max 1) Fosmet (max 2)	4
	Neonicotinoids	Acetamiprid	1
	Total		**5**
2016	Organophosphates	Clorpirifos methyl (max 2) Clorpirifos (max 1) Fosmet (max 2)	5
	Neonicotinoids	Acetamiprid	1
	Total		**7**
2017	Organophosphates	Clorpirifos methyl (max 1) Clorpirifos (max 1) Fosmet (max 2)	4
	Neonicotinoids	Acetamiprid (max 2)	2
	Total		**6**
2018–19	Organophosphates	Clorpirifos methyl (max 2) Clorpirifos (max 1) Fosmet (max 2)	4
	Neonicotinoids	Acetamiprid (max 2)	2
	Pyrethroids	Deltamethrin Tau-fluvalinate Etofenprox	2
	Total		**8**
2020	Organophosphates	Clorpirifos methyl (max 2) Fosmet (max 2)	4
	Neonicotinoids	Acetamiprid (max 2)	2
	Pyrethroids	Deltamethrin Tau-fluvalinate Lambda cihalotrina Etophenprox	3
	IGR	Triflumuron	2
	Total		**11**
2021–22	Organophosphates	Fosmet	2
	Neonicotinoids	Acetamiprid (max 2)	2
	Pyrethroids	Deltamethrin Tau-fluvalinate Lambda cihalotrina Etophenprox	3
	IGR	Triflumuron	2
	Total		**9**
2023	IGR (ecdisone mimic)	Tebufenozide	2
	Neonicotinoids	Acetamiprid	2
	Pyrethroids	Deltamethrin Tau-fluvalinate Lambda cihalotrina Etophenprox	3
	Total		**7**

codling moth and pear psylla were the main pests, a maximum of five treatments were allowed per crop cycle, with organophosphates and neonicotinoids. As BMSB pressure increased, pyrethroids had to be included and the number of allowed treatments increased (up to 11 in 2020) between 2019 and 2022. From 2023, the banning of certain active ingredients resulted in a reduction of authorised insecticide types and numbers. The situation was similar for other crops affected by BMSB in all regions of northern Italy. It was therefore immediately necessary to identify other strategies to rationalise the management of BMSB.

7.1 The Importance of the Landscape

BMSB has a strong dispersal capacity and spreads at landscape level. Bosco et al. (2020) investigated the dispersal capacity and behaviour of overwintered BMSB adults in relation to host plants before the colonisation of hazelnut crops in four polyculture areas in Piedmont (north western Italy). They found that BMSB adults responded to pheromones after overwintering and exhibited both short- and long-range dispersal from overwintering sites and/or early host plants to wild and crop plants. Hybrid plane, European spindle tree, walnut, oak, and European elder were the first plants on which overwintered BMSB were observed to feed.

Tamburini et al. (2023) found that landscapes rich in semi-natural habitats and vineyards and poor in annual crops supported a higher abundance of BMSB juveniles in traps, whereas adult abundance was not affected by landscape. The study confirmed a high dispersal capacity and suggests that landscapes influences BMSB populations and crop colonisation dynamics.

7.2 Physical Barriers

The use of physical barriers, such as exclusion nets, is one of the most readily available tools for crop protection and an environmentally friendly alternative to pesticides. Exclusion nets, which have proven effective in controlling codling moth in France and Italy (Alaphilippe et al. 2016), have been tested by Caruso et al. (2019) in commercial pear orchards. The results show that different types of netting can significantly reduce BMSB damage and the need for insecticide treatments. These positive results have led to the implementation of exclusion netting in orchards with the support of regional administrations in all regions affected by BMSB.

In a field trial in north western Italy (Candian et al. 2018), exclusion nets significantly reduced damage by 45% in peaches compared to untreated trees or insecticide treated trees, and by 20% in apples compared to insecticide treatments. The nets had no negative effect on the arthropod fauna or fruit quality. According to Pozzebon et al. (2023), white netting and early deployment are recommended for optimal effectiveness.

7.3 Behaviour-Based Strategies

Given that most BMSB damage is mainly found in the edges of cultivated fields, a method has been developed that profitably exploits this behavior. The Integrated Pest Management—Crop Perimeter Restructuring (IPM-CPR) strategy was developed in peach orchards in the USA (Blaauw et al. 2015), using border spraying to intercept and eliminate BMSB in the orchard perimeter, preventing it from damaging the central part. This method was applied in field trials on pear orchards in Emilia in 2016–2017 (Caruso et al. 2018). The results show no significant differences in the fruit damaged between IPM-CPR and conventionally treated orchards, but the use of insecticides was reduced by 44–50% with the adoption of IPM-CPR.

Attract-and-kill (AK) strategies are used to control pests while minimizing contact between pesticides and crops. These strategies involve using an attractant and an insecticide to kill pests. A first AK approach, using barriers of pyrethroids-impregnated nets, was effective in killing BMSB in orchards in Emilia Romagna (Sabbatini Peverieri et al. 2018a). A second approach, using pyramid-like structures covered with a pyrethroid-impregnated net, reduced BMSB abundance but did not reduce fruit damage (Masetti et al. 2023). Technical issues and the ban on insecticide-treated nets highlighted the need for alternative methods. A third approach involves black glued panels baited with aggregation pheromones over a bottom tank filled with soapy water. According to Scaccini et al. (2023) these d-evices reduced infestation levels, suggesting their potential for monitoring and mass trapping strategies.

Field trials (Batistič et al. 2023) tested the effectiveness of attracting BMSB to different trap crops in Slovenia to protect orchards and showed that sorghum was the most attractive to adults, followed by soybean and sunflower.

The employment of repellent compounds shows promise to control BMSB, when used alongside specific attractants like the aggregation pheromone. This can help drive pests away from fields and implement a push-pull approach (Cook et al. 2007). Bulgarini et al. (2021a, b, c) found that turmeric and clove oils were effective in repelling adult BMSB insects when the concentration was over 3%. Sulphur-based products can also be used as a tool in integrated pest management strategies for fruit crops (Mirandola et al. 2023).

7.4 Use of Natural Substances

Among the natural alternatives to synthetic insecticides, geomaterials and plant extracts were tested. Results showed that kaolin was more effective in reducing pest damage than zeolite materials (Preti et al. 2023).

The development and application of polyphenol-based microencapsulated extracts of *Stevia rebaudiana* and *Aronia melanocarpa* as green insecticides against BMSB was studied; microparticles loaded with *Stevia* extract showed high pesticidal efficacy in both contact and oral treatments, especially on nymphs, whereas those loaded with *Aronia* extract showed some toxic effects on all growth stages of contact application (Pajač-Živković et al. 2020). Gokturk (2021) demonstrated *Satureja spicigera* essential oil toxicity against nymphs and adults BMSB. Essential oils and their terpenic constituents have potential as fumigants (Finetti et al. 2021a).

7.5 Modeling and Innovative Approaches

7.5.1 Modeling Approaches

Bout et al. (2022) developed a model based on the concept of degree days, which predicts egg-laying periods and the appearance of early nymphal stages.

Lazar et al. (2023) proposes a simple model for the development of BMSB, which used quadratic equations to calculate the duration of each instar under variable temperatures, based on the dataset from Costi et al. (2017). The results show that the model can reasonably simulate the observed phenology of BMSB, with a high correlation between observed and simulated start days of different developmental stages.

Rossini et al. (2023) developed a mathematical model using ordinary differential equations to describe the population dynamics of BMSB and its interaction with the environment in order to predict infestations in hazelnut orchards.

Grilo (2023) used data collected over 3 years in Emilia Romagna through the CIMICE.NET project (OpenStreetMap Contributors 2020) to develop a spatial-temporal model. The model is based on the SSD algorithm, which is typically used to produce estimation models in geological or climatic contexts. The geostatistical tools utilized by the model provide a simplified 3D representation of the location and density of BMSB over time, which was also valuable in validating the forecast model.

After the first records in the sea port of Odessa, a study was performed to determine the potential suitable bioclimatic range of BMSB in Ukraine (Tytar and Kozynenko 2020). The bioclimatic modeling results indicate that the pest has diverse likelihoods of invasion in the near future under Ukraine's current climate conditions, with southern regions particularly vulnerable to invasion.

7.5.2 RNA-Based Treatments

Finetti et al. (2021b) has characterized HhTAR1, a gene that belongs to type 1 tyramine receptor and plays a crucial role in detecting the alarm pheromone (E)-2-decenal through olfactory reception. RNAi silencing of HhTAR1 in nymphs decreases their sensitivity to (E)-2-decenal, and this represents a promising novel approach to managing BMSB. The same group tested the efficiency of double-stranded (ds) RNA-based treatment by applying it to BSMB topically (Finetti et al. 2023). The study revealed that specific genes, including IAP and ATPase, caused a more significant mortality rate in the nymphs, making them practical for controlling the BMSB population with minimal non-target effects.

7.5.3 Symbiotic Control

An innovative approach to the management of BMSB is symbiotic control by targeting their obligate bacterial symbionts, specifically '*Candidatus* (*Pantoea*) *carbekii*', which is vertically transmitted through maternal secretions and acquired by newborns from the egg surface. Gonella et al. (2019) tested micronutrient fertilizers with antimicrobial activity on BMSB egg masses and found that zinc, copper, and citric acid biocomplexes were most effective in disrupting symbiont acquisition, resulting in high nymph mortality. Orrù et al. (2023) found no natural symbiont acquisition by native and exotic egg parasitoids after Dentamet® micronutrient biocomplex treatment, indicating that parasitoid fitness is unlikely to be altered by dysbiotic effects and supporting further implementation of symbiotic control in combination with biological control. Gonella et al. (2022) found significant mortality in BMSB nymphs treated with Amylo-X® (a commercial formulation containing *Bacillus velezensis*) due to its potential antagonistic activity and elimination of the symbiont from the egg surface.

7.5.4 Irradiation

Roselli et al. (2023b) conducted irradiation experiments on overwintered male BMSB, achieving sterility rates of over 95% without significant longevity effects. The study provides insights into the use of Sterile Insect Technique as a component of area-wide BMSB management strategies. Irradiation can induce complete sterility in BMSB eggs and the performance of the sterile eggs as a substrate for oviposition and larval development of the parasitoid *T. japonicus* varies with the age of the eggs (Roselli et al. 2023a).

7.5.5 Precision Agriculture

Liakos et al. (2023) studied BMSB populations on kiwi orchards using precision agriculture techniques. They found higher populations in areas with greener, healthy canopies with high leaf density and high moisture content. The study also shows a correlation between altitude and population density, with valuable information for predicting and managing BMSB populations.

8 Biological Control

8.1 Use of Native Parasitoids

As classical biological control with exotic agents was initially not feasible in European countries, the use of the native egg parasitoid *A. bifasciatus* for augmentative biological control of BMSB was investigated. Stahl et al. (2019b) studied the reproductive parameters, longevity, phenology and temperature requirements for the development of this species and found that it completes three generations between June and October, with an average lifetime fecundity of 52 eggs over 46 days. Over 3 years, *A. bifasciatus* was released in four orchards in Switzerland and northern Italy, with release densities ranging from 11,000 to 26,000 females per hectare. Egg parasitism on frozen sentinel egg masses was 6%, not high enough to effectively suppress the pest. In addition, parasitism of non-target lepidopteran species reached an average rate of 8% (Stahl et al. 2019c). Maistrello et al. (2020) conducted a study in Emilia, Italy, using weekly inundative releases of *A. bifasciatus* at a density of 10,000 females per hectare. The results showed an average parasitisation of 9.4% in the release area, which was not significantly different from the control (6.6%). Bernardinelli et al. (2020) conducted a similar trial in Friuli Venezia Giulia with 10,400 adult females released in hedgerows. The results showed very low parasitisation and no differences between treated and untreated areas. In the study by Iacovone et al. (2022), augmentative releases of *A. bifasciatus* were carried out at 11 sites in Trentino-Alto Adige (northern Italy) at a release density of 1000 individuals per hectare of orchard (325,000 in total). The results showed that the parasitism rate of *A. bifasciatus* on BMSB naturally laid eggs was 1.2% in the control sites whereas it increased to 16.7% in the release sites.

The release of *A. bifasciatus* in agroecosystems can increase parasitism on BMSB eggs, but is not sufficient to significantly reduce populations. This is probably due to the fact that these polyphagous parasitoids, once released, can also parasitise the eggs of other species. A risk assessment study confirmed the ability of *A. bifasciatus* to attack non-target species, including Lepidoptera of conservation interest (Stahl et al. 2018).

Other native parasitoids have also been considered as candidates for BMSB biocontrol. During a 4-year study, Moraglio et al. (2021a) collected the egg parasitoids emerging from native Pentatomidae and Scutelleridae eggs in north western Italy and tested their ability to parasitise fresh BMSB eggs in laboratory no-choice experiments. The results showed that all *Trissolcus* species induced abortion of host eggs, but only *T. kozlovi* successfully produced offspring from fresh BMSB eggs. Moraglio et al. (2021b) found that *T. kozlovi* showed a physiological host range quite similar to that of *T. japonicus*. After field releases in hazelnut orchards, *T. kozlovi* parasitised BMSB eggs, but only immediately after field release.

Tunca et al. (2019) suggests that BMSB is a suitable host for *O. pityocampae*, supporting its potential use as biocontrol agent.

8.2 Classical Biological Control with Exotic Parasitoids

Classical biological control is seen as a feasible and environmentally sustainable long-term solution for managing invasive pests. However, strict international regulations (EPPO 2014; FAO and IPPC 2021) are in place to limit the potential impact on biodiversity. In Europe, new exotic biocontrol agents must undergo approval after a non-target risk analysis, requiring studies on their physiological host range.

Haye et al. (2020) conducted behavioral tests to evaluate the acceptance and suitability of European non-target host species for *T. japonicus*. In no-choice tests, eleven out of thirteen non-target species were suitable for development, with four species (*Arma custos, Palomena prasina, Pentatoma rufipes, Rhaphigaster nebulosa*) showing acceptance levels similar to BMSB. In large-arena choice tests, only parasitism on *Palomena prasina* was not significantly different from BMSB. The simplified laboratory protocol proposed by Sabbatini-Peverieri et al. (2021) confirmed *T. japonicus* oligophagy and allowed host egg morphology to be correlated with acceptance rates. Haye et al. (2023) examined the host range of *T. japonicus* by exposing egg masses of BMBS and 18 non-target species in Switzerland and northern Italy, in areas where adventive populations of this species occur. Results show that 15 out of 18 non-target species were successfully parasitised, however, at lower rates than BMSB, and that *Pentatoma rufipes* eggs was the most parasitised non-target species.

Giovannini et al. (2022b) showed that in southern Europe *T. mitsukurii* has an oligophagous host range restricted to certain species within the Pentatomoidea superfamily, comparable to that of *T. japonicus*, and with high parasitisation rates on eggs of BMSB, *Acrosternum heegeri* and *Dolycoris baccarum*.

Optimizing laboratory rearing procedures is crucial for successful biological control programs. Sabbatini-Peverieri et al. (2020a) provides guidelines for the large-scale production of *T. japonicus* and *T. mitsukurii* to control BMSB and suggest that offering host eggs to host-deprived females once a week for three weeks can optimize host resources and maximize progeny production. Bittau et al. (2021) suggest that to improve efficiency in *T. japonicus* mass rearing, BMSB egg masses can be stored effectively at 6 °C for several months, while adult parasitoids can be stored at 16 °C for up to 90 days with minimal decrease in survival.

The collection of eggs at release sites is crucial for verifying the effectiveness of biological control programs on target species and non-target species. This is typically done using naturally laid eggs or sentinel eggs from laboratory rearing.

8.2.1 Classical Biocontrol Programs

Due to its specificity and high parasitisation rate, *T. japonicus* was identified as the most promising BMSB biocontrol agent for the classical biocontrol in Italy, France and Turkey.

In Italy, following the devastating economic impact of BMSB on fruit production and the repercussions on the entire agro-food chain, a decree by the Italian Ministry of the Environment and Protection of the Land and Sea finally authorised *T. japonicus* for the biocontrol of BMSB (MATTM 2020), leading to one of the largest biocontrol projects ever attempted in Italy and Europe. Coordinated by the Central Phytosanitary Service and the CREA-DC (Council for Agricultural Research and Economics—Division of Plant Protection and Certification), the "National Action Programme for the Biological Control of BMSB" was created by a technical committee with members of regional phytosanitary services and scientists. It outlines the procedures for mass rearing of the parasitoid, the criteria for selecting release sites, the number of parasitoids per release unit, the criteria for evaluating the success of the programme and the impact on non-target insects (Servizio Fitosanitario Centrale 2020). Launched in the summer of 2020, the program has been implemented for 3 years, exclusively in the Italian regions/autonomous provinces that were authorized: initially Emilia Romagna, Veneto, Lombardy, Piedmont, Friuli Venezia Giulia, Trento and Bolzano ("Group 1"), followed by Liguria, Marche, Umbria, Sardinia, Campania ("Group 2") in 2021 and Tuscany ("Group 3") in 2022. The programme, developed as a concerted effort of several institutions (regional and local plant protection services, IPM operators, universities and other research organisations, companies, farmers' associations) aimed to establish a stable population of *T. japonicus* and achieve an ecological balance to keep BMSB at manageable levels. The parasitoids (100 females +10 males/each release, 2 releases/site) were released during the peak BMSB egg laying periods (late June and mid-August) at sites (selected by the regional plant protection services) close to productive crops, in environments free of chemical inputs, such as ecological corridors, hedgerows, riverbanks and forests. In total, several thousand *T. japonicus* were released (Fig. 15.7).

To verify the parasitization of *T. japonicus* on target and non-target insects, naturally laid eggs of both BMSB and other Pentatomoidea were collected from at least 10% of the release sites. Each Region/autonomous Province sent the results to the CREA DC and an annual report was produced, which was shared with the Central Plant Protection Service and the Ministries of Agriculture and Environment. A summary of the results obtained after 3 years in "Group" 1 regions (AA.VV. 2023) is shown in Fig. 15.7, after permission was obtained from all parties involved in the project.

The overall results show a progressive increase in the number of sites with *T. japonicus* as well as in its parasitisation on BMSB. It should be noted that parasitisation, also due to the adventive populations already present in northern Italy, was extremely variable among the different sites (e.g. for 2022 in Emilia Romagna from 0.3% to 48%, in Veneto from 0.01% to 85%, in Piedmont from 4.5% to 69%). The parasitoid was also found in sites where it had only been released in the first year, indicating successful overwintering and establishment. The total impact of *T. japonicus* on non-target species was 0.62% in 2020, 0.47% in 2021 and 0.72% in 2022, mainly on *Palomena prasina* and *Pentatoma rufipes*, especially in the autonomous provinces of Bolzano and Trento. Parasitisation of BMSB eggs by other

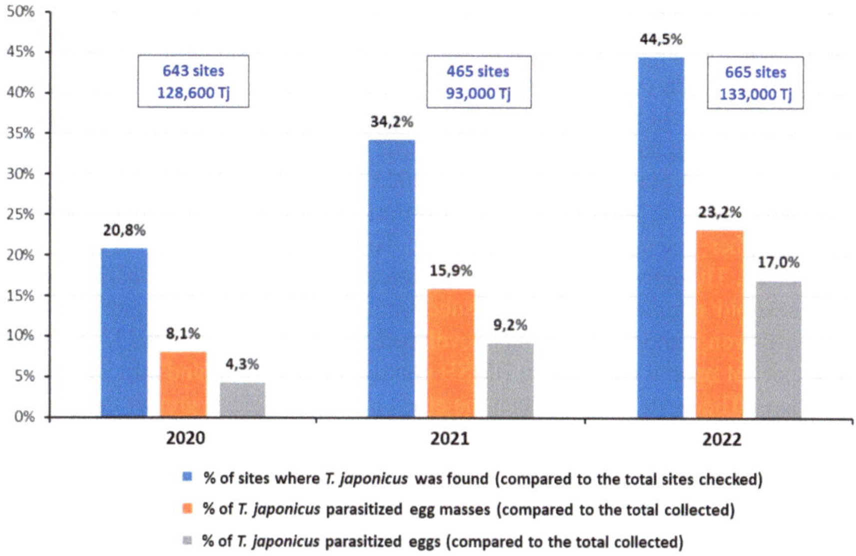

Fig. 15.7 Results of the Italian 3-year biocontrol programme of *Halyomorpha halys* for Group 1 regions. In the boxes framed in blue, the total number of release sites and released females of *Trissolcus japonicus* (Tj) for each year. Data collected and managed by the Regional Phytosanitary Services of Friuli Venezia Giulia, Veneto, Lombardia, Piemonte, Emilia-Romagna, Province Autonome di Bolzano e Trento and Scientific Support Institutions (CREA, Università di Torino, Centro di Sperimentazione Laimburg, Fondazione Edmund Mach, Università di Padova, Università di Modena e Reggio Emilia) identified in DG DISR—DISR 05—Prot. N.0036832, 18/11/2019

species was also recorded during these 3 years (see in Table 15.5 the sites where *T. japonicus* is marked with an asterisk).

In conjunction with the national project, each region has carried out further surveys, collecting egg masses also from sites where releases have never been carried out. In Trentino Alto Adige, *T. japonicus* was the most abundant parasitoid of BMSB with minimal impact on non-target insects. The efficacy of *T. mitsukurii* was lower in sites where *T. japonicus* was established. BMSB and *T. japonicus* were more common at lower altitudes and in permanent crops. Predation and parasitisation resulted in up to 50% BMSB mortality at release sites (Falagiarda et al. 2023). A similar mortality was also observed by Maistrello et al. (2022) in Emilia Romagna where the most abundant parasitoid was *A. bifasciatus* followed by *T. mitsukurii, A. sinicus* and then by *T. japonicus.*

In France, the INRAE (National Research Institute for Agriculture, Food and the Environment) received legal authorisation from the Ministries of Agriculture and Environment in early 2023 to release a strain of *T. japonicus* in the field (the same strain used for the Italian releases). The first releases will take place in 2024 in the PACA and Nouvelle Aquitaine regions, and should be extended to other sites and regions in subsequent years (A. Bout p.c.).

In Turkey, the Ministry of Agriculture authorities have decided to introduce the classical biological control method for BMSB in the long term by importing *T. japonicus* into the country. The authorised institution is the Black Sea Agricultural Research Institute in Samsun, which established facilities for mass rearing of parasitoids in 2021 and imported the first parasitoid species from Italy in 2022. The first field releases took place in the summer of 2023 (Tuncer and Ozdemir 2023).

A study using the MaxEnt algorithm (Tortorici et al. 2023), found that *T. japonicus* has higher potential suitability on a global scale and at lower latitudes than *T. mtsukurii*. These results could enhance biological control effectiveness by targeting parasitoid releases in the most successful areas. The physiologically based weather-driven demographic model developed by Gutierrez et al. (2023) evaluates the impact of natural enemies on BMSB control under current and climate change conditions. The model suggests that the effect of *T. japonicus* is greater than that of *T. mitsukurii*, but their parasitism rates are insufficient for economic control of BMSB populations.

9 Final Considerations

BMSB has rapidly colonized almost all of continental Europe in less than 20 years. Its arrival in Italy had a devastating impact on agriculture and it became one of the most serious phytosanitary emergencies ever faced. This led to the disruption of previous IPM programmes and, therefore, to the need to carry out research aimed at better understanding its bioecology and identifying a series of strategies for its management, culminating in the implementation of the largest biocontrol projects ever attempted in Italy and Europe. Other European countries are currently experiencing increasing crop losses and have initiated their own research studies, monitoring and management programmes.

The purpose of this work was to assess the presence and distribution of BMSB in Europe and its impact on crops. It also covers various bio-ecological aspects, with a focus on egg parasitoids, and provides an overview of current and past efforts in European countries to improve monitoring and management. Research is progressing and several innovative technologies are on the horizon that could help enhance the tools and strategies available to manage this invasive pest in an increasingly sustainable way.

Acknowledgements The following people are thanked for providing support, data and updates from different European countries. Albania: Ejup Çota; Austria: Georg Friebe, Christa Lethmayer; Belgium: Tim Belien, Olivier Berteloot, Stéphane Claerebout; Bosnia-Herzegovina: Nedzad Karic; Bulgaria: Nikolay Simov; Croatia: Lucija Šerić Jelaska, Ivana Pajač Živković; Czech Republic: Petr Kment, Jiříslav Skuhrovec; Finland: Sari Haikola; France: Bertrand Alison, Alxandre Bout, Lionel Delbac, Jean-Claude Streito; Georgia: George Japoshvili, Natalia Kharabadze, Manana Kereselidze, Maka Murvanidze; Germany: Christine Dieckhoff, Ricardo Bauer Pilla, Christian Scheer; Greece: Stefanos Andreadis, Petros Damos; Hungary: Katalin

Radácsiné Hári, Jozsef Fail, David Redei; Iceland: Matthías Alfreðsson, Thorhildur Isberg; Italy: Gianfranco Anfora, Gianni Anselmo, Loredana Antoniacci, Massimo Bariselli, Walter Baruzzo, Luca Benvenuto, Iris Bernardinelli, Pietro Bertolotto, Alessandra Bianchi, Stefano Boncompagni, Rita Bonfanti, Roberta Bonsignore, Andrea Bordoni, Giovanni Bosio, Maurizio Bottura, Giacomo Bulgarini, Giovanni Burgio, Alda Butturini, Giuseppe Campo, Arturo Caponero, Stefano Caruso, Beniamino Cavagna, Michele Cesari, Eliana Consolani, Stefano Colazza, Mario Contarini, Eric Conti, Elena Costi, Paolo Culatti, Fabrizio Dagostin, Sandro Dallou, Domenico D'Ascenzo, Dalia Del Nista, Emanuele Di Bella, Davide Di Domenico, Lorenzo Di Martino, Paride Dioli, Stefano Endrizzi, Martina Falagiarda, Luca Fagioli, Bruno Caio Faraglia, Stefanie Fischnaller, Tiziano Galassi, Massimiliano Ghironi, Daniele Giannetti, Roberto Giordano, Massimo Giorgini, Massimo Gragnani, Donato Antonio Grasso, Roberto Guidetti, Claudio Ioriatti, Laura Loru, Vincenzo Maione, Alan Maistrello, Flavio Pietro Maistrello, Giorgio Malossini, Antonio Masetti, Federica Manucci, Valerio Mazzoni, Alberto Mele, Teresa Mirandoli, Elena Monari, Silvia Moraglio, Patrizia Nappa, Sandro Nardi, Claudio Panizza, Niccolò Patelli, Roberto Pantaleoni, Tarasca Pardo, Lucia Piemontese, Sabrina Pintus, Michele Preti, Alberto Pozzebon, Gabriele Rondoni, Pio Federico Roversi, Giuseppino Sabbatini-Peverieri, Martino Salvetti, Angela Sanchioni, Agostino Santomauro, Davide Scaccini, Silvia Schmidt, Davide Serra, Stefano Speranza, Ivana Stella, Luciana Tavella, Lorenza Tessari, Rocchina Tiso, Flaviagrazia Tropiano, Grazia Tommasini, Francesco Tortorici, Michael Unterthurner, Giacomo Vaccari, Michele Violi, Robert Wiedmer, Giovanni Zanini; Netherlands: Herman Helsen, Karin Winkler; Luxembourg: Tiago De Sousa, Julien Reiners, Francesco Vitali, Alexander Weigand; Montenegro: Zorka Prljević; Portugal: Silvia Castro, Hugo Gaspar, Luis Miguel Moleiro Grilo, João Loureiro; Romania: Ioana Grozea; Russia: Natalia Karpun, Ilya M. Mityushev, Dmitrii Musolin, Sergei Reznik; Serbia: Aleksandra Konjevic, S. Šestić; Slovenia: Mojca Rot, Stanislav Trdan, Ivan Žežlina; Spain: Adriana Escudero; Sweden: Axel Benediktsson, Johanna Jansson, Philip Zandelin; Switzerland: Barbara Egger, Tim Haye, Cristina Marazzi, Cornelia Sauer; Turkey: İsmail Oğuz Özdemir, Celal Tuncer; Ukraine: Tatiana Stefanovska; United Kingdom: Michelle Fountain, Francis Warmonje.

The author thanks the following projects and funders: the Grant "Innovative tools and protocols for monitoring and sustainable control of the alien stink bug *Halyomorpha halys*, a new phytosanitary threat, and of other harmful heteropterans for the fruit crops of the territory of Modena" (2013.065) of "Fondazione Cassa di Risparmio di Modena" Modena, Italy; the projects HALYS, HALY.BIO, A&K, ALIEN.STOP, CIMICE.NET, funded by the Emilia Romagna Rural Development Plan 2014-2020 (EIP-Agri Programme); the HALY.ID project, as part of ERA-NET Cofund ICT-AGRI-FOOD, with funding provided by national sources (Ministero delle Politiche Agricole e Forestali, MIPAAF) and co-funding by the European Union's Horizon 2020 research and innovation program, Grant Agreement number 862665 ERA-NET ICT-AGRI-FOOD (HALY-ID 862671). From 2023 further research is ongoing under the project INTEGR.HALYS funded by Emilia Romagna Region (L.R. 17/2022).

References

AA VV (2023) Report 2022 Lotta biologica alla cimice asiatica *Halyomorpha halys*: risultati rilascio *Trissolcus japonicus* e integrazioni in riferimento alla delibera SNPA 158/2022. GRUPPO 1: Regioni Friuli Venezia Giulia, Veneto, Lombardia, Piemonte, Emilia-Romagna e Province Autonome di Bolzano e Trento. Internal Report

Abram PK, Hoelmer KA, Acebes-Doria A, Andrews H, Beers EH, Bergh JC, Bessin R, Biddinger D, Botch P, Buffington ML, Cornelius ML, Costi E, Delfosse ES, Dieckhoff C, Dobson R, Donais Z, Grieshop M, Hamilton G, Haye T, Hedstrom C, Herlihy MV, Hoddle MS, Hooks CRR, Jentsch P, Joshi NK, Kuhar TP, Lara J, Lee JC, Legrand A, Leskey TC, Lowenstein D, Maistrello L, Mathews CR, Milnes JM, Morrison WR, Nielsen AL, Ogburn EC, Pickett CH,

Poley K, Pote J, Radl J, Shrewsbury PM, Talamas E, Tavella L, Walgenbach JF, Waterworth R, Weber DC, Welty C, Wiman NG (2017) Indigenous arthropod natural enemies of the invasive brown marmorated stink bug in North America and Europe. J Pest Sci 90:1009–1020

Ak K, Uluca M, Aydin Ö, Gokturk T (2019) Important invasive species and its pest status in Turkey: *Halyomorpha halys* (Stål) (Heteroptera: Pentatomidae). J Plant Dis Prot 126:401–408

Ak K, Uluca M, Tuncer C (2023a) Current status of brown marmorated stink bug in Turkey. Acta Hortic 1379:417–424

Ak K, Uluca M, Tuncer C (2023b) Distribution and population density of *Halyomorpha halys* (Stål, 1855) (Hemiptera: Pentatomidae) in Black Sea Region of Türkiye. Turk J Zool 47:120–129

Alaphilippe A, Capowiez Y, Severac G, Simon S, Sandreau M, Caruso S, Vergnani S (2016) Codling moth exclusion netting: an overview of French and Italian experiences. IOBC-WPRS Bull 112:31–35

Aletta Ő, Szabolcs S, Attila M, Antal N (2021) Adatok az invazív márványos poloska (*Halyomorpha halys* Stal, 1855) telelési sikeréről városi környezetben (Heteroptera). Acta Nat Pannonica 22:59–69

Alison B, Berud M, Bout A, Drouzy N, Mercier A-S (2020) Les punaises phytophages en verger de fruits à pépins – Enseignements et perspectives du projet SUPOR. Infos CTIFL 367:37–43

Almstedt L, Baltieri D, Betti Sorbelli F, Cattozzi D, Giannetti D, Kargar A, Maistrello L, Navarra A, Niederprüm D, O'Flynn B, Palazzetti L, Patelli N, Piccinini L, Pinotti C, Wolf L, Zorbas D (2023) Technological Innovations in Agriculture for Scouting Halyomorpha halys in Orchards. In: 2023 19th International Conference on Distributed Computing in Smart Systems and the Internet of Things (DCOSS-IoT). IEEE, Pafos, Cyprus, pp 1–8

Altanlar E, Kiliç M, Altaş K, Talamas E, Tuncer C (2023) First record of native egg parasitoid, *Anastatus bifasciatus*, on *Halyomorpha halys* (Stål, 1855) (Hemiptera: Pentatomidae) eggs in Türkiye. Kahramanmaraş Sütçü İmam Üniversitesi Tarım Ve Doğa Derg 26:545–551

Andreadis SS, Navrozidis EI, Farmakis A, Pisalidis A (2018) First evidence of *Halyomorpha halys* (Hemiptera: Pentatomidae) infesting kiwi fruit (Actinidia chinensis) in Greece. J Entomol Sci 53:402–405

Andreadis SS, Gogolashvili NE, Fifis GT, Navrozidis EI, Thomidis T (2021) First report of native parasitoids of *Halyomorpha halys* (Hemiptera: Pentatomidae) in Greece. Insects 12:984

Andreadis SS, Moysiadis T, Gogolashvili NE, Koutsogeorgiou EI, Fifis GT, Tsaliki E (2022) Quality degradation of industrial hemp due to infestation by *Halyomorpha halys* (Hemiptera: Pentatomidae). Int J Pest Manag. https://doi.org/10.1080/09670874.2022.2089769

Arakawa R, Namura Y (2002) Effects of temperature on development of three *Trissolcus* spp. (Hymenoptera: Scelionidae), egg parasitoids of the brown marmorated stink bug, *Halyomorpha halys* (Hemiptera: Pentatomidae). Entomol Sci 5:215–218

Arif MA, Alınç T, Guarino S, Colazza S, Cusumano A, Peri E (2021) Only females oviposit: Chemical discrimination of adult stink bug sex by the egg parasitoid *Trissolcus japonicus*. Front Ecol Evol 9:652561

Arnold K (2009) *Halyomorpha halys* (Stål, 1855), eine für die europäische Fauna neu nachgewiesene Wanzenart (Insecta: Heteroptera, Pentatomidae, Pentatominae, Cappaeini). Mitteilungen Thüring Entomol 16:19

Arter project group (2023) Species recordings from the Danish National portal Arter.dk. Miljøstyrelsen / The Danish Environmental Protection Agency. https://www.gbif.org/dataset/963a6b96-4d22-4428-86e4-afee52cf4a8e. Accessed 1 Nov 2023

Aşkin AK, Yiğit Ş, Saruhan İ, Akça İ (2022) Biyolojik Preparatların Halyomorpha halys (Stål, 1885) (Hemiptera: Pentatomidae)'a Karşı Etkinliğinin Belirlenmesi. Kahramanmaraş Sütçü İmam Üniversitesi Tarım Ve Doğa Derg 25:100–104

Aukema B, den Bieman K, Lommen G, Maat G, Troisfontaine L, Vossen P (2019) Nieuwe en interessante Nederlandse wantsen IX (Hemiptera: Heteroptera) [New and intersting Dutch bugs IX (Hemiptera: Heteroptera)]. 52:25–40

Bariselli M, Bugiani R, Maistrello L (2016) Distribution and damage caused by *Halyomorpha halys* in Italy. EPPO Bull 46:332–334

Batistič L, Bohinc T, Trdan S (2023) Seasonal dynamics and abundance of brown marmorated stink bug *Halyomorpha halys* (Stål) on four trap crops. Plant Prot Sci 59:264–277

Bauer Pilla R, Scheer C, Trautmann M (2022) Die Marmorierte Baumwanze. Aktuelle Situation in der Bodensee-Region. Obstbau Fachmag 4:416–421

Bauer Pilla R, Trautmann M, Bischoff R, Scheer C (2023) Parasitization of *Halyomorpha halys* Stål (Heteroptera: Pentatomidae) by *Trissolcus japonicus* (Ashmead) (Hymenoptera: Scelionidae) and other parasitoids in the Lake Constance region (Germany). Pheromones Semiochem Integr Prod Integr Prot Fruit Crops IOBC-WPRS Bull 166:154–158

Benvenuto L, Malossini G, Bernardinelli I, Colussi G, Masino F, Antonelli A, Anaclerio F, De Candido M (2020) Indagine sugli effetti della presenza di Halyomorpha halys nel processo di vinificazione delle varietà pinot grigio, ribolla gialla e merlot in Friuli Venezia Giulia. Atti Giornate Fitopatol 1:195–202

Bernardinelli I, Benvenuto L, Malossini G, Raso R, Oian B, Zampa C (2020) Valutazione dell'efficacia di Anastatus bifasciatus per il controllo biologico della cimice marmorata asiatica in Friuli Venezia Giulia. Atti Giornate Fitopatol 1:177–184

Betti Sorbelli F, Palazzetti L, Pinotti C (2023a) Preliminary Results for Halyomorpha halys Monitoring Relying on a Custom Dataset. In: 2023 5th IEEE International Workshop on Metrology for Agriculture and Forestry (MetroAgriFor). IEEE, Pisa, pp 1–6

Betti Sorbelli F, Palazzetti L, Pinotti CM (2023b) YOLO-based detection of Halyomorpha halys in orchards using RGB cameras and drones. Comput Electron Agric 213:108228

Betti-Sorbelli F, Corò F, Das SK, Di Bella E, Maistrello L, Palazzetti L, Pinotti CM (2022) A Drone-based application for scouting *Halyomorpha halys* bugs in orchards with multifunctional nets. In: 2022 IEEE International Conference on Pervasive Computing and Communications Workshops and other Affiliated Events (PerCom Workshops). IEEE, Pisa, pp 127–129

Bittau B, Dindo ML, Burgio G, Sabbatini-Peverieri G, Hoelmer KA, Roversi PF, Masetti A (2021) Implementing mass rearing of *Trissolcus japonicus* (Hymenoptera: Scelionidae) on cold-stored host eggs. Insects 12:840

Blaauw BR, Polk D, Nielsen AL (2015) IPM-CPR for peaches: incorporating behaviorally-based methods to manage *Halyomorpha halys* and key pests in peach. Pest Manag Sci 71:1513–1522

Bohinc T, Jelnikar J, Strauss G, Sinkovič T, Trdan S (2022) Occurrence of brown marmorated stink bug (*Halyomorpha halys*, Hemiptera, Pentatomidae) in urban area (Ljubljana, Slovenia). IOBC-WPRS Bull – Landsc Manag Funct Biodivers 156:109–112

Bosco L, Moraglio ST, Tavella L (2018) *Halyomorpha halys*, a serious threat for hazelnut in newly invaded areas. J Pest Sci 91:661–670

Bosco L, Nardelli M, Tavella L (2020) First insights on early host plants and dispersal behavior of *Halyomorpha halys* (Hemiptera: Pentatomidae) from overwintering to crop colonization. Insects 11:866

Bout A, Tortorici F, Hamidi R, Warot S, Tavella L, Thomas M (2021) First detection of the adventive egg parasitoid of *Halyomorpha halys* (Stål) (Hemiptera: Pentatomidae) *Trissolcus mitsukurii* (Ashmead) (Hymenoptera: Scelionidae) in France. Insects 12:761

Bout A, Jail R, Hamidi R, Gachein EA (2022) Prédire les pics de présence d'*Halyomorpha halys*. Phytoma 755:9–14

Bozsik G, Kerezsi V, Kontschán J (2021) Attraction of adults of *Halyomorpha halys* (Stål, 1855) and *Nezara viridula* (Linnaeus, 1758) (Hemiptera: Pentatomidae). Acta Zool Bulg 73:451–455

Bulgarini G, Badra Z, Leonardi S, Maistrello L (2021a) Predatory ability of generalist predators on eggs, young nymphs and adults of the invasive *Halyomorpha halys* in southern Europe. BioControl 66:355–366

Bulgarini G, Bortolini S, Maistrello L (2021b) Repellent activity of essential oils on adults of *Halyomorpha halys* (Heteroptera: Pentatomidae) in different physiological-behavioural phases. J Appl Entomol 145:575–586

Bulgarini G, Castracani C, Mori A, Grasso DA, Maistrello L (2021c) Searching for new predators of the invasive *Halyomorpha halys*: the role of the black garden ant Lasius Niger. Entomol Exp Appl 169:799–806

Bulgarini G, Piemontese L, Guidetti R, Cesari M, di Bella E, Maistrello L (2022) Identification of predatory arthropods of the invasive *Halyomorpha halys* through molecular gut content analysis. Agric For Entomol 24:219–228

Burjanadze M, Gorgadze O, De Luca F, Troccoli A, Lortkipanidze M, Kharabadze N, Arjevanidze M, Fanelli E, Tarasco E (2020a) Potential of native entomopathogenic nematodes for the control of brown marmorated stink bug *Halyomorpha halys* in Georgia. Biocontrol Sci Tech 30:962–974

Burjanadze M, Kharabadze N, Chkhidze N (2020b) Testing local isolates of entomopathogenic microorganisms against brown marmorated stink bug *Halyomorpha halys* in Georgia. BIO Web Conf 18:00006

Bury J (2021) Pierwsze stwierdzenie *Halyomorpha halys* (Stål, 1855) (Hemiptera: Heteroptera: Pentatomidae) w Beskidzie Wschodnim / First records of *Halyomorpha halys* (Stål, 1855) (Heteroptera: Heteroptera: Pentatomidae) in the Eastern Beskids. Heteroptera Pol—Acta Faun 15:11–14

Bury J, Mazepa J (2022) Nowe dane o rozmieszczeniu lądowych pluskwiaków różnoskrzydłych (Hemiptera: Heteroptera) wybranych lokalizacji południowo-wschodniej i północno-wschodniej Polski—część 2. Heteroptera Pol—Acta Faun 16:15–36

Callot H-J, Brua C (2013) *Halyomorpha halys* (Stål, 1855), la punaise diabolique, nouvelle espèce pour la faune de France (Heteroptera Pentatomidae). L'Entomologiste 69:69–71

Candian V, Pansa MG, Briano R, Peano C, Tedeschi R, Tavella L (2018) Exclusion nets: a promising tool to prevent Halyomorpha halys from damaging nectarines and apples in NW Italy. Bull Insectology 71:21–30

Caorsi V, Cornara D, Wells KE, Moser D, Berardo A, Miselli R, Torriani M, Pugno NM, Tasin M, Maistrello L, Mazzoni V (2021) Design of ideal vibrational signals for stinkbug male attraction through vibrotaxis experiments. Pest Manag Sci 77:5498–5508

Carapezza A, Lo Verde G (2017) First record of Halyomorpha halys (Stål, 1855) (Hemiptera Pentatomidae) in Sicily. Nat Sicil 6:139–145

Caruso S, Vaccari G, Zanetti G, Maistrello L (2018) Gestione del perimetro del frutteto per il controllo integrato di *Halyomorpha halys*. CLUEB srl 1:320

Caruso S, Vergnani S, Vaccari G, Maistrello L (2019) Managing *Halyomorpha halys* by means of exclusion netting: trials 2016-2018. Pheromones Semiochem Integr Prod Integr Prot Fruit Crops IOBC-WPRS Bull 146:108–110

Castracani C, Bulgarini G, Giannetti D, Spotti FA, Maistrello L, Mori A, Grasso DA (2017) Predatory ability of the ant *Crematogaster scutellaris* on the brown marmorated stink bug *Halyomorpha halys*. J Pest Sci 90:1181–1190

Çerçi B (2021) First record of *Halyomorpha halys* (Stål, 1855) (Pentatomidae: Heteroptera) in Aegean Region of Turkey. Acta Biol Turc 34:35–37

Çerçi B, Koçak Ö (2017) Further contribution to the Heteroptera (Hemiptera) fauna of Turkey with a new synonymy. Acta Biol Turc 30:121–127

Cesari M, Maistrello L, Ganzerli F, Dioli P, Rebecchi L, Guidetti R (2015) A pest alien invasion in progress: potential pathways of origin of the brown marmorated stink bug *Halyomorpha halys* populations in Italy. J Pest Sci 88:1–7

Cesari M, Maistrello L, Piemontese L, Bonini R, Dioli P, Lee W, Park C-G, Partsinevelos GK, Rebecchi L, Guidetti R (2017) Genetic diversity of the brown marmorated stink bug *Halyomorpha halys* in the invaded territories of Europe and its patterns of diffusion in Italy. Biol Invasions 20:1073–1092

Chartois M, Streito J-C, Pierre É, Armand J-M, Gaudin J, Rossi J-P (2021) A crowdsourcing approach to track the expansion of the brown marmorated stinkbug *Halyomorpha halys* (Stål, 1855) in France. Biodivers Data J 9:e66335

Cherpitel T, Casset L (2018) *Halyomorpha halys* (Stål, 1855), la punaise diabolique, atteint la façade atlantique (Heteroptera Pentatomidae). L'Entomologiste 75:59–60

Chierici E, Sabbatini-Peverieri G, Roversi PF, Rondoni G, Conti E (2023) Phenotypic plasticity in an egg parasitoid affects olfactory response to odors from the plant–host complex. Front Ecol Evol 11:1233655

Cianferoni F, Graziani F, Dioli P, Ceccolini F (2018) Review of the occurrence of *Halyomorpha halys* (Hemiptera: Heteroptera: Pentatomidae) in Italy, with an update of its European and World distribution. Biologia (Bratisl) 73:599–607

Cianferoni F, Graziani F, Ceccolini F (2019) The unstoppable march of *Halyomorpha halys*: new first country records (Hemiptera, Pentatomidae). Spixiana 42:60

Ciceoi R, Mardare E, Teodorescu E, Dobrin I (2016) The status of brown marmorated stink bug, *Halyomorpha halys*, in Bucharest area. J Hortic For Biotechnol 20:18–25

Ciceoi R, Badulescu L, Minodora G, Ivan E, Pomohaci C-M (2017a) Citizen-generated data on invasive alien species in Romania: trends and challenges. Acta Zool Bulg 9:255–260

Ciceoi R, Bolocan IG, Ionela D (2017b) The spread of brown marmorated stink bug, Halyomorpha halys, in Romania. J Hortic For Biotechnol 21:15–20

Ciceoi R, Dumbrava M, Jerca IO, Pomohaci CM, Dobrin I (2017c) Assessment of the damages on maize crop by the invasive stink bugs *Halyomorpha halys* (Stål, 1855) and *Nezara viridula* (Linnaeus, 1758) (Hemiptera: Pentatomidae). Acta Zool Bulg 9:211–217

Ciceoi R, Gutue C, Gutue M, Roşca I (2017d) Current status of pests associated with urban vegetation in Bucharest area. Acta Zool Bulg 9:181–190

Claerebout S, Haye T, Ólafsson E, Pannier É, Bultot J (2018) Premières occurrences de *Halyomorpha halys* (Stål, 1855) pour la Belgique et actualisation de sa répartition en Europe (Hemiptera: Heteroptera: Pentatomidae). Bull Société R Belge D'Entomologie 154:2015–2227

Conti E, Avila G, Barratt B, Cingolani F, Colazza S, Guarino S, Hoelmer K, Laumann RA, Maistrello L, Martel G, Peri E, Rodriguez-Saona C, Rondoni G, Rostas M, Roversi P, Sforza R, Tavella L, Wajnberg E (2021) Biological control of invasive stink bugs: review of global state and future prospects. Entomol Exp Appl 169:28–51

Cook SM, Khan ZR, Pickett JA (2007) The use of push-pull strategies in integrated pest management. Annu Rev Entomol 52:375–400

Corsini-Foka M, Zava B, Insacco G (2021) Occurrence of *Halyomorpha halys* (Stål, 1855) (Hemiptera: Pentatomidae) in Rhodes Island, Greece. J Insect Biodivers 29:12–15

Costi E, Haye T, Maistrello L (2017) Biological parameters of the invasive brown marmorated stink bug, *Halyomorpha halys*, in southern Europe. J Pest Sci 90:1059–1067

Costi E, Haye T, Maistrello L (2018) Surveying native egg parasitoids and predators of the invasive *Halyomorpha halys* in Northern Italy. J Appl Entomol 143:299–307

Costi E, Di Bella E, Iotti D, Maistrello L (2022) Biocontrol implications of multiparasitism by *Trissolcus mitsukurii* and *Trissolcus japonicus* on the invasive brown marmorated stink bug. Entomol Exp Appl 170:772–781

CSO Italy – Centro Servizi Ortofrutticoli (2019) Estimation of damage from brown marmorated stink bug and pathologies related to climate change. www.csoservizi.com

Daher E, Chierici E, Urbani S, Cinosi N, Rondoni G, Servili M, Famiani F, Conti E (2023) Characterization of olive fruit damage induced by invasive *Halyomorpha halys*. Insects 14:848

Damos P, Soulopoulou P, Thomidis T (2020) First record and current status of the brown marmorated sting bug *Halyomorpha halys* damaging peaches and olives in northern Greece. J Plant Prot Res 60:323–326

De Benedetta F, Giaccone M, Pica F, Lisanti MT, Vinale F, Turrà D, Giacca GM, Bernardo U (2023) Fruit phenology of two hazelnut cultivars and incidence of damage by *Halyomorpha halys* in treated and untreated hazel groves. Horticulturae 9:727

De Michele A, Grozea I (2018) Review of the spreading of *Halyomorpha halys* in Italy and confirmation of presence in Romania. Res J Agric Sci 50:111–115

Delbac L, Rouzes R, Hamidi R, Thiéry D (2022) First occurrence of *Halyomorpha halys* in Bordeaux vineyards. OENO One 56:253–257

Delbac L, Laflaquière L, Achilli P, Rouzes R (2023) Evaluation of risk related to *Halyomorpha halys* in the Libournais wine region. In: Proceedings of IOBC-WPRS Meeting, Working Group Integrated Protection in Viticulture (IOBC-WPRS 2023), Logroño (Spain), October 3–5

Derjanschi V, Chimişliu C (2019) Ploşniţa marmorată *Halyomorpha halys* (Stål, 1855) (Heteroptera, Pentatomidae)—specie alogenă invazivă nouă în fauna Republicii Moldova. Bul Ştiinţ Rev Etnografie Ştiinţ Nat Şi Muzeol 30:18–22

Devetak M, Hoblaj S, Rot M, Jančar M (2023) Ugotavljanje vpliva marmorirane smrdljivke (*Halyomorpha halys*) na odpadanje plodov oljk. Oljka Novice Druš Oljkarjev 28:13–15

Dieckhoff C, Wenz S, Renninger M, Reißig A, Rauleder H, Zebitz CPW, Reetz J, Zimmermann O (2021) Add Germany to the list—Adventive population of *Trissolcus japonicus* (Ashmead) (Hymenoptera: Scelionidae) emerges in Germany. Insects 12:414

Dioli P, Leo P, Maistrello L (2016) Prime segnalazioni in Spagna e in Sardegna della specie aliena *Halyomorpha halys* (Stål, 1855) e note sulla sua distribuzione in Europa (Hemiptera, Pentatomidae). Rev Gaditana Entomol 7:539–548

Dumbadze G, Gokturk T, Jgenti L, Chelidze N (2019) Distribution of brown marmorated stink bug *Halyomorpha halys*. Bioecological features and control mechanisms in Batumi (Georgia). In: 4th international symposium on innovative approaches in engineering and natural sciences proceedings. SETSCI, pp 539–542

Egri Á, Mészáros Á, Kriska G, Fail J (2023) Dichromacy in the brown marmorated stink bug? Spectral sensitivity of the compound eyes and phototaxis of *Halyomorpha halys*. J Pest Sci. https://doi.org/10.1007/s10340-023-01655-z

EPPO (2013) EPPO Alert List. In: Eur. Mediterr. Plant Prot. Organ. EPPO. http://www.eppo.int/QUARANTINE/Alert_List/alert_list.htm. Accessed 25 Nov 2013

EPPO (2014) PM 6/2 (3) import and release of non-indigenous biological control agents. EPPO Bull 44:320–329

Escudero-Colomar LA (2020) Biologia, ecologia, danys i monitoratge de les poblacions de bernat marró marbrejat. In: Guia tècnica. El bernat marró marbrejat *Halyomorpha halys* (Hemiptera, Heteroptera, Pentatomidae). Departament d'Agricultura, Ramaderia, Pesca i Alimentació, Generalitat de Catalunya, Barcelona, pp 23–30

Falagiarda M, Carnio V, Chiesa SG, Pignalosa A, Anfora G, Angeli G, Ioriatti C, Mazzoni V, Schmidt S, Zapponi L (2023) Factors influencing short-term parasitoid establishment and efficacy for the biological control of *Halyomorpha halys* with the samurai wasp *Trissolcus japonicus*. Pest Manag Sci 79:2397–2414

FAO, IPPC (2021) Pest risk analysis for quarantine pests. International Standard for Phytosanitary Measures 11. Available online: https://www.ippc.int/en/publications/639

Faúndez EI, Rider DA (2017) The brown marmorated stink bug *Halyomorpha halys* (Stål, 1855) (Heteroptera: Pentatomidae) in Chile. Arq Entomolóxicos 17:305–307

Ferrari V, Calvini R, Boom B, Menozzi C, Rangarajan AK, Maistrello L, Offermans P, Ulrici A (2023) Evaluation of the potential of near infrared hyperspectral imaging for monitoring the invasive brown marmorated stink bug. Chemom Intell Lab Syst 234:104751

Finetti L, Civolani S, Bernacchia G (2021a) Monoterpenes-induced toxicity in nymphal stages of *Halyomorpha halys*. J Plant Dis Prot 128:1371–1375

Finetti L, Pezzi M, Civolani S, Calò G, Scapoli C, Bernacchia G (2021b) Characterization of *Halyomorpha halys* TAR1 reveals its involvement in (E)-2-decenal pheromone perception. J Exp Biol 224:jeb238816

Finetti L, Benetti L, Leyria J, Civolani S, Bernacchia G (2023) Topical delivery of dsRNA in two hemipteran species: Evaluation of RNAi specificity and non-target effects. Pestic Biochem Physiol 189:105295

Finnish Biodiversity Information Facility (2023) Hemiptera Eastern Fennoscandia (Luomus). https://www.gbif.org/dataset/53f222a9-d397-42a3-b05b-ee4ee1bcefb6. Accessed 1 Nov 2023

Forresi C, Gallinucci E, Golfarelli M, Maistrello L, Preti M, Vaccari G (2024) A data platform for real-time monitoring and analysis of the brown marmorated stink bug in northern Italy. Ecol Inform 82:102713

Francati S, Masetti A, Martinelli R, Mirandola D, Anteghini G, Busi R, Dalmonte F, Spinelli F, Burgio G, Dindo ML (2021) *Halyomorpha halys* (Hemiptera: Pentatomidae) on kiwifruit in Northern Italy: Phenology, infestation, and natural enemies assessment. J Econ Entomol 114:1733–1742

Friebe JG, Zimmermann K, Ritter E, Gort M (2019) Streudaten zur Fauna Vorarlbergs. I. Zugewanderte und selten dokumentierte einheimische Wanzen (Heteroptera). Inatura Forschung Online 65:1–10

Gandjaeva LA, Hudayberdieva MO, Abdullaev II, Mirzayeva GS, Yusupboev EK (2022) First record of *Halyomorpha halys* (Heteroptera: Pentatomidae) from Uzbekistan. Zoosyst Ross 31:329–331

Gapon DA (2016) First records of the brown marmorated stink bug *Halyomorpha halys* (Stål, 1855) (Heteroptera, Pentatomidae) in Russia, Abkhazia, and Georgia. Entomol Rev 96:1086–1088

Gariepy TD, Haye T, Fraser H, Zhang J (2014) Occurrence, genetic diversity, and potential pathways of entry of *Halyomorpha halys* in newly invaded areas of Canada and Switzerland. J Pest Sci 87:17–28

Gariepy TD, Bruin A, Haye T, Milonas P, Vétek G (2015) Occurrence and genetic diversity of new populations of *Halyomorpha halys* in Europe. J Pest Sci 88:451–460

Gariepy TD, Musolin DL, Konjević A, Karpun NN, Zakharchenko VY, Zhuravleva EN, Tavella L, Bruin A, Haye T (2021) Diversity and distribution of cytochrome oxidase I (COI) haplotypes of the brown marmorated stink bug, *Halyomorpha halys* Stål (Hemiptera, Pentatomidae), along the eastern front of its invasive range in Eurasia. NeoBiota 68:53–77

Garrouste R, Nel P, Nel A, Horellou A, Pluot-Sigwalt D (2015) *Halyomorpha halys* (Stål 1855) in Ile de France (Hemiptera: Pentatomidae: Pentatominae): Monitoring the brown marmorated stink bug. Ann Soc Entomol Fr 50:257–259

Gaspar H, Castro S, Grosso-Silva JM, van der Heyden T, Loureiro J (2023) Exponential outspread of *Halyomorpha halys* (Stål, 1855) (Hemiptera: Pentatomidae) in Portugal. Arq Entomolóxicos 26:13–18

GBIF (2023a) Occurrence of *Halyomorpha halys* (Stål, 1855). https://www.gbif.org/species/4485843. Accessed 4 Nov 2023

GBIF (2023b) Occurrence of *Halyomorpha halys* in Belgium. https://www.gbif.org/occurrence/search?country=BE&taxon_key=4485843. Accessed 4 Nov 2023

Gennaro M, Giorcelli A (2019) The biotic adversities of poplar in Italy: a reasoned analysis of factors determining the current state and future perspectives. Ann Silvic Res 43:41–51

Giannetti D, Patelli N, Palazzetti L, Betti Sorbelli F, Pinotti C, Maistrello L (2024) First use of unmanned aerial vehicles to monitor Halyomorpha halys and recognize it using Artificial Intelligence. Pest Manag Sci 80(9):4074–4084

Gierlasiński G, Fiedor M, Dorda A, Taszakowski A (2020a). Pluskwiaki różnoskrzydłe (Hemiptera: Heteroptera) Cieszyna (Beskidy Zachodnie, południowa Polska) / True-bugs (Hemiptera: Heteroptera) of Cieszyn (Western Beskidy Mts., southern Poland)) Heteroptera Pol—Acta Faun 14:127–150

Gierlasiński G, Kolago G, Pacuk B, Taszakowski A, Syratt M, Regner J, Itczak A, Żóralski R, Rutkowski T, Radzimkiewicz D, Kucza W, Ogłaza B (2020b) Przyczynek do rozmieszczenia pluskwiaków różnoskrzydłych (Hemiptera: Heteroptera) w Polsce. II. Heteroptera Pol—Acta Faun 14:53–108

Gierlasiński G, Kolago G, Pacuk B, Taszakowski A, Syratt M, Regner J, Żóralski R, Rutkowski T, Kanja J, Kucza W, Milkowski M, Maslowski A (2021) Przyczynek do rozmieszczenia pluskwiaków różnoskrzydłych Hemiptera Heteroptera w Polsce III. Heteroptera Pol—Acta Faun 15:31–68

Gierlasiński G, Kolago G, Taszakowski A, Miłkowski M, Syratt M, Masłowski A, Grzywocz J, Rutkowski T (2022) Przyczynek do rozmieszczenia pluskwiaków różnoskrzydłych (Hemiptera: Heteroptera) w Polsce – IV. Heteroptera Poloniae 16:37–62

Giovannini L, Sabbatini-Peverieri G, Tillman PG, Hoelmer KA, Roversi PF (2021) Reproductive and developmental biology of *Acroclisoides sinicus*, a hyperparasitoid of scelionid parasitoids. Biology 10:229

Giovannini L, Mazza G, Chitarra W, Sabbatini-Peverieri G, Sonnati C, Roversi PF, Nerva L (2022a) New insights from the virome of *Halyomorpha halys* (Stål, 1855). Virus Res 316:198802

Giovannini L, Sabbatini-Peverieri G, Marianelli L, Rondoni G, Conti E, Roversi PF (2022b) Physiological host range of *Trissolcus mitsukurii*, a candidate biological control agent of *Halyomorpha halys* in Europe. J Pest Sci 95:605–618

Giovannini L, Sabbatini-Peverieri G, Simoni S, Cervo R, Hoelmer KA, Roversi PF (2022c) Interspecific competition between *Trissolcus japonicus* and *Trissolcus mitsukurii*, two promising candidates for biocontrol of *Halyomorpha halys*. Biol Control 176:105068

Gokturk T (2021) Chemical composition of *Satureja spicigera* essential oil and its insecticidal effectiveness on *Halyomorpha halys* nymphs and adults. Z Naturforsch C J Biosci 76:451–457

Gokturk T, Tozlu G (2019) An important agricultural pest for Turkey: invasive species *Halyomorpha halys*. In: International Black Sea Coastline Countries Symposium – Proceedings. Batumi (Georgia), pp 283–297

Gokturk T, Burjanadze M, Tozlu G, Tarihi G (2023) First record of *Cylindromyia bicolor* (Diptera: Tachinidae) as an adult parasitoid of *Halyomorpha halys* (Heteroptera: Pentatomidae) in the world / *Halyomorpha halys*'ın (Hemiptera: Pentatomidae) ergin parazitoiti olarak Cylindromyia bicolor'un (Diptera: Tachinidae) dünyadaki ilk kaydı. Artvin Çoruh Üniversitesi Orman Fakültesi Derg 24:249–253

Gonella E, Orrù B, Alma A (2019) Egg masses treatment with micronutrient fertilizers has a suppressive effect on newly-emerged nymphs of the brown marmorated stink bug *Halyomorpha halys*. Entomol Gen 39:231–238

Gonella E, Orrù B, Alma A (2022) Symbiotic control of *Halyomorpha halys* using a microbial biopesticide. Entomol Gen 42:901–909

Gorgadze O, Bakhtadze G, Kereselidze M, Lortkhipanidze M (2017) The efficacy of entomopathogenic agents against *Halyomorpha halys* (Hemiptera: Pentatomidae). Int J Curr Res 9:62177–62180

Grilo LMM (2023) Spatio-temporal modeling of *Halyomorpha halys* (Hemiptera, Pentatomidae) populations on Emilia Romagna (Italy). Master Thesis. Universidade de Evora, Portugal

Grisafi F, Papa G, Barbato M, Tombesi S, Negri I (2021) Influence of microclimate factors on *Halyomorpha halys* dehydration. Insects 12:897

Grosso-Silva JM, Gaspar H, Castro S, Loureiro J, Amorim F, van der Heyden T (2020) Confirmation of the presence of *Halyomorpha halys* (Stål, 1855) (Hemiptera: Pentatomidae) in mainland Portugal. Arq Entomolóxicos 22:373–376

Grozea I, Costea SA (2020) Apricot trees, a new attraction for the brown marmorated stink bug. Res J Agric Sci 52:122–127

Güncan A, Gümüş E (2019) Brown marmorated stink bug, *Halyomorpha halys* (Stål, 1855) (Hemiptera: Heteroptera, Pentatomidae), a new and important pest in Turkey. Entomol News 128:204–210

Gutierrez AP, Sabbatini Peverieri G, Ponti L, Giovannini L, Roversi PF, Mele A, Pozzebon A, Scaccini D, Hoelmer KA (2023) Tritrophic analysis of the prospective biological control of brown marmorated stink bug, *Halyomorpha halys*, under extant weather and climate change. J Pest Sci 96:921–942

Hajek AE, Solter LF, Maddox JV, Huang W-F, Estep AS, Krawczyk G, Weber DC, Hoelmer KA, Sanscrainte ND, Becnel JJ (2018) *Nosema maddoxi* sp. nov. (Microsporidia, Nosematidae), a widespread pathogen of the green stink bug *Chinavia hilaris* (Say) and the brown marmorated stink bug *Halyomorpha halys* (Stål). J Eukaryot Microbiol 65:315–330

HALY.ID (2023) HALY.ID – HALYomorpha halys IDentification: Innovative ICT tools for targeted monitoring and sustainable management of the brown marmorated stink bug and other pests. https://www.haly-id.eu/about/. Accessed 13 Nov 2023

Hamilton GC, Ahn JJ, Bu W, Leskey TC, Nielsen AL, Park Y-L, Rabitsch W, Hoelmer KA (2017) *Halyomorpha halys* (Stål). In: McPherson JE (ed) Invasive Stink Bugs and Related Species (Pentatomoidea): Biology, Higher Systematics, Semiochemistry, and Management. CRC Press, Boca Raton, pp 243–292

Hanselmann D (2017) Aliens and citizens in Germany: *Halyomorpha halys* (Stål, 1855) and *Nezara viridula* (Linnaeus, 1758) new to Rhineland-Palatinate, *Oxycarenus lavaterae* (Fabricius, 1787) new to Saxony, *Arocatus longiceps* Stål, 1872 new to Hesse. Mainz Naturwissenschaftliches Arch 53:159–177

Hartung V, Heller M, Husemann M (2022) First record of the invasive true bug (Heteroptera) *Halyomorpha halys* (Stål, 1855) from Hamburg with other records from Northern Germany and possible implications for its range expansion. Evol Syst 6:57–64

Haye T, Weber DC (2017) Special issue on the brown marmorated stink bug, *Halyomorpha halys*: an emerging pest of global concern. J Pest Sci 90:987–988

Haye T, Zimmermann O (2017) Establishment of the brown marmorated stink bug, *Halyomorpha halys* (Stål, 1855), in Germany (original title: Etablierung der Marmorierten Baumwanze, Halyomorpha halys (STÅL, 1855), in Deutschland). Heteropteron 48:34–37

Haye T, Abdallah S, Gariepy T, Wyniger D (2014a) Phenology, life table analysis and temperature requirements of the invasive brown marmorated stink bug, *Halyomorpha halys*, in Europe. J Pest Sci 87:407–418

Haye T, Wyniger D, Gariepy T (2014b) Recent range expansion of brown marmorated stink bug in Europe. In: Proceedings of the Eighth International Conference of Urban Pests. Zurich, Switzerland, pp 309–314

Haye T, Fischer S, Zhang J, Gariepy T (2015a) Can native egg parasitoids adopt the invasive brown marmorated stink bug, *Halyomorpha halys* (Heteroptera: Pentatomidae), in Europe? J Pest Sci 88:693–705

Haye T, Gariepy T, Hoelmer K, Rossi J-P, Streito J-C, Tassus X, Desneux N (2015b) Range expansion of the invasive brown marmorated stinkbug, *Halyomorpha halys*: an increasing threat to field, fruit and vegetable crops worldwide. J Pest Sci 88:665–673

Haye T, Moraglio ST, Stahl J, Visentin S, Gregorio T, Tavella L (2020) Fundamental host range of *Trissolcus japonicus* in Europe. J Pest Sci 93:171–182

Haye T, Moraglio ST, Tortorici F, Marazzi C, Gariepy TD, Tavella L (2023) Does the fundamental host range of *Trissolcus japonicus* match its realized host range in Europe? J Pest Sci 97:299–321

Heckmann R (2012) Erster Nachweis von *Halyomorpha halys* (Stål, 1855) (Heteroptera: Pentatomidae) für Deutschland. Heteropteron Heft 36:17–18

Hemala V, Kment P (2017) First record of *Halyomorpha halys* and mass occurrence of *Nezara viridula* in Slovakia (Hemiptera: Heteroptera: Pentatomidae). Plant Prot Sci 53:247–253

Hess B, Baufeld P, Dominic AR, Menz C, Reißig A, Strassemeyer J, Waldau T, Wilstermann A, Zimmermann O, Schrader G (2020) Modellierung klimasensitiver Schadorganismen in der Pflanzengesundheit. J Für Kult 72:435–439

Hess B, Zimmermann O, Baufeld P, Reißig A, Lutsch B, Schrader G (2022) Current distribution and spatial spread patterns of *Halyomorpha halys* in Germany. EPPO Bull 52:164–174

Hoblaj S, Rot M, Devetak M, Jančar M, Žežlina J (2022) Razširjenost marmorirane smrdljivke (*Halyomorpha halys*) v oljčnikih in njeno obvladovanje. Oljka Novice Druš Oljkarjev 27:18–20

Hoebeke ER, Carter ME (2003) *Halyomorpha halys* (Stål) (Heteroptera: Pentatomidae): a polyphagous plant pest from Asia newly detected in North America. Proc Entomol Soc Wash 105:225–237

Hristozova M, Harizanova A (2022) Laboratory study of the lifecycle parameters of the brown marmorated stink bug (*Halyomorpha halys* Stal)—An alien pest established in Bulgaria. Agric Sci 14:23–30

Iacovone A, Masetti A, Mosti M, Conti E, Burgio G (2022) Augmentative biological control of *Halyomorpha halys* using the native European parasitoid *Anastatus bifasciatus:* Efficacy and ecological impact. Biol Control 172:104973

Ibrahim A, Giovannini I, Anfora G, Stacconi MVR, Malek R, Maistrello L, Guidetti R, Romani R (2019) A closer look at the antennae of the invasive *Halyomorpha halys*: fine structure of the sensilla. Bull Insectology 72:187–199

iNaturalist contributors (2023a) Observations of brown marmorated stink bug. In: iNaturalist. https://www.inaturalist.org/observations?taxon_id=81923. Accessed 4 Nov 2023

iNaturalist contributors (2023b) BMSB in Greece. In: iNaturalist. https://www.inaturalist.org/projects/bmsb-in-greece. Accessed 4 Nov 2023

ISIP (2022) Obstbau—Früher aktiv: Erstnachweis 2022 der Eiablage der invasiven Marmorierten Baumwanze am Bodensee bereits Ende Mai – Aufruf zu Meldungen!! | ISIP. https://www.isip.de/isip/servlet/isip-de/regionales/baden-wuerttemberg/obstbau/allgemeines/obstbau%2D%2D-frueher-aktiv%2D%2Derstnachweis-2022-der-eiablage-der-invasiven-marmorierten-baumwanze-am-bodensee-bereits-ende-mai%2D%2D-aufruf-zu-meldungen%2D%2D-342686. Accessed 14 Nov 2023

Ivancic T, Grohar MC, Jakopic J, Veberic R, Hudina M (2022) Effect of brown marmorated stink bug (*Halyomorpha halys* Stål). Infestation on the phenolic response and quality of olive fruits (*Olea europaea* L.). Agronomy 12:2200

Izeltlabuak.hu (2023) *Halyomorpha halys* – ázsiai márványospoloska. https://www.izeltlabuak.hu/faj/azsiai-marvanyospoloska/terkep. Accessed 4 Nov 2023

Japoshvili G, Arabuli T, Salakaia M, Tskaruashvili Z, Kirkitadze G, Talamas E (2022) Surveys for *Halyomorpha halys* (Stål)(Hemiptera: Pentatomidae) and its biocontrol potential by parasitic wasps in the Republic of Georgia (Sakartvelo). Phytoparasitica 50:127–137

Kalashian MY, Ghrejyan TL, Karagyan GH (2022) Brown marmorated stink bug *Halyomorpha halys* (Stål, 1855) (Heteroptera: Pentatomidae) penetrated into Armenia. Russ J Biol Invasions 13:305–308

Kapantaidaki DE, Evangelou VI, Morrison WR, Leskey TC, Brodeur J, Milonas P (2019) *Halyomorpha halys* (Hemiptera: Pentatomidae) genetic diversity in North America and Europe. Insects 10:174

Kavran M, Konjević A, Petrić D, Ćupina AI (2023) The introduction and establishment of four invasive insect species in Serbia. Insects 14:728

Kehrli P, Rösti J, Lorenzini F, Deneulin P, Linder C (2021) Influence of processed *Halyomorpha halys* bugs on the aroma and taste of "Chardonnay" and "Merlot" musts and wines. VITIS – J Grapevine Res 60:43–50

Kereselidze M, Aleksidze G, Haye T (2018) First record native parasitoid attacking *Halyomorpha halys* (Heteroptera: Pentatomidae) in Georgia. Bull Georgian Acad Agric Sci 39:127–129

Kereselidze M, Pilarska D, Linde A (2019) First record of a microsporidium in the population of brown marmorated stink bug *Halyomorpha halys* (Stål, 1855) (Heteroptera: Pentatomidae) in the Republic of Georgia. Acta Zool Bulg 71:427–432

Kereselidze M, Pilarska D, Guntadze N, Linde A (2022) *Halyomorpha halys* Stål, (Hemiptera:Pentatomidae) feeding effects on some agricultural fruits in Georgia. Turk J Zool 46:298–303

Kharabadze N (2022) Review of the *Halyomorpha halys* (Stal, 1855) (Hemiptera: Heteroptera: Pentatomidae) in Georgia: Distribution, biology and management. Ann Agrar Sci 20:131–147

Kharabadze N, Chkhaidze N, Abramishvili T, Göktürk T, Lobjanidze M, Burjanadze M (2022) First report of *Hierodula transcaucasica* (Brunner von Wattenwyl, 1878) predation on the *Halyomorpha halys* (Stål, 1855) in Georgia. Int J Trop Insect Sci 42:3283–3292

Kment P, Březíková M (2018) First record of the invasive brown marmorated stink bug (*Halyomorpha halys*) (Hemiptera: Heteroptera: Pentatomidae) in The Czech Republic První nález invazní kněžice mramorované (*Halyomorpha halys*) (Hemiptera: Heteroptera: Pentatomidae) v České republice. Klapalekiana 54:221–232

Konjević A (2020) First records of the brown marmorated stink bug *Halyomorpha halys* (Stål, 1855) (Hemiptera: Pentatomidae) in Republic of North Macedonia. Acta Zool Bulg 72:687–690

Konjevic A (2022) Importance of monitoring of brown marmorated stink bug (*Halyomorpha halys*) in blueberry orchards. Biodivers Online J 2:1–2

Konopka JK, Haye T, Gariepy TD, McNeil JN (2017) Possible coexistence of native and exotic parasitoids and their impact on control of *Halyomorpha halys*. J Pest Sci 90:1119–1125

Kott P (2019) *Halyomorpha halys* (Stål, 1855)—Beobachtungen zur Ausbreitung der Art im Kölner Raum. Heteropteron Heft 56:33–38

Koutsogeorgiou EI, Kouloussis NA, Koveos DS, Andreadis SS (2022) Females of *Halyomorpha halys* (Hemiptera: Pentatomidae) experience a facultative reproductive diapause in Northern Greece. Insects 13:866

Koutsogeorgiou EI, Kouloussis NA, Sarrou E, Andreadis SS (2023) Headspace determination of the volatile organic compounds (VOCs) emitted by host plants of the brown marmorated stink bug *Halyomorpha halys*. Anal Lett 56:2898–2909

Lazar C, Popescu D, Maistrello L, Costi E, Ichim L, Georgescu EIV (2023) A Model for simulation of developmental instars of *Halyomorpha halys*. In: 5th IEEE Int Workshop Metrol Agric For MetroAgriFor Pisa Italy (in press)

Lee D-H, Short BD, Joseph SV, Bergh JC, Leskey TC (2013) Review of the biology, ecology, and management of *Halyomorpha halys* (Hemiptera: Pentatomidae) in China, Japan, and the Republic of Korea. Environ Entomol 42:627–641

Lee W, Guidetti R, Cesari M, Gariepy TD, Park Y-L, Park C-G (2018) Genetic diversity of *Halyomorpha halys* (Hemiptera, Pentatomidae) in Korea and comparison with COI sequence datasets from East Asia, Europe, and North America. Fla Entomol 101:49–54

Leskey TC, Short BD, Butler BR, Wright SE (2012) Impact of the invasive brown marmorated stink bug, *Halyomorpha halys* (Stål), in Mid-Atlantic tree fruit orchards in the United States: Case studies of commercial management. Psyche J Entomol 2012:535062

Leskey TC, Andrews H, Bády A, Benvenuto L, Bernardinelli I, Blaauw B, Bortolotti PP, Bosco L, Di Bella E, Hamilton G, Kuhar T, Ludwick D, Maistrello L, Malossini G, Nannini R, Nixon LJ, Pasqualini E, Preti M, Short BD, Spears L, Tavella L, Vétek G, Wiman N (2021) Refining pheromone lures for the invasive *Halyomorpha halys* (Hemiptera: Pentatomidae) through collaborative trials in the United States and Europe. J Econ Entomol 114:1666–1673

Liakos V, Navrozidis IE, Koutsogeorgiou EI, Gogolashvili NE, Samourgkanidou E, Faraslis I, Gravalos I, Thomidis T, Andreadis SS (2023) Analyzing on-farm spatiotemporal distribution of *Halyomorpha halys* (Stål) (Hemiptera: Pentatomidae) populations from a precision agriculture perspective. Plan Theory 12:2282

Liljeblad J (2023) Artportalen (Swedish Species Observation System). https://www.gbif.org/dataset/38b4c89f-584c-41bb-bd8f-cd1def33e92f. Accessed 1 Nov 2023

Limonta L, Dioli P, Selva J, Salvetti M (2016) *Halyomorpha halys* (Stål, 1855) (Insecta: Heteroptera) in Valtellina and Northern Lake of Como (Lombardy, Italy). Il Nat Valtellin Atti Mus Civ Storia Nat Morbegno 27:5–14

Loru L, Marras PM, Nannini M, Pantaleoni RA (2018) The occult spread of the invasive brown marmorated stink bug, *Halyomorpha halys* (Stål,1855) (Hemiptera Pentatomidae), in Sardinia (Italy). Biodivers J 9:447–450

Lupi D, Dioli P, Limonta L (2017) First evidence of *Halyomorpha halys* (Stål) (Hemiptera Heteroptera, Pentatomidae) feeding on rice (*Oryza sativa* L.). J Entomol Acarol Res 49:6679

Macavei L, Băeţan R, Oltean I, Florian T, Varga M, Costi E, Maistrello L (2015) First detection of *Halyomorpha halys*, a new invasive species with a high potential of damage on agricultural crops in Romania. Lucr Stiintifice Ser Agron 58:105–108

Maistrello L, Dioli P (2014) *Halyomorpha halys* Stål 1855, trovata per la prima volta nelle Alpi centrali italiane (Insecta: Heteroptera: Pentatomidae). Il Nat Valtellin Atti Mus Civ Storia Nat Morbegno 25:51–57

Maistrello L, Dioli P, Vaccari G, Nannini R, Bortolotti P, Caruso S, Costi E, Montermini A, Casoli L, Bariselli M (2014) Primi rinvenimenti in Italia della cimice esotica *Halyomorpha halys*, una nuova minaccia per la frutticoltura. In: Atti delle Giornate Fitopatologiche. CLUEB, Bologna, pp 283–288

Maistrello L, Dioli P, Bariselli M, Mazzoli GL, Giacalone-Forini I (2016a) Citizen science and early detection of invasive species: phenology of first occurrences of *Halyomorpha halys* in Southern Europe. Biol Invasions 18:3109–3116
Maistrello L, Vaccari G, Bortolini S, Costi E, Guidetti R, Bortolotti PP, Caruso S, Nannini R, Montermini A, Casoli L (2016b) Field monitoring and damage by the alien bug *Halyomorpha halys* in Emilia Romagna: a threat becoming a concrete problem. Atti Giornate Fitopatol 1:171–178
Maistrello L, Vaccari G, Caruso S, Costi E, Bortolini S, Macavei L, Foca G, Ulrici A, Bortolotti PP, Nannini R, Casoli L, Fornaciari M, Mazzoli GL, Dioli P (2017) Monitoring of the invasive *Halyomorpha halys*, a new key pest of fruit orchards in northern Italy. J Pest Sci 90:1231–1244
Maistrello L, Dioli P, Dutto M, Volani S, Pasquali S, Gilioli G (2018) Tracking the spread of sneaking aliens by integrating crowdsourcing and spatial modeling: the Italian invasion of *Halyomorpha halys*. Bioscience 68:979–989
Maistrello L, Costi E, Di Bella E, Vaccari G, Caruso S, Casoli L, Preti M, Landi M, Bombardini E, Moretti C, Tommasini MG, Bariselli M, Butturini A, Haye T (2020) Rilasci inondativi del parassitoide autoctono *Anastatus bifasciatus* per il controllo della cimice invasiva *Halyomorpha halys* in Italia. Atti Delle Giornate Fitopatol 1:167–176
Maistrello L, Costi E, Di Bella E, Masetti A, Bittau B, Lami F, Zaniboni A, Torreggiani D, Vaccari G, Caruso S, Preti M, Rufolo E, Landi M, Fagioli L, Manucci F, Fabbri G, Ferrari R, Zoboli F, Bariselli M, Tommasini MG (2022) Indagini sulla parasitizzazione di *Halyomorpha halys* in Emilia Romagna nell'ambto del progetto "Haly.Bio.". Atti Giornate Fitopatol 1:205–214
Malek R, Tattoni C, Ciolli M, Corradini S, Andreis D, Ibrahim A, Mazzoni V, Eriksson A, Anfora G (2018) Coupling traditional monitoring and citizen science to disentangle the invasion of *Halyomorpha halys*. ISPRS Int J Geo Inf 7:171
Malek R, Zapponi L, Eriksson A, Ciolli M, Mazzoni V, Anfora G, Tattoni C (2019) Monitoring 2.0: Update on the *Halyomorpha halys* invasion of Trentino. ISPRS Int J Geo Inf 8:564
Malek R, Kaser JM, Anfora G, Ciolli M, Khrimian A, Weber DC, Hoelmer KA (2021) *Trissolcus japonicus* foraging behavior: Implications for host preference and classical biological control. Biol Control 161:104700
Malossini G, Benvenuto L, Bernardinelli I (2020) *Halyomorpha halys* in Friuli Venezia Giulia: primi segnali di un'invesione di tendenza. Atti Giornate Fitopatol 1:153–160
Malumphy C (2014) Second interception of *Halyomorpha halys* (Stål) (Hemiptera: Pentatomidae) in Britain. Het News 3rd Ser 21:4–5
Martinez-Sañudo I, Perotti MA, Scaccini D, Pozzebon A, Marri L, Mazzon L (2020) Co-haplotyping symbiont and host to unravel invasion pathways of the exotic pest *Halyomorpha halys* in Italy. Sci Rep 10:18441
Martorana L, Foti MC, Rondoni G, Conti E, Colazza S, Peri E (2017) An invasive insect herbivore disrupts plant volatile-mediated tritrophic signalling. J Pest Sci 90:1079–1085
Masetti A, Preti M, Albertini M, Capriotti M, Scagliarini O, Tommasini MG (2023) Evaluation of an attract-and-kill strategy for the management of the brown marmorated stink bug in Northern Italy. Pheromones Semiochem Integr Prod Integr Prot Fruit Crops IOBC-WPRS Bull 166:104–107
Matsuo K, Hirose Y, Johnson NF (2014) A taxonomic issue of two species of *Trissolcus* (Hymenoptera: Platygastridae) parasitic on eggs of the brown-winged green bug, *Plautia stali* (Hemiptera: Pentatomidae): resurrection of *T. plautiae*, a cryptic species of *T. japonicus* revealed by morphology, reproductive isolation and molecular evidence. Appl Entomol Zool 49:385–394
MATTM (2020) Decreto del Ministero dell'Ambiente e dela Tutela del territorio e del mare – Direzione generale per il patrimonio naturalistico, 09 giugno 2020, n. 42967, Immissione in natura della specie non autoctona *Trissolcus japonicus* quale agente di controllo biologico del fitofago *Halyomorpha halys* sensi del Decreto del Presidente della Repubblica 8 settembre 1997, n. 357, art. 12, comma 4

Mazzoni V, Polajnar J, Baldini M, Stacconi MVR, Anfora G, Guidetti R, Maistrello L (2017) Use of substrate-borne vibrational signals to attract the brown marmorated stink bug, *Halyomorpha halys*. J Pest Sci 90:1219–1229

Mele A, Scaccini D, Pozzebon A (2021) Hyperparasitism of *Acroclisoides sinicus* (Huang and Liao) (Hymenoptera: Pteromalidae) on two biological control agents of *Halyomorpha halys*. Insects 12:617

Mele A, Scaccini D, Zanolli P, Pozzebon A (2022) Semi-natural habitats promote biological control of *Halyomorpha halys* (Stål) by the egg parasitoid *Trissolcus mitsukurii* (Ashmead). Biol Control 166:104833

Meshkova V (2022) Alien phytophagous insects in forest and urban stands of Ukraine. Bucov For 22:29–40

Milonas PG, Partsinevelos GK (2014) First report of brown marmorated stink bug *Halyomorpha halys* Stål (Hemiptera: Pentatomidae) in Greece. EPPO Bull 44:183–186

Mirandola E, Scaccini D, Chama A, Fornasiero D, Galli G, Lombardo V, Simoni F, Duso C, Pozzebon A (2023) Repellence of different sulphur-based compounds on *Halyomorpha halys* under controlled conditions. Pheromones Semiochem Integr Prod Integr Prot Fruit Crops IOBC-WPRS Bull 166:111–112

Mityushev IM (2016) First record of marmorated bug detection in Russia. Zas Karantin Rast 3:48

Mityushev IM (2020) Invasion and trophic relations of the brown marmorated stink bug *Halyomorpha halys* (Stål) (Heteroptera) in South Russia. Euroasian Entomol J 19:356–358

Mityushev IM (2021) Host plants of *Halyomorpha halys* in the urban ecosystem on the Azov Sea Coast of Russia. EPPO Bull 51:305–310

Mohekar P, Osborne J, Wiman NG, Walton V, Tomasino E (2017) Influence of winemaking processing steps on the amounts of (E)-2-decenal and tridecane as off-odorants caused by brown marmorated stink bug (*Halyomorpha halys*). J Agric Food Chem 65:872–878

Moraglio ST, Tortorici F, Pansa MG, Castelli G, Pontini M, Scovero S, Visentin S, Tavella L (2019) A 3-year survey on parasitism of *Halyomorpha halys* by egg parasitoids in northern Italy. J Pest Sci 93:183–194

Moraglio ST, Tortorici F, Giromini D, Pansa MG, Visentin S, Tavella L (2021a) Field collection of egg parasitoids of Pentatomidae and Scutelleridae in Northwest Italy and their efficacy in parasitizing *Halyomorpha halys* under laboratory conditions. Entomol Exp Appl 169:52–63

Moraglio ST, Tortorici F, Visentin S, Pansa MG, Tavella L (2021b) *Trissolcus kozlovi* in North Italy: Host specificity and augmentative releases against *Halyomorpha halys* in hazelnut orchards. Insects 12:464

Morrison WR, Milonas P, Kapantaidaki DE, Cesari M, Bella ED, Guidetti R, Haye T, Maistrello L, Moraglio ST, Piemontese L, Pozzebon A, Ruocco G, Short BD, Tavella L, Vétek G, Leskey TC (2017) Attraction of *Halyomorpha halys* (Hemiptera: Pentatomidae) haplotypes in North America and Europe to baited traps. Sci Rep 7:1–11

Mueller G, Landau Luescher I, Schmidt M (2011) New data on the incidence of household arthropod pests and new invasive pests in Zurich (Switzerland). In: Robinson WH, Campos AEC (eds) Proceedings of the Seventh International Conference on Urban Pests. Instituto Biológico, São Paulo, pp 99–104

Muntean AC, Grozea I (2021) Abundance of insect species harmful to ornamental plants in urban ecosystems. Sci Pap Ser B Hortic LXV:645–650

Musolin D (2015) *Halyomorpha halys* crawls across Eurasia. Studying the spread of the invasive brown marmorated bug. https://citizenscience.ru/projects/rasprostranenie-invazionnogo-korichnevogo-mramornogo-klopa-halyomorpha-halys-v-evrazii.html. Accessed 4 Nov 2023

Musolin DL, Konjević A, Karpun NN, Protsenko VY, Ayba LY, Saulich AK (2018) Invasive brown marmorated stink bug *Halyomorpha halys* (Stål) (Heteroptera: Pentatomidae) in Russia, Abkhazia, and Serbia: History of invasion, range expansion, early stages of establishment, and first records of damage to local crops. Arthropod Plant Interact 12:517–529

Musolin DL, Dolgovskaya MY, Protsenko VY, Karpun NN, Reznik SY, Saulic AK (2019) Photoperiodic and temperature control of nymphal growth and adult diapause induction in the invasive Caucasian population of the brown marmorated stink bug, *Halyomorpha halys*. J Pest Sci 92:621–631

Musolin DL, Dolgovskaya MY, Zakharchenko VY, Karpun NN, Haye T, Saulich AK, Reznik SY (2022a) Flying over Eurasia: Geographic variation of photoperiodic control of nymphal development and adult diapause induction in native and invasive populations of the brown marmorated stink bug, *Halyomorpha halys* (Hemiptera: Heteroptera: Pentatomidae). Insects 13:522

Musolin DL, Kirichenko NI, Karpun NN, Aksenenko EV, Golub VB, Kerchev IA, Mandelshtam MY, Vasaitis R, Volkovitsh MG, Zhuravleva EN, Selikhovkin AV (2022b) Invasive insect pests of forests and urban trees in Russia: Origin, pathways, damage, and management. Forests 13:521

Nannini R, Bortolotti P, Casoli L, Montevecchi G, Masino F, Antonelli A, Preti M (2018) Survey on *Halyomorpha halys* infestations and damage in grapevine productions. Atti Giornate Fitopatol 1:295–300

Neacsu M, Grozea I, Stef R (2019) Pre-wintering behavior of *Halyomorpha halys* (Insecta: Hemiptera: Pentatomidae). Res J Agric Sci 51:97–103

Neimorevets VV (2018) Brown marmorated stink bug *Halyomorpha halys* (Heteroptera: Pentatomidae): morphology, biology, distribution and threats to agriculture in The Russian Federation (Analytical review). Plant Prot News 1:11–16

Nuriyeva IA (2018) Azərbaycanda bitkiçilik üçün yeni, təhlükəli növ olan Mərmər taxtabiti (Halyo*morpha halys* Stal.). AMEA "Elm Və Həyat" 1:66–68

Nuriyeva IA, Nadirova GI (2020) *Halyomorpha halys* Stal (Hemiptera, Pentatomidae) and *Metcalfa pruinosa* Say (Hemiptera, Flatidae) – the new dangerous invasive pests of walnut trees (*Juglans regia* L. and *Corylus maxima* Mill.) in Azerbaijan. Trans Pedagog Univ Ser Math Nat Sci 68:187–201

OC Media (2017) Stink bug devastates hazelnut crops in Samegrelo and Abkhazia. OC Media. https://oc-media.org/stink-bug-devastates-hazelnut-crops-in-samegrelo-abkhazia/. Accessed 12 Apr 2023

OpenStreetMap Contributors (2020) CIMICE.NET – Monitoring *Halyomorpha halys* in Emilia-Romagna. https://big.csr.unibo.it/projects/cimice/monitoring.php?lan=EN. Accessed 13 Nov 2023

Orrù B, Moraglio ST, Tortorici F, Gonella E, Tavella L, Alma A (2023) No adverse effects of symbiotic control on the parasitism of *Halyomorpha halys* by egg parasitoids. J Pest Sci 96:1091–1102

Özdemír İO, Yildirim E, Uluca M, Tuncer C (2022) Efficacy of native *Beauveria bassiana* and *B. pseudobassiana* isolates against invasive brown marmorated stink bug, *Halyomorpha halys* (Stål) (Hemiptera: Pentatomidae). Black Sea J Agric 5:227–233

Özdemir İO, Doğan F, Tunçer C (2023a) The preliminary study on the biology of an invasive species, *Halyomorpha halys* (Hemiptera: Pentatomidae) in Northwest Türkiye. Turk J Agric – Food Sci Technol 11:1380–1385

Özdemir IO, Karakaya O, Ates U, Ozturk B, Uluca M, Tuncer C (2023b) Characterization of hazelnut kernel responses to brown marmorated stink bug [*Halyomorpha halys* Stål (Hemiptera: Pentatomidae)] infestations: Changes in bioactive compounds and fatty acid composition. J Food Compos Anal 124:105696

Öztemiz S, Sağut M, Adak Y (2019) *Halyomorpha halys* (Hemiptera: Pentatomidae): a new invasive species in north-western of Turkey. Munis Entomol Zool 14:634–637

Pagola-Carte S, Zabalegui I (2019) Dos hemípteros asiáticos nuevos para Gipuzkoa, norte de la Península Ibérica (Hemiptera: Pentatomidae, Cicadellidae). Heteropterus Rev Entomol 19:355–360

Pajač Živković I, Jurić S, Vinceković M, Galešić MA, Marijan M, Vlahovićek-Kahlina K, Mikac KM, Lemic D (2020) Polyphenol-based microencapsulated extracts as novel green insecticides for sustainable management of polyphagous brown marmorated stink bug. Sustainability 12:10079

Pajač Živković I, Skendžić S, Lemić D (2021) Rapid spread and first massive occurrence of *Halyomorpha halys* (Stål, 1855) in agricultural production in Croatia. J Cent Eur Agric 22:531–538

Pajač Živković I, Mulamehmedović J, Göldel B, Lemić D (2022) Sexual dimorphism of brown marmorated stink bug. J Cent Eur Agric 23:62–68

Paltrinieri MG, Francati S, Dindo ML, Bertaccini A (2016) Cimice Asiatica confermata come insetto vettore di fitoplasmi. Inf Agrar 45:60–61

Pansa MG, Tavella L, Asteggiano L, Vittone G, Costamagna C (2013) Primo ritrovamento di *Halyomorpha halys* nei pescheti piemontesi. Inf Agrar 60:60–61

Papa G, Negri I (2020) Cannibalism in the brown marmorated stink bug *Halyomorpha halys* (Stål). Insects 11:643

Papa G, Abbà S, Galetto L, Parise C, Marzachì C, Negri I (2023) Distribution and prevalence of viral genomes in Italian populations of the invasive brown marmorated stink bug *Halyomorpha halys*. J Invertebr Pathol 200:107977

Peri E, Foti MC, Martorana L, Cusumano A, Colazza S (2021) The invasive stink bug *Halyomorpha halys* affects the reproductive success and the experience-mediated behavioural responses of the egg parasitoid *Trissolcus basalis*. BioControl 66:329–342

Peusens G, Belien T (2012) Life cycle and control of the forest bug *Pentatoma rufipes* L. in organically managed pear orchards. Commun Agric Appl Biol Sci 77:663–666

Piemontese L (2020) Detection and population structure assessment of the invasive alien species *Halyomorpha halys* (Hemiptera, Pentatomidae) using molecular approaches. PhD Thesis. Università di Modena e Reggio Emilia, Italy

Polajnar J, Maistrello L, Bertarella A, Mazzoni V (2016) Vibrational communication of the brown marmorated stink bug (*Halyomorpha halys*). Physiol Entomol 41:249–259

Polajnar J, Maistrello L, Ibrahim A, Mazzoni V (2019) Can vibrational playback improve control of an invasive stink bug? In: Hill PSM, Lakes-Harlan R, Mazzoni V, Narins PM, Virant-Doberlet M, Wessel A (eds) Biotremology: Studying Vibrational Behavior. Springer International Publishing, Cham, pp 375–398

Powell G (2020) The biology and control of an emerging shield bug pest, *Pentatoma rufipes* (L.) (Hemiptera: Pentatomidae). Agric For Entomol 22:298–308

Powell G, Barclay M, Couch Y, Evans A (2021) Current invasion status and potential for UK establishment of the brown marmorated stink bug, *Halyomorpha halys* (Hemiptera: Pentatomidae). Br J Entomol Nat Hist 34:9–21

Pozzebon A, Scaccini D, Lombardo V, Galli G, Fornasiero D (2023) Influence of exclusion net employment time and colour on the brown marmorated stink bug, *Halyomorpha halys* (Hemiptera: Pentatomidae) infestation and damage in pear and apple orchards. Pheromones Semiochem Integr Prod Integr Prot Fruit Crops IOBC-WPRS Bull 166:161–162

PPC (2021) Brown marmorated stink bug spotted in UK. PPC-Prof Pest Control 103:7

Preston CE, Agnello AM, Vermeylen F, Hajek AE (2019) Impact of *Nosema maddoxi* on the survival, development, and female fecundity of *Halyomorpha halys*. J Invertebr Pathol 169:107303

Preston CE, Agnello AM, Hajek AE (2020) *Nosema maddoxi* (Microsporidia: Nosematidae) in brown marmorated stink bug, *Halyomorpha halys* (Hemiptera: Pentatomidae), populations in the United States. Biol Control 144:104213

Preti M, Fagioli L, Capriotti M, Coltorti M, Tommasini MG, Pozzebon A (2023) *Halyomorpha halys* fruit injury on pear can be reduced by applying kaolin and zeolites. Pheromones Semiochem Integr Prod Integr Prot Fruit Crops IOBC-WPRS Bull 166:159–160

Rabitsch W, Friebe GJ (2015) From the west and from the east? First records of *Halyomorpha halys* (Stål, 1855) (Hemiptera: Heteroptera: Pentatomidae) in Vorarlberg and Vienna. Beitr Entomof 16:115–139

Rancati M, Pansa MG, Ghiglione M, Tavella L (2017) Danni da cimice asiatica su mais di secondo raccolto. Inf Agrar 14:66–68

Regione Emilia Romagna (2022) Il Sistema agroalimentare dell'Emilia-Romagna – Rapporto 2022. In: Agric. Caccia E Pesca. https://agricoltura.regione.emilia-romagna.it/agricoltura-incifre/rapporto-agroalimentare/rapporto-2022/sistema-agroalimentare-emilia-romagna-rapporto-2022. Accessed 17 Nov 2023

Reznik SY, Karpun NN, Zakharchenko VY, Shoshina YI, Dolgovskaya MY, Saulich AK, Musolin DL (2022) To everything there is a season: Phenology and photoperiodic control of seasonal development in the invasive caucasian population of the brown marmorated stink bug, *Halyomorpha halys* (Hemiptera: Heteroptera: Pentatomidae). Insects 13:580

Roca-Cusachs M, Fernandez D, Goula M, Escudero-Colomar L-A (2018) New records of the invasive alien plant pest *Halyomorpha halys* (Stål, 1855) in the Iberian Peninsula (Heteroptera: Pentatomidae). Butlletí Inst Catalana D'Història Nat 82:73–77

Rondoni G, Bertoldi V, Malek R, Foti MC, Peri E, Maistrello L, Haye T, Conti E (2017) Native egg parasitoids recorded from the invasive *Halyomorpha halys* successfully exploit volatiles emitted by the plant–herbivore complex. J Pest Sci 90:1087–1095

Rondoni G, Bertoldi V, Malek R, Djelouah K, Moretti C, Buonaurio R, Conti E (2018) *Vicia faba* plants respond to oviposition by invasive *Halyomorpha halys* activating direct defences against offspring. J Pest Sci 91:671–679

Rondoni G, Chierici E, Giovannini L, Sabbatini-Peverieri G, Roversi PF, Conti E (2022a) Olfactory responses of *Trissolcus mitsukurii* to plants attacked by target and non-target stink bugs suggest low risk for biological control. Sci Rep 12:1880

Rondoni G, Chierici E, Marchetti E, Nasi S, Ferrari R, Conti E (2022b) Improved captures of the invasive brown marmorated stink bug, *Halyomorpha halys*, using a novel multimodal trap. Insects 13:527

Roselli G, Anfora G, Sasso R, Zapponi L, Musmeci S, Cemmi A, Suckling DM, Hoelmer KA, Ioriatti C, Cristofaro M (2023a) Combining irradiation and biological control against brown marmorated stink bug: Are sterile eggs a suitable substrate for the egg parasitoid *Trissolcus japonicus*? Insects 14:654

Roselli G, Anfora G, Suckling DM, Mazzoni V, Vanoni V, Menegotti L, Fellin L, Rossi Stacconi MV, Ioriatti C, Cristofaro M (2023b) Effects of irradiation on biology and mating behaviour of wild males of brown marmorated stink bug using a 6 MV medical linear accelerator. Insects 14:460

Rossini L, Contarini M, Bono Rosselló N, Garone E, Speranza S (2023) Prediction of infestations by true bugs in hazelnut orchards: feasibility and preliminary approaches in the case of *Halyomorpha halys*. Acta Hortic 1379:463–472

Rot M, Devetak M, Carlevaris B, Žežlina J, Žežlina I (2018) First record of brown marmorated stink bug *Halyomorpha halys* (Stål, 1855) (Hemiptera: Pentatomidae) in Slovenia. Acta Entomol Slov 26:5–12

Rot M, Devetak M, Žigon P, Ferlež Ruz A, Matko B, Peterlin A (2019) Marmorirana smrdljivka (*Halyomorpha halys* (Stal, 1855) [Hemiptera, Pentatomidae]); pojav in razširjenost nove invazivne, tujerodne stenice v Sloveniji. In: Proceedings of the Zbornik Predavanj in Referatov 14. Slovenskega Posvetovanja o Varstvu Rastlin z Mednarodno Udeležbo, Maribor, Slovenia, 5–6 March 2019, Društvo za Varstvo Rastlin Slovenije: Ljubljana, Slovenia. Trdan, S., Maribor, Slovenia, pp 134–141

Rot M, Maistrello L, Costi E, Bernardinelli I, Malossini G, Benvenuto L, Trdan S (2021) native and non-native egg parasitoids associated with brown marmorated stink bug (*Halyomorpha halys* [Stål, 1855]; Hemiptera: Pentatomidae) in Western Slovenia. Insects 12:505

Rot M, Maistrello L, Costi E, Trdan S (2022a) biological parameters, phenology and temperature requirements of *Halyomorpha halys* (Hemiptera: Pentatomidae) in the sub-mediterranean climate of Western Slovenia. Insects 13:956

Rot M, Žežlina I, Carlevaris B, Devetak M, Žežlina J, Dariž J, Juretič V, Trdan S (2022b) Prvi korak na poti k biotičnemu varstvu marmorirane smrdljivke (Halyomorpha halys [Stål, 1855], Hemiptera, Pentatomidae) v Sloveniji. In: Proceedings of the Zbornik predavanj in referatov 15. slovenskega posvetovanja o varstvu rastlin z mednarodno udeležbo, Portorož, 1.-2. marec 2022. Portorož, Slovenia, pp 40–51

Roversi PF, Binazzi F, Marianelli L, Costi E, Maistrello L, Sabbatini Peverieri G (2016) Searching for native egg-parasitoids of the invasive alien species *Halyomorpha halys* Stål (Heteroptera, Pentatomidae) in Southern Europe. Redia 99:63–70

Sabbatini Peverieri G, Bortolotti PP, Nannini R, Marianelli L, Roversi PF (2018a) Efficacy of long-lasting insecticide nets in killing *Halyomorpha halys* in pear orchards. Outlooks Pest Manag 29:70–74

Sabbatini Peverieri G, Talamas E, Bon MC, Marianelli L, Bernardinelli I, Malossini G, Benvenuto L, Roversi PF, Hoelmer K (2018b) Two Asian egg parasitoids of *Halyomorpha halys* (Stål) (Hemiptera, Pentatomidae) emerge in northern Italy: *Trissolcus mitsukurii* (Ashmead) and *Trissolcus japonicus* (Ashmead) (Hymenoptera, Scelionidae). J Hymenopt Res 67:37–53

Sabbatini Peverieri G, Mitroiu M-D, Bon M-C, Balusu R, Benvenuto L, Bernardinelli I, Fadamiro H, Falagiarda M, Fusu L, Grove E, Haye T, Hoelmer K, Lemke E, Malossini G, Marianelli L, Moore MR, Pozzebon A, Roversi P-F, Scaccini D, Shrewsbury P, Tillman G, Tirello P, Waterworth R, Talamas EJ (2019) Surveys of stink bug egg parasitism in Asia, Europe and North America, morphological taxonomy, and molecular analysis reveal the Holarctic distribution of *Acroclisoides sinicus* (Huang & Liao) (Hymenoptera, Pteromalidae). J Hymenopt Res 74:123–151

Sabbatini-Peverieri G, Dieckhoff C, Giovannini L, Marianelli L, Roversi PF, Hoelmer K (2020a) Rearing *Trissolcus japonicus* and *Trissolcus mitsukurii* for biological control of *Halyomorpha halys*. Insects 11:787

Sabbatini-Peverieri G, Giovannini L, Benvenuti C, Madonni L, Hoelmer K, Roversi PF (2020b) Characteristics of the meconia of European egg parasitoids of *Halyomorpha halys*. J Hymenopt Res 77:187–201

Sabbatini-Peverieri G, Boncompagni L, Mazza G, Paoli F, Dapporto L, Giovannini L, Marianelli L, Hoelmer K, Roversi PF (2021) Combining physiological host range, behavior and host characteristics for predictive risk analysis of *Trissolcus japonicus*. J Pest Sci 94:1003–1016

Šapina I, Jelaska LŠ (2018) First report of invasive brown marmorated stink bug *Halyomorpha halys* (Stål, 1855) in Croatia. EPPO Bull 48:138–143

Sargent C, Martinson HM, Raupp MJ (2014) Traps and trap placement may affect location of brown marmorated stink bug (Hemiptera: Pentatomidae) and increase injury to tomato fruits in home gardens. Environ Entomol 43:432–438

Sauer C (2012) Die Marmorierte Baumwanze tritt neu im Deutschschweizer Gemüsebau auf. Gemüsebau Info 28:4–5

Sauer C, Total R (2019) Wanzen im Gemüsebau—Befallssituation 2019 und Schadbilder. Gemüsebau Info 30:3–7

Saulich AK, Musolin DL (2017) Seasonal cycles of Pentatomoidea. In: Invasive Stink Bugs and Related Species (Pentatomoidea). CRC Press, p 44

Scaccini D, Duso C, Pozzebon A (2019) Lethal effects of high temperatures on brown marmorated stink bug adults before and after overwintering. Insects 10:355

Scaccini D, Falagiarda M, Tortorici F, Martinez-Sañudo I, Tirello P, Reyes-Domínguez Y, Gallmetzer A, Tavella L, Zandigiacomo P, Duso C, Pozzebon A (2020a) An insight into the role of *Trissolcus mitsukurii* as biological control agent of *Halyomorpha halys* in Northeastern Italy. Insects 11:306

Scaccini D, Vanishvili L, Tirello P, Walton VM, Duso C, Pozzebon A (2020b) Lethal and sub-lethal effects of low-temperature exposures on *Halyomorpha halys* (Hemiptera: Pentatomidae) adults before and after overwintering. Sci Rep 10:15231

Scaccini D, Fornasiero D, Bertucco R, Lombardo V, Galli G, Pozzebon A (2023) Developing pheromone-based traps for mass capture of the brown marmorated stink bug, *Halyomorpha halys* (Hemiptera: Pentatomidae). Pheromones Semiochem Integr Prod Integr Prot Fruit Crops IOBC-WPRS Bull 166:102–103

Scala M, Fouani JM, Zapponi L, Mazzoni V, Wells KE, Biondi A, Baser N, Verrastro V, Anfora G (2022) Attraction of egg parasitoids *Trissolcus mitsukurii* and *Trissolcus japonicus* to the chemical cues of *Halyomorpha halys* and *Nezara viridula*. Insects 13:439

Schifani E, Giannetti D, Costi E, Franconi G, Campostrini A, Maistrello L, Grasso DA (2023) Interactions between egg parasitoids and predatory ants for the biocontrol of the invasive brown marmorated stink bug *Halyomorpha halys*. J Appl Entomol 147:868–874
Schuler H, Elsler D, Fischnaller S (2021) Population genetics of the brown marmorated stink bug *Halyomorpha halys* in the early phase of invasion in South Tyrol (Northern Italy). Bull Entomol Res 111:394–401
Schweizer Obstverband SOV in Bauernzeitung (2019) Marmorierte Baumwanze verursacht Millionenschäden im Obstbau. In: bauernzeitung.ch. https://www.bauernzeitung.ch/artikel/pflanzen/marmorierte-baumwanze-verursacht-millionenschaeden-im-obstbau-380118. Accessed 4 Nov 2023
Šeat J (2015) *Halyomorpha halys* (Stål, 1855) (Heteroptera: Pentatomidae) a new invasive species in Serbia. Acta Entomol Serbica 20:167–171
Serteyn L, Ponnet L, Saive M, Fauconnier M-L, Francis F (2020) Changes of feeding behavior and salivary proteome of brown marmorated stink bug when exposed to insect-induced plant defenses. Arthropod Plant Interact 14:101–112
Servizio Fitosanitario Centrale (2020) Cimice asiatica – Decreto 29 aprile 2020. In: Prot. Delle Piante. https://www.protezionedellepiante.it/cimice-asiatica-decreto-20-luglio-2020/. Accessed 27 Sep 2023
Signor M, Benvenuto L, Bernardinelli I, Malossini G, Barbiani G (2018) Cimice asiatica su soia, come contenere i danni. Inf Agrar 29:43–44
Simov N (2016) The invasive brown marmorated stink bug *Halyomorpha halys* (Stål,1855) (Heteroptera: Pentatomidae) already in Bulgaria. Ecol Montenegrina 9:51–53
Skuhrovec J, Zeman Š, Křížová K, Platková H (2022) Výskyt kněžice mramorované (Halyomorpha halys) v České republic. Praha, Výzkumný ústav rostlinné výroby, v.v.i., 2022. Specializovaná mapa s odborným obsahem
Smolis A, Kadej M (2022) Pierwsze stwierdzenia tarczówki marmurkowatej Halyomorpha halys (Stål, 1855) (Hemiptera, Heteroptera, Pentatomidae) w zachodniej Polsce. / First records of the brown marmorated stink bug *Halyomorpha halys* (Stål, 1855) (Hemiptera, Heteroptera, Pentatomidae) form Western Poland
Stäheli N, Bravin E, Kambor J, Egger B (2022) Umfrage zu Schäden und Massnahmen zur Bekämpfung der Marmorierten Baumwanze. Obst-und Weinbau 14:26–29
Stahl JM, Babendreier D, Haye T (2018) Using the egg parasitoid *Anastatus bifasciatus* against the invasive brown marmorated stink bug in Europe: can non-target effects be ruled out? J Pest Sci 91:1005–1017
Stahl J, Tortorici F, Pontini M, Bon M-C, Hoelmer K, Marazzi C, Tavella L, Haye T (2019a) First discovery of adventive populations of *Trissolcus japonicus* in Europe. J Pest Sci 92:371–379
Stahl JM, Babendreier D, Haye T (2019b) Life history of *Anastatus bifasciatus*, a potential biological control agent of the brown marmorated stink bug in Europe. Biol Control 129:178–186
Stahl JM, Babendreier D, Marazzi C, Caruso S, Costi E, Maistrello L, Haye T (2019c) Can *Anastatus bifasciatus* be used for augmentative biological control of the brown marmorated stink bug in fruit orchards? Insects 10:108
Stoeckli S, Felber R, Haye T (2020) Current distribution and voltinism of the brown marmorated stink bug, *Halyomorpha halys*, in Switzerland and its response to climate change using a high-resolution CLIMEX model. Int J Biometeorol 64:2019–2032
Streito J-C, Chartois M, Pierre É, Dusoulier F, Armand J-M, Gaudin J, Rossi J-P (2021) Citizen science and niche modeling to track and forecast the expansion of the brown marmorated stinkbug *Halyomorpha halys* (Stål, 1855). Sci Rep 11:11421
Stryukova NM, Stryukov AA (2022) Alien insects—phytophages of agricultural, ornamental and forest plantations of the Crimea. Plant Health Quar 4:16–32
Suckling DM, Levy MC, Roselli G, Mazzoni V, Ioriatti C, Deromedi M, Cristofaro M, Anfora G (2019a) Live traps for adult brown marmorated stink bugs. Insects 10:376
Suckling DM, Mazzoni V, Roselli G, Levy MC, Ioriatti C, Stringer LD, Zeni V, Deromedi M, Anfora G (2019b) Trapping brown marmorated stink bugs: "The Nazgûl" lure and kill nets. Insects 10:433

Szanyi S, Ősz A, Szanyi K, Potish L, Nagy A (2022) Distribution of brown marmorated stink bug *Halyomorpha halys* (Stål, 1855) (Hemiptera: Pentatomidae) in the Northeast part of the Carpathian Lowland (West Ukraine). Acta Agrar Debr 1:191–194

Talamas EJ, Japoshvili G (2022) List of species of Pentatomids (Hemiptera: Pentatomidae) in the hazelnut orchards of Sakartvelo (Georgia) and their potential parasitoids. Ann Agrar Sci 20:225–231

Talamas EJ, Buffington ML, Hoelmer KA (2017) Revision of palearctic *Trissolcus* Ashmead (Hymenoptera, Scelionidae). J Hymenopt Res 56:3–185

Tamburini G, Laterza I, Nardi D, Mele A, Mori N, Pasini M, Scaccini D, Pozzebon A, Marini L (2023) Effect of landscape composition on the invasive pest *Halyomorpha halys* in fruit orchards. Agric Ecosyst Environ 353:108530

Tanco J (2022) *Halyomorpha halys* (Stål, 1855) (Hemiptera: Pentatomidae), nueva cita para Aragón (NE España). Arq Entomolóxicos 25:291–292

Tanco J (2023) *Halyomorpha halys* (Stål, 1855) (Hemiptera, Pentatomidae), nueva cita para la Comunidad Foral de Navarra. Arq Entomolóxicos 26:23–24

Tanco J, Canyelles X (2023) Confirmació de la presència *d'Halyomorpha halys* (Stål, 1855) (Hemiptera: Pentatomidae) a les Illes Balears SHNB. Bolleti Soc Hist Nat Balears 66:69–74

Tassini C, Mifsud D (2019) The brown marmorated stink bug, *Halyomorpha halys* (Hemiptera: Heteroptera: Pentatomidae) in Malta. EPPO Bull 49:132–136

Temreshev II, Esenbekova PA, MU A (2018) New records of a dangerus invasive pests – brown marmorated stink bug *Halyomorpha halys* Stal, 1855 (Heteroptera, Pentatomidae) in Kazakhstan. Acta Biol Sib 4:94–101

Terrestrial Heteroptera Recording Scheme—Shieldbugs & allied species (2023) Shieldbug & allied species records, from iRecord. In: Shield. Allied Species Rec. IRecord. https://www.gbif.org/dataset/ad07ac5e-718d-40cc-8093-b61169b745ef. Accessed 1 Nov 2023

Tommasini MG, Maistrello L, Vaccari G, Nannini R, Bortolotti P, Caruso S, Casoli L, Vergnani S, Preti M, Montanari M, Landi M, Simoni M, Costi E, Di Bella E, Bulgarini G, Masino F, Antonelli A (2019) Approccio multidisciplinare per contenere la cimice asiatica. Inf Agrar 13:38–43

Tortorici F, Talamas EJ, Moraglio ST, Pansa MG, Asadi-Farfar M, Tavella L, Caleca V (2019) A morphological, biological and molecular approach reveals four cryptic species of *Trissolcus* Ashmead (Hymenoptera Scelionidae), egg parasitoids of Pentatomidae (Hemiptera). J Hymenopt Res 93:153–200

Tortorici F, Bombi P, Loru L, Mele A, Moraglio ST, Scaccini D, Pozzebon A, Pantaleoni RA, Tavella L (2023) *Halyomorpha halys* and its egg parasitoids *Trissolcus japonicus* and *T. mitsukurii*: the geographic dimension of the interaction. NeoBiota 85:197–221

Tozlu E, Saruhan I, Tozlu G, Kotan R, Dadaşoğlu F, Tekiner N (2019) Potentials of some entomopathogens against the brown marmorated stink bug, *Halyomorpha halys* (Stål, 1855) (Hemiptera: Pentatomidae). Egypt J Biol Pest Control 29:76

Tunca H, Venard M, Colombel E-A, Capelli M, Tabone E (2019) Life history traits of *Ooencyrtus pityocampae* (Hymenoptera: Encyrtidae) reared on *Halyomorpha halys* eggs (Hemiptera: Pentatomidae). Entomol Gen 39:93–101

Tuncer C, Ozdemir IO (2023) Current status and importance of the brown marmorated stink bug in Turkey. In: The 4th Asian Horticultural Congress 28–31 August 2023Tokyo, Japan, Abstract book, pp 69–70

Tytar V, Kozynenko II (2020) Bioclimatic modeling of the distribution of brown marmorated stink bug *Halyomorpha halys* (Stål, 1855), with special reference to Ukraine. Rep Natl Acad Sci Ukr 2:82–86

USDA-NIFA SCRI (2021) State-by-State – StopBMSB.org. https://www.stopbmsb.org/where-is-bmsb/state-by-state/. Accessed 3 Nov 2023

Uzhevskaya SF (2017) Invasive insect species in Odessa region. Proc Mus Fund Named Brauner 14:57–64

Vaccari G, Caruso S, Pozzebon A, Maistrello L (2018) Pheromone traps for *Halyomorpha halys* monitoring: evaluation of side-effects in fruit orchards. In: Book of Abstracts, ECE 2018—XIth European Congress of Entomology, Napoli, Italy, July 2–6, 2018, p 167

Vaccari G, Caruso S, Pozzebon A, Maistrello L (2019) Pheromone traps for *Halyomorpha halys*: a three-year comparison in pear orchards. Pheromones Semiochem Integr Prod Integr Prot Fruit Crops IOBC-WPRS Bull 146:125–126

Vaccari G, Preti M, Bariselli M, Fagioli L, Civolani S, Tommasini MG, Golfarelli M, Gallinucci E, Forresi C, Maistrello L (2023) A territorial monitoring system for *Halyomorpha halys* in Emilia-Romagna Region. Pheromones Semiochem Integr Prod Integr Prot Fruit Crops IOBC-WPRS Bull 166:151–153

van der Heyden T, Petrovan S (2023) New records of Heteroptera from the Canary Islands (Spain), II. Arq Entomolóxicos 26:109–111

van der Heyden T, Saci A, Dioli P (2021) First record of the brown marmorated stink bug *Halyomorpha halys* (Stål, 1855) in Algeria and its presence in North Africa (Heteroptera: Pentatomidae). Rev Gaditana Entomol 12:147–154

Vétek G (2017) Severe damage to vegetables by the invasive brown marmorated stink bug, *Halyomorpha halys* (Hemiptera: Pentatomidae), in Hungary. Period Biol 119:131–135

Vétek G, Papp V, Haltrich A, Redei D (2014) First record of the brown marmorated stink bug, *Halyomorpha halys* (Hemiptera: Heteroptera: Pentatomidae), in Hungary, with description of the genitalia of both sexes. Zootaxa 3780:194–200

Vétek G, Károlyi B, Mészáros Á, Horváth D, Korányi (2018) The invasive brown marmorated stink bug (*Halyomorpha halys*) is now widespread in Hungary. Entomol Gen 38:3–14

Vidović V, Nikolić P, Hrnčić S, Nježić B (2023) Monitoring of the brown marmorated stink bug *Halyomorpha halys* (Stål, 1855) in Banja Luka, Bosnia and Herzegovina. AGRO – Knowl J 24:27–36

Waarneming.nl (2023) *Halyomorpha halys* (Stål, 1855). Waarneming.nl. https://waarneming.nl/species/197583/maps/?start_date=2013-09-26&interval=315360000&end_date=2023-09-24&map_type=grid10k. Accessed 4 Nov 2023

Weber NC, Razinger J, Jakopič J, Schmitzer V, Hudina M, Slatnar A, Veberič R, Štampar F, Zamljen T (2021) Brown marmorated stink bug (*Halyomorpha halys* Stål.) attack induces a metabolic response in strawberry (Fragaria × ananassa Duch.). Fruit Horticulturae 7:561

Wermelinger B, Wyniger D, Forster B (2008) First records of an invasive bug in Europe: *Halyomorpha halys* Stal (Heteroptera: Pentatomidae), a new pest on woody ornamentals and fruit trees? Mitt-Schweiz Entomol Ges 81:1–8

Yan J, Pal C, Anderson D, Vétek G, Farkas P, Burne A, Fan Q-H, Zhang J, Gunawardana DN, Balan RK, George S, Li D (2021) Genetic diversity analysis of brown marmorated stink bug, *Halyomorpha halys* based on mitochondrial COI and COII haplotypes. BMC Genomic Data 22:7

Yanovsky Y (2021) Brown-marble bug (*Halyomorpha halys*): peculiarities of biology and measures to control its harmfulness in apple plantations of Ukraine (in ukr.). AgroTimes Httpsagro timesuaarticlekorychnevo-Marmurovyj-Klop

Yonow T, Kriticos DJ, Ota N, Avila GA, Hoelmer KA, Chen H, Caron V (2021) Modelling the potential geographic distribution of two *Trissolcus* species for the brown marmorated stink bug, *Halyomorpha halys*. Insects 12:491

Zakharchenko V, Karpun N, Borisov B (2020) Trophic connections of the brown marmorated stink bug *Halyomorpha halys* Stål in the conditions of the invasive area on the Black Sea coast of the Caucasus. BIO Web Conf 21:00007

Zamljen T, Medič A, Veberič R, Hudina M, Štampar F, Slatnar A (2021a) Apple fruit (*Malus domestica* Borkh.) metabolic response to infestation by invasive brown marmorated stink bug (*Halyomorpha halys* Stal.). Horticulturae 7:212

Zamljen T, Veberic R, Hudina M, Slatnar A (2021b) The brown marmorated stink bug (*Halyomorpha halys* Stål) influences pungent and non-pungent *Capsicum* cultivars' pre- and post-harvest quality. Agronomy 11:2252

Zapponi L, Bon MC, Fouani JM, Anfora G, Schmidt S, Falagiarda M (2020) Assemblage of the egg parasitoids of the invasive stink bug *Halyomorpha halys*: insights on plant host associations. Insects 11:588

Zapponi L, Tortorici F, Anfora G, Bardella S, Bariselli M, Benvenuto L, Bernardinelli I, Butturini A, Caruso S, Colla R, Costi E, Culatti P, Di Bella E, Falagiarda M, Giovannini L, Haye T, Maistrello L, Malossini G, Marazzi C, Marianelli L, Mele A, Michelon L, Moraglio ST, Pozzebon A, Preti M, Salvetti M, Scaccini D, Schmidt S, Szalatnay D, Roversi PF, Tavella L, Tommasini MG, Vaccari G, Zandigiacomo P, Sabbatini-Peverieri G (2021) Assessing the distribution of exotic egg parasitoids of *Halyomorpha halys* in Europe with a large-scale monitoring program. Insects 12:316

Zapponi L, Nieri R, Zaffaroni-Caorsi V, Pugno NM, Mazzoni V (2023) Vibrational calling signals improve the efficacy of pheromone traps to capture the brown marmorated stink bug. J Pest Sci 96:587–597

Zhang J, Zhang F, Gariepy T, Mason P, Gillespie D, Talamas E, Haye T (2017) Seasonal parasitism and host specificity of *Trissolcus japonicus* in northern China. J Pest Sci 90:1127–1141

Zovko M, Ostojić I, Jurković D, Karić N (2019) First report of the brown marmorated stink bug, *Halyomorpha halys* (Stål, 1855) in Bosnia and Herzegovina. Rad Poljopr Prehrambenog Fak Univ U Sarajev Fac Agric Food Sci Univ Sarajevo 64:68–78

Chapter 16
Case Study 3: Current Pest Status and Management of the Invasive Brown Marmorated Stink Bug in the USA

Laura J. Nixon, Anne L. Nielsen, and Tracy C. Leskey

1 Introduction

The invasive brown marmorated stink bug (BMSB), *Halyomorpha halys* (Stål), was officially detected in Allentown, PA, USA, in 2001, though aggregations of adults were observed in the area beginning in the mid-1990s. Since that first detection, BMSB has spread throughout the USA, Canada, a number of European countries, and Chile. When BMSB populations exploded in the mid-Atlantic USA in 2010, the list of crops with significant damage was lengthy and included tree fruit, especially apples, peaches and pears, fruiting vegetable such as tomatoes and peppers, grapes, ornamental trees, and row crops including corn and soybean. Apple alone incurred $37 million in losses and estimations indicated half the peach crop was lost.

Additionally, homeowners and businesses experienced extreme nuisance pest problems as large numbers of adults entered anthropogenic structures in the fall to overwinter. Concerns for similar devasting impacts across the USA was acute at that time, as no management recommendations or monitoring tools were in place for this pest, little was known about its biology, ecology and behavior, and the potential for establishment in new regions, impacts on additional crops, or long-term prospects

L. J. Nixon
USDA ARS, Kearneysville, WV, USA

Rutgers University, Bridgeton, NJ, USA

A. L. Nielsen
Rutgers University, Bridgeton, NJ, USA

T. C. Leskey (✉)
USDA ARS, Kearneysville, WV, USA
e-mail: tracy.leskey@usda.gov

A. F. Bueno, A. R. Panizzi (eds.), *Stink Bugs (Hemiptera: Pentatomidae) Research and Management*, Entomology in Focus 9,
https://doi.org/10.1007/978-3-031-69742-5_16

for landscape level biological control efforts were uncertain. Here, we summarize the knowledge gained regarding BMSB biology, ecology and behavior, current pest status and management tools, and lingering issues that remain in the USA nearly 15 years following the 2010 outbreak.

2 Factors Affecting Current Distribution

In early 2011, BMSB populations were confirmed in seven Eastern states: DE, MD, NC, NJ, NY, PA, VA, WV, as well as known isolated populations in OR and WA. Since that time, BMSB has been detected in nearly every other state in the USA, but significant impacts from established populations have been confined to Eastern states and some regions on the West Coast.

Abiotic and possibly biotic barriers appear to be limiting their spread and establishment in portions of the USA, leading to regions with confirmed detections, but no established populations, and therefore no subsequent agricultural or nuisance pest issues. Moreover, while bivoltinism has been documented in some regions and is predicted for much of the United States (Leskey et al. 2012a; Nielsen et al. 2016), differences in abiotic conditions impact the size of each generation which combined with overlapping generations makes separation of each generation difficult in some locations. In a recent life table study comparing populations from NJ and OR, similar minimum and maximum temperatures of 14.3 °C and 35.9 °C for development were recorded, with optimal development occurring between 25 °C and 30 °C, however stark differences were observed in the reproductive potential for each population (Mermer et al. 2023), in alignment with previous studies in NJ and in MN (Nielsen et al. 2008b; Govindan and Hutchison 2020). Temperatures higher than these optimal conditions appear to be a limiting factor for BMSB. For example, in NC, despite the potential for bivoltinism in warmer coastal plain areas, populations have remained low; recent studies found higher summer temperatures (>30 °C) induced heat stress and limited successful hatch of F2 eggs laid by F1 adults (~26% hatch), whereas eggs laid by overwintered adults in the spring in the same region had ~90% hatch when temperatures averaged 26 °C (Ogburn et al. 2023). Indeed, high adult BMSB mortality has been recorded for temperatures >35 °C, with >90% at temperatures of 42 °C for 4 h (Aigner and Kuhar 2016). More evidence for high temperatures limiting BMSB populations is also apparent in a recent study in CA. Here, high temperatures (40 °C and 42 °C) and low humidity (17%) were used to simulate CA conditions; results revealed reduced survivorship of early instar BMSB nymphs under these conditions. Interestingly, high temperatures and high humidity (79%) reduced older instar nymphal survivorship, and high temperatures reduced overall reproduction (Fisher et al. 2020). Based on these results, high temperatures coupled with low and/or high humidity (depending on the life stage) may be responsible for minimal or lack of establishment in portions of the USA such as southeastern coastal plains and arid regions of the West.

Conversely, low temperature thresholds also may constrict BMSB establishment and geographic range. BMSB is a chill intolerant species, succumbing to low temperatures before it freezes, i.e., reaching its supercooling point. In studies involving MN and Mid-Atlantic (VA and WV) adults, high mortality occurred when temperatures were below −10 °C and prior to BMSB reaching supercooling points of −17.0 °C and − 13.9 °C for MN and mid-Atlantic, respectively, with acclimation location likely responsible for the supercooling temperature differences (Cira et al. 2016). Certainly, BMSB likely buffer lethal low temperatures by overwintering in anthropogenic structures or beneath the bark of trees in nature (Inkley 2012; Lee et al. 2014b). In colder, northern climes such as those in Ontario, Canada, it is clear these anthropogenic structures are critical for survivorship. Indeed, while Canadian populations from Ontario were found to be chill intolerant with similar supercooling temperatures as USA populations, adult survivorship from simulated overwintering shelters held outdoors compared with those held in a garage were significantly different (Ciancio et al. 2021). In colder climates, it is likely that BMSB will have more difficulty establishing as their populations cannot tolerate low temperatures without access to robust overwintering refugia.

Photoperiod may also contribute to capacity of BMSB to establish in new regions. While early estimates as low as 12.7 h photoperiod for diapause termination were identified (Nielsen et al. 2017), other more recent studies have cited photoperiods between 13.0 and 13.5 h, with at least 50% of females initiating reproductive development in Eastern US, the region where densities have historically been greatest (McDougall et al. 2021). Interestingly, in more southerly US regions such as FL that have fewer days exceeding these critical photoperiod lengths (13 + h light), only very small populations have been detected at several sites near Orlando to date (Penca and Hodges 2018). In this case, photoperiod, high humidity, and high temperatures all may be reducing the capacity of BMSB to establish in these regions despite many reports of accidental introductions from the belongings of human visitors traveling to the state. Diapause induction in an agent-based stochastic model also relied on a 13.5 h photoperiod as well (Nielsen et al. 2016) with overwintering females remaining reproductively immature until the following spring (Nielsen et al. 2017).

Overall, abiotic factors including temperature, humidity and photoperiod have likely influenced establishment of BMSB. A number of models have been generated to predict current and future BMSB distribution in the USA and around the globe, though quality of prediction appears to vary (Fraser et al. 2017; Kriticos et al. 2017; Zhu et al. 2017; Gutiérrez Illán et al. 2022). For example, in the CLIMEX model published by Kriticos et al. (2017), abiotic parameters included a photoperiod parameter of 12 h, thereby producing predictions of strong suitability of places like FL, though now it is clear that the state is not very suitable. More recently, however, predictions generated by a maximum entropy model that included a photoperiod/solar radiation parameter were well aligned with current BMSB distribution and abundance based on a nationwide trapping network providing ground truthing for model predictions (Gutiérrez Illán et al. 2022). Based on collective results to date, regions like FL and the arid west are not very suitable currently.

Host plant availability also likely affects BMSB distribution. While BMSB can readily complete development on single hosts of peach, they can complete development on multi-host diets of uncultivated and cultivated hosts such as the invasive tree of heaven paired with apple (Acebes-Doria et al. 2016a). Indeed, these cultivated tree fruit hosts and uncultivated hosts are commonly often in close proximity of each other at orchard-woodland borders in the mid-Atlantic (Acebes-Doria et al. 2017) possibly leading to larger relative densities and likely supporting establishment of populations (Bergh et al. 2021; Hadden et al. 2022; Park et al. 2024). On the other hand, multiple uncultivated plants species present in regions of the Pacific Northwest did not support strong survivorship and development of BMSB when compared with Eastern host plant diets (Hepler et al. 2023) also possibly contributing to differences in establishment and size of BMSB populations in these two production regions.

Ultimately, both abiotic and biotic factors and their interactions likely contribute to BMSB distribution and establishment. Across the US, the most recent results from a national monitoring network revealed that largest relative densities were present in the mid-Atlantic and inland Southeast locations, while relative densities in the Great Lakes and PNW were ~50% lower, and sites in more arid western regions were >90% lower (Acebes-Doria et al. 2020). These generalized patterns have not changed substantially since this study was conducted.

3 Current Pest Status

3.1 Crops

When populations exploded in the USA, BMSB was recorded on numerous cultivated species including deciduous tree fruit, fruiting vegetables, row crops, small fruit and ornamentals (Leskey et al. 2012d), though it was unclear if BMSB would become a damaging pest and if other issues such as contamination at harvest would lead to additional impacts. Since that time, a list of confirmed hosts for BMSB has been maintained on StopBMSB.org with over 180 plant species listed to date (StopBMSB.org). Among cultivated crops, deciduous tree fruit have been found to be extremely susceptible to feeding injury from BMSB. For apples and peaches, both adults and nymphs can inflict feeding injury (Leskey et al. 2012c; Acebes-Doria et al. 2016b) with greatest injury incurred in the early season for peaches (much of it internal) and in the late season for apple (Joseph et al. 2015). Both European and Asian pear were also identified as susceptible host plants (Nielsen and Hamilton 2009). In the mid-Atlantic, damage from BMSB on tree fruit has declined substantially since the major outbreak in 2010, but legacy effects remain.

Growers must still treat for this pest, with most applying between 1 and 3 insecticide applications annually. If growers do not treat or delay treatments, they are still vulnerable to serious injury. Management also becomes challenging closer to harvest as growers must follow EPA pre-harvest insecticide application intervals, contend with large BMSB populations, and devote more labor to harvest activities rather than pest management. Indeed, BMSB remains an important tree fruit pest and is still considered one of the most serious for Eastern growers (Blaauw et al. 2015). In other parts of the USA, however, BMSB has yet to produce long-term, serious pest issues.

Orchard nut crops are also vulnerable to BMSB damage. In OR, hazelnut was identified as a susceptible host (Hedstrom et al. 2014), and BMSB has continued to inflict economic injury with early estimates of up to 6% injury annually (Walton et al. 2022) with most injury occurring late in the season (Handbooks 2024) and recent estimates as high as 18% at harvest (Walton unpubl). This pest continues to be a very serious problem for hazelnut growers in the region. In CA, BMSB was able to complete development on almond and pistachio as single host diets, though mortality was high under abiotic conditions experienced in the field (Stahl et al. 2021). While injury to almond has been detected in one commercial farm in 2017 (Rijal and Gyawaly 2018), widespread injury on almond and no injury to pistachio has not been reported to date. In all likelihood, abiotic factors are contributing to low populations in CA, and thus, lower pest pressure.

BMSB also can certainly cause significant damage on a wide variety of vegetables including sweet corn, tomatoes, peppers, eggplant, okra, and beans (Kuhar et al. 2012; Cissel et al. 2015; Zobel et al. 2016; Philips et al. 2017) as well as row crops (Rice et al. 2014; Aigner et al. 2017). During the early years of the outbreak, damage on mid-Atlantic farms was common and often very high (Leskey et al. 2012c; Rice et al. 2014). Organic growers were at the greatest disadvantage (Morehead and Kuhar 2017). And like tree fruit, the greatest damage was often present near wooded borders and landscapes (Rice et al. 2016; Aigner et al. 2017). However, while BMSB still continues to attack vegetable crops, widespread problems generated solely by BMSB are less common, and this pest is often lumped into management strategies for the greater stink bug complex (Aigner et al. 2016).

Finally, small fruit and grapes appear to have escaped serious problems. While BMSB will feed on and damage blueberry (Wiman et al. 2015), widespread damage and contamination issues from mechanical harvesting have not occurred. Moreover, following the introduction of spotted wing drosophila and the continued devastating consequences of its presence, the sprays being directed at this invasive fly likely mitigated any potential effects from BMSB. Conversely, the presence of BMSB in grape clusters can lead to taint in wine (Mohekar et al. 2017, 2018) and it is recommended that every effort is made to reduce presence of BMSB in clusters at harvest. To date, there are no widespread reports of taint from BMSB presence, though scouting for the pest prior to harvest is strongly recommended with 3 BMSB/cluster recommended as an action threshold (Mohekar et al. 2017).

3.2 Nuisance and Biosecurity Issues

When BMSB populations exploded in the mid-Atlantic in 2010, homeowners and businesses were subjected to large numbers of adults aggregating on the exterior of homes and buildings and entering to seek overwintering shelter in concealed areas. Press coverage by major news outlets such as the New York Times, Washington Post and National Public Radio raised the awareness of these additional issues (Leskey et al. 2012d). Since that time, more information regarding the dispersal process to overwintering sites, settling behavior within structures, and dispersal from them the following spring has contributed to our understanding of the phenomenon and strategies for reducing invasion into anthropogenic spaces.

BMSB adults begin moving toward potential overwintering sites several weeks before the fall equinox in the USA. In counts made over five consecutive years on a large building in rural VA, over 88% of this dispersal activity occurred between 15 September-15 October with peak numbers present between 1500 and 1600 h on north- and east-facing walls (Bergh and Quinn 2018). In a study involving citizen scientists across 20 states, BMSB numbers on the sides of homes were also greatest on north and east sides. Additionally, greater numbers were recorded by homeowners in rural, forested landscapes compared with other habitat types, and on homes that were brown or grey in color and with wood, cement, or stone exteriors (Hancock et al. 2019). Elevation appears to influence relative abundance of BMSB during the fall dispersal period with larger numbers present at higher elevations in the Eastern US compared with lower elevations (Cullum et al. 2020) though similar studies have not been performed at higher elevation regions, such as the PNW.

When BMSB enter structures, adults can easily pass through circular openings of ≥7 mm and slits ≥3 mm wide with preferences for overwintering in aggregations in cavities 4.5–5.0 mm wide within dry locations (Chambers et al. 2019, 2020; Cullum et al. 2020). The sheer numbers present in homes can be astounding. One homeowner collected data on the numbers removed from each floor of their home and their locations daily; over 26,000 adults were removed over a 181-day period with greatest numbers recovered from an attic crawlspace (Inkley 2012). As BMSB overwinter, they slowly deplete nutrient reserves with those that may no longer have adequate resources leaving overwintering sites earlier in search of food compared with more robust individuals (Skillman et al. 2018). It is during this period when BMSB become an apparent interior nuisance pest for homeowners. A survey of homeowners revealed their perceptions on the magnitude of the problem. Homeowners who estimated they had >1000 BMSB in their homes (~13% of respondents) were five times as likely to characterize their problem as 'bad' or 'horrible' compared with homeowners who had <100 (~45% of respondents) (Ludwick et al. 2020a). Thus, the best approach to reduce these problems is to prevent BMSB entry in the fall.

This propensity for settling in anthropogenic structures has led to shipments of BMSB being of goods harboring hitchhiking BMSB adults in countries trying to exclude them from entry, such as Australia and New Zealand. Overwintering BMSB aggregations have been identified as a primary risk of introduction in these countries. Both Australia and New Zealand require phytosanitary treatments during the highest risk period, i.e., when overwintering adults from the Northern Hemisphere may be present in shipments. This has culminated in ongoing research between the US and Australia and New Zealand on optimal treatments for at-risk cargo (Kawagoe et al. 2022; Corbett et al. 2023) and novel detection methods for overwintering aggregations (Nixon et al. 2018, 2019; Bedoya et al. 2020). Phytosanitary treatments for BMSB include heat treatments, which can be very expensive, and sulfuryl fluoride fumigations (Abrams et al. 2020), which is effective but not labelled in all relevant countries. Recent studies have shown that ethyl formate dilutions in carbon dioxide are effective at managing BMSB in shipments and labelled more widely in receiving countries (Kawagoe et al. 2022, Corbett et al. 2023). Studies on overwintering BMSB in the US have identified volatile organic compounds released by BMSB that act as alarm and defense pheromones (Nixon et al. 2018, 2022), and vibrational signals produced by male BMSB (Bedoya et al. 2020). However, neither stimuli source was reliable enough to be used for biosecurity detection (Nixon et al. 2019, Bedoya et al. 2020).

4 Monitoring and Management

4.1 Insecticide-Based Management

Growers in invaded regions in the mid-Atlantic region of the USA were forced to increase insecticide applications by up to four-fold for BMSB (Leskey et al. 2012c), which resulted in the disruption of many long-standing IPM programs for specialty crops and generated secondary pest problems. Insecticide programs previously recommended for native stink bugs in the US were not effective against BMSB (Leskey et al. 2012b) and materials which were reported to provide excellent management of BMSB in Asia were either not registered for crops in the US or performed poorly in further experiments (Lee et al. 2013b). Extensive laboratory, semi-field, and field trials assessing insecticidal efficacy for vulnerable crops including tree fruit, vegetables, and row crops (Kuhar and Kamminga 2017) demonstrated that several chemicals in the pyrethroid, organophosphate, carbamate, organochlorine, and neonicotinoid classes were effective against BMSB.

Major issues identified for labelled materials, however, included short residual activity, as well as knockdown and recovery responses particularly for some pyrethroids and neonicotinoids (Nielsen et al. 2008a; Leskey et al. 2012b; Lee et al. 2013a). Because of these shortcomings, as well as existing regulatory constraints, two additional insecticides, bifenthrin and dinotefuran, were identified and approved for emergency use in apple and peach in seven eastern states (DE, MD, NJ, NY, PA, VA, WV). However, every effective insecticide against BMSB is also considered broad-spectrum with negative effects on beneficial arthropods including pollinators and predators. Some materials such as the organochlorine, endosulfan, have had their registrations cancelled. Neonicotinoids continue to face regulatory challenges in many locations due to negative pollinator impacts, though they are still used against BMSB. In the US, the most commonly recommended materials for BMSB are pyrethroids and include bifenthrin, permethrin, fenpropathrin, lambdacyhalothrin, and zeta-cypermethrin, depending on the crop.

Other commonly used materials include neonicotinoids dinotefuran, clothianidin, thiamethoxam and the carbamate, methomyl. For organic production, very few options are available; pyrethrins and spinosad were shown to be effective in laboratory assays, however, do not hold up for regular field applications (Lee et al. 2014a; Kuhar and Kamminga 2017; Morehead and Kuhar 2017).

How insecticides are applied can improve impact against BMSB and reduce crop injury. For example, alternate row middle (ARM) sprays (Hull and VanStarner 1983) have been utilized to increase fresh residues in orchard blocks to combat short residual activity of materials against BMSB (Leskey et al. 2014, 2020); because only half the block is sprayed 1 week (every other row), and the other half a week later both "half-treatments" equal a single full block treatment, but provide improved residual activity over the 2 week period. Moreover, exploiting BMSB behavioral attributes has been used to tailor insecticide applications. BMSB are present in large numbers in crop-woodland border interfaces, leading to a strong edge effect exhibited by BMSB into cropping systems based on both trap captures and injury patterns (Joseph et al. 2014; Rice et al. 2016; Aigner et al. 2017; Hadden et al. 2022; Hadden et al. 2023). IPM Crop Perimeter Restructuring (CPR), is a concept based on this behavior that utilizes perimeter sprays rather than standard full block or ARM treatments, resulting in injury at harvest statistically equal to standard insecticide management programs, but reducing insecticide inputs by up to 60% (Blaauw et al. 2015; Akotsen-Mensah et al. 2020; Leskey et al. 2020).

Field crops, such as corn and soybean, also experience increased edge effects with BMSB injury, with severity increased when edges are bordered by woods, other crops, or buildings (Venugopal et al. 2014; Aigner et al. 2017). Recommendations for these crops are often few well-timed perimeter sprays and avoiding planting different host crops (e.g. soybean and wheat) adjacent to one another (Venugopal et al. 2015). In vegetable crops, targeted soil drenches with systemic neonicotinoids have provided significant reduction of BMSB damage in tomato and pepper crops, replacing labor-intensive foliar applications (Aigner et al. 2015).

4.2 Monitoring Tools

In the absence of effective monitoring tools, growers were applying insecticides against BMSB frequently to ensure they mitigated the chances of catastrophic injury. Effective monitoring systems are essential for decision-based IPM strategies. Unfortunately for BMSB, the only decision support tool in the early days of the outbreak, black pyramid traps baited with methyl (2E,4E,6Z)-2,4,6-decatrienoate, the pheromone of another Asian stink bug species, *Plautia stali*, could not be used to guide season-long management as adults did not respond to this stimulus until late in the growing season (Leskey et al. 2012a). The discovery of the male-produced BMSB aggregation pheromone ((3S,6S,7R,10S)-10,11-epoxy-1-bisabolen-3-ol and (3R,6S,7R,10S)-10,11-epoxy-1-bisabolen 3-ol) (Khrimian et al. 2014) and that MDT (methyl (2E,4E,6Z)-2,4,6-decatrienoate) serves as a powerful synergist, has made season-long monitoring of BMSB viable (Weber et al. 2014; Leskey et al. 2015). Both ground-deployed pyramid and sticky traps reliably capture BMSB adults and nymphs, reflecting presence, relative abundance, and seasonal activity accurately across the US (Leskey et al. 2012a; Acebes-Doria et al. 2018, 2020). While other traps have been evaluated (Morrison et al. 2015; Rice et al. 2018; Nixon et al. 2024), they have not been as effective and reliable. Those traps deployed in trees are most problematic because while foraging BMSB respond to the baited trap, they may remain on branches near the trap itself, never entering the collection device (Rice et al. 2018).

Using calculations for estimate plume reach, maximum dispersive distance, and trapping area as per Miller et al. (2015) and fluorescent marking methods (Kirkpatrick et al. 2020), the trapping area of ground deployed sticky cards baited with pheromone + MDT has been calculated for BMSB in open fields and apple orchards. The trapping areas in for nymphs and adults in open fields was calculated at 0.58 ha and 4.83–5.56 ha, respectively, and in apple orchards was reduced to 1.67 ha for adults (Kirkpatrick et al. 2019). This is critical information to inform how many traps are necessary to reliably capture BMSB and measure presence, relative abundance and seasonal activity.

Moreover, baited traps can be used as a decision support tool. In trials conducted over 2 years, using a cumulative threshold of 10 adults/trap in baited black pyramid traps (Fig. 16.1) to trigger insecticide applications (each application includes two consecutive ARM sprays) resulted in effective management of BMSB; injury at harvest was equal to that of weekly insecticide treatments, but resulted in 40% reduction in insecticide applications. Lower and higher thresholds resulted in increased insecticide inputs with no measurable reduction in injury or fewer insecticide applications but with increased injury, respectively (Short et al. 2017). The drawback of this monitoring method is that the bulky, ground-deployed traps are labor-intensive to install and difficult to integrate into commercial orchards. However, baited clear sticky panels (Fig. 16.2) deployed atop of 1.2 m wooden posts (Acebes-Doria et al. 2018, 2020; Nixon et al. 2024) have proven effective, and less time-intensive to install. A cumulative threshold of 4 adults/trap used to trigger insecticide treatment has proven effective in experimental and commercial apple orchard plots with insecticide applications reduced insecticides by over 50% compared with weekly treatments, but with similar fruit injury at harvest (Leskey et al. unpubl). Currently, no other thresholds for tree fruit or nut crops have been developed to guide BMSB management.

Fig. 16.1 Black pyramid traps baited with pheromone + MDT to trigger insecticide applications agains the brown marmorated stink bug, *Halyomorpha halys*

Fig. 16.2 Baited sticky panel to sample the brown marmorated stink bug, *Halyomorpha halys*

4.3 Border-Based Management Strategies

Several iterations of attract-and-kill (AK) strategies across cropping systems have been investigated for managing BMSB populations while reducing the amount of insecticide applied. The first published study involved baiting selected border trees with high dose experimental pheromone lures in combination with MDT then spraying baited trees plus those on either side on a weekly basis (Morrison III et al. 2019; Leskey et al. 2020). When compared with grower standard spray programs in commercial orchards, this AK protocol decreased BMSB fruit injury by 2–7× and reduced insecticide application areas by over 90%. However, this approach also was more expensive than standard management programs due to the cost of experimental pheromone lures used (now commercially available and more affordable) and due to weekly labor inputs (Morrison III et al. 2019). To reduce the labor involved based on weekly treatment of AK sites, deltamethrin-impregnated long-lasting insecticide netting (LLIN), which is manufactured as mosquito netting were used to replace insecticide application. This approach reduces labor inputs, as nets remain effective season-long (Kuhar et al. 2017; Ibrahim et al. 2020). Recent trials conducted with nets placed next to border row apple trees (Fig. 16.3). resulted in damage that was statistically equivalent to grower standard treatments. However, because LLINs are not registered for agricultural use, adoption of this promising technology is not possible.

Fig. 16.3 Long-lasting insecticide netting used as killing agent at attract-and-kill sites in apple orchards against the brown marmorated stink bug, *Halyomorpha halys*

4.4 Nuisance Pest Management

The first step to managing BMSB in homes is preventing their entry. Exclusion recommendations include sealing gaps around window, doors, and other openings, repair any damaged screens or pipe openings, and checking attic and basement vents. Once inside, BMSB can be mechanically removed with vacuums or by hand with empty containers (Raupp 2018). Despite the discovery and synthesis of the BMSB aggregation pheromone, semiochemical-baited traps were not effective at capturing BMSB during diapause, as even those that may leave overwintering locations and roaming inside structures through the winter were not responsive to this olfactory cue (Morrison III et al. 2017b). Conversely, an aluminum foil pan filled with soapy water and paired with a simple desk lamp proved to be effective at reducing those that have become active inside homes (Aigner Jr and Kuhar 2014). There are products available to homeowners such as insecticidal soaps, oils, and capsaicin; however, these are generally ineffective against BMSB adults (Bergmann and Raupp 2014). Much like cropping systems, nuisance problems have become far less severe since the 2010 outbreak.

5 Biological Control

Generalist predators native to the USA have been reported for all lifestages of BMSB through field surveys, and semi-field and laboratory bioassays (Morrison III et al. 2016, 2017a; Ogburn et al. 2016; Pote and Nielsen 2017). Egg masses are the most common target of generalist feeding with effective predators including ground beetles (Carabidae), katydids (Tettigoniidae), earwigs (Forficulidae), jumping spiders (Salticidae), grasshopper (Acrididae), and crickets (Gryllidae) (Morrison III et al. 2016; Pote and Nielsen 2017). There are also three principal genera of Hymenoptera parasitoids that have been found to target BMSB eggs in North America: *Telenomus* spp., *Trissolcus* spp., and *Anastatus* spp., with *Anastatus* being the most frequent to emerge from BMSB egg masses from the field (Cornelius et al. 2016; Herlihy et al. 2016; Abram et al. 2017; Dieckhoff et al. 2017; Jones et al. 2017). Parasitism and predation rates of egg masses by indigenous species is generally reported at <10% in field surveys (Abram et al. 2017). Laboratory and semi-field studies have shown that generalist predators such as assassin bugs (Reduviidae), grasshoppers, damsel bugs (*Nabis* spp.), and soldier spine bugs (*Podisus maculiventris* Say) will predate on BMSB nymphs (Pote and Nielsen 2017), spiders in the families Theridiidae, Pholcidae, or Agelenidae will predate on adults (Morrison III et al. 2017a), and the wheel bug (*Arilus cristatus* L.) will predate on all mobile lifestages (Jones et al. 2017). Although there has not been any meaningful quantification of overall predator impacts on BMSB mobile lifestages in the field. Analysis of landscape elements influencing natural enemies of BMSB identified herbaceous crops, specifically soybean, to be significantly associated with natural enemies, specifically Scelionid wasps (McDougall et al. unpub).

Trissolcus japonicus (Ashmead) is an Asian parasitoid wasp which co-evolved with BMSB (Zhang et al. 2017) and can parasitize 50–90% of BMSB egg masses. This species was being researched for consideration as a biocontrol agent against BMSB (Buffington et al. 2018) when, in 2014, an adventive population was discovered in Maryland (Talamas et al. 2015; Herlihy et al. 2016). Since then, adventive populations of *T. japonicus* have established in ten Eastern states and three states in the Pacific Northwest (stopBMSB). Models suggest that *T. japonicus* can establish in any of the regions that BMSB has invaded in the US (Avila and Charles 2018). *Trissolcus japonicus* was originally recovered from wooded and non-managed areas (Talamas et al. 2015; Milnes et al. 2016), but further detections included agricultural sites (Jentsch 2017; Kaser et al. 2018). *T. japonicus* has been consistently monitored for in tree of heaven, a common wild host of BMSB, which is one of the most common trees bordering crops and transport corridors in the mid-Atlantic US. Unbaited yellow sticky cards can be hung throughout the canopy of tree of heaven for effective monitoring of relative abundance of *T. japonicus* (Dyer et al. 2022a, 2022b, 2024). This monitoring method was successfully used to track *T. japonicus* establishment at 11 sites throughout Virginia tree fruit production regions over a 5-year period of 2017–2022 (Bergh et al. 2023). These surveys showed that augmentative releases of *T. japonicus* in a production region led to detection of populations within 2–4 years.

A recent study also showed that *T. japonicus* as a classical biocontrol agent can be integrated into orchard pest management for BMSB (Ludwick et al. 2020b); emergence from parasitized egg masses was not affected after deployment in orchards treated with bifenthrin, thiamethoxam + λ-cyhalothrin, clothianidin, or methomyl. Additionally, egg masses from orchards treated with either clothianidin or methomyl were able to be parasitized post-treatment without significant effects to female wasps or emergence (Ludwick et al. 2020b). These findings are positives for the future of *T. japonicus* biocontrol for managing BMSB populations.

The microsporidian *Nosema maddoxi* (Becnel, Solter, Hajek, Huang, Sanscrainte & Estep) is an entomopathogenic fungus that infects BMSB populations and causes decreased fitness in infected insects. The first infected BMSB in the US were detected in declining colonies and subsequently in field populations (Hajek et al. 2018). In 2018, *N. maddoxi* infected BMSB were present in 11 states, with an average of 18.9% infection rates in field populations (Preston et al. 2020a), and since has been detected in a further 6 states in overwintering populations (Hajek et al. 2023). This naturally occurring pathogen does not generally lead to immediate insect death, but rather decreases insect longevity, egg production, and egg viability (Preston et al. 2020b).

6 Legacy Issues, Lessons Learned and Future Directions

While BMSB is no longer the devastating pest that first attacked crops and invaded homes in tremendous numbers back in 2010, legacy effects remain. Orchard crop growers (deciduous tree fruit and nuts) continue to make insecticide applications against this serious pest in the late season, affecting their overall financial bottom line. Additionally, available IPM tools have not been fully integrated into management programs, and growers may be applying too many insecticides simply to provide them with the confidence they will not incur significant crop damage. The need for increased efforts to demonstrate the benefits of IPM tools is still a high priority in the USA. Conversely, IPM tools do not exist for many vulnerable crops; trap-based thresholds, for example, still are not available for stone fruit and nut crops. Focusing on their development would likely reduce unnecessary treatment. Additionally, while AK is an effective management tool, LLINs are lacking a registration for this specific use leaving this technology as purely experimental currently.

Furthermore, we do not know if the impacts of landscape-level biological control agents such as *T. japonicus* have reached their maximum. While evidence for density-dependent regulation of BMSB has recently been documented (Dyer et al. 2024), BMSB still requires management intervention in vulnerable orchard crops. Conducting research on the impact of augmentative releases of *T. japonicus* near vulnerable cropping systems would be beneficial as this could determine if their release increases their establishment, ultimately reducing pest pressure from BMSB and subsequently crop damage and the need for insecticide inputs.

Ultimately, the successful response to the invasive BMSB by researchers, Extension specialists, regulators and industry provides a template for responses to future invasive species (Ludwick et al. 2020a). Ensuring that a comprehensive strategy is developed in the early stages of the invasion that includes short, medium, and longer-term goals and links all critical stakeholders are essential.

References

Abram PK, Hoelmer KA, Acebes-Doria A, Andrews H, Beers EH, Bergh JC, Bessin R, Biddinger D, Botch P, Buffington ML (2017) Indigenous arthropod natural enemies of the invasive brown marmorated stink bug in North America and Europe. J Pest Sci 90:1009–1020

Abrams AE, Kawagoe JC, Najar-Rodriguez A, Walse SS (2020) Sulfuryl fluoride fumigation to control brown marmorated stinkbug (Hempitera: Pentatomidae). Postharvest Biol Technol 163:111111

Acebes-Doria AL, Leskey TC, Bergh JC (2016a) Host plant effects on *Halyomorpha halys* (Hemiptera: Pentatomidae) nymphal development and survivorship. Environ Entomol 45:663–670

Acebes-Doria AL, Leskey TC, Bergh JC (2016b) Injury to apples and peaches at harvest from feeding by *Halyomorpha halys* (Stål) (Hemiptera: Pentatomidae) nymphs early and late in the season. Crop Prot 89:58–65

Acebes-Doria AL, Leskey TC, Bergh JC (2017) Temporal and directional patterns of nymphal *Halyomorpha halys* (Hemiptera: Pentatomidae) movement on the trunk of selected wild and fruit tree hosts in the mid-Atlantic region. Environ Entomol 46:258–267

Acebes-Doria AL, Morrison WR III, Short BD, Rice KB, Bush HG, Kuhar TP, Duthie C, Leskey TC (2018) Monitoring and biosurveillance tools for the brown marmorated stink bug, *Halyomorpha halys* (Stål)(Hemiptera: Pentatomidae). Insects 9:82

Acebes-Doria AL, Agnello AM, Alston DG, Andrews H, Beers EH, Bergh JC, Bessin R, Blaauw BR, Buntin GD, Burkness EC, Chen S, Cottrell TE, Daane KM, Fann LE, Fleischer SJ, Guédot C, Gut LJ, Hamilton GC, Hilton R, Hoelmer KA, Hutchison WD, Jentsch P, Krawczyk G, Kuhar TP, Lee JC, Milnes JM, Nielsen AL, Patel DK, Short BD, Sial AA, Spears LR, Tatman K, Toews MD, Walgenbach JD, Welty C, Wiman NG, Van Zoeren J, Leskey TC (2020) Season-long monitoring of the brown marmorated stink bug (Hemiptera: Pentatomidae) throughout the United States using commercially available traps and lures. J Econ Entomol 113:159–171

Aigner JD Jr, Kuhar TP (2014) Using citizen scientists to evaluate light traps for catching brown marmorated stink bugs in homes in Virginia. J Extension 52:16

Aigner J, Kuhar T (2016) Lethal high temperature extremes of the brown marmorated stink bug (Hemiptera: Pentatomidae) and efficacy of commercial heat treatments for control in export shipping cargo. J Agric Urban Entomol 32:1–6

Aigner J, Walgenbach J, Kuhar T (2015) Toxicities of neonicotinoid insecticides for systemic control of brown marmorated stink bug (Hemiptera: Pentatomidae) in fruiting vegetables. J Agric Urban Entomol 31:70–80

Aigner BL, Herbert DA, Dively GP, Venugopal D, Whalen J, Cissel B, Kuhar TP, Brewster CC, Hogue JW, Seymore E (2016) Comparison of two sampling methods for assessing *Halyomorpha halys* (Hemiptera: Pentatomidae) numbers in soybean fields. J Econ Entomol 109:2586–2589

Aigner BL, Kuhar TP, Herbert D, Brewster CC, Hogue J, Aigner JD (2017) Brown marmorated stink bug (Hemiptera: Pentatomidae) infestations in tree borders and subsequent patterns of abundance in soybean fields. J Econ Entomol 110:487–490

Akotsen-Mensah C, Blaauw B, Short B, Leskey TC, Bergh JC, Polk D, Nielsen AL (2020) Using IPM-CPR as a management program for apple orchards. J Econ Entomol 113:1894–1902

Avila G, Charles J (2018) Modelling the potential geographic distribution of *Trissolcus japonicus*: a biological control agent of the brown marmorated stink bug, *Halyomorpha halys*. BioControl 63:505–518

Bedoya C, Brockerhoff EG, Hayes M, Leskey TC, Morrison WR III, Rice KB, Nelson XJ (2020) Brown marmorated stink bug overwintering aggregations are not regulated through vibrational signals during autumn dispersal. R Soc Open Sci 7:201371

Bergh JC, Quinn NF (2018) Can the dispersal behavior of *Halyomorpha halys* (Hemiptera: Pentatomidae) inform the use of insecticide-treated netting to mitigate homeowner issues from its fall invasion? Environ Entomol 47:1501–1508

Bergh JC, Morrison WR, Stallrich JW, Short BD, Cullum JP, Leskey TC (2021) Border habitat effects on captures of *Halyomorpha halys* (Hemiptera: Pentatomidae) in pheromone traps and fruit injury at harvest in apple and peach orchards in the mid-Atlantic, USA. Insects 12:419

Bergh JC, Talamas EJ, Brandt SN, Edwards A, MacRae L, Monger G, Bowen J, Lawrence K, Reed K, Pottorff S (2023) Releasing and tracking the distribution of adventive *Trissolcus japonicus* (hymenoptera: Scelionidae) in Virginia. Environ Entomol 52:583–592

Bergmann EJ, Raupp MJ (2014) Efficacies of common ready to use insecticides against *Halyomorpha halys* (Hemiptera: Pentatomidae). Fla Entomol 97:791–800

Blaauw BR, Polk D, Nielsen AL (2015) IPM-CPR for peaches: incorporating behaviorally-based methods to manage Halyomorpha halys and key pests in peach. Pest Manag Sci 71:1513–1522

Buffington ML, Talamas EJ, Hoelmer KA (2018) Team Trissolcus: integrating taxonomy and biological control to combat the brown marmorated stink bug. Am Entomol 64:224–232

Chambers BD, Kuhar TP, Reichard G, Leskey TC, Pearce AR (2019) Size restrictions on the passage of overwintering *Halyomorpha halys* (Hemiptera: Pentatomidae) through openings. J Econ Entomol 112:1343–1347

Chambers BD, Leskey TC, Cullum JP, Pearce AR, Kuhar TP (2020) Cavity tightness preferences of overwintering *Halyomorpha halys* (Hemiptera: Pentatomidae). J Econ Entomol 113:1572–1575

Ciancio JJ, Turnbull KF, Gariepy TD, Sinclair BJ (2021) Cold tolerance, water balance, energetics, gas exchange, and diapause in overwintering brown marmorated stink bugs. J Insect Physiol 128:104171

Cira TM, Venette RC, Aigner J, Kuhar T, Mullins DE, Gabbert SE, Hutchison WD (2016) Cold tolerance of *Halyomorpha halys* (Hemiptera: Pentatomidae) across geographic and temporal scales. Environ Entomol 45:484–491

Cissel WJ, Mason CE, Whalen J, Hough-Goldstein J, Hooks CRR (2015) Effects of brown marmorated stink bug (Hemiptera: Pentatomidae) feeding injury on sweet corn yield and quality. J Econ Entomol 108:1065–1071

Corbett SMA, Cullum JP, Abrams AE, Rodriguez MS, Leskey TC, Walse SS (2023) Confirmatory evaluation of eFUME for control of brown marmorated stink bug, *Halyomorpha halys* (Hemiptera: Pentatomidae). J Econ Entomol 116:771–778

Cornelius ML, Dieckhoff C, Hoelmer KA, Olsen RT, Weber DC, Herlihy MV, Talamas EJ, Vinyard BT, Greenstone MH (2016) Biological control of sentinel egg masses of the exotic invasive stink bug *Halyomorpha halys* (Stål) in mid-Atlantic USA ornamental landscapes. Biol Control 103:11–20

Cullum JP, Nixon LJ, Morrison WR, Raupp MJ, Shrewsbury PM, Venugopal PD, Martinson H, Bergh JC, Leskey TC (2020) Influence of landscape factors and abiotic conditions on dispersal behavior and overwintering site selection by *Halyomorpha halys* (Hemiptera: Pentatomidae). J Econ Entomol 113:2016–2021

Dieckhoff C, Tatman KM, Hoelmer KA (2017) Natural biological control of *Halyomorpha halys* by native egg parasitoids: a multi-year survey in northern Delaware. J Pest Sci 90:1143–1158

Dyer J, Talamas EJ, Leskey TC, Bergh JC (2022a) Evaluating chemical cues associated with *Halyomorpha halys* toward enhanced sensitivity of surveillance for *Trissolcus japonicus*. Environ Entomol 51:679–687

Dyer J, Talamas EJ, Leskey TC, Bergh JC (2022b) Influence of trap location in the tree canopy on captures of adventive *Trissolcus japonicus* (hymenoptera: Scelionidae). J Econ Entomol 115:904–908

Dyer J, Talamas EJ, Leskey TC, Bergh JC (2024) Season-long trapping of *Trissolcus japonicus* and *Halyomorpha halys*: temporal relationships between a parasitoid and its host. Environ Entomol 12(2):118

Fisher JJ, Rijal JP, Zalom FG (2020) Temperature and humidity interact to influence brown marmorated stink bug (Hemiptera: Pentatomidae) survival. Environ Entomol 50:390–398

Fraser D, Kumar S, Aguilar G (2017) Mapping the potential global range of the brown marmorated stink bug, *Halyomorpha halys*, with particular reference to New Zealand. Climate 5:75

Govindan BN, Hutchison WD (2020) Influence of temperature on age-stage, two-sex life tables for a Minnesota-acclimated population of the brown marmorated stink bug (*Halyomorpha halys*). Insects 11:108

Gutiérrez Illán J, Zhu G, Walgenbach JF, Acebes-Doria A, Agnello AM, Alston DG, Andrews H, Beers EH, Bergh JC, Bessin RT, Blaauw B, Buntin GD, Burkness EC, Cullum J, Daane KM, Fann LE, Fisher J, Girod P, Gut L, Hamilton G, Hepler J, Hilton R, Hoelmer KA, Hutchison W, Jentsch P, Joseph SV, Kennedy G, Krawczyk G, Kuhar T, Lee J, Leskey T, Marsha A, Milnes JM, Nielsen A, Patel DK, Peterson H, Reisig D, Rijal J, Sial AA, Spears L, Stahl J, Tatman KM, Taylor S, Tillman G, Toews MD, Villanueva R, Welty C, Wiman N, Wilson J, Zalom FG, Crowder D (2022) Evaluating invasion risk and population dynamics of the brown marmorated stink bug across the contiguous United States. Pest Manag Sci 78:4929–4938

Hadden WT, Nixon LJ, Leskey TC, Bergh JC (2022) Seasonal distribution of *Halyomorpha halys* (Hemiptera: Pentatomidae) captures in woods-to-orchard pheromone trap transects in Virginia. J Econ Entomol 115:109–115

Hadden W, Brewster CC, Leskey TC, Bergh JC (2023) *Halyomorpha halys* (Hemiptera: Pentatomidae) trap captures at orchard and nonorchard sites and the influence of uncultivated woody host plants in adjoining woodlots. J Econ Entomol 116:2076–2084

Hajek AE, Solter LF, Maddox JV, Huang WF, Estep AS, Krawczyk G, Weber DC, Hoelmer KA, Sanscrainte ND, Becnel JJ (2018) *Nosema maddoxi* sp. nov.(microsporidia, Nosematidae), a widespread pathogen of the green stink bug *Chinavia hilaris* (say) and the brown marmorated stink bug *Halyomorpha halys* (Stål). J Eukaryot Microbiol 65:315–330

Hajek AE, Brandt SN, González JB, Bergh JC (2023) Entomopathogens infecting brown marmorated stink bugs before, during, and after overwintering. J Insect Sci 23. https://doi.org/10.1093/jisesa/iead033

Hancock TJ, Lee DH, Bergh JC, Morrison WR III, Leskey TC (2019) Presence of the invasive brown marmorated stink bug *Halyomorpha halys* (Stål)(Hemiptera: Pentatomidae) on home exteriors during the autumn dispersal period: results generated by citizen scientists. Agric For Entomol 21:99–108

Handbooks PNPM (2024) Hazelnut-Brown marmorated stink bug. https://pnwhandbooks.org/insect/nut/hazelnut/hazelnut-brown-marmorated-stink-bug

Hedstrom CS, Shearer PW, Miller JC, Walton VM (2014) The effects of kernel feeding by *Halyomorpha halys* (Hemiptera: Pentatomidae) on commercial hazelnuts. J Econ Entomol 107:1858–1865

Hepler JR, Mill NJ, Smytheman P, Beers EH (2023) Suitability of the shrub-steppe habitat of eastern Washington for *Halyomorpha halys* (Hemiptera: Pentatomidae). Environ Entomol 52:1152–1161

Herlihy MV, Talamas EJ, Weber DC (2016) Attack and success of native and exotic parasitoids on eggs of *Halyomorpha halys* in three Maryland habitats. PLoS One 11:e0150275

Hull LA, VanStarner R (1983) Impact of four synthetic pyrethroids on major natural enemies and pests of apple in Pennsylvania. J Econ Entomol 76:122–130

Ibrahim A, Kirkpatrick D, Nixon LJ, Ludwick DC, Anfora G, Leskey TC (2020) Effect of deltamethrin-incorporated nets on mobility and survivorship of *Halyomorpha halys* (Hemiptera: Pentatomidae) adults and nymphs in the laboratory. J Appl Entomol 144:589–597

Inkley DB (2012) Characteristics of home invasion by the brown marmorated stink bug (Hemiptera: Pentatomidae). J Entomol Sci 47:125–130
Jentsch P (2017) Expanding the range of the samurai wasp, *Trissolcus japonicus*, in New York orchards. Fruit Quart 25:31–35
Jones AL, Jennings DE, Hooks CR, Shrewsbury PM (2017) Field surveys of egg mortality and indigenous egg parasitoids of the brown marmorated stink bug, *Halyomorpha halys*, in ornamental nurseries in the mid-Atlantic region of the USA. J Pest Sci 90:1159–1168
Joseph SV, Stallings JW, Leskey TC, Krawczyk G, Polk D, Butler B, Bergh JC (2014) Spatial distribution of brown marmorated stink bug (Hemiptera: Pentatomidae) injury at harvest in mid-Atlantic apple orchards. J Econ Entomol 107:1839–1848
Joseph SV, Nita M, Leskey TC, Bergh JC (2015) Temporal effects on the incidence and severity of brown marmorated stink bug (Hemiptera: Pentatomidae) feeding injury to peaches and apples during the fruiting period in Virginia. J Econ Entomol 108:592–599
Kaser JM, Akotsen-Mensah C, Talamas EJ, Nielsen AL (2018) First report of *Trissolcus japonicus* parasitizing *Halyomorpha halys* in north American agriculture. Fla Entomol 101:680–683
Kawagoe JC, Abrams AE, Lourie AP, Walse SS (2022) Ethyl formate dilution in carbon dioxide for fumigation control of the brown marmorated stink bug *Halyomorpha halys*, Stål (Hemiptera: Pentatomidae). Pest Manag Sci 78:3090–3097
Khrimian A, Zhang A, Weber DC, Ho HY, Aldrich JR, Vermillion KE, Siegler MA, Shirali S, Guzman F, Leskey TC (2014) Discovery of the aggregation pheromone of the brown marmorated stink bug *(Halyomorpha halys)* through the creation of stereoisomeric libraries of 1-bisabolen-3-ols. J Nat Prod 77:1708–1717
Kirkpatrick DM, Rice KB, Ibrahim A, Fleischer SJ, Tooker JF, Tabb A, Medeiros H, Morrison WR III, Leskey TC (2020) The influence of marking methods on mobility, survivorship, and field recovery of *Halyomorpha halys* (Hemiptera: Pentatomidae) adults and nymphs. Environ Entomol 49:1026–1031
Kirkpatrick DM, Acebes-Doria AL, Rice KB, Short BD, Adams CG, Gut LJ, Leskey TC (2019) Estimating monitoring trap plume reach and trapping area for nymphal and adult *Halyomorpha halys* (Hemiptera: Pentatomidae) in crop and non-crop habitats. Environ Entomol 48:1104–1112
Kriticos DJ, Kean JM, Phillips CB, Senay SD, Acosta H, Haye T (2017) The potential global distribution of the brown marmorated stink bug, *Halyomorpha halys*, a critical threat to plant biosecurity. J Pest Sci 90:1033–1043
Kuhar TP, Kamminga K (2017) Review of the chemical control research on *Halyomorpha halys* in the USA. J Pest Sci 90:1021–1031
Kuhar TP, Kamminga KL, Whalen J, Dively GP, Brust G, Hooks CRR, Hamilton G, Herbert DA (2012) The pest potential of brown marmorated stink bug on vegetable crops. Plant Health Prog 13:41
Kuhar TP, Short BD, Krawczyk G, Leskey TC (2017) Deltamethrin-incorporated nets as an integrated pest management tool for the invasive *Halyomorpha halys* (Hemiptera: Pentatomidae). J Econ Entomol 110:543–545
Lee DH, Wright SE, Leskey TC (2013a) Impact of insecticide residue exposure on the invasive pest, *Halyomorpha halys* (Hemiptera: Pentatomidae): analysis of adult mobility. J Econ Entomol 106:150–158
Lee DH, Short BD, Joseph SV, Bergh JC, Leskey TC (2013b) Review of the biology, ecology, and management of *Halyomorpha halys* (Hemiptera: Pentatomidae) in China, Japan, and the Republic of Korea. Environ Entomol 42:627–641
Lee DH, Short BD, Nielsen AL, Leskey TC (2014a) Impact of organic insecticides on the survivorship and mobility of *Halyomorpha halys* (Stål)(Hemiptera: Pentatomidae) in the laboratory. Fla Entomol 97:414–421
Lee DH, Cullum JP, Anderson JL, Daugherty JL, Beckett LM, Leskey TC (2014b) Characterization of overwintering sites of the invasive brown marmorated stink bug in natural landscapes using human surveyors and detector canines. PLoS One 9:e91575

Leskey TC, Wright SE, Short BD, Khrimian A (2012a) Development of behaviorally-based monitoring tools for the brown marmorated stink bug (Heteroptera: Pentatomidae) in commercial tree fruit orchards. J Entomol Sci 47:76–85

Leskey TC, Lee DH, Short BD, Wright SE (2012b) Impact of insecticides on the invasive *Halyomorpha halys* (Hemiptera: Pentatomidae): analysis of insecticide lethality. J Econ Entomol 105:1726–1735

Leskey TC, Short BD, Butler BR, Wright SE (2012c) Impact of the invasive brown marmorated stink bug, *Halyomorpha halys* (Stål), in mid-Atlantic tree fruit orchards in the United States: case studies of commercial management. Psyche J Entomol 2012:14

Leskey TC, Hamilton GC, Nielsen AL, Polk DF, Rodriguez-Saona C, Bergh JC, Herbert DA, Kuhar TP, Pfeiffer D, Dively GP, Hooks CRR, Raup MJ, Shrewsbury PM, Krawcyzck G, Shearer PW, Whalen J, Koplinka-Loehr C, Myers E, Inkley D, Hoelmer KA, Lee DH, Wright SE (2012d) Pest status of the brown marmorated stink bug, *Halyomorpha halys* in the USA. Outlooks Pest Manag 23:218–226

Leskey TC, Short BD, Lee DH (2014) Efficacy of insecticide residues on adult *Halyomorpha halys* (Stål)(Hemiptera: Pentatomidae) mortality and injury in apple and peach orchards. Pest Manag Sci 70:1097–1104

Leskey TC, Agnello A, Bergh JC, Dively GP, Hamilton GC, Jentsch P, Khrimian A, Krawczyk G, Kuhar TP, Lee DH (2015) Attraction of the invasive *Halyomorpha halys* (Hemiptera: Pentatomidae) to traps baited with semiochemical stimuli across the United States. Environ Entomol 44:746–756

Leskey TC, Short BD, Ludwick D (2020) Comparison and refinement of integrated pest management tactics for *Halyomorpha halys* (Hemiptera: Pentatomidae) management in apple orchards. J Econ Entomol 113:1725–1734

Ludwick D, Morrison WR III, Acebes-Doria AL, Agnello AM, Bergh JC, Buffington ML, Hamilton GC, Harper JK, Hoelmer KA, Krawczyk G, Kuhar TP, Pfeiffer DG, Nielsen AL, Rice KB, Rodriguez-Saona C, Shearer PW, Shrewsbury PM, Talamas EJ, Walgenbach JF, Wiman NG, Leskey TC (2020a) Invasion of the brown marmorated stink bug (Hemiptera: Pentatomidae) into the United States: developing a national response to an invasive species crisis through collaborative research and outreach efforts. J Integr Pest Manag 11. https://doi.org/10.1093/jipm/pmaa001

Ludwick DC, Patterson J, Leak LB, Carper L, Leskey TC (2020b) Integrating *Trissolcus japonicus* (Ashmead, 1904) (hymenoptera: Scelionidae) into management programs for *Halyomorpha halys* (Stål, 1855) (Hemiptera: Pentatomidae) in apple orchards: impact of insecticide applications and spray patterns. Insects 11:833

McDougall RN, Ogburn EC, Walgenbach JF, Nielsen AL (2021) Diapause termination in invasive populations of the brown marmorated stink bug (Hemiptera: Pentatomidae) in response to photoperiod. Environ Entomol 50:1400–1406

Mermer S, Maslen EA, Dalton DT, Nielsen AL, Rucker A, Lowenstein D, Wiman N, Bhattarai M, Soohoo-Hui A, Harris ET, Pfab F, Walton VM (2023) Temperature-dependent life table parameters of brown marmorated stink bug, *Halyomorpha halys* (Stål) (Hemiptera: Pentatomidae) in the United States. Insects 14:248

Miller JR, Adams CG, Weston PA, Schenker JH (2015) Trapping of small organisms moving randomly: principles and applications to Pest monitoring and management. Springer, Dordrecht

Milnes JM, Wiman NG, Talamas EJ, Brunner JF, Hoelmer KA, Buffington ML, Beers EH (2016) Discovery of an exotic egg parasitoid of the brown marmorated stink bug, *Halyomorpha halys* (Stål) in the Pacific Northwest. Proc Entomol Soc Wash 118:466–470

Mohekar P, Osborne J, Wiman NG, Walton V, Tomasino E (2017) Influence of winemaking processing steps on the amounts of (*E*)-2-decenal and tridecane as off-odorants caused by brown marmorated stink bug (*Halyomorpha halys*). J Agric Food Chem 65:872–878

Mohekar P, Osborne J, Tomasino E (2018) Effects of fining agents, reverse osmosis and wine age on brown marmorated stink bug (*Halyomorpha halys*) taint in wine. Beverages 4:17

Morehead JA, Kuhar TP (2017) Efficacy of organically approved insecticides against brown marmorated stink bug, *Halyomorpha halys* and other stink bugs. J Pest Sci 90:1277–1285
Morrison WR, Cullum JP, Leskey TC (2015) Evaluation of trap designs and deployment strategies for capturing *Halyomorpha halys* (Hemiptera: Pentatomidae). J Econ Entomol 108:1683–1692
Morrison WR III, Mathews CR, Leskey TC (2016) Frequency, efficiency, and physical characteristics of predation by generalist predators of brown marmorated stink bug (Hemiptera: Pentatomidae) eggs. Biol Control 97:120–130
Morrison WR III, Bryant AN, Poling B, Quinn NF, Leskey TC (2017a) Predation of *Halyomorpha halys* (Hemiptera: Pentatomidae) from web-building spiders associated with anthropogenic dwellings. J Insect Behav 30:70–85
Morrison WR III, Acebes-Doria A, Ogburn E, Kuhar TP, Walgenbach JF, Bergh JC, Nottingham L, Dimeglio A, Hipkins P, Leskey TC (2017b) Behavioral response of the brown marmorated stink bug (Hemiptera: Pentatomidae) to semiochemicals deployed inside and outside anthropogenic structures during the overwintering period. J Econ Entomol 110:1002–1009
Morrison WR III, Blaauw BR, Short BD, Nielsen AL, Bergh JC, Krawczyk G, Park YL, Butle B, Khrimian A, Leskey TC (2019) Successful management of *Halyomorpha halys* (Hemiptera: Pentatomidae) in commercial apple orchards with an attract-and-kill strategy. Pest Manag Sci 75:104–114
Nielsen AL, Hamilton GC (2009) Life history of the invasive species *Halyomorpha halys* (Hemiptera: Pentatomidae) in northeastern United States. Ann Entomol Soc Am 102:608–616
Nielsen AL, Shearer PW, Hamilton GC (2008a) Toxicity of insecticides to *Halyomorpha halys* (Hemiptera: Pentatomidae) using glass-vial bioassays. J Econ Entomol 101:1439–1442
Nielsen AL, Hamilton GC, Matadha D (2008b) Developmental rate estimation and life table analysis for *Halyomorpha halys* (Hemiptera: Pentatomidae). Environ Entomol 37:348–355
Nielsen AL, Chen S, Fleischer SJ (2016) Coupling developmental physiology, photoperiod, and temperature to model phenology and dynamics of an invasive heteropteran, *Halyomorpha halys*. Front Physiol 7:177600
Nielsen AL, Fleischer S, Hamilton GC, Hancock T, Krawczyk G, Lee JC, Ogburn E, Pote JM, Raudenbush A, Rucker A, Saunders M, Skillman VP, Sullivan J, Timer J, Walgenbach J, Wiman NG, Leskey TC (2017) Phenology of brown marmorated stink bug described using female reproductive development. Ecol Evol 7:6680–6690
Nixon LJ, Morrison WR, Rice KB, Brockerhoff EG, Leskey TC, Guzman F, Khrimian A, Goldson S, Rostás M (2018) Identification of volatiles released by diapausing brown marmorated stink bug, *Halyomorpha halys* (Hemiptera: Pentatomidae). PLoS One 13:e0191223
Nixon LJ, Tabb A, Morrison WR, Rice KB, Brockerhoff EG, Leskey TC, van Koten C, Goldson S, Rostás M (2019) Volatile release, mobility, and mortality of diapausing *Halyomorpha halys* during simulated shipping movements and temperature changes. J Pest Sci 92:633–641
Nixon L, Morrison WR III, Rice KB, Goldson S, Brockerhoff EG, Khrimian A, Rostás M, Leskey TC (2022) Behavioural responses of diapausing *Halyomorpha halys* (Hemiptera: Pentatomidae) to conspecific volatile organic compounds. J Appl Entomol 146:319–327
Nixon LJ, Acebes-Doria A, Kirkpatrick D, Leskey TC (2024) Influence of deployment strategy and maintenance on efficacy of sticky card traps for *Halyomorpha halys* (Hemiptera: Pentatomidae). J Econ Entomol. https://doi.org/10.1093/jee/tov159
Ogburn EC, Bessin R, Dieckhoff C, Dobson R, Grieshop M, Hoelmer KA, Mathews C, Moore J, Nielsen AL, Poley K (2016) Natural enemy impact on eggs of the invasive brown marmorated stink bug, *Halyomorpha halys* (Stål)(Hemiptera: Pentatomidae), in organic agroecosystems: a regional assessment. Biol Control 101:39–51
Ogburn EC, Ohmen TM, Huseth AS, Reisig DD, Kennedy GG, Walgenbach JF (2023) Temperature-driven differences in phenology and habitat suitability for brown marmorated stink bug, *Halyomorpha halys,* in two ecoregions of North Carolina. J Pest Sci 96:373–387
Park YL, Choi K, Cullum J, Hoelmer KA, Weber DC, Morrison WR III, Rice KB, Krawczyk G, Fleischer SJ, Hamilton G, Ludwick D, Nielsen A, Kaser JM, Polk D, Shrewsbury PM, Bergh CJ, Kuhar T, Leskey T (2024) Landscape-scale spatiotemporal dynamics of *Halyomorpha*

halys (Stål)(Hemiptera: Pentatomidae) populations: implications for spatially-based pest management. Pest Manag Sci 80:953–966

Penca C, Hodges A (2018) First report of brown marmorated stink bug (Hemiptera: Pentatomidae) reproduction and localized establishment in Florida. Fla Entomol 101:708–711

Philips C, Kuhar T, Dively G, Hamilton G, Whalen J, Kamminga K (2017) Seasonal abundance and phenology of *Halyomorpha halys* (Hemiptera: Pentatomidae) on different pepper cultivars in the Mid-Atlantic (United States). J Econ Entomol 110:192–200

Pote JM, Nielsen AL (2017) Life stage specific predation of *Halyomorpha halys* (Stål) by generalist predators. Biol Control 114:1–7

Preston CE, Agnello AM, Hajek AE (2020a) *Nosema maddoxi* (Microsporidia: Nosematidae) in brown marmorated stink bug, *Halyomorpha halys* (Hemiptera: Pentatomidae), populations in the United States. Biol Control 144:104213

Preston CE, Agnello AM, Vermeylen F, Hajek AE (2020b) Impact of *Nosema maddoxi* on the survival, development, and female fecundity of *Halyomorpha halys*. J Invertebr Pathol 169:107303

Raupp MJ (2018) Brown marmaroted stink bug: keeping stink bugs out of your house. University of Maryland Home and Garden Information Center. https://www.youtube.com/watch?v=0kG-2fetbZA. Last accessed 14 May 2024

Rice KB, Bergh CJ, Bergmann EJ, Biddinger DJ, Dieckhoff C, Dively G, Fraser H, Gariepy T, Hamilton G, Haye T, Herbert TA, Holmer KA, Cooks CR, Jones A, Krawczyk G, Kuhar T, Martinson H, Mitchell W, Nielsen A, Pfeiffer D, Raupp MJ, Rodriguez-Saona C, Shearer P, Shrewsbury PM, Venugopal PD, Whalen J, Wiman N, Leskey TC, Tooker JF (2014) Biology, ecology, and management of brown marmorated stink bug (Hemiptera: Pentatomidae). J Integr Pest Manag 5:A1–A13

Rice KB, Troyer RR, Watrous KM, Tooker JF, Fleischer SJ (2016) Landscape factors influencing stink bug injury in mid-Atlantic tomato fields. J Econ Entomol 110:94–100

Rice KB, Morrison WR III, Short BD, Acebes-Doria A, Bergh JC, Leskey TC (2018) Improved trap designs and retention mechanisms for *Halyomorpha halys* (Hemiptera: Pentatomidae). J Econ Entomol 111:2136–2142

Rijal J, Gyawaly S (2018) Characterizing brown marmorated stink bug injury in almond, a new host crop in California. Insects 9:126

Short BD, Khrimian A, Leskey TC (2017) Pheromone-based decision support tools for management of *Halyomorpha halys* in apple orchards: development of a trap-based treatment threshold. J Pest Sci 90:1191–1204

Skillman VP, Wiman NG, Lee JC (2018) Nutrient declines in overwintering *Halyomorpha halys* populations. Entomol Exp Appl 166:778–789

Stahl JM, Scaccini D, Daane KM (2021) Field survival of the brown marmorated stink bug *Halyomorpha halys* (Hemiptera: Pentatomidae) on California tree crops. Environ Entomol 50:1187–1193

Talamas EJ, Herlihy MV, Dieckhoff C, Hoelmer KA, Buffington M, Bon MC, Weber DC (2015) *Trissolcus japonicus* (Ashmead) (Hymenoptera, Scelionidae) emerges in North America. J Hymenopt Res 43:119–128

Venugopal PD, Coffey PL, Dively GP, Lamp WO (2014) Adjacent habitat influence on stink bug (Hemiptera: Pentatomidae) densities and the associated damage at field corn and soybean edges. PLoS One 9:e109917

Venugopal PD, Dively GP, Lamp WO (2015) Spatiotemporal dynamics of the invasive *Halyomorpha halys* (Hemiptera: Pentatomidae) in and between adjacent corn and soybean fields. J Econ Entomol 108:2231–2241

Walton V, Dalton D, Miller B, Harris E, Mirandola E, Tait G, Krawczyk G, Mermer S (2022) Economic impact and management of brown marmorated stink bug, *Halyomorpha halys* (Stål) (Hemiptera: Pentatomidae). In: Oregon hazelnut orchards, X International Congress on Hazelnut, pp 425–432

Weber DC, Leskey TC, Walsh GC, Khrimian A (2014) Synergy of aggregation pheromone with methyl (E, E, Z)-2, 4, 6-decatrienoate in attraction of *Halyomorpha halys* (Hemiptera: Pentatomidae). J Econ Entomol 107:1061–1068

Wiman NG, Parker JE, Rodriguez-Saona C, Walton VM (2015) Characterizing damage of brown marmorated stink bug (Hemiptera: Pentatomidae) in blueberries. J Econ Entomol 108:1156–1163

Zhang J, Zhang F, Gariepy T, Mason P, Gillespie D, Talamas E, Haye T (2017) Seasonal parasitism and host specificity of *Trissolcus japonicus* in northern China. J Pest Sci 90:1127–1141

Zhu G, Gariepy TD, Haye T, Bu W (2017) Patterns of niche filling and expansion across the invaded ranges of *Halyomorpha halys* in North America and Europe. J Pest Sci 90:1045–1057

Zobel ES, Hooks CRR, Dively GP (2016) Seasonal abundance, host suitability, and feeding injury of the brown marmorated stink bug, *Halyomorpha halys* (Heteroptera: Penatomidae), in selected vegetables. J Econ Entomol 109:1289–1302

Chapter 17
Perspectives on Pest Stink Bugs Research and Management in Agriculture

Adeney Freitas Bueno and Antônio Ricardo Panizzi

1 Introduction

Stink bugs (Hemiptera: Pentatomidae) form one of the largest families within the suborder Heteroptera (true bugs) (Panizzi 2008). Being polyphagous pests, they have brought significant economic losses to a variety of crops (Grabarczyk et al. 2021). Despite varying greatly among different stink bug species and plant attacked (Panizzi 2008), these insects are considered major pests around the world (Panizzi et al. 2000), causing damage mainly to soybean (Bueno et al. 2023a) and maize (Gomes et al. 2020) in the Neotropics.

At least 54 different stink bug species are reported damaging soybean (Panizzi and Slansky Jr 1985) and maize (Bueno et al. 2015a). Feeding directly on soybean pods or maize seedlings, stink bugs seriously reduce yields (Bueno et al. 2015b; Gomes et al. 2020) by impairing the physiological and sanitary quality of the soybean seeds (Corrêa-Ferreira and Azevedo 2002) or maize growth (Duarte et al. 2015) if not properly managed.

Not only can stink bugs damage Leguminosae (soybean) and Gramineae (maize) but also several different other hosts, many of which are also cash crops. Stink bugs are usually generalist feeders, frequently with over 100 host plant species (Lye and Story 1988; Panizzi et al. 2000; Siebert et al. 2005; Rice et al. 2014). For instance, *Edessa meditabunda* (F.) is a pest of many vegetable crops, particularly Solanaceae such as tomato and potato, and Leguminosae, such as peas, soybean, and alfalfa. It also feeds on cotton, eggplant, tobacco, sunflower, papaya, and grapes.

A. F. Bueno (✉)
Embrapa Soja, Londrina, PR, Brazil
e-mail: adeney.bueno@embrapa.br

A. R. Panizzi
Embrapa Trigo, Passo Fundo, RS, Brazil

A. F. Bueno, A. R. Panizzi (eds.), *Stink Bugs (Hemiptera: Pentatomidae) Research and Management*, Entomology in Focus 9,
https://doi.org/10.1007/978-3-031-69742-5_17

Pests of trees include the spined citrus bug, *Biprorulus bibax* Breddin, on *Citrus* spp. in Australia; *Lincus* spp. on coconut in South America; *Bathycoelia thalassina* (Herrich-Schaeffer) on cocoa in Africa; and *Plautia* spp. on orchard plants in the Orient (Panizzi 2008).

In South America, the Neotropical brown stink bug, *Euschistus heros* (F.) is the most common species (Panizzi and Corrêa-Ferreira 1997) followed by *Diceraeus* spp. (Bueno et al. 2015a). *Nezara viridula* (L.), is a pest of grain, legumes and vegetables worldwide. *Piezodorus* spp. feed mostly on legumes, being a severe pest of soybean in South America, in Africa, and in the Orient. Members of the genus *Euschistus* are major pests of several crops, mostly damaging legumes such as soybean in the New World. *Eurydema* spp. in the Orient attack grain legumes and vegetables. Members of the genera *Oebalus*, *Mormidea*, and *Aelia* are major pests of Gramineae, particularly on rice and wheat (Panizzi 2008).

In order to mitigate losses caused by those pests and consequently increase food production, farmers have frequently controlled these phytophagous arthropods (Zalucki et al. 2009) spraying chemical insecticides as the first line of defense (Dara 2019; Bueno et al. 2021), which when overused will select resistant stink bug populations (Tibola et al. 2021), harm biological control (Torres and Bueno 2018), and the environment (Jacquet et al. 2022), besides other possible negative impacts. Therefore, in order to manage stink bugs more sustainably, the traditional use of insecticides should be mitigated, giving place to more ecological sound strategies. These certainly will include biological control (both conservation and augmentative use of parasitoids, entomopathogens and predators); host plant resistance; plant derived insecticides; RNAi and CRISPR based control strategies; as well as the use of any other low impact alternative solution applied based on action levels inside an integrated pest management (IPM) approach. In this concluding chapter those perspectives previously discussed in the book chapters will be presented briefly and interconnected, bringing future perspectives and demands.

2 Main Strategies on Integrated Pest Management (IPM) Programs

When crops are cultivated, not only is biodiversity into the agroecosystem reduced to one single plant species of economic interest but also temperature, humidity, as well as, the whole environment at the local level are impacted. This favors pest species that are adapted to the new formed local environment and feed on the cultivated plant, triggering outbreaks of those pest species (Biswas et al. 2023). Adopting traditional chemical insecticides to control those outbreaks significantly reduce natural biological control (Torres and Bueno 2018). It favors pest resurgence as well as outbreaks of secondary pest species (e.g., Panizzi et al. 1977; Yadav et al. 2023) in a whole cycle of causes of imbalance that certainly further disrupt the equilibrium of the entomophauna of the agroecosystem. That is the reason IPM adoption is mandatory (Maciel and Bueno 2022).

Integrated pest management is an approach designed taking the whole landscape into consideration, used to sustainably manage pests in the crop system by grouping different technologies which are intended to disturb the least possible the ecological equilibrium of the entomophauna into the agroecosystems. That is the reason future researches on stink bug must focus on alternatives to be integrated inside IPM and help mitigating or eliminating traditional insecticide use at medium and long-terms goals, respectively (Bueno and Bueno 2012).

Overall, the adoption of IPM reduces the use of insecticides (costs) through the use of economic thresholds (ETs) as well as increase yields by managing better pests (reducing losses caused by pests), bringing significant economic and ecological benefits. Not only is the adoption of IPM capable of reducing costs, eliminating unnecessary sprays (Bueno et al. 2023b), but also increasing natural biological control by adopting more selective insecticides (Carvalho et al. 2021), prioritizing less impactful pest control tools (Bueno et al. 2023a), increasing flowering plants on the periphery of crops (Gurr et al. 2017), among other alternatives which can help to preserve natural stink bug suppression, as well as, any other ecological services provided by beneficial arthropods, an ecosystem service valued at U$4.5 billion annually (Losey and Vaughan 2006).

Integrated pest management is based upon the premise that crop plants can tolerate certain levels of injury with no economically significant yield reduction (Higley and Peterson 1996). Pest control should not be preventively adopted and only is justifiable when the pest population is equal to or greater than the economic thresholds (ETs) or is expected to surpass this level within few hours or days (Bueno et al. 2013; Onstad et al. 2020). Therefore, IPM's success is dependable on an easy, cheap and efficient pest sampling that forms the foundation of this technology. Correct management decisions require reliable, accurate and rapid assessment of the pest population density (Bueno et al. 2021).

Currently strategies for samplings stink bugs usually have the aid of a sweep net or a sample cloth. Despite being the most widely used device to sample insects worldwide, sweep net results have been significantly variable accordingly several causes including environmental variables (temperature, humidity, wind speed, sun position, etc.) and plant characteristics (size, density, etc.) (De Long 1932). Sampling cloth requires a great amount of time and qualified work which can be not only limiting due to the difficulties of hiring field samplers in enough numbers but also expensive due to the well-trained work required (Bueno et al. 2021). That is one of the reasons future research projects in stink bug management should also include new innovative sampling strategies to make it faster, simpler and cheaper.

Several studies in this area have been carried out in order to help monitoring stink bugs as well as other insects. Pheromones in baited traps were efficient in attracting and capturing stink bugs (e.g., Borges et al. 1998; Pires et al. 2000; Schmidt et al. 2003) but still without a simple correlation between the number of insects collected in the traps and the size of the pest population in the field, limiting, so far, its use for ET decision.

The use of automated traps, with sensors that perform stink bug counting, would revolutionize stink bugs' sampling and might be closer to come into the reality with the aid of artificial intelligence (AI). Imaging methods to identify captured insects, combining texture, color and shape information or by species wing beat pattern (Potamitis et al. 2015; Potamitis and Rigakis 2015; Wen et al. 2015) is really disruptive. This is certainly on the edge of science but it is still far from farmers' reality and requires investments and researches to develop such innovations as medium or long-term goals.

Satellite or drone images have been also mentioned as possible pest sampling tools of the future for quite long time. According to Nansen et al. (2014) plants under abiotic or biotic stress have spectral behaviors that differ from healthy plants. Studies of aphid-infested sorghum and soybean plants show reductions in near infrared reflectance (NIR) and normalized difference vegetation index (NDVI) (Alves et al. 2015; Elliot et al. 2015). However, using such data to pest management decisions in IPM programs is still a challenge. However, if solved it can revolutionize stink bug management.

2.1 Biological Control

Biological control, also called biocontrol, is the action of one or more living species (called biocontrol agents) to control another living species (called pests) for diverse purposes to provide human benefits (Stenberg et al. 2021). In general, biological control is divided in four main categories (Bettiol 2022):

(a) *Natural biological control*: Control is performed by biocontrol agents of natural occurrence without any human intervention.
(b) *Conservation biological control*: Control is performed by biocontrol agents of natural occurrence but with human intervention. Human actions are to protect and stimulate preservation and increase of natural populations of beneficial agents.
(c) *Classical biological control*: Control is performed by biocontrol agents which are introduced from another area (different country, usually from the area of origin of the pest) which may or not result in a permanent population. Usually there are limited releases of the biocontrol agent and control can be achieved in medium or long time frames.
(d) *Augmentative biological control*: Control is performed by biocontrol agents massively released in the area to rapidly control target pests. Biocontrol agents (bioinsecticides) will be bought at the market or produced on farm before releasing at the field.

Despite a large and diverse complex of natural enemies of stink bugs being recorded around the world, egg parasitoids are considered the most important biocontrol agents (Koppel et al. 2009; Laumann et al. 2010; Bueno et al. 2022). For instance, *Telenomus podisi* Ashmead can parasitize more than 100 eggs of stink bugs such as

Euschistus heros (F.) (Silva et al. 2018) and *Diceraeus melacanthus* (Dallas) (Taguti et al. 2019) during adult parasitoid lifespan. Overall, eggs, late nymphal stages and adults of stink bugs are most commonly attacked by different parasitoid species; eggs and early nymph stages can also be attacked by predators (Ademokoya et al. 2022).

Preserving such biocontrol agents naturally occurring in the agroecosystems or massively releasing one or more species to rapidly reduce stink bug populations are certainly helpful to manage this important pest. Some of those alternatives are previously discussed in Chap. 4 (egg parasitoids), Chap. 5 (entomopathogens) and Chap. 6 (conservation biological control). As mentioned in each of these chapters, the existent alternatives of biological control are still insufficient. Investments in disruptive research in those themes are urgently needed to sustainably manage stink bugs in different agroecosystems.

Alternatives of conservation biological control which are effective, cheap and easily adopted are of great theoretical and practical interests in the management of stink bugs. Adopting IPM and ETs to avoid unnecessary insecticide sprays to control stink bugs and the resulting further unbalance in the agroecosystem is important as discussed in the case study of Chap. 14.

From 2015 to 2021, the adoption of IPM in soybean in Brazil not only reduced 40% the use of insecticides to control stink bugs but also increased 15 days the time for the first insecticide to control stink bug in the fields. This clearly benefited the preservation of biological control agents into the agroecosystem triggering yield increase of 75 kg compared with non-IPM fields due to better pest control. Despite such economic and ecological benefits still traditional chemical insecticides are the first line of defense to manage stink bugs. It calls the attention to the challenges to the massive adoption of IPM that includes the increase of number of well-trained extensionists in the field besides more sustainable tools to sample and manage the pest in a simple, cheap and efficient way.

Augmentative biological control is certainly a major tool to manage stink bugs in a sustainable way. In the Chap. 4 (egg parasitoids) and in the Chap. 5 (entomopathogens) great examples of augmentative biological control (ABC) recommendations were discussed but there still is a limitation in the number of biocontrol agents available to manage the pest. Despite important advances in the last few years, more research and innovation are certainly needed to make ABC greatly used for stink bug management. This will likely occur not only in Leguminosae or Gramineae but also in the several other crops where stink bugs are a pest problem to food production around the world.

2.2 *Host Plant Resistance*

Host resistance or plant resistance, which are characterized by plants with genetically inherited attributes manifested by being less damaged by pests than susceptible counterparts (Painter 1951; Baldin et al. 2019), is one of several tools inside IPM with crucial importance for providing stink bug management (Arias et al. 2022).

Plant resistance to insects was initially classified by Painter (1951) into three mechanisms: (1) antibiosis; (2) non-preference (later called antixenosis by Kogan and Ortman 1978); and (3) tolerance. Antibiosis and antixenosis are defined as the types of resistance in which the plant presents at least one characteristic that negatively affects the biology of the insect. In the case of tolerance, there is only plant's response making host less susceptible to pest injury, in which the insect is not affected (Peterson et al. 2017). Thus, insect-resistant crops should suppress stink bug abundance or elevate the damage tolerance level of cultivated plants altering the relationship stink bugs have with their host plants.

For instance, in Brazil, soybean cultivars tolerant to stink bugs, carrying the called Block® technology, were developed by scientists from Embrapa Soybean (see Chap. 10). Soybean block cultivars combine desirable agronomic characteristics and preserve low leaf retention and grains with little damage by stink bugs. It is possible because those cultivars have genetic characteristics that allow the plant under attack by high stink bug infestations to have lower yield loss compared to cultivars without the technology. Soybean cultivars with these genes tolerate the double of stink bug numbers compared to susceptible cultivars, providing greater flexibility to adopt control measures against them in situations of high pest pressure. This ensures greater production stability in situations of delayed applications or large stink bug population outbreaks.

Soybean cultivars with Block® technology have in their genealogy PI 274454, one of the wild plants introductions (PIs), originating from centers of soybean origin. This PI was also used in the crosses that gave rise to IAC-100, which was the first soybean cultivar resistant to insects in Brazil, including stink bugs (Rossetto et al. 1990; Veiga et al. 1999). PI 227687, PI 274454 and IAC-100 have higher concentrations of defense metabolites, such as flavonoids that confer resistance to different soybean insect pests (Hoffmann-Campo et al. 2001; Piubelli et al. 2003; Graça et al. 2016; Vieira et al. 2016).

The adoption of insect-resistant plants in the IPM strategies, in which concept the need to use multiple tactics to maintain stink bug abundance and damage below economic injury levels is economically, ecologically, and environmentally advantageous (Warghat et al. 2023). Economic benefits occur because crop yields are saved from loss caused by insect pests and money is saved by not applying insecticides that would have been applied on susceptible plants (Table 17.1).

In most cases, seed of resistant plants to stink bugs will costs no more, or little more, than susceptible seeds of counterparts. Ecological and environmental benefits arise from increases in species diversity in the agroecosystem, in part because of reduced use of insecticides. Increases in species diversity and ecosystem stability which promote a more sustainable agroecosystem far less polluted and detrimental to natural resources. Despite such crucial benefits of adopting resistant plants to stink bugs that are of great theoretical and practical interests of millions of farmers impacted by this pest group around the world, commercial seeds of stink bug-resistant plants are still limited to a few crops such as soybean (Arias et al. 2022), cowpea (Schalk and Fery 1986) and rice (Almeida et al. 2021; Bhavanam and Stout 2022).

Table 17.1 Yield and seed quality of soybean cultivars susceptible and resistant to stink bug damage from fields in where the economic threshold (ET) used was 4 stink bugs/meter to decide for insecticide spray (crop season 2014/15)

County, State	Cultivar	Number of insecticide sprays	Yield (kg/ha)	Tetrazolium test (%) Vigor	Viability	Seeds Stink bug damaged seeds	Non-viable seeds
Andirá, Paraná	BRS 391 (Block)	2	4638 a	78.9 a	91.5 a	65.3 b	7.3 b
	BRS 232 (susceptible)	2	3222 b	19.2 b	19.2 b	98.3 a	37.5 a
Cândido Mota, São Paulo	BRS 391 (Block)	2	4485 a	73.9 a	90.7 a	69.9 a	7.5 a
	BRS 232 (susceptible)	4	4258 b	73.7 a	90.1 a	63.7 a	7.3 a

Means followed by the same letter in the collum for each county did not differ significantly (Tukey test at 5% of probability)

Plant resistance to pests has advantages over other direct control tactics. For example, plant resistance to insects is compatible with all insecticide use, while biological control is not when a non-selective insecticide is used. Plant resistance to insects is not density dependent, whereas biological control is. Plant resistance is specific, only affecting the target pest. Often effects of using insect-resistant cultivars are cumulative over time. In addition, usually the effectiveness of resistant cultivars is long-lasting. Therefore, future efforts in stink bug research should be dedicated to overcome the challenges of releasing commercial cultivars resistant to stink bugs.

Recent new tools related to biotechnology genetic transformation, rDNA methods, or genetic engineering can help to improve the availability of stink bug-resistant plants in near future, what certainly will require investments and a great effort of scientists to bring innovation and disruption into the area of resistant plants. Plant resistance should always be used together with other IPM strategies. To use plant resistance as the sole control strategy for an insect pest rather than a component of IPM will certainly result in pest resistance, repeating the mistake done during the insecticide era.

2.3 Use of Alternative Products to Traditional Commercial Chemical Insecticides

Very few products have been explored as potential insecticides alternatives to synthetic chemicals to be use in integrated pest management (IPM) programs. Among them, botanical insecticides are been considered a potential alternative.

Despite the tremendous diversity of plant species in nature to be explored with the aim to find suitable insect toxic compounds for commercialization, few products with lethal or sublethal effects have been tested and used to control phytophagous stink bugs.

So far, the plant families Piperaceae, Meliaceae, Lamiaceae and Annonaceae have been investigated with potential to provide toxic compounds to stink bugs. Compounds such as dilapiol, azadirachtin and acetogenins are among the substances tested with favourable results. They may cause several impacts to stink bugs biology, including lethal, repellent, ovicidal, cytological, and reproduction negative effects (see details in Chap. 7).

There is no doubt that active ingredients extracted from plants with insecticidal properties are a potential revenue to be explored. However, there are some challenges that need to be overcome to make this a reality, which include, among others: further evaluations of the mechanisms of action; development of methodological protocols; and feasibility of use on a commercial level.

3 Novel Approaches with Potential to Manage Stink Bugs

As the knowledge of new technologies advance fast in present time, soon there will be available novel powerful tools to be used in integrated pest management (IPM) programs for stink bugs. To illustrate this, several advances' research lines, previously presented in specific chapters of this book, are here mentioned such as studies on RNAi of topical use to prevent development of stink bugs, behavioral manipulation of these pests, genetically modified plants using promising tools, such as the CRISPR/Cas9 genome editing toolbox, and even developing genetically modified stink bugs less harmful to the plants.

As these novel approaches to manage pest stink bugs are developed and proved to be effective, a second major step is needed, which is to validate these new tools inside the frame of present IPM's programs. Certainly, this will change the current IPM programs. Therefore, the modified IPM including these new tools should be tested in different agroecosystems in order to validate its implementation at the grower's level. Once adopted, the goal is to spread its use in order to mitigate the impact of these insects and promote a more sustainable food production globally.

4 Concluding Remarks

It is clear the importance of stink bugs (pentatomids) as pests worldwide. It is known that traditional commercial insecticides (chemicals) are used as the main control strategy with detrimental effects. These include contamination of the environment, impact on human health, and elimination of biocontrol agents and pollinators. As the pressure for food production increases, and agriculture becomes every year more and more intense, the need for sustainable control measurements of pests become mandatory.

The elaboration of holistic integrated pest management (IPM) programs is the way to be followed. After they initial success in past decades, these programs suffer a decline in use by farmers for several reasons, which have been presented and discussed elsewhere. However, with the introduction of the new tools here presented and discussed, the implementation and adoption of IPM programs will, much likely, get back on track. This surely will revolutionize the future of stink bug management around the world with favorable and long-lasting results.

References

Ademokoya B, Athey K, Ruberson J (2022) Natural Enemies and Biological Control of Stink Bugs (Hemiptera: Heteroptera) in North America. Insects 13:932

Almeida ACSA, Jesus FG, Lanna AC, Barrigossi JAF (2021) Evidence for rice tolerance to *Tibraca limbativentris* (Hemiptera: Pentatomidae). Pest Manag Sci 77:4181–4191

Alves TM, Macrae IV, Koch RL (2015) Soybean aphid (Hemiptera: Aphididae) affects soybean spectral reflectance. J Econ Entomol 108:2655–2664

Arias CAA, Hoffmann-Campo CB, Soares RM, Meyer MC (2022) Contribuição do melhoramento genético de soja para o manejo de doenças e pragas. In: Meyer MC, Bueno AF, Mazaro SMM, Silva JC (eds) Bioinsumos na Cultura da Soja. Embrapa, Brasília, pp 53–72

Baldin ELL, Vendramim JD, Lourenção AL (2019) Introdução a resistência de plantas a insetos: fundamentos e aplicações. In Resistência de Plantas a Insetos: Fundamentos e Aplicações. FEALQ, Piracicaba, SP, Brasil

Bettiol W (2022) Pesquisa, desenvolvimento e inovação com bioinsumos. In: Meyer MC, Bueno AF, Mazaro SMM, Silva JC (eds) Bioinsumos na Cultura da Soja. Embrapa, Brasília, pp 21–38

Bhavanam S, Stout MJ (2022) Varietal resistance and chemical ecology of the Rice Stink Bug, *Oebalus pugnax*, on rice, *Oryza sativa*. Plants (Basel) 11:3169

Biswas SK, Jaisval GK, Shukla P, Kumar S (2023) What is pest, categories of pests and causes of pest outbreak. In: Mishra PK, Kumar A, Kumar P, Aman AS, Bajpeyi (eds) Modern aspects of entomology. Integrated Publications TM, Delhi, pp 93–102

Borges M, Schmidt FGV, Sujii ER, Medeiros MA, Mori K, Zarbin PHG, Ferreira JTB (1998) Field responses of stink bugs to the natural and synthetic pheromone of the Neotropical brown stink bug, *Euschistus heros* (Heteroptera: Pentatomidae). Physiol Entomol 23:202–207

Bueno AF, Bueno RCOF (2012) Integrated pest management as a tool to mitigate the pesticide negative impact into the agroecosystem: the soybean example. In: The impact of pesticides, 1st edn. AcademyPublish.Org®, Cheyenne, pp 164–190

Bueno AF, Paula-Moraes SV, Gazzoni DL, Pomari AF (2013) Economic thresholds in soybean integrated pest management: Old concepts, current adoption, and adequacy. Neotrop Entomol 42:439–447

Bueno AF, Corrêa-Ferreira BS, Roggia S, Bianco R (2015a) Silenciosos e daninhos. Rev Cult 196:25–27

Bueno AF, Bortolotto OC, Pomari-Fernandes A, França-Neto JB (2015b) Assessment of a more conservative stink bug economic threshold for managing stink bugs in Brazilian soybean. Crop Prot 71:132–137

Bueno AF, Panizzi AR, Hunt TE, Dourado PM, Pitta RM, Gonçalves J (2021) Challenges for adoption of integrated pest management (IPM): the soybean example. Neotrop Entomol 50:5–20

Bueno AF, Parra JRP, Colombo FC, Colmenarez YC, Narde BVF, Pereira FF (2022) Manejo de pragas com parasitoides. In: Meyer MC, Bueno AF, Mazaro SMM, Silva JC (eds) Bioinsumos na Cultura da Soja. Embrapa, Brasília, pp 417–434

Bueno AF, Sutil WP, Jahnke SM, Carvalho GA, Cingolani MF, Colmenarez YC, Corniani N (2023a) Biological control as part of the soybean integrated pest management (IPM): potential and challenges. Agronomy (Basel) 13:2532

Bueno AF, Colmenarez YC, Carnevalli RA, Sutil WP (2023b) Benefits and perspectives of adopting soybean-IPM: the success of a Brazilian programme. Plant Health Cases 2023:1–16

Carvalho GA, Reis PR, Grutzmacher AD, Degrande PE, Yamamoto PT, Bueno AF (2021) Seletividade de produtos fitossanitários: uma estratégia viável para a agricultura sustentável. In: Controle Biológico com Parasitoides e Predadores na Agricultura Brasileira, 1st edn. FEALQ, Piracicaba, pp 481–510

Corrêa-Ferreira BS, Azevedo J (2002) Soybean seed damage by different species of stink bugs. Agric For Entomol 4:145–150

Dara SK (2019) The new integrated pest management paradigm for the modern age. J Integr Pest Manag 10:1–9

De Long DM (1932) Some problems encountered in the estimation of insect populations by the sweeping method. Ann Entomol Soc Am 25:13–17

Duarte MM, Avila CJ, Santos V (2015) Danos e nível de dano econômico do percevejo barriga verde na cultura do milho. Rev Bras Milho Sorgo 14:291–299

Elliot NC, Backoulou GF, Brewer MJ, Giles KL (2015) NDVI to detect sugarcane aphid injury to grain sorghum. J Econ Entomol 108:1452–1455

Gomes EC, Hayashida R, Bueno AF (2020) *Dichelops melacanthus* and *Euschistus heros* injury on maize: basis for re-evaluating stink bug thresholds for IPM decisions. Crop Prot 130:105050

Grabarczyk EE, Cottrell TE, Tillman G (2021) Characterizing the spatiotemporal distribution of three native stink bugs (Hemiptera: Pentatomidae) across an agricultural landscape. Insects 12:1–15

Graça JP, Ueda TE, Janegitz T, Vieira SS, Salvador MC, Oliveira MCN, Zingaretti SM, Powers SJ, Pickett JA, Birkett MA, Hoffmann-Campo CB (2016) The natural plant stress elicitor *cis*-jasmone causes cultivar-dependent reduction in growth of the stink bug, *Euschistus heros* and associated changes in flavonoid concentrations in soybean, *Glycine max*. Phytochemistry 131:84–91

Gurr GM, Wratten SD, Landis DA, You M (2017) Habitat management to suppress pest populations: progress and prospects. Annu Rev Entomol 62:91–109

Higley LG, Peterson RKD (1996) The biological basis of the EIL. In: Higley LG, Pedigo LP (eds) Economic thresholds for integrated pest management. University of Nebraska Press, Lincoln, pp 22–40

Hoffmann-Campo CB, Harbone JB, Mcaffery AR (2001) Pre-ingestive and post-ingestive effects of soya bean extracts and rutin on *Trichoplusia ni* growth. Entomol Exp Appl 98:181–194

Jacquet F, Jeuffroy M-H, Jouan J, Le Cadre E, Litrico I, Malausa T, Reboud X, Huyghe C (2022) Pesticide-free agriculture as a new paradigm for research. Agron Sustain Dev 42:8

Kogan M, Ortman E (1978) Antixenosis: a new term proposed to define Painter's "nonpreference" modality of resistance. Bull Entomol Soc Am 24:175–176

Koppel AL, Herbert DA, Kuhar TP, Kamminga K (2009) Survey of stink bug (Hemiptera: Pentatomidae) egg parasitoids in wheat, soybean, and vegetable crops in southeast Virginia. Environ Entomol 38:375–379

Laumann RA, Moraes MCB, Silva JPD, Vieira AMC, Silveira SD, Borges M (2010) Egg parasitoid wasps as natural enemies of the neotropical stink bug *Dichelops melacanthus*. Pesq Agrop Brasileira 45:442–449

Losey JE, Vaughan M (2006) The economic value of ecological services provided by insects. BioScience 56:311–323

Lye BH, Story RN (1988) Feeding preference of the southern green stink bug (Hemiptera: Pentatomidae) on tomato fruit. J Econ Entomol 81:522–526

Maciel RMA, Bueno AF (2022) The role of integrated pest management for sustainable food production: the soybean example. In: Galanakis CM (ed) Biodiversity, Functional Ecosystems and Sustainable Food Production, Springer Nature, Switzerland, p 117–139

Nansen C, Zhang A, Yan G (2014) Use of variogram analysis to classify field peas with and without internal defects caused by weevil infestation. J Food Eng 123:17–22

Onstad DW, Bueno AF, Favetti BM (2020) Economic thresholds and sampling in integrated pest management. In: Onstad DW, Crain PR (eds) The Economics of Integrated Pest Management of Insects, 1st edn. CABI, Wallingford, pp 122–139

Painter R (1951) Insect resistance in crop plants. University of Kansas Press, Lawrence

Panizzi AR (2008) Stink bugs (Hemiptera: Pentatomidae), emphasizing economic importance. In: Capinera JL (ed) Encyclopedia of Entomology. Springer, Dordrecht, pp 3567–3570

Panizzi AR, Corrêa-Ferreira BS (1997) Dynamics in the insect fauna adaptation to soybean in the tropics. Trends Entomol 1:71–88

Panizzi AR, Slansky F Jr (1985) Review of phytophagous pentatomids (Hemiptera: Pentatomidae) associated with soybean in the Americas. Fla Entomol 68:184–203

Panizzi AR, Corrêa BS, Newman GG, Turnipseed SG (1977) Efeito de inseticidas na população das principais pragas da soja. An Soc Entomol Brasil 6:264–275

Panizzi AR, McPherson JE, James DG, Javahery M (2000) Economic importance of stink bugs (Pentatomidae). In: Shaefer CW, Panizzi AR (eds) Heteroptera of economic importance. CRC Press, Boca Raton, FL, pp 421–474

Peterson RKD, Varella AC, Higley LG (2017) Tolerance: the forgotten child of plant resistance. Peer J 5:e3934

Pires CSS, Sujii ER, Schmidt FGV, Santos HM, Pais JSO, Borges M (2000) Potencial de utilização de armadilhas iscadas com o feromônio sexual do percevejo marrom, *Euschistus heros*: uma nova metodologia para o monitoramento populacional de percevejos praga da soja. Embrapa-Cenargen, Brasília, DF, Brasil, Circular Técnica 7:1–14

Piubelli GC, Hoffmann-Campo CB, Arruda IC, Franchini JC, Lara FM (2003) Flavonoid increase in soybean genotypes as response of *Nezara viridula* injury and its effect on insect-feeding preference. J Chem Ecol 29:1223–1233

Potamitis I, Rigakis I (2015) Novel noise-robust optoacoustic sensors to identify insects through wingbeats. IEEE Sensors J 15:4621–4631

Potamitis I, Rigakis I, Fysarakis K (2015) Insect biometrics: Optoacoustic signal processing and its applications to remote monitoring of McPhail type traps. PLoS One 10:1–33

Rice KB, Bergh CJ, Bergmann EJ, Biddinger DJ, Dieckhoff C, Dively G, Fraser H, Gariepy T, Hamilton G, Haye T, Herbert A, Hoelmer K, Hooks CR, Jones A, Krawczyk G, Kuhar T, Martinson H, Mitchell W, Nielsen AL, Pfeiffer DG, Raupp MJ, Rodriguez-Saona C, Shearer P, Shrewsbury P, Venugopal PD, Whalen J, Wiman NG, Leskey TC, Tooker JF (2014) Biology, ecology, and management of brown marmorated stink bug (Hemiptera: Pentatomidae). J Integr Pest Manag 5:A1–A13

Rossetto CJ, Tisselli Filho O, Cione J, Gallo PB, Razera LF, Teixeira JPF, Bertoletto N (1990) Cultivar de soja IAC 100. IAC Folder, Campinas, SP, Brasil

Schalk JM, Fery RL (1986) Resistance in cowpea to the southern green stink bug. HortScience 21:1189–1190

Schmidt FGV, Pires CSS, Sujii ER, Borges M, Pantaleão DC, Lacerda ALM, Azevedo VCR (2003) Comportamento e captura das fêmeas de *Euschistus heros* em armadilhas iscadas com feromônio sexual. Embrapa-Cenargen, Brasília, DF, Brazil. Comunicado Técnico 93:1–4

Siebert MW, Leonard BR, Gable RH, LaMotte LR (2005) Cotton boll age influences feeding preference by brown stink bug (Heteroptera: Pentatomidae). J Econ Entomol 98:82–87

Silva GV, Bueno AF, Neves PMOJ, Favetti BM (2018) Biological characteristics and parasitism capacity of *Telenomus podisi* (Hymenoptera: Platygastridae) on eggs of *Euschistus heros* (Hemiptera: Pentatomidae). J Agric Sci 10:210–220

Stenberg JA, Sundh I, Becher PG, Björkman C, Dubey M, Egan PA, Friberg H, Gil JF, Jensen DF, Jonsson M, Karlsson M, Khalil S, Ninkovic V, Rehermann G, Vetukuri RR, Viketoft M (2021) When is it biological control? A framework of definitions, mechanisms, and classifications. J Pest Sci 94:665–676

Taguti EA, Goncalvez J, Bueno AF, Marchioro ST (2019) *Telenomus podisi* parasitism on *Dichelops melacanthus* and *Podisus nigrispinus* eggs at different temperatures. Fla Entomol 102:607–613

Tibola CM, Silva L, Sgubin F, Omoto C (2021) Monitoring resistance of *Euschistus heros* (Fabricius) (Hemiptera: Pentatomidae) to insecticides by using encapsulated artificial diet bioassay. Insects 12:599, 1–17

Torres JB, Bueno AF (2018) Conservation biological control using selective insecticides: A valuable tool for IPM. Biol Control 126:53–64

Veiga RFA, Razera LF, Gallo PB, Bertoletto N, Medina PF, Tisselli Filho O, Cione J (1999) Caracterização morfológica e agronômica da cultivar IAC 100. Instituto Agronômico de Campinas, Boletim Técnico, 177, Campinas, SP, Brasil, 23p

Vieira SS, Lourenção AL, Graça JP, Janegitz T, Salvador MC, Oliveira MCN, Hoffmann-Campo CB (2016) Biological aspects of *Bemisia tabaci* biotype B and the chemical causes of resistance in soybean genotypes. Arthropod-Plant Inter 10:525–534

Warghat AN, Kumar A, Raghuvanshi HR, Aman AS, Kumar A (2023) Recent advancements in plant protection. In: Kumar N, Purushotham P, Kumar A, Sahu A, Nandeesha SV (eds) Recent advances in plant protection. Golden Leaf Publishers, Uttar Pradesh, pp 1–25

Wen C, Wu D, Hu H, Pan W (2015) Pose estimation-dependent identification method for field moth images using deep learning architecture. Biosyst Eng 136:117–128

Yadav A, Aman AS, Kumar S (2023) Resurgence and pest outbreak of destructive pest by mismanagement tactics. In: Mishra PK, Kumar A, Kumar P, Aman AS, Bajpeyi (eds) Modern aspects of entomology. Integrated Publications TM, Delhi, pp 77–92

Zalucki MP, Adamson D, Furlong MJ (2009) The future of IPM: whither or wither? Aust J Entomol 48:85–96

Insect Index

A
Acledra bonariensis, 9
Acledra fraterna, 9, 11
Acroclisoides, 320
Acroclisoides sinicus, 313, 314, 318, 337
Acrosternum, 17, 41
Acrosternum heegeri, 335
Acrosternum hilare, 41, 98, 99, 106, 109
Acyrthosiphon pisum, 172
Adalia bipunctata, 323
Aelia, 384
Aethemenes, 17
Alcaeorrhynchus grandis, 32, 34
Anastatus, 373
Anastatus bifasciatus, 313, 314, 318, 320, 334, 337
Anopheles gambiae, 217
Antheminia, 6
Anticarsia, 66
Anticarsia gemmatalis, 230
Arachnocephalus vestitus, 322
Arilus cristatus, 373
Arma custos, 32, 33, 321, 335
Arocera apta, 3
Arvelius albopunctatus, 10, 11
Ascra bífida, 9, 11
Aspavia armigera, 10, 11

B
Bagrada hilaris, 7, 247
Banasa dimidiata, 10, 11
Banasa euchlora, 10, 11
Bathycoelia thalassina, 384
Bemisia argentifolii, 172
Bemisia tabaci, 172
Biprorulus bibax, 384
Blattella germanica, 216
B. nigrolimbatus, 34
Bombyx mori, 67
Brontocoris, 30, 34
Brontocoris tabidus, 32, 34

C
Callosobruchus chinensis, 217
Camponotus rufipes, 3
Carpocoris, 6, 12
Chinavia, 17
Chinavia acuta, 8
Chinavia armigera, 10, 11
Chinavia aseada, 10, 11
Chinavia erythrocnemis, 10, 11
Chinavia hilare, 41
Chinavia hilaris, 7, 10, 11, 17, 19, 23
Chinavia impicticornis, 10, 11, 17–19
Chinavia longicorialis, 10, 11
Chinavia marginata, 8
Chinavia nigridorsata, 10, 11
Chinavia obstinata, 10, 11
Chinavia pengue, 10, 11
Chinavia ubica, 10, 11, 17–19, 73
Chlorochroa, 17
Chlorochroa ligata, 121, 125, 127
Chlorochroa persimilis, 8, 10, 11, 23
Chrysoperla carnea, 323

A. F. Bueno, A. R. Panizzi (eds.), *Stink Bugs (Hemiptera: Pentatomidae) Research and Management*, Entomology in Focus 9,
https://doi.org/10.1007/978-3-031-69742-5

Chrysoperla externa, 265
C. melanoticus, 3
Coccinella septempunctata, 323
Coenus delius, 9, 11
Cosmopepla lintneriana, 8, 9, 11, 23
Cotesia flavipes, 66, 261
Crematogaster scutellaris (Olivier), 324, 325
Culex pipiens pallens, 159
Cylindromyia bicolor, 320

D
Dalbulus maidis, 202
Diabrotica virgifera virgifera, 153, 158
Diaphorina citri, 159, 172
Diatraea saccharalis, 66, 261
Diceraeus, 8, 12, 97, 170, 214, 228, 244, 384
Diceraeus furcatus, 7, 9, 11–14, 23, 73, 74, 89, 170, 204, 206, 207, 223, 241, 244–247
Diceraeus melacanthus, 9, 11–14, 23, 43, 45–47, 52, 70, 73, 139, 142, 153, 170, 171, 206, 221, 223, 224, 243–247
Diceraeus (Dichelops) melacanthus, 199
Dichelops, 12, 97
Dictyotus coenosus, 10
Dolycoris, 6, 12
Dolycoris baccarum, 9, 11, 335
Dysdercus koenigii, 98, 99, 106, 111

E
Edessa, 11, 12, 32, 97
Edessa loxdalii, 8, 9, 11, 12
Edessa meditabunda, 4, 7, 9, 11–14, 214, 383
Edessa ruformarginata, 8
Epilachna varivestis, 33
Eupholidoptera chabrieri, 322
Eurydema, 384
Eurystethus microlobatus, 3
Euschistus, 6, 12–16, 23, 97, 384
Euschistus conspersus, 121, 125, 127
Euschistus cornutus, 9, 11
Euschistus crassus, 8, 9, 11, 23
Euschistus crenator, 9, 11
Euschistus heros, 7, 8, 11–14, 16, 23, 43, 46, 47, 52–57, 60, 65, 67, 69, 70, 73, 74, 85, 88, 98, 99, 103–105, 109–111, 119–128, 130, 131, 133, 135, 137–140, 142, 145, 147–149, 152, 153, 155, 160, 170–172, 174, 176, 181–195, 199, 204–206, 209, 214, 217, 219, 220, 223, 228, 230, 241, 243–246, 255–267, 384, 387
Euschistus ictericus, 9, 11
Euschistus picticornis, 3, 9, 11
Euschistus quadrator, 9, 11
Euschistus servus, 7, 9, 11–13, 15, 23, 99, 106, 109
Euschistus taurulus, 8, 9, 11, 23
Euschistus tristigmus, 7, 9, 11–13, 15, 23
Euschistus variolarius, 9, 11–13, 15, 23

F
Forficula auricularia, 322

G
Geocoris, 87, 171
Grazia tincta, 3

H
Halyomorpha, 16
Halyomorpha halys, 7–9, 11, 13, 16–17, 23, 48, 89, 97, 99, 107, 108, 110, 111, 119–122, 124–128, 130, 132–136, 138, 139, 142, 144–146, 148–150, 152–155, 157, 158, 170, 171, 189, 241–243, 271–338, 361, 370–372
Haplostegus nigricrus, 34
Harmonia axyridis, 321, 323
Helicoverpa, 33
Helicoverpa armigera, 76
Heteroperreyia hubrichi, 34
Hexacladia smithii, 264
Hierodula transcaucasica, 322
Himacerus mirmicoides, 323
Hippodamia variegata, 323
Holcostethus limbolarius, 9, 11
Homalodisca vitripennis, 172
Hymenarcis nervosa, 9, 11
Hyphantria cunea, 32

L
Lasius niger, 324
Lattinidea geographica, 3
Leptinotarsa decemlineata, 31, 32, 158, 159
Lethocerus, 2
Lincus, 384
Loxa deducta, 10, 11
Lygus hesperus, 172

M
Mayrinia curvidens, 10, 11
Mayrinia variegata, 3
Mecidea major, 8, 10, 11, 23

Microtheca ochroloma, 33
Mormidea, 384
Mormidea lugens, 9, 11
Mormidea v-luteum, 9, 11
Murgantia histriônica, 10, 11, 123, 125, 132–136, 144, 146, 152, 153, 155, 248
Musca domestica, 35, 36, 216, 217
Myzus persicae, 173

N
Nabis, 87, 373
Nagusta goedelii, 323
Neotiglossa undata, 8, 9, 11, 23
Nezara, 6, 17–20, 23, 97
Nezara antennata, 10, 11, 19
Nezara viridula, 7, 8, 10, 11, 16, 18, 19, 23, 36, 41, 43, 46, 47, 73, 85, 88, 90, 99–103, 109–111, 120–122, 125, 127, 131–136, 139, 143–146, 148, 152–155, 170, 171, 204, 214, 223, 228, 230, 241, 243, 244, 247, 299, 320, 321, 384
Nilaparvata lugens, 159, 173
N. yunnana, 20

O
Oebalus, 384
Oebalus poecilus, 9, 11
Oebalus pugnax, 9, 11
Oebalus ypsilogriseus, 9, 11
Oenopia conglobata, 323
Oncopeltus fasciatus, 172
Ooencyrtus, 313, 314
Ooencyrtus pityocampae (Mercet), 313, 314, 319, 334
Ooencyrtus telenomicida, 90, 313, 314, 320
Orius, 87

P
Palomena prasina, 335, 336
Patanius vitattus, 3
Pentatoma rufipes, 308, 335, 336
Peregrinus maidis, 173
Perillus, 32
Perillus bioculatus, 31–33, 36
Phaneroptera falcata, 322
Picromerus bidens, 32–34
Piezodorus, 20–22, 97, 384
Piezodorus guildinii, 7, 8, 10, 11, 18, 20, 22, 23, 36, 43, 46, 47, 54, 73, 120–128, 131, 145, 152, 160, 170, 171, 204, 214, 219, 220, 228, 244, 247
Piezodorus hybneri, 7, 8, 10, 11, 18, 20, 22, 23
Piezodorus lituratus, 10, 11, 20
Piezodorus oceanicus, 10, 11, 20
Plautia, 384
Plautia affinis, 50
Plautia stali, 9, 11, 131, 132, 134, 136–138, 140–150, 152, 153, 155, 369
Podisus, 30, 87
Podisus fuscescens, 36
Podisus maculiventris, 31–32, 373
Podisus nigrispinus, 31, 32, 35
Podisus sagitta, 36
Proxys albopunctulatus, 9, 11
Proxys punctulatus, 9, 11
Pyrrhocoris apterus, 173, 174

R
Reduviolus roseipennis, 171
Rhaphigaster nebulosa, 335
Rhodnius prolixus, 158
Rhycholepta grandicallosa, 3
Rhynocoris iracundus, 323
Rhyzopertha dominica, 217
Runibia decorata, 3

S
Scaptocoris, 97
Scaptocoris castanea, 70
Sitophilus oryzae, 217
Sogatella furcifera, 174
Solenopsis invicta, 171
Spodoptera, 33
Spodoptera eridania, 31, 35
Stiretrus, 33
Stiretrus anchorago, 32, 33
Stiretrus decemguttatus, 33
Supputius, 30
Supputius cincticeps, 32, 35

T
Telenomus, 313, 314, 373
Telenomus chloropus, 313, 314
Telenomus podisi, 51–61, 65, 67, 85, 110, 264, 266, 386
Telenomus turesis, 313, 314
Tenebrio molitor, 35, 36
Thaumastocoris peregrinus, 35
Thyanta, 11, 17, 20
Thyanta calceata, 10, 11
Thyanta custator, 10, 11, 18, 20, 21
Thyanta pallidovirens, 10, 11, 121, 125, 127

Thyanta perditor, 10, 11, 18, 20, 21
Tibraca limbativentris, 9, 11, 98, 99, 107, 109, 110, 214
Tribolium castaneum, 124, 126, 128, 217
Trichogramma galloi, 66
Trichogramma pretiosum, 57
Trichopepla semivittata, 8, 23
Trissolcus, 314, 334, 373
Trissolcus basalis, 85, 88, 90, 171, 264, 313, 314, 319, 320
Trissolcus belenus, 313, 314, 319
Trissolcus cultratus, 313, 314
Trissolcus japonicus, 171, 313, 314, 318, 320, 321, 333–338, 373, 374
Trissolcus kozlovi, 313, 314, 334
Trissolcus mitsukurii, 313, 314, 319–321, 325, 335, 337, 338
Trissolcus semistriatus, 313, 314, 319
Trissolcus urichi, 264
Tylopsis liliifolia, 322
Tynacantha marginata, 32, 35–36

V
Vatiga illudens, 171
Vatiga manihotae, 171

Y
Yersinella raymondi, 322

Subject Index

A
Active ingredients, 60, 85, 106, 203, 204, 207–210, 214–218, 224, 242, 248, 330, 390
Asopinae, 1, 2, 4, 7, 29–36
Augmentative biological control (ABC), 53–56, 65–78, 258, 260, 261, 264, 265, 267, 334, 386, 387

B
Biocontrol agents, 32, 52, 54, 57, 66, 84, 90, 258, 261, 263–265, 321, 325, 334, 335, 373, 374, 386, 387, 390
Bioecology, 82, 338
Biological control, 29–36, 65–78, 81–91, 95, 112, 158, 171, 181, 207, 215, 231, 241, 256, 258, 261, 263–267, 321, 333–338, 362, 373–374, 384, 386–387, 389
Biotechnologies, 72, 171, 177, 389
Brown marmorated stink bug (BMSB), 48, 98, 119, 170, 171, 189, 242–243, 271–294, 298–301, 308, 311–313, 320, 321, 325–328, 330–338, 361–375

C
Chemical control, 98, 156, 170, 199–210, 214–216, 250, 258, 260
Citizen science, 272–292
Control, 30, 49, 65, 81, 96, 117, 181, 202, 213, 241, 256, 298, 362, 384

D
Distributions, 1–23, 31, 35, 47, 58, 75, 76, 99, 169, 171, 203, 214, 225, 229, 230, 241, 243, 272, 292, 294, 299, 300, 338, 362–364
Double stranded RNA (dsRNA), 117, 118, 120, 121, 125–129, 144–158, 160

E
Economic thresholds (ET), 60, 66, 83, 84, 89, 90, 182, 213, 223, 226, 228, 256, 260–263, 266, 385, 389
Egg parasitoids, 43, 51, 52, 55–57, 59, 65, 67, 82, 85, 88–90, 110, 171, 249, 263, 313, 320, 325, 333, 334, 338, 386, 387
Electropenetrography (EPG), 182–184, 189, 194–196, 301
Entomopathogenic fungi, 65–70, 72–77, 220, 231–233, 325, 374

F
Fruits, 1, 8, 96, 98, 107, 110, 169, 170, 241, 271, 272, 276, 301–308, 311–312, 330, 331, 336, 361, 364, 365, 367, 369, 371, 373, 374

A. F. Bueno, A. R. Panizzi (eds.), *Stink Bugs (Hemiptera: Pentatomidae) Research and Management*, Entomology in Focus 9,
https://doi.org/10.1007/978-3-031-69742-5

G
Gene editing, 171–174, 176–177
Gene silencing, 117–120, 123, 129, 149–151, 153, 154, 157, 158, 160

H
Habitat manipulations, 86, 89
Hemiptera, 2, 33, 41, 70, 73, 74, 82, 85, 87, 89, 97, 118, 154, 169, 171–174, 199, 213, 215, 217, 219, 223, 233, 260, 271, 383
Heteroptera, 1, 2, 41, 73, 158, 241, 287, 288, 301, 323, 326, 383
Host plant resistance, 181–197, 384, 387–389
Host plants, 17, 83, 98, 181–197, 222, 243, 300, 301, 320, 330, 364, 383, 387–389

I
Insect feeding, 109
Insecticide resistance management (IRM), 200, 203, 204, 207, 210, 250
Insecticides, 48, 66, 81, 95, 158, 170, 181, 200, 214, 241, 258, 320, 365, 384
Insecticide-susceptibility, 247
Integrated pest management (IPM), 31, 41, 42, 47–49, 60, 65–67, 70, 71, 82, 84, 96, 112, 117, 158, 159, 177, 200, 201, 228, 231, 249, 256–263, 265, 267, 271, 328, 329, 331, 336, 338, 367, 368, 374, 384–391
Invasive, 7, 152, 183, 241, 292, 293, 298, 325, 361–375
Invasive pests, 335, 338

M
Maize, 42, 43, 45, 158, 203, 214, 218, 220–228, 234, 244–246, 248–250, 303, 305, 306, 308, 311, 383
Management, 31, 42, 65, 82, 95, 129, 170, 181, 199, 214, 249, 255, 272, 365, 385
Mode of action, 67, 68, 70, 109–111, 203, 204, 208, 214, 215, 246, 248

N
Natural enemies efficiency, 83
Natural enemies preservation, 264
Neotropical brown stink bug, 43, 98, 170, 181, 199, 214, 228, 241, 243–244, 255, 384

P
Pentatomidae, 1, 2, 4–7, 10, 12, 23, 29, 36, 41, 73, 74, 82, 85, 89, 97–99, 117, 120, 125, 127, 130, 145, 151–153, 155, 157, 158, 160, 169, 171, 172, 199, 213, 215, 217, 219, 223, 233, 241, 271, 334, 383
Pest management, 31, 82, 84, 129, 149, 150, 158, 160, 171, 174, 176–177, 181, 182, 189, 231, 232, 234, 256, 258, 262, 263, 267, 365, 372, 374, 386
Pests, 2, 30, 41, 81, 95, 117, 169, 181, 199, 213, 241, 258, 271, 361, 385
Plant extracts, 97, 111, 331

R
Resistance, 71, 113, 117, 144, 150, 158–160, 173, 181–197, 200, 201, 204, 206, 207, 210, 241–250, 256, 387–389
RNAi machinery, 118–129, 150, 158, 160

S
Scelionidae, 43, 51, 82, 85, 171, 264, 313, 314
Selective insecticides, 61, 84, 258, 265, 385
Soybean, 7, 41, 65, 82, 98, 151, 181, 199, 214, 241, 255, 308, 361, 383
Species identification, 12, 16, 17, 20, 264
Spray nozzles, 230, 232
Stink bug management, 61, 171, 182, 200–203, 207, 213–234, 248, 256, 258, 263, 265, 267, 385–387, 391
Stink bugs, 1–23, 29–36, 41–49, 51–61, 65–78, 81–91, 95–112, 117–160, 169–177, 181–197, 213–234, 241–250, 255, 258, 260–267, 312, 320, 321, 325, 326, 365, 367, 369, 383–391
Sustainable agriculture, 31

U
Updated taxonomy, 1–23

MIX
Papier aus verantwortungsvollen Quellen
Paper from responsible sources
FSC® C105338

If you have any concerns about our products,
you can contact us on
ProductSafety@springernature.com

In case Publisher is established outside the EU,
the EU authorized representative is:
Springer Nature Customer Service Center GmbH
Europaplatz 3, 69115 Heidelberg, Germany

Printed by Libri Plureos GmbH
in Hamburg, Germany